MW00639759

PRINCIPLES OF CHARGED PARTICLE ACCELERATION

Stanley Humphries, Jr.

Professor Emeritus
University of New Mexico
Albuquerque, New Mexico

President
Field Precision LLC
Albuquerque, New Mexico

DOVER PUBLICATIONS, INC.
MINEOLA, NEW YORK

Bibliographical Note

This Dover edition, first published in 2012, is an unabridged republication of the work originally published in 1986 by John Wiley & Sons, Inc., New York. The author has provided a new Preface for this Dover edition.

Library of Congress Cataloging-in-Publication Data

Humphries, Stanley.
   Principles of charged particle acceleration / Stanley Humphries.
      p. cm.
   Originally published: New York : J. Wiley, c1986.
   Summary: "This authoritative text offers a unified, programmed summary of the principles underlying all charged particle accelerators — it also doubles as a reference collection of equations and material essential to accelerator development and beam applications. The only text that covers linear induction accelerators, the work contains straightforward expositions of basic principles rather than detailed theories of specialized areas"—Provided by publisher.
   Includes bibliographical references and index.
   ISBN-13: 978-0-486-49818-8 (pbk.)
   ISBN-10: 0-486-49818-2 (pbk.)
      1. Particle acceleration. I. Title.

QC787.P3H86 2012
539.7'3—dc213

                                              2012015806

www.doverpublications.com

To my parents, Katherine and Stanley Humphries

Preface to the Dover Edition

This book evolved from the first term of a two-term course on the physics of charged particle acceleration that I taught at the University of New Mexico and at Los Alamos National Laboratory. The first term covered conventional accelerators in the single particle limit. The second term covered collective effects in charged-particle beams, including high-current transport and instabilities. The material was selected to make the course accessible to graduate students in physics and electrical engineering with no previous background in accelerator theory. Nonetheless, I sought to make the course relevant to accelerator researchers by including complete derivations and essential formulas.

The organization of the book reflects my outlook as an experimentalist. I followed a building-block approach, starting with basic material and adding new techniques and insights in a programmed sequence. I included extensive review material in areas that would not be familiar to the average student and in areas where my own understanding needed reinforcement. I tried to make the derivations as simple as possible by making physical approximations at the beginning of the derivation rather than at the end. Because the text was intended as an introduction to the field of accelerators, I felt that it was important to preserve a close connection with the physical basis of the derivations; therefore, I avoided treatments that required advanced methods of mathematical analysis. Accelerator specialists will no doubt find many important areas that are not covered. I apologize in advance for the inevitable consequence of writing a book of finite length.

I want to express my appreciation to my students at Los Alamos and the University of New Mexico for the effort they put into the course and for their help in resolving ambiguities in the material. In particular, I would like to

v

thank Alan Wadlinger, Grenville Boicourt, Steven Wipf, and Jean Berlijn of Los Alamos National Laboratory for lively discussions on problem sets and for many valuable suggestions.

I am grateful to Francis Cole of Fermilab, Wemer Joho of the Swiss Nuclear Institute, William Herrmannsfeldt of the Stanford Linear Accelerator Center, Andris Faltens of Lawrence Berkeley Laboratory, Richard Cooper of Los Alamos National Laboratory, Daniel Prono of Lawrence Livermore Laboratory, Helmut Milde of Ion Physics Corporation, and George Fraser of Physics International Company for contributing material and commenting on the manuscript. I was aided in the preparation of the manuscript by lecture notes developed by James Potter of LANL and by Francis Cole. I would like to take this opportunity to thank David W. Woodall, L. K. Len, David Straw, Robert Jameson, Francis Cole, James Benford, Carl Ekdahl, Brendan Godfrey, William Rienstra, and McAllister Hull for their encouragement of and contributions towards the creation of an accelerator research program at the University of New Mexico. I appreciate the generosity of the original publisher, John Wiley and Sons, for transferring the copyright to make this Dover edition possible.

                                        STANLEY HUMPHRIES, JR.

Albuquerque, New Mexico 2012

Preface

This book evolved from the first term of a two-term course on the physics of charged particle acceleration that I taught at the University of New Mexico and at Los Alamos National Laboratory. The first term covered conventional accelerators in the single particle limit. The second term covered collective effects in charged particle beams, including high current transport and instabilities. The material was selected to make the course accessible to graduate students in physics and electrical engineering with no previous background in accelerator theory. Nonetheless, I sought to make the course relevant to accelerator researchers by including complete derivations and essential formulas.

The organization of the book reflects my outlook as an experimentalist. I followed a building block approach, starting with basic material and adding new techniques and insights in a programmed sequence. I included extensive review material in areas that would not be familiar to the average student and in areas where my own understanding needed reinforcement. I tried to make the derivations as simple as possible by making physical approximations at the beginning of the derivation rather than at the end. Because the text was intended as an introduction to the field of accelerators, I felt that it was important to preserve a close connection with the physical basis of the derivations; therefore, I avoided treatments that required advanced methods of mathematical analysis. Most of the illustrations in the book were generated numerically from a library of demonstration microcomputer programs that I developed for the courses. Accelerator specialists will no doubt find many important areas that are not covered. I apologize in advance for the inevitable consequence of writing a book of finite length.

I want to express my appreciation to my students at Los Alamos and the University of New Mexico for the effort they put into the course and for their help in resolving ambiguities in the material. In particular, I would like to thank Alan Wadlinger, Grenville Boicourt, Steven Wipf, and Jean Berlijn of Los Alamos National Laboratory for lively discussions on problem sets and for many valuable suggestions.

I am grateful to Francis Cole of Fermilab, Werner Joho of the Swiss Nuclear Institute, William Herrmannsfeldt of the Stanford Linear Accelerator Center, Andris Faltens of Lawrence Berkeley Laboratory, Richard Cooper of Los Alamos National Laboratory, Daniel Prono of Lawrence Livermore Laboratory, Helmut Milde of Ion Physics Corporation, and George Fraser of Physics International Company for contributing material and commenting on the manuscript. I was aided in the preparation of the manuscript by lecture notes developed by James Potter of LANL and by Francis Cole. I would like to take this opportunity to thank David W. Woodall, L. K. Len, David Straw, Robert Jameson, Francis Cole, James Benford, Carl Ekdahl, Brendan Godfrey, William Rienstra, and McAllister Hull for their encouragement of and contributions towards the creation of an accelerator research program at the University of New Mexico. I am grateful for support that I received to attend the 1983 NATO Workshop on Fast Diagnostics.

Finally, I want to express my love and appreciation to my wife, Sandra, for her help on this book and her support over the years. She gave me the inspiration and opportunity to complete the project, despite the demands on her time as a mother and as an artist.

STANLEY HUMPHRIES, JR.

University of New Mexico
December, 1985

Contents

PRINCIPLES OF CHARGED PARTICLE ACCELERATION

1

Introduction

This book is an introduction to the theory of charged particle acceleration. It has two primary roles:

1. A unified, programmed summary of the principles underlying all charged particle accelerators.
2. A reference collection of equations and material essential to accelerator development and beam applications.

The book contains straightforward expositions of basic principles rather than detailed theories of specialized areas.

Accelerator research is a vast and varied field. There is an amazingly broad range of beam parameters for different applications, and there is a correspondingly diverse set of technologies to achieve the parameters. Beam currents range from nanoamperes (10^{-9} A) to megaamperes (10^6 A). Accelerator pulselengths range from less than a nanosecond to steady state. The species of charged particles range from electrons to heavy ions, a mass difference factor approaching 10^6. The energy of useful charged particle beams ranges from a few electron volts (eV) to almost 1 TeV (10^{12} eV).

Organizing material from such a broad field is inevitably an imperfect process. Before beginning our study of beam physics, it is useful to review the order of topics and to define clearly the objectives and limitations of the book. The goal is to present the theory of accelerators on a level that facilitates the design of accelerator components and the operation of accelerators for applications. In order to accomplish this effectively, a considerable amount of potentially interesting material must be omitted:

1.   Accelerator theory is interpreted as a mature field. There is no attempt to review the history of accelerators.

2.   Although an effort has been made to include the most recent developments in accelerator science, there is insufficient space to include a detailed review of past and present literature.

3.   Although the theoretical treatments are aimed toward an understanding of real devices, it is not possible to describe in detail specific accelerators and associated technology over the full range of the field.

These deficiencies are compensated by the books and papers tabulated in the bibliography.

We begin with some basic definitions. A *charged particle* is an elementary particle or a macroparticle which contains an excess of positive or negative charge. Its motion is determined mainly by interaction with electromagnetic forces. *Charged particle acceleration* is the transfer of kinetic energy to a particle by the application of an electric field. A *charged particle beam* is a collection of particles distinguished by three characteristics: (1) beam particles have high kinetic energy compared to thermal energies, (2) the particles have a small spread in kinetic energy, and (3) beam particles move approximately in one direction. In most circumstances, a beam has a limited extent in the direction transverse to the average motion. The antithesis of a beam is an assortment of particles in thermodynamic equilibrium.

Most applications of charged particle accelerators depend on the fact that beam particles have high energy and good directionality. Directionality is usually referred to as *coherence*. Beam coherence determines, among other things, (1) the applied force needed to maintain a certain beam radius, (2) the maximum beam propagation distance, (3) the minimum focal spot size, and (4) the properties of an electromagnetic wave required to trap particles and accelerate them to high energy.

The process for generating charged particle beams is outlined in Table 1.1. Electromagnetic forces result from mutual interactions between charged particles. In accelerator theory, particles are separated into two groups: (1) particles in the beam and (2) charged particles that are distributed on or in surrounding materials. The latter group is called the external charge. Energy is required to set up distributions of external charge; this energy is transferred to the beam particles via electromagnetic forces. For example, a power supply can generate a voltage difference between metal plates by subtracting negative charge from one plate and moving it to the other. A beam particle that moves between the plates is accelerated by attraction to the charge on one plate and repulsion from the charge on the other.

Electromagnetic forces are resolved into electric and magnetic components. Magnetic forces are present only when charges are in relative motion. The

TABLE 1.1   Acceleration Process

Single-Particle Theory	Multiparticle Theory

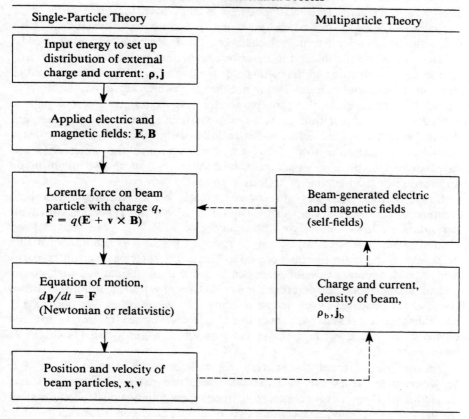

ability of a group of external charged particles to exert forces on beam particles is summarized in the *applied electric and magnetic fields*. Applied forces are usually resolved into those aligned along the average direction of the beam and those that act transversely. The axial forces are acceleration forces; they increase or decrease the beam energy. The transverse forces are confinement forces. They keep the beam contained to a specific cross-sectional area or bend the beam in a desired direction. Magnetic forces are always perpendicular to the velocity of a particle; therefore, magnetic fields cannot affect the particle's kinetic energy. Magnetic forces are confinement forces. Electric forces can serve both functions.

The distribution and motion of external charge determines the fields, and the fields determine the force on a particle via the Lorentz force law, discussed in Chapter 3. The expression for force is included in an appropriate equation of motion to find the position and velocity of particles in the beam as a function of time. A knowledge of representative particle orbits makes it possible to estimate average parameters of the beam, such as radius, direction, energy, and current. It is also possible to sum over the large number of particles in the beam to find charge density ρ_b and current density j_b. These quantities act as source terms for beam-generated electric and magnetic fields.

This procedure is sufficient to describe low-current beams where the contribution to total electric and magnetic fields from the beam is small compared to those of the external charges. This is not the case at high currents. As shown in Table 1.1, calculation of beam parameters is no longer a simple linear procedure. The calculation must be *self-consistent*. Particle trajectories are determined by the total fields, which include contributions from other beam particles. In turn, the total fields are unknown until the trajectories are calculated. The problem must be solved either by successive iteration or application of the methods of collective physics.

Single-particle processes are covered in this book. Although theoretical treatments for some devices can be quite involved, the general form of all derivations follows the straight-line sequence of Table 1.1. Beam particles are treated as test particles responding to specified fields. A continuation of this book addressing collective phenomena in charged particle beams is in preparation. A wide variety of useful processes for both conventional and high-power pulsed accelerators are described by collective physics, including (1) beam cooling, (2) propagation of beams injected into vacuum, gas, or plasma, (3) neutralization of beams, (4) generation of microwaves, (5) limiting factors for efficiency and flux, (6) high-power electron and ion guns, and (7) collective beam instabilities.

An outline of the topics covered in this book is given in Table 1.2. Single-particle theory can be subdivided into two categories: *transport* and *acceleration*. Transport is concerned with beam confinement. The study centers on methods for generating components of electromagnetic force that localize beams in space. For steady-state beams extending a long axial distance, it is sufficient to consider only transverse forces. In contrast, particles in accelerators with time-varying fields must be localized in the axial direction. Force components must be added to the accelerating fields for longitudinal particle confinement (phase stability).

Acceleration of charged particles is conveniently divided into two categories: electrostatic and electromagnetic acceleration. The accelerating field in electrostatic accelerators is the gradient of an electrostatic potential. The peak energy of the beam is limited by the voltage that can be sustained without breakdown. Pulsed power accelerators are included in this category because pulselengths are almost always long enough to guarantee simple electrostatic acceleration.

In order to generate beams with kinetic energy above a few million electron volts, it is necessary to utilize time-varying electromagnetic fields. Although particles in an electromagnetic accelerator experience continual acceleration by an electric field, the field does not require prohibitively large voltages in the

TABLE 1.2 Organization of Topics

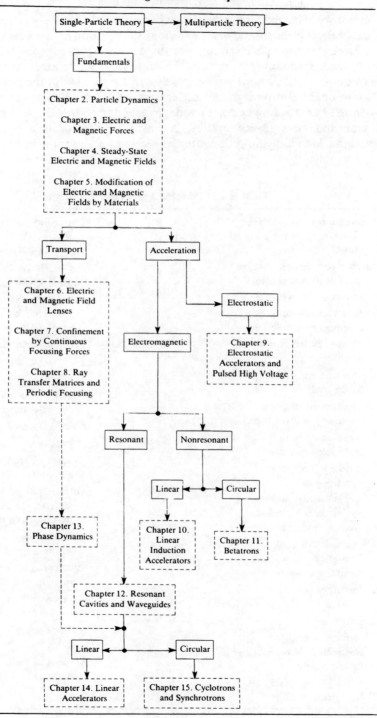

laboratory. The accelerator geometry is such that inductively generated electric fields cancel electrostatic fields except at the position of the beam.

Electromagnetic accelerators are divided into two subcategories: nonresonant and resonant accelerators. Nonresonant accelerators are pulsed; the motion of particles need not be closely synchronized with the pulse waveform. Nonresonant electromagnetic accelerators are essentially step-up transformers, with the beam acting as a high-voltage secondary. The class is subdivided into linear and circular accelerators. A linear accelerator is a straight-through machine. Generally, injection into the accelerator and transport is not difficult;

TABLE 1.3   Modern Accelerator Applications

Production of short-lived radioisotopies for medical diagnosis	Cross-linking of thermoplastics
Intense ion beams to drive inertial fusion reactors	Image intensifiers and fast streak tubes
Electron beam welding	Pulsed X-ray radiography
Pulsed neutron sources for uranium borehole logging	Surface modification of materials by ion implantation
Cathode ray tubes and fast digitizers	Nuclear structure studies
Electronuclear breeding of fissile fuels	Plasma heating for fusion reactors
Measurement of cross sections for atomic physics	Assay of nuclear materials for safeguard applications
Processing of semi-conductor circuits	Sterilization of food products
Secondary ion mass spectroscopy	Studies of transuranic elements
Generation of synchrotron radiation for solid-state physics research	Intense pulsed neutron sources for radiography and materials studies
Diagnostics of rock formations in oil and natural gas wells	Generation of X rays and pions for radiation therapy
Elementary particle physics	Electron and ion surface microprobes
Electron microscopy	Analysis of trace elements for biology and archeology
Materials testing for controlled thermonuclear fusion reactors	Calibration of radiation detectors
Drivers for gas lasers and free electron lasers	Studies of radiation damage in fission reactors

linear accelerators are useful for initial acceleration of low-energy beams or the generation of high-flux beams. In circular machines, the beam is recirculated many times through the acceleration region during the pulse. Circular accelerators are well suited to the production of beams with high kinetic energy.

The applied voltage in a resonant accelerator varies harmonically at a specific frequency. The word *resonant* characterizes two aspects of the accelerator: (1) electromagnetic oscillations in resonant cavities or waveguides are used to transform input microwave power from low to high voltage and (2) there is close coupling between properties of the particle orbits and time variations of the accelerating field. Regarding the second property, particles must always be at the proper place at the proper time to experience a field with accelerating polarity. Longitudinal confinement is a critical issue in resonant accelerators. Resonant accelerators can also be subdivided into linear and circular machines, each category with its relative virtues.

In the early period of accelerator development, the quest for high kinetic energy, spurred by nuclear and elementary particle research, was the overriding goal. Today, there is increased emphasis on a diversity of accelerator applications. Much effort in modern accelerator theory is devoted to questions of current limits, beam quality, and the evolution of more efficient and cost-effective machines. The best introduction to modern accelerators is to review some of the active areas of research, both at high and low kinetic energy. The list in Table 1.3 suggests the diversity of applications and potential for future development.

2

Particle Dynamics

Understanding and utilizing the response of charged particles to electromagnetic forces is the basis of particle optics and accelerator theory. The goal is to find the time-dependent position and velocity of particles, given specified electric and magnetic fields. Quantities calculated from position and velocity, such as total energy, kinetic energy, and momentum, are also of interest.

The nature of electromagnetic forces is discussed in Chapter 3. In this chapter, the response of particles to general forces will be reviewed. These are summarized in laws of motion. The Newtonian laws, treated in the first sections, apply at low particle energy. At high energy, particle trajectories must be described by relativistic equations. Although Newton's laws and their implications can be understood intuitively, the laws of relativity cannot since they apply to regimes beyond ordinary experience. Nonetheless, they must be accepted to predict particle behavior in high-energy accelerators. In fact, accelerators have provided some of the most direct verifications of relativity.

This chapter reviews particle mechanics. Section 2.1 summarizes the properties of electrons and ions. Sections 2.2–2.4 are devoted to the equations of Newtonian mechanics. These are applicable to electrons from electrostatic accelerators of in the energy range below 20 kV. This range includes many useful devices such as cathode ray tubes, electron beam welders, and microwave tubes. Newtonian mechanics also describes ions in medium energy accelerators used for nuclear physics. The Newtonian equations are usually simpler to solve than relativistic formulations. Sometimes it is possible to describe transverse motions of relativistic particles using Newtonian equations with a relativistically corrected mass. This approximation is treated in Section

8

2.10. In the second part of the chapter, some of the principles of special relativity are derived from two basic postulates, leading to a number of useful formulas summarized in Section 2.9.

2.1 CHARGED PARTICLE PROPERTIES

In the theory of charged particle acceleration and transport, it is sufficient to treat particles as dimensionless points with no internal structure. Only the influence of the electromagnetic force, one of the four fundamental forces of nature, need be considered. Quantum theory is unnecessary except to describe the emission of radiation at high energy.

This book will deal only with ions and electrons. They are simple, stable particles. Their response to the fields applied in accelerators is characterized completely by two quantities: mass and charge. Nonetheless, it is possible to apply much of the material presented to other particles. For example, the motion of macroparticles with an electrostatic charge can be treated by the methods developed in Chapters 6–9. Applications include the suspension of small objects in oscillating electric quadrupole fields and the acceleration and guidance of inertial fusion targets. At the other extreme are unstable elementary particles produced by the interaction of high-energy ions or electrons with targets. Beamlines, acceleration gaps, and lenses are similar to those used for stable particles with adjustments for different mass. The limited lifetime may influence hardware design by setting a maximum length for a beamline or confinement time in a storage ring.

An electron is an elementary particle with relatively low mass and negative charge. An ion is an assemblage of protons, neutrons, and electrons. It is an atom with one or more electrons removed. Atoms of the isotopes of hydrogen have only one electron. Therefore, the associated ions (the proton, deuteron, and triton) have no electrons. These ions are bare nucleii consisting of a proton with 0, 1, or 2 neutrons. Generally, the symbol Z denotes the atomic number of an ion or the number of electrons in the neutral atom. The symbol Z^* is often used to represent the number of electrons removed from an atom to create an ion. Values of Z^* greater than 30 may occur when heavy ions traverse extremely hot material. If $Z^* = Z$, the atom is fully stripped. The atomic mass number A is the number of nucleons (protons or neutrons) in the nucleus. The mass of the atom is concentrated in the nucleus and is given approximately as Am_p, where m_p is the proton mass.

Properties of some common charged particles are summarized in Table 2.1. The meaning of the rest energy in Table 2.1 will become clear after reviewing the theory of relativity. It is listed in energy units of million electron volts (MeV). An electron volt is defined as the energy gained by a particle having one fundamental unit of charge ($q = \pm e = \pm 1.6 \times 10^{-19}$ coulombs) passing

TABLE 2.1   Charged Particle Properties

Particle	Charge (coulomb)	Mass (kg)	Rest Energy (MeV)	A	Z	Z*
Electron (β particle)	-1.60×10^{-19}	9.11×10^{-31}	0.511	—	—	—
Proton	$+1.60 \times 10^{-19}$	1.67×10^{-27}	938	1	1	1
Deuteron	$+1.60 \times 10^{-19}$	3.34×10^{-27}	1875	2	1	1
Triton	$+1.60 \times 10^{-19}$	5.00×10^{-27}	2809	3	1	1
He$^+$	$+1.60 \times 10^{-19}$	6.64×10^{-27}	3728	4	2	1
He^{++} (α particle)	$+3.20 \times 10^{-19}$	6.64×10^{-27}	3728	4	2	2
C$^+$	$+1.6 \times 10^{-19}$	1.99×10^{-26}	1.12×10^4	12	6	1
U$^+$	$+1.6 \times 10^{-19}$	3.95×10^{-25}	2.22×10^5	238	92	1

through a potential difference of one volt. In mks units,[†] the electron volt is

$$1 \text{ eV} = (1.6 \times 10^{-19} \text{ C})(1 \text{ V}) = 1.6 \times 10^{-19} \text{ J}.$$

Other commonly used metric units are keV (10^3 eV) and GeV (10^9 eV). Relativistic mechanics must be used when the particle kinetic energy is comparable to or larger than the rest energy.

There is a factor of 1843 difference between the mass of the electron and the proton. Although methods for transporting and accelerating various ions are similar, techniques for electrons are quite different. Electrons are relativistic even at low energies. As a consequence, synchronization of electron motion in linear accelerators is not difficult. Electrons are strongly influenced by magnetic fields; thus they can be accelerated effectively in a circular induction accelerator (the betatron). High-current electron beams (~ 10 kA) can be focused effectively by magnetic fields. In contrast, magnetic fields are ineffective for high-current ion beams. On the other hand, it is possible to neutralize the charge and current of a high-current ion beam easily with light electrons, while the inverse is usually impossible.

2.2 NEWTON'S LAWS OF MOTION

The charge of a particle determines the strength of its interaction with the electromagnetic force. The mass indicates the resistance to a change in velocity. In Newtonian mechanics, mass is constant, independent of particle motion.

[†] MKS units are used throughout the book unless specifically noted.

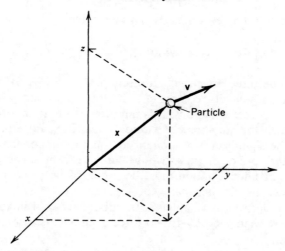

Figure 2.1 Position and velocity vectors of a particle in Cartesian coordinates.

The Newtonian mass (or *rest mass*) is denoted by a subscript: m_e for electrons, m_p for protons, and m_0 for a general particle. A particle's behavior is described completely by its position in three-dimensional space and its velocity as a function of time. Three quantities are necessary to specify position; the position x is a vector. In the Cartesian coordinates (Figure 2.1), x can be written

$$\mathbf{x} = (x, y, z). \tag{2.1}$$

The particle velocity is[†]

$$\mathbf{v} = (v_x, v_y, v_z) = (dx/dt, dy/dt, dz/dt) = d\mathbf{x}/dt. \quad \boxed{2.2}$$

Newton's first law states that a moving particle follows a straight-line path unless acted upon by a force. The tendency to resist changes in straight-line motion is called the momentum, **p**. Momentum is the product of a particle's mass and velocity,

$$\mathbf{p} = m_0\mathbf{v} = (p_x, p_y, p_z). \quad \boxed{2.3}$$

Newton's second law defines force **F** through the equation

$$d\mathbf{p}/dt = \mathbf{F}. \quad \boxed{2.4}$$

[†] Equations that have general applicability or demonstrate an important result are designated with a boxed number.

In Cartesian coordinates, Eq. (2.4) can be written

$$dp_x/dt = F_x, \qquad dp_y/dt = F_y, \qquad dp_z/dt = F_z. \qquad (2.5)$$

Motions in the three directions are decoupled in Eq. (2.5). With specified force components, velocity components in the x, y, and z directions are determined by separate equations. It is important to note that this decoupling occurs only when the equations of motion are written in terms of Cartesian coordinates. The significance of straight-line motion is apparent in Newton's first law, and the laws of motion have the simplest form in coordinate systems based on straight lines. Caution must be exercised using coordinate systems based on curved lines. The analog of Eq. (2.5) for cylindrical coordinates (r, θ, z) will be derived in Chapter 3. In curvilinear coordinates, momentum components may change even with no force components along the coordinate axes.

2.3 KINETIC ENERGY

Kinetic energy is the energy associated with a particle's motion. The purpose of particle accelerators is to impart high kinetic energy. The kinetic energy of a particle, T, is changed by applying a force. Force applied to a static particle cannot modify T; the particle must be moved. The change in T (work) is related to the force by

$$\Delta T = \int \mathbf{F} \cdot d\mathbf{x}. \qquad \boxed{2.6}$$

The integrated quantity is the vector dot product; $d\mathbf{x}$ is an incremental change in particle position.

In accelerators, applied force is predominantly in one direction. This corresponds to the symmetry axis of a linear accelerator or the main circular orbit in a betatron. With acceleration along the z axis, Eq. (2.6) can be rewritten

$$\Delta T = \int F_z \, dz = \int F_z (dz/dt) \, dt. \qquad (2.7)$$

The chain rule of derivatives has been used in the last expression. The formula for T in Newtonian mechanics can be derived by (1) rewriting F_z using Eq. (2.4), (2) taking $T = 0$ when $v_z = 0$, and (3) assuming that the particle mass is not a function of velocity:

$$T = \int m_0 v_z (dv_z/dt) \, dt = m_0 v_z^2 / 2. \qquad \boxed{2.8}$$

The differential relationship $d(m_0 v_z^2/2)/dt = m_0 v_z \, dv_z/dt$ leads to the last expression. The differences of relativistic mechanics arise from the fact that assumption 3 is not true at high energy.

When static forces act on a particle, the *potential energy U* can be defined. In this circumstance, the sum of kinetic and potential energies, $T + U$, is a constant called the total energy. If the force is axial, kinetic and potential energy are interchanged as the particle moves along the z axis, so that $U = U(z)$. Setting the total time derivative of $T + U$ equal to 0 and assuming $\partial U/\partial t = 0$ gives

$$m_0 v_z (dv_z/dt) = -(\partial U/\partial z)(dz/dt). \tag{2.9}$$

The expression on the left-hand side equals $F_z v_z$. The static force and potential energy are related by

$$F_z = -\partial U/\partial z, \qquad \mathbf{F} = -\nabla U, \qquad \boxed{2.10}$$

where the last expression is the general three-dimensional form written in terms of the vector gradient operator,

$$\nabla = \hat{x}\,\partial/\partial x + \hat{y}\,\partial/\partial y + \hat{z}\,\partial/\partial z. \tag{2.11}$$

The quantities $\hat{x}$, $\hat{y}$, and $\hat{z}$ are unit vectors along the Cartesian axes.

Potential energy is useful for treating electrostatic accelerators. Stationary particles at the source can be considered to have high U (potential for gaining energy). This is converted to kinetic energy as particles move through the acceleration column. If the potential function, $U(x, y, z)$, is known, focusing and accelerating forces acting on particles can be calculated.

2.4 GALILEAN TRANSFORMATIONS

In describing physical processes, it is often useful to change the viewpoint to a frame of reference that moves with respect to an original frame. Two common frames of reference in accelerator theory are the *stationary frame* and the *rest frame*. The stationary frame is identified with the laboratory or accelerating structure. An observer in the rest frame moves at the average velocity of the beam particles; hence, the beam appears to be at rest. A *coordinate transformation* converts quantities measured in one frame to those that would be measured in another moving with velocity **u**. The transformation of the properties of a particle can be written symbolically as

$$(\mathbf{x}, \mathbf{v}, m, \mathbf{p}, T) \rightarrow (\mathbf{x}', \mathbf{v}', m', \mathbf{p}', T'),$$

where primed quantities are those measured in the moving frame. The oper-

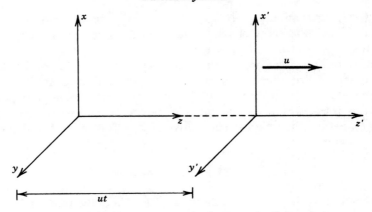

Figure 2.2   Galilean transformation between coordinate systems.

ation that transforms quantities depends on **u**. If the transformation is from the stationary to the rest frame, **u** is the particle velocity **v**.

The transformations of Newtonian mechanics (Galilean transformations) are easily understood by inspecting Figure 2.2. Cartesian coordinate systems are defined so that the z axes are colinear with **u** and the coordinates are aligned at $t = 0$. This is consistent with the usual convention of taking the average beam velocity along the z axis. The position of a particle transforms as

$$x' = x, \qquad y' = y, \qquad z' = z - ut. \tag{2.12}$$

Newtonian mechanics assumes inherently that measurements of particle mass and time intervals in frames with constant relative motion are equal: $m' = m$ and $dt' = dt$. This is not true in a relativistic description. Equations (2.12) combined with the assumption of invariant time intervals imply that $dx' = dx$ and $dx'/dt' = dx/dt$. The velocity transformations are

$$v'_x = v_x, \qquad v'_y = v_y, \qquad v'_z = v_z - u. \tag{2.13}$$

Since $m' = m$, momenta obey similar equations. The last expression shows that velocities are additive. The axial velocity in a third frame moving at velocity w with respect to the x' frame is related to the original quantity by $v''_z = v_z - u - w$.

Equations (2.13) can be used to determine the transformation for kinetic energy,

$$T' = T + \tfrac{1}{2}m_0\left(-2uv_z + u^2\right). \tag{2.14}$$

Measured kinetic energy depends on the frame of reference. It can be either larger or smaller in a moving frame, depending on the orientation of the velocities. This dependence is an important factor in beam instabilities such as the two-stream instability.

2.5 POSTULATES OF RELATIVITY

The principles of special relativity proceed from two postulates:

1. The laws of mechanics and electromagnetism are identical in all inertial frames of reference.
2. Measurements of the velocity of light give the same value in all inertial frames.

Only the theory of special relativity need be used for the material of this book. General relativity incorporates the gravitational force, which is negligible in accelerator applications. The first postulate is straightforward; it states that observers in any *inertial frame* would derive the same laws of physics. An inertial frame is one that moves with constant velocity. A corollary is that it is impossible to determine an absolute velocity. Relative velocities can be measured, but there is no preferred frame of reference.

The second postulate follows from the first. If the velocity of light were referenced to a universal stationary frame, tests could be devised to measure absolute velocity. Furthermore, since photons are the entities that carry the electromagnetic force, the laws of electromagnetism would depend on the absolute velocity of the frame in which they were derived. This means that the forms of the Maxwell equations and the results of electrodynamic experiments would differ in frames in relative motion. Relativistic mechanics, through postulate 2, leaves Maxwell's equations *invariant* under a coordinate transformation. Note that invariance does not mean that measurements of electric and magnetic fields will be the same in all frames. Rather, such measurements will always lead to the same governing equations.

The validity of the relativistic postulates is determined by their agreement with experimental measurements. A major implication is that no object can be induced to gain a measured velocity faster than that of light,

$$c = 2.998 \times 10^8 \text{ m/s.} \tag{2.15}$$

This result is verified by observations in electron accelerators. After electrons gain a kinetic energy above a few million electron volts, subsequent acceleration causes no increase in electron velocity, even into the multi-GeV range. The constant velocity of relativistic particles is important in synchronous accelerators, where an accelerating electromagnetic wave must be matched to the

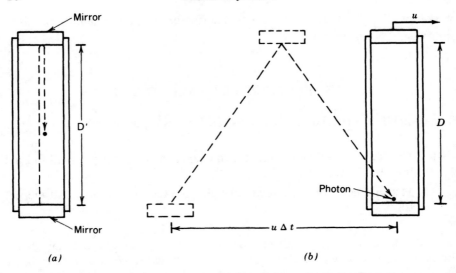

Figure 2.3 Effect of time dilation on the observed rates of a photon clock. (*a*) Clock rest frame. (*b*) Stationary frame.

motion of the particle.

2.6 TIME DILATION

In Newtonian mechanics, observers in relative motion measure the same time interval for an event (such as the decay of an unstable particle or the period of an atomic oscillation). This is not consistent with the relativistic postulates. The variation of observed time intervals (depending on the relative velocity) is called *time dilation*. The term *dilation* implies extending or spreading out.

The relationship between time intervals can be demonstrated by the clock shown in Figure 2.3, where double transits (back and forth) of a photon between mirrors with known spacing are measured. This test could actually be performed using a photon pulse in a mode-locked laser. In the rest frame (denoted by primed quantities), mirrors are separated by a distance D', and the photon has no motion along the z axis. The time interval in the clock rest frame is

$$\Delta t' = 2D'/c. \tag{2.16}$$

If the same event is viewed by an observer moving past the clock at a velocity $-u$, the photon appears to follow the triangular path shown in Figure 2.3*b*. According to postulate 2, the photon still travels with velocity c but follows a longer path in a double transit. The distance traveled in the laboratory frame is

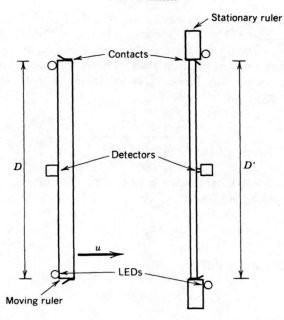

Figure 2.4 Experiment to demonstrate invariance of transverse lengths between frames in relative motion.

$$c \, \Delta t = 2 \left[D^2 + (u \, \Delta t / 2)^2 \right]^{1/2},$$

or

$$\Delta t = \frac{2D/c}{\left(1 - u^2/c^2\right)^{1/2}}. \tag{2.17}$$

In order to compare time intervals, the relationship between mirror spacing in the stationary and rest frames (D and D') must be known. A test to demonstrate that these are equal is illustrated in Figure 2.4. Two scales have identical length when at rest. Electrical contacts at the ends allow comparisons of length when the scales have relative motion. Observers are stationed at the centers of the scales. Since the transit times of electrical signals from the ends to the middle are equal in all frames, the observers agree that the ends are aligned simultaneously. Measured length may depend on the magnitude of the relative velocity, but it cannot depend on the direction since there is no preferred frame or orientation in space. Let one of the scales move; the observer in the scale rest frame sees no change of length. Assume, for the sake of argument, that the stationary observer measures that the moving scale has shortened in the transverse direction, $D < D'$. The situation is symmetric, so that the roles of stationary and rest frames can be interchanged. This leads to conflicting conclusions. Both observers feel that their clock is the same length but the other is shorter. The only way to resolve the conflict is to take $D = D'$. The key to the argument is that the observers agree on simultaneity of the comparison events (alignment of the ends). This is not true in tests to compare axial length, as discussed in the next section. Taking $D = D'$, the relationship between time intervals is

$$\Delta t = \frac{\Delta t'}{\left(1 - u^2/c^2\right)^{1/2}}. \qquad (2.18)$$

Two dimensionless parameters are associated with objects moving with a velocity u in a stationary frame:

$$\beta = u/c, \qquad \gamma = \left(1 - u^2/c^2\right)^{-1/2}. \qquad \boxed{2.19}$$

These parameters are related by

$$\gamma = \left(1 - \beta^2\right)^{-1/2}, \qquad \boxed{2.20}$$

$$\beta = \left(1 - 1/\gamma^2\right)^{1/2}. \qquad \boxed{2.21}$$

A time interval Δt measured in a frame moving at velocity u with respect to an object is related to an interval measured in the rest frame of the object, $\Delta t'$, by

$$\Delta t = \gamma \Delta t'. \qquad \boxed{2.22}$$

For example, consider an energetic π^+ pion (rest energy 140 MeV) produced by the interaction of a high-energy proton beam from an accelerator with a target. If the pion moves at velocity 2.968×10^8 m/s in the stationary frame, it has a β value of 0.990 and a corresponding γ value of 8.9. The pion is unstable, with a decay time of 2.5×10^{-8} s at rest. Time dilation affects the decay time measured when the particle is in motion. Newtonian mechanics predicts that the average distance traveled from the target is only 7.5 m, while relativistic mechanics (in agreement with observation) predicts a decay length of 61 m for the high-energy particles.

2.7 LORENTZ CONTRACTION

Another familiar result from relativistic mechanics is that a measurement of the length of a moving object along the direction of its motion depends on its velocity. This phenomenom is known as *Lorentz contraction*. The effect can be

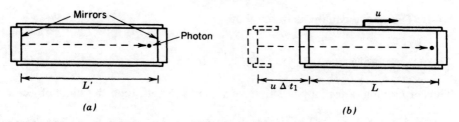

Figure 2.5   Lorentz contraction of a photon clock. (*a*) Clock rest frame. (*b*) Stationary frame.

demonstrated by considering the clock of Section 2.6 oriented as shown in Figure 2.5.

The detector on the clock measures the double transit time of light between the mirrors. Pulses are generated when a photon leaves and returns to the left-hand mirror. Measurement of the single transit time would require communicating the arrival time of the photon at the right-hand mirror to the timer at the left-hand mirror. Since the maximum speed with which this information can be conveyed is the speed of light, this is equivalent to a measurement of the double transit time. In the clock rest frame, the time interval is $\Delta t' = 2L'/c$.

To a stationary observer, the clock moves at velocity u. During the transit in which the photon leaves the timer, the right-hand mirror moves away. The photon travels a longer distance in the stationary frame before being reflected. Let Δt_1 be the time for the photon to travel from the left to right mirrors. During this time, the right-hand mirror moves a distance $u \, \Delta t_1$. Thus,

$$c \, \Delta t_1 = (L + u \, \Delta t_1),$$

where L is the distance between mirrors measured in the stationary frame. Similarly, on the reverse transit, the left-hand mirror moves toward the photon. The time to complete this leg is

$$\Delta t_2 = (L - u \, \Delta t_2)/c.$$

The total time for the event in the stationary frame is

$$\Delta t = \Delta t_1 + \Delta t_2 = \frac{L}{c - u} + \frac{L}{c + u}$$

or

$$\Delta t = \frac{2L/c}{1 - u^2/c^2}.$$

Time intervals cannot depend on the orientation of the clock, so that Eq. (2.22) holds. The above equations imply that

$$L = L'/\gamma. \qquad \boxed{2.23}$$

Thus, a moving object appears to have a shorter length than the same object at rest.

The acceleration of electrons to multi-GeV energies in a linear accelerator provides an interesting example of a Lorentz contraction effect. Linear accelerators can maintain longitudinal accelerating gradients of, at most, a few megavolts per meter. Lengths on the kilometer scale are required to produce high-energy electrons. To a relativistic electron, the accelerator appears to be rushing by close to the speed of light. The accelerator therefore has a contracted apparent length of a few meters. The short length means that focusing lenses are often unnecessary in electron linear accelerators with low-current beams.

2.8 LORENTZ TRANSFORMATIONS

Charged particle orbits are characterized by position and velocity at a certain time, $(\mathbf{x}, \mathbf{v}, t)$. In Newtonian mechanics, these quantities differ if measured in a frame moving with a relative velocity with respect to the frame of the first measurement. The relationship between quantities was summarized in the Galilean transformations.

The Lorentz transformations are the relativistic equivalents of the Galilean transformations. In the same manner as Section 2.4, the relative velocity of frames is taken in the z direction and the z and z' axes are colinear. Time is measured from the instant that the two coordinate systems are aligned ($z = z' = 0$ at $t = t' = 0$). The equations relating position and time measured in one frame (unprimed quantities) to those measured in another frame moving with velocity u (primed quantities) are

$$x' = x, \qquad (2.24)$$

$$y' = y, \qquad (2.25)$$

$$z' = \frac{z - ut}{\left(1 - u^2/c^2\right)^{1/2}} = \gamma(z - ut), \qquad (2.26)$$

$$t' = \frac{t - uz/c^2}{\left(1 - u^2/c^2\right)^{1/2}} = \gamma\left(t - \frac{uz}{c^2}\right). \qquad (2.27)$$

The primed frame is not necessarily the rest frame of a particle. One major difference between the Galilean and Lorentz transformations is the presence of the γ factor. Furthermore, measurements of time intervals are different in frames in relative motion. Observers in both frames may agree to set their clocks at $t = t' = 0$ (when $z = z' = 0$), but they will disagree on the subsequent passage of time [Eq. (2.27)]. This also implies that events at different locations in z that appear to be simultaneous in one frame of reference may not be simultaneous in another.

Equations (2.24)–(2.27) may be used to derive transformation laws for particle velocities. The differentials of primed quantities are

$$dx' = dx, \qquad dy' = dy, \qquad dz' = \gamma(dz - u\,dt),$$

$$dt' = \gamma\,dt\left(1 - uv_z/c^2\right).$$

In the special case where a particle has only a longitudinal velocity equal to u, the particle is at rest in the primed frame. For this condition, time dilation and Lorentz contraction proceed directly from the above equations.

Velocity in the primed frame is dx'/dt'. Substituting from above,

$$v'_x = \frac{v_x}{\gamma\left(1 - uv_z/c^2\right)}. \tag{2.28}$$

When a particle has no longitudinal motion in the primed frame (i.e., the primed frame is the rest frame and $v_z = u$), the transformation of transverse velocity is

$$v'_x = \gamma v_x. \tag{2.29}$$

This result follows directly from time dilation. Transverse distances are the same in both frames, but time intervals are longer in the stationary frame.

The transformation of axial particle velocities can be found by substitution for dz' and dt',

$$\frac{dz'}{dt'} = \frac{\gamma\,dt\left(dz/dt - u\right)}{\gamma\,dt\left(1 - uv_z/c^2\right)},$$

or

$$v'_z = \frac{v_z - u}{1 - uv_z/c^2}. \tag{2.30}$$

This can be inverted to give

$$v_z = \frac{v_z' + u}{1 + uv_z'/c^2}.$$

$$\boxed{2.31}$$

Equation (2.31) is the relativistic velocity addition law. If a particle has a velocity v_z' in the primed frame, then Eq. (2.31) gives observed particle velocity in a frame moving at $-u$. For $v_z' \to c$, inspection of Eq. (2.31) shows that v_z also approaches c. The implication is that there is no frame of reference in which a particle is observed to move faster than the velocity of light. A corollary is that no matter how much a particle's kinetic energy is increased, it will never be observed to move faster than c. This property has important implications in particle acceleration. For example, departures from the Newtonian velocity addition law set a limit on the maximum energy available from cyclotrons. In high-power, multi-MeV electron extractors, saturation of electron velocity is an important factor in determining current propagation limits.

2.9 RELATIVISTIC FORMULAS

The motion of high-energy particles must be described by relativistic laws of motion. Force is related to momentum by the same equation used in Newtonian mechanics

$$d\mathbf{p}/dt = \mathbf{F}.$$

$$(2.32)$$

This equation is consistent with the Lorentz transformations if the momentum is defined as

$$\mathbf{p} = \gamma m_0 \mathbf{v}.$$

$$\boxed{2.33}$$

The difference from the Newtonian expression is the γ factor. It is determined by the total particle velocity v observed in the stationary frame, $\gamma = (1 - v^2/c^2)^{-1/2}$. One interpretation of Eq. (2.33) is that a particle's effective mass increases as it approaches the speed of light. The *relativistic* mass is related to the rest mass by

$$m = \gamma m_0.$$

$$\boxed{2.34}$$

The relativistic mass grows without limit as $v_z \to c$. Thus, the momentum increases although there is a negligible increase in velocity.

In order to maintain Eq. (2.6), relating changes of energy to movement under the influence of a force, particle energy must be defined as

$$E = \gamma m_0 c^2.$$

$$\boxed{2.35}$$

The energy is not zero for a stationary particle, but approaches m_0c^2, which is called the *rest energy*. The kinetic energy (the portion of energy associated with motion) is given by

$$T = E - m_0c^2 = m_0c^2(\gamma - 1). \qquad \boxed{2.36}$$

Two useful relationships proceed directly from Eqs. (2.20), (2.33), and (2.35):

$$E = \left(c^2p^2 + m_0^2c^4\right)^{1/2}, \qquad \boxed{2.37}$$

where $p^2 = \mathbf{p} \cdot \mathbf{p}$, and

$$\mathbf{v} = c^2\mathbf{p}/E. \qquad \boxed{2.38}$$

The significance of the rest energy and the region of validity of Newtonian mechanics is clarified by expanding Eq. (2.35) in limit that $v/c \ll 1$.

$$E = \frac{m_0c^2}{\left(1 - v^2/c^2\right)^{1/2}}$$

$$= m_0c^2\left(1 + v^2/2c^2 + \cdots \right). \qquad (2.39)$$

The Newtonian expression for T [Eq. (2.8)] is recovered in the second term. The first term is a constant component of the total energy, which does not affect Newtonian dynamics. Relativistic expressions must be used when $T \geq m_0c^2$. The rest energy plays an important role in relativistic mechanics.

Rest energy is usually given in units of electron volts. Electrons are relativistic when T is in the MeV range, while ions (with a much larger mass) have rest energies in the GeV range. Figure 2.6 plots β for particles of interest for accelerator applications as a function of kinetic energy. The Newtonian result is also shown. The graph shows saturation of velocity at high energy and the energy range where departures from Newtonian results are significant.

2.10 NONRELATIVISTIC APPROXIMATION FOR TRANSVERSE MOTION

A relativistically correct description of particle motion is usually more difficult to formulate and solve than one involving Newtonian equations. In the study of the transverse motions of charged particle beams, it is often possible to express the problem in the form of Newtonian equations with the rest mass replaced by the relativistic mass. This approximation is valid when the beam is well directed so that transverse velocity components are small compared to the axial velocity of beam particles.

Particle Dynamics

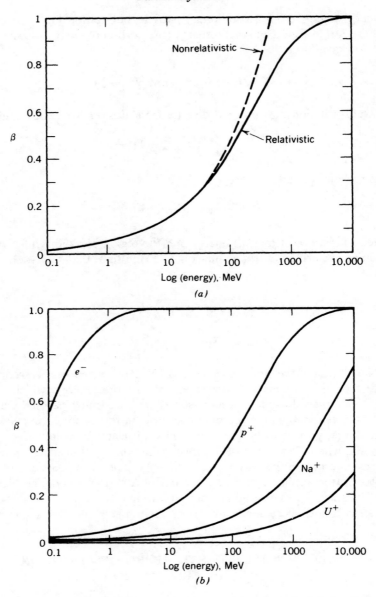

Figure 2.6 Particle velocity normalized to the speed of light as a function of kinetic energy. (*a*) Protons: solid line, relativistic prediction; dashed line, Newtonian prediction. (*b*) Relativistic predictions for various particles.

Consider the effect of focusing forces applied in the x direction to confine particles along the z axis. Particles make small angles with this axis, so that v_x is always small compared to v_z. With $\mathbf{F} = \hat{x} F_x$, Eq. (2.32) can be written in the form

$$\gamma m_0 v_x (\dot{\gamma}/\gamma + \dot{v}_x/v_x) = F_x, \qquad (2.40)$$

where the dot indicates a time derivative. Equation (2.39) can be rewritten as

$$E = m_0 c^2 \left(1 + \left(v_z^2 + v_x^2 \right)/2c^2 + 3 \left(v_z^2 + v_x^2 \right)^2/8c^4 + \cdots \right).$$

When $v_x \ll v_z$, relative changes in γ due to the transverse motion are small. In Eq. (2.40), the first term in parenthesis is much less than the second, so that the equation of motion is approximately

$$\gamma m_0 (dv_x/dt) \cong F_x. \qquad (2.41)$$

This has the form of a Newtonian expression with m_0 replaced by γm_0.

3

Electric and Magnetic Forces

Electromagnetic forces determine all essential features of charged particle acceleration and transport. This chapter reviews basic properties of electromagnetic forces. Advanced topics, such as particle motion with time-varying forces, are introduced throughout the book as they are needed.

It is convenient to divide forces between charged particles into electric and magnetic components. The relativistic theory of electrodynamics shows that these are manifestations of a single force. The division into electric and magnetic interactions depends on the frame of reference in which particles are observed.

Section 3.1 introduces electromagnetic forces by considering the mutual interactions between pairs of stationary charges and current elements. Coulomb's law and the law of Biot and Savart describe the forces. Stationary charges interact through the electric force. Charges in motion constitute currents. When currents are present, magnetic forces also act.

Although electrodynamics is described completely by the summation of forces between individual particles, it is advantageous to adopt the concept of fields. Fields (Section 3.2) are mathematical constructs. They summarize the forces that could act on a test charge in a region with a specified distribution of other charges. Fields characterize the electrodynamic properties of the charge distribution. The Maxwell equations (Section 3.3) are direct relations between electric and magnetic fields. The equations determine how fields arise from distributed charge and current and specify how field components are related to each other.

Electric and magnetic fields are often visualized as vector lines since they obey equations similar to those that describe the flow of a fluid. The field magnitude (or strength) determines the density of lines. In this interpretation, the Maxwell equations are fluidlike equations that describe the creation and flow of field lines. Although it is unnecessary to assume the physical existence of field lines, the concept is a powerful aid to intuit complex problems.

The Lorentz law (Section 3.2) describes electromagnetic forces on a particle as a function of fields and properties of the test particle (charge, position and velocity). The Lorentz force is the basis for all orbit calculations in this book. Two useful subsidiary functions of field quantities, the electrostatic and vector potentials, are discussed in Section 3.4. The electrostatic potential (a function of position) has a clear physical interpretation. If a particle moves in a static electric field, the change in kinetic energy is equal to its charge multiplied by the change in electrostatic potential. Motion between regions of different potential is the basis of electrostatic acceleration. The interpretation of the vector potential is not as straightforward. The vector potential will become more familiar through applications in subsequent chapters.

Section 3.6 describes an important electromagnetic force calculation, motion of a charged particle in a uniform magnetic field. Expressions for the relativistic equations of motion in cylindrical coordinates are derived in Section 3.5 to facilitate this calculation.

3.1 FORCES BETWEEN CHARGES AND CURRENTS

The simplest example of electromagnetic forces, the mutual force between two stationary point charges, is illustrated in Figure 3.1a. The force is directed along the line joining the two particles, $\mathbf{r}$. In terms of $\hat{r}$ (a vector of unit length aligned along $\mathbf{r}$), the force on particle 2 from particle 1 is

$$\mathbf{F}(1 \rightarrow 2) = (1/4\pi\varepsilon_0)q_1 q_2 \hat{r}/r^2 \text{ (newtons).} \qquad \boxed{3.1}$$

The value of ε_0 is

$$\varepsilon_0 = 8.85 \times 10^{-12} \text{ A-s/V-m}$$

In Cartesian coordinates, $\mathbf{r} = (x_2 - x_1)\hat{x} + (y_2 - y_1)\hat{y} + (z_2 - z_1)\hat{z}$. Thus, $r^2 = (x_2 - x_1)^2 + (y_2 - y_1)^2 + (z_2 - z_1)^2$. The force on particle 1 from particle 2 is equal and opposite to that of Eq. (3.1). Particles with the same polarity of charge repel one another. This fact affects high-current beams. The electrostatic repulsion of beam particles causes beam expansion in the absence of strong focusing.

Currents are charges in motion. Current is defined as the amount of charge in a certain cross section (such as a wire) passing a location in a unit of time. The mks unit of current is the ampere (coulombs per second). Particle beams

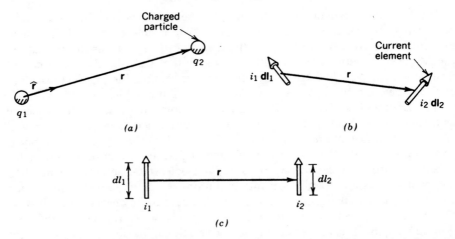

Figure 3.1 Forces between charged particles. (*a*) Electrostatic force. (*b*) Magnetostatic force. (*c*) Magnetostatic force between parallel current elements.

may have charge and current. Sometimes, charge effects can be neutralized by adding particles of opposite-charge sign, leaving only the effects of current. This is true in a metal wire. Electrons move through a stationary distribution of positive metal ions. The force between currents is described by the law of Biot and Savart. If $i_2\,\mathbf{dl}_1$ and $i_2\,\mathbf{dl}_2$ are current elements (e.g., small sections of wires) oriented as in Figure 3.1*b*, the force on element 2 from element 1 is

$$d\mathbf{F} = (\mu_0/4\pi)[(i_2\,\mathbf{dl}_2 \times (i_2\,\mathbf{dl}_1 \times \hat{r}))]/r^2, \qquad \boxed{3.2}$$

where

$$\mu_0 = 4\pi \times 10^{-7} = 1.26 \times 10^{-6} \text{ V-s/A-m}.$$

Equation (3.2) is more complex than (3.1); the direction of the force is determined by vector cross products. Resolution of the cross products for the special case of parallel current elements is shown in Figure 3.1*c*. Equation (3.2) becomes

$$d\mathbf{F}(1 \to 2) = -(\mu_0/4\pi)(i_1 i_2\, dl_1\, dl_2/r^2)\hat{r}.$$

Currents in the same direction attract one another. This effect is important in high-current relativistic electron beams. Since all electrons travel in the same direction, they constitute parallel current elements, and the magnetic force is attractive. If the electric charge is neutralized by ions, the magnetic force dominates and relativistic electron beams can be self-confined.

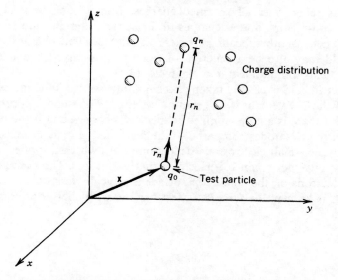

Figure 3.2 Electrical force exerted on a test particle by a distribution of charged particles.

3.2 THE FIELD DESCRIPTION AND THE LORENTZ FORCE

It is often necessary to calculate electromagnetic forces acting on a particle as it moves through space. Electric forces result from a specified distribution of charge. Consider, for instance, a low-current beam in an electrostatic acceleration. Charges on the surfaces of the metal electrodes provide acceleration and focusing. The electric force on beam particles at any position is given in terms of the specified charges by

$$\mathbf{F} = \sum (1/4\pi\varepsilon_0) q_0 q_n \hat{r}_n / r_n^2,$$

where q_0 is the charge of a beam particle and the sum is taken over all the charges on the electrodes (Fig. 3.2).

In principle, particle orbits can be determined by performing the above calculation at each point of each orbit. A more organized approach proceeds from recognizing that (1) the potential force on a test particle at any position is a function of the distribution of external charges and (2) the net force is proportional to the charge of the test particle. The function $\mathbf{F}(x)/q_0$ characterizes the action of the electrode charges. It can be used in subsequent calculations to determine the orbit of any test particle. The function is called the *electric field* and is defined by

$$\mathbf{E}(\mathbf{x}) = \sum (1/4\pi\varepsilon_0) q_n \hat{r}_n / r_2^2. \qquad \boxed{3.3}$$

The sum is taken over all specified charges. It may include freely moving charges in conductors, bound charges in dielectric materials, and free charges in space (such as other beam particles). If the specified charges move, the electric field may also be a function of time; in this case, the equations that determine fields are more complex than Eq. (3.3).

The electric field is usually taken as a smoothly varying function of position because of the $1/r^2$ factor in the sum of Eq. (3.3). The smooth approximation is satisfied if there is a large number of specified charges, and if the test charge is far from the electrodes compared to the distance between specified charges. As an example, small electrostatic deflection plates with an applied voltage of 100 V may have more than 10^{10} electrons on the surfaces. The average distance between electrons on the conductor surface is typically less than 1 μm.

When **E** is known, the force on a test particle with charge q_0 as a function of position is

$$\mathbf{F(x)} = q\mathbf{E(x)}. \qquad \boxed{3.4}$$

This relationship can be inverted for measurements of electric fields. A common nonperturbing technique is to direct a charged particle beam through a region and infer electric field by the acceleration or deflection of the beam.

A summation over current elements similar to Eq. (3.3) can be performed using the law of Biot and Savart to determine forces that can act on a differential test element of current. This function is called the *magnetic field* **B**. (Note that in some texts, the term magnetic field is reserved for the quantity **H**, and **B** is called the magnetic induction.) In terms of the field, the magnetic force on $i\,\mathbf{dl}$ is

$$\mathbf{dF} = i\,\mathbf{dl} \times \mathbf{B}. \qquad (3.5)$$

Equation (3.5) involves the vector cross product. The force is perpendicular to both the current element and magnetic field vector.

An expression for the total electric and magnetic forces on a single particle is required to treat beam dynamics. The differential current element, $i\,\mathbf{dl}$, must be related to the motion of a single charge. The correspondence is illustrated in Figure 3.3. The test particle has charge q and velocity **v**. It moves a distance dl in a time $dt = |dl|/|v|$. The current (across an arbitrary cross section) represented by this motion is $q/(|dl|/|v|)$. A moving charged particle acts like a current element with

$$i\,\mathbf{dl} = \frac{q\,\mathbf{dl}}{|dl|/|v|} = q\mathbf{v}.$$

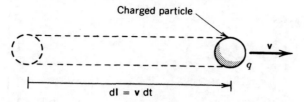

Figure 3.3 Representation of a moving charged particle as a current element.

The magnetic force on a charged particle is

$$\mathbf{F} = q\mathbf{v} \times \mathbf{B}.$$

3.6

Equations (3.4) and (3.6) can be combined into a single expression (the *Lorentz force law*)

$$\mathbf{F}(\mathbf{x}, t) = q(\mathbf{E} + \mathbf{v} \times \mathbf{B}).$$

3.7

Although we derived Equation (3.7) for static fields, it holds for time-dependent fields as well. The Lorentz force law contains all the information on the electromagnetic force necessary to treat charged particle acceleration. With given fields, charged particle orbits are calculated by combining the Lorentz force expression with appropriate equations of motion.

In summary, the field description has the following advantages.

1. Fields provide an organized method to treat particle orbits in the presence of large numbers of other charges. The effects of external charges are summarized in a single, continuous function.

2. Fields are themselves described by equations (Maxwell equations). The field concept extends beyond the individual particle description. Chapter 4 will show that field lines obey geometric relationships. This makes it easier to visualize complex force distributions and to predict charged particle orbits.

3. Identification of boundary conditions on field quantities sometimes makes it possible to circumvent difficult calculations of charge distributions in dielectrics and on conducting boundaries.

4. It is easier to treat time-dependent electromagnetic forces through direct solution for field quantities.

The following example demonstrates the correspondence between fields and charged particle distributions. The parallel plate capacitor geometry is shown in Figure 3.4. Two infinite parallel metal plates are separated by a distance d. A battery charges the plates by transferring electrons from one plate to the other. The excess positive charge and negative electron charge spread uniformly on the inside surfaces. If this were not true, there would be electric fields inside the metal. The problem is equivalent to calculating the electric

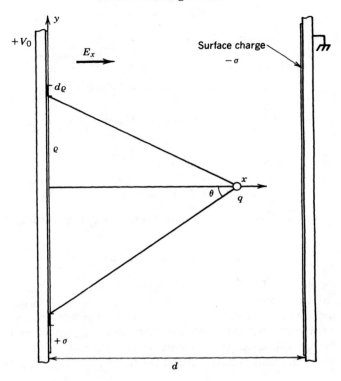

Figure 3.4   Calculation of force on a test particle between uniformly charged plates.

fields from two thin sheets of charge, as shown in Figure 3.4. The surface charge densities, $\pm\sigma$ (in coulombs per square meter), are equal in magnitude and opposite in sign.

A test particle is located between the plates a distance x from the positive electrode. Figure 3.4 defines a convenient coordinate system. The force from charge in the differential annulus illustrated is repulsive. There is force only in the x direction; by symmetry transverse forces cancel. The annulus has charge $(2\pi\rho\,d\rho\,\sigma)$ and is a distance $(\rho^2 + x^2)^{1/2}$ from the test charge. The total force [from Eq. (3.1)] is multiplied by $\cos\theta$ to give the x component.

$$dF_x = \frac{2\pi\rho\,d\rho\,\sigma q_0\cos\theta}{4\pi\varepsilon_0(\rho^2 + x^2)},$$

where $\cos\theta = x/(\rho^2 + x^2)^{1/2}$. Integrating the above expression over ρ from 0 to ∞ gives the net force

$$F^+ = \int \frac{\rho \, d\rho \, \sigma q_0 x}{2\varepsilon_0 (\rho^2 + x^2)^{3/2}} = \frac{q_0 \sigma}{2\varepsilon_0}. \tag{3.8}$$

A similar result is obtained for the force from the negative-charge layer. It is attractive and adds to the positive force. The electric field is found by adding the forces and dividing by the charge of the test particle

$$E_x(x) = x(F^+ - F^-)/q = \sigma/\varepsilon_0. \qquad \boxed{3.9}$$

The electric field between parallel plates is perpendicular to the plates and has uniform magnitude at all positions. Approximations to the parallel plate geometry are used in electrostatic deflectors; particles receive the same impulse independent of their position between the plates.

3.3   THE MAXWELL EQUATIONS

The Maxwell equations describe how electric and magnetic fields arise from currents and charges. They are continuous differential equations and are most conveniently written if charges and currents are described by continuous functions rather than by discrete quantities. The source functions are the *charge density*, $\rho(x, y, z, t)$ and *current density* $\mathbf{j}(x, y, z, t)$.

The charge density has units of coulombs per cubic meters (in mks units). Charges are carried by discrete particles, but a continuous density is a good approximation if there are large numbers of charged particles in a volume element that is small compared to the minimum scale length of interest. Discrete charges can be included in the Maxwell equation formulation by taking a charge density of the form $\rho = q\delta[\mathbf{x} - \mathbf{x}_0(t)]$. The delta function has the following properties:

$$\delta(\mathbf{x} - \mathbf{x}_0) = 0, \qquad \text{if} \quad \mathbf{x} \neq \mathbf{x}_0,$$

$$\int dx \int dy \int dz \, \delta(\mathbf{x} - \mathbf{x}_0) = 1. \tag{3.10}$$

The integral is taken over all space.

The current density is a vector quantity with units amperes per square meter. It is defined as the differential flux of charge, or the charge crossing a small surface element per second divided by the area of the surface. Current density can be visualized by considering how it is measured (Fig. 3.5). A small current probe of known area is adjusted in orientation at a point in space until the current reading is maximized. The orientation of the probe gives the

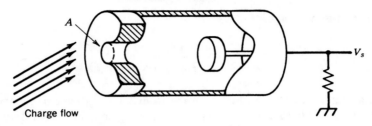

Figure 3.5   Probe to measure current density.

direction, and the current divided by the area gives the magnitude of the current density.

The general form of the Maxwell equations in mks units is

$$\nabla \times \mathbf{E} = -\partial \mathbf{B}/\partial t, \qquad \boxed{3.11}$$

$$\nabla \times \mathbf{B} = (1/c^2)(\partial \mathbf{E}/\partial t) + \mu_0 \mathbf{j}, \qquad \boxed{3.12}$$

$$\nabla \cdot \mathbf{E} = \rho/\varepsilon_0, \qquad \boxed{3.13}$$

$$\nabla \cdot \mathbf{B} = 0. \qquad \boxed{3.14}$$

Although these equations will not be derived, there will be many opportunities in succeeding chapters to discuss their physical implications. Developing an intuition and ability to visualize field distributions is essential for understanding accelerators. Characteristics of the Maxwell equations in the static limit and the concept of field lines will be treated in the next chapter.

No distinction has been made in Eqs. (3.11)–(3.14) between various classes of charges that may constitute the charge density and current density. The Maxwell equations are sometimes written in terms of vector quantities **D** and **H**. These are subsidiary quantities in which the contributions from charges and currents in linear dielectric or magnetic materials have been extracted. They will be discussed in Chapter 5.

3.4   ELECTROSTATIC AND VECTOR POTENTIALS

The electrostatic potential is a scalar function of the electric field. In other words, it is specified by a single value at every point in space. The physical meaning of the potential can be demonstrated by considering the motion of a charged particle between two parallel plates (Fig. 3.6). We want to find the change in energy of a particle that enters that space between the plates with kinetic energy T. Section 3.2 has shown that the electric field E_x is uniform.

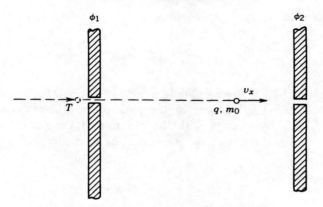

Figure 3.6 Acceleration of a particle between charged parallel plates.

The equation of motion is therefore

$$dp_x/dt = F_x = qE_x.$$

The derivative can be rewritten using the chain rule to give

$$(dp_x/dx)(dx/dt) = v_x\, dp_x/dx = qE_x.$$

The relativistic energy E of a particle is related to momentum by Eq. (2.37). Taking the derivative in x of both sides of Eq. (2.37) gives

$$c^2 p_x(dp_x/dx) = E(dE/dx).$$

This can be rearranged to give

$$dE/dx = \left[c^2 p_x/E\right] dp_x/dx = v_x\, dp_x/dx. \qquad (3.15)$$

The final form on the right-hand side results from substituting Eq. (2.38) for the term in brackets. The expression derived in Eq. (3.15) confirms the result quoted in Section 2.9. The right-hand side is dp_x/dt which is equal to the force F_x. Therefore, the relativistic form of the energy [Eq. (2.35)] is consistent with Eq. (2.6). The integral of Eq. (3.15) between the plates is

$$\Delta E = q\int dx\, E_x. \qquad (3.16)$$

The *electrostatic potential* ϕ is defined by

$$\phi = - \int \mathbf{E} \cdot \mathbf{dx}. \qquad \boxed{3.17}$$

The change in potential along a path in a region of electric fields is equal to the integral of electric field tangent to the path times differential elements of pathlength. Thus, by analogy with the example of the parallel plates [Eq. (3.16)] $\Delta E - q \Delta \phi$. If electric fields are static, the total energy of a particle can be written

$$E = m_0 c^2 + T_0 - q(\phi - \phi_0), \qquad \boxed{3.18}$$

where T_0 is the particle kinetic energy at the point where $\phi = \phi_0$.

The potential in Eq. (3.18) is not defined absolutely; a constant can be added without changing the electric field distribution. In treating electrostatic acceleration, we will adopt the convention that the zero point of potential is defined at the particle source (the location where particles have zero kinetic energy). The potential defined in this way is called the *absolute potential* (with respect to the source). In terms of the absolute potential, the total energy can be written

$$E = m_0 c^2 - q\phi$$

or

$$\gamma = \left(1 - q\phi/m_0 c^2\right). \qquad \boxed{3.19}$$

Finally, the static electric field can be rewritten in the differential form,

$$\mathbf{E} = -\nabla\phi = (\partial\phi/\partial x)\hat{x} + (\partial\phi/\partial y)\hat{y} + (\partial\phi/\partial z)\hat{z}$$

$$= E_x\hat{x} + E_y\hat{y} + E_z\hat{z}. \qquad \boxed{3.20}$$

If the potential is known as a function of position, the three components of electric field can be found by taking spatial derivatives (the gradient operation). The defining equation for electrostatic fields [Eq. (3.3)] can be combined with Eq. (3.20) to give an expression to calculate potential directly from a specified distribution of charges

$$\phi(\mathbf{x}) = \sum \frac{q_n/4\pi\varepsilon_0}{|\mathbf{x} - \mathbf{x}_n|}. \qquad (3.21)$$

The denominator is the magnitude of the distance from the test charge to the

nth charge. The integral form of this equation in terms of charge density is

$$\phi(\mathbf{x}) = \frac{1}{4\pi\varepsilon_0} \iiint d^3x' \frac{\rho(\mathbf{x}')}{|\mathbf{x} - \mathbf{x}'|} \tag{3.22}$$

Although Eq. 3.22 can be used directly to find the potential, we will usually use differential equations derived from the Maxwell equations combined with boundary conditions for such calculations (Chapter 4).

The *vector potential* **A** is another subsidiary quantity that can be valuable for computing magnetic fields. It is a vector function related to the magnetic field through the vector curl operation

$$\mathbf{B} = \nabla \times \mathbf{A}. \qquad \boxed{3.23}$$

This relationship is general, and holds for time-dependent fields. We will use **A** only for static calculations. In this case, the vector potential can be written as a summation over source current density

$$\mathbf{A}(\mathbf{x}) = \frac{\mu_0}{4\pi} \iiint d^3x' \frac{\mathbf{j}(\mathbf{x}')}{|\mathbf{x} - \mathbf{x}'|}. \tag{3.24}$$

Compared to the electrostatic potential, the vector potential does not have a straightforward physical interpretation. Nonetheless, it is a useful computational device and it is helpful for the solution of particle orbits in systems with geometry symmetry. In cylindrical systems it is proportional to the number of magnetic field lines encompassed within particle orbits (Section 7.4).

3.5 INDUCTIVE VOLTAGE AND DISPLACEMENT CURRENT

The static concepts already introduced must be supplemented by two major properties of time-dependent fields for a complete and consistent theory of electrodynamics. The first is the fact that time-varying magnetic fields lead to electric fields. This is the process of *magnetic induction*. The relationship between inductively generated electric fields and changing magnetic flux is stated in Faraday's law. This effect is the basis of betatrons and linear induction accelerators. The second phenomenon, first codified by Maxwell, is that a time-varying electric field leads to a virtual current in space, the *displacement current*. We can verify that displacement currents "exist" by measuring the magnetic fields they generate. A current monitor such as a Rogowski loop enclosing an empty space with changing electric fields gives a current reading. The combination of inductive fields with the displacement current leads to predictions of electromagnetic oscillations. Propagating and stationary electromagnetic waves are the bases for RF (radio-frequency) linear accelerators.

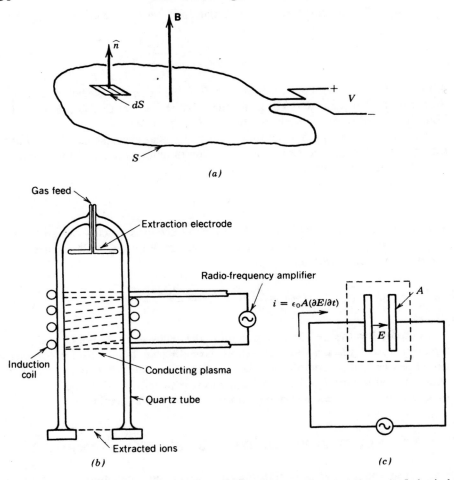

Figure 3.7 Inductive voltage and displacement currents. (*a*) Faraday's law. (*b*) Inductively coupled plasma source. (*c*) Alternating-current circuit with a parallel-plate capacitor.

Faraday's law is illustrated in Figure 3.7*a*. A wire loop defines a surface S. The magnetic flux Ψ passing through the loop is given by

$$\Psi = \int\int \mathbf{B} \cdot \hat{n}\, dS, \tag{3.25}$$

where $\hat{n}$ is a unit vector normal to S and dS is a differential element of surface area. Faraday's law states that a voltage is induced around the loop when the magnetic flux changes according to

$$V = -d\Psi/dt. \qquad \boxed{3.26}$$

The time derivative of Ψ is the total derivative. Changes in Ψ can arise from a time-varying field at constant loop position, motion of the loop to regions of different field magnitude in a static field, or a combination of the two.

The term *induction* comes from induce, to produce an effect without a direct action. This is illustrated by the example of Figure 3.7*b*, an inductively coupled plasma source. (A plasma is a conducting medium of hot, ionized gas.) Such a device is often used as an ion source for accelerators. In this case, the plasma acts as the loop. Currents driven in the plasma by changing magnetic flux ionize and heat the gas through resistive effects. The magnetic flux is generated by windings outside the plasma driven by a high-frequency ac power supply. The power supply couples energy to the plasma through the intermediary of the magnetic fields. The advantage of inductive coupling is that currents can be generated without immersed electrodes that may introduce contaminants.

The sign convention of Faraday's law implies that the induced plasma currents flow in the direction opposite to those of the driving loop. Inductive voltages always drive reverse currents in conducting bodies immersed in the magnetic field; therefore, oscillating magnetic fields are reduced or canceled inside conductors. Materials with this property are called *diamagnetic*. Inductive effects appear in the Maxwell equations on the right-hand side of Eq. (3.11). Application of the Stokes theorem (Section 4.1) shows that Eqs. (3.11) and (3.26) are equivalent.

The concept of displacement current can be understood by reference to Figure 3.7*c*. An electric circuit consists of an ac power supply connected to parallel plates. According to Eq. 3.9, the power supply produces an electric field E_x between the plates by moving an amount of charge

$$Q = \varepsilon_0 E_x A,$$

where A is the area of the plates. Taking the time derivative, the current through the power supply is related to the change in electric field by

$$i = \varepsilon_0 A (\partial E_x / \partial t). \qquad (3.27)$$

The partial derivative of Eq. (3.27) signifies that the variation results from the time variation of E_x with the plates at constant position. Suppose we considered the plate assembly as a black box without knowledge that charge was stored inside. In order to guarantee continuity of current around the circuit, we could postulate a virtual current density between the plates given by

$$j_d = \varepsilon_0 (\partial E_x / \partial t). \qquad (3.28)$$

This quantity, the displacement current density, is more than just an abstraction to account for a change in space charge inside the box. The experimentally observed fact is that there are magnetic fields around the plate assembly that are identical to those that would be produced by a real wire connecting the plates and carrying the current specified by Eq. (3.27) (see Section 4.6). There is thus a parallelism of time-dependent effects in electromagnetism. Time-varying magnetic fields produce electric fields, and changing electric fields produce magnetic fields. The coupling and interchange of electric and magnetic field energy is the basis of electromagnetic oscillations.

Displacement currents or, equivalently, the generation of magnetic fields by time-varying electric fields, enter the Maxwell equations on the right side of Eq. (3.12). Noting that

$$c = 1/(\varepsilon_0\mu_0)^{1/2}, \qquad \boxed{3.29}$$

we see that the displacement current is added to any real current to determine the net magnetic field.

3.6   RELATIVISTIC PARTICLE MOTION IN CYLINDRICAL COORDINATES

Beams with cylindrical symmetry are encountered frequently in particle accelerators. For example, electron beams used in applications such as electron microscopes or cathode ray tubes have cylindrical cross sections. Section 3.7 will introduce an important application of the Lorentz force, circular motion in a uniform magnetic field. In order to facilitate this calculation and to derive useful formulas for subsequent chapters, the relativistic equations of motion for particles in cylindrical coordinates are derived in this section.

Cylindrical coordinates, denoted by (r, θ, z), are based on curved coordinate lines. We recognize immediately that equations of the form $dp_r/dt = F_r$ are incorrect. This form implies that particles subjected to no radial force move in a circular orbit ($r = $ constant, $dp_r/dt = 0$). This is not consistent with Newton's first law. A simple method to derive the proper equations is to express $d\mathbf{p}/dt = \mathbf{F}$ in Cartesian coordinates and make a coordinate transformation by direct substitution.

Reference to Figure 3.8 shows that the following equations relate Cartesian coordinates to cylindrical coordinates sharing a common origin and a common z axis, and with the line $(r, 0, 0)$ lying on the x axis:

$$x = r\cos\theta, \qquad y = r\sin\theta, \qquad z = z, \qquad (3.30)$$

and

$$r = (x^2 + y^2)^{1/2}, \qquad \theta = \tan^{-1}(y/x). \qquad (3.31)$$

Motion along the z axis is described by the same equations in both frames, $dp_z/dt = F_x$. We will thus concentrate on equations in the (r, θ) plane. The Cartesian equation of motion in the x direction is

$$dp_x/dt = F_x. \qquad (3.32)$$

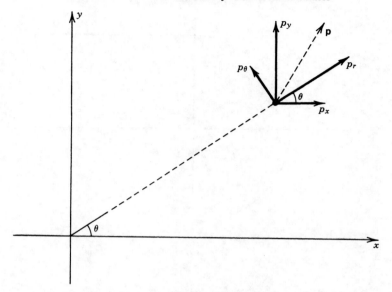

Figure 3.8 Relationship between cylindrical and Cartesian coordinates.

Figure 3.8 shows that

$$p_x = p_r\cos\theta - p_\theta\sin\theta, \qquad F_x = F_r\cos\theta - F_\theta\sin\theta.$$

Substituting in Eq. (3.32),

$$\dot{p}_r\cos\theta - p_r\sin\theta\,\dot{\theta} - \dot{p}_\theta\sin\theta - p_\theta\cos\theta\,\dot{\theta} = F_r\cos\theta - F_\theta\sin\theta$$

where dotted quantities are time derivatives. The equation must hold at all positions, or at any value of θ. Thus, terms involving $\cos\theta$ and $\sin\theta$ must be separately equal. This yields the cylindrical equations of motion

$$dp_r/dt = F_r + [\, p_\theta\, d\theta/dt\,], \qquad \boxed{3.33}$$

$$dp_\theta/dt = F_\theta - [\, p_r\, d\theta/dt\,]. \qquad \boxed{3.34}$$

The quantities in brackets are correction terms for cylindrical coordinates. Equations (3.33) and (3.34) have the form of the Cartesian equations if the bracketed terms are considered as *virtual forces*. The extra term in the radial equation is called the *centrifugal force*, and can be rewritten

$$\text{Centrifugal force} = -\gamma m_0 v_\theta^2/r, \qquad \boxed{3.35}$$

noting that $v_\theta = r\, d\theta/dt$. The bracketed term in the azimuthal equation is the

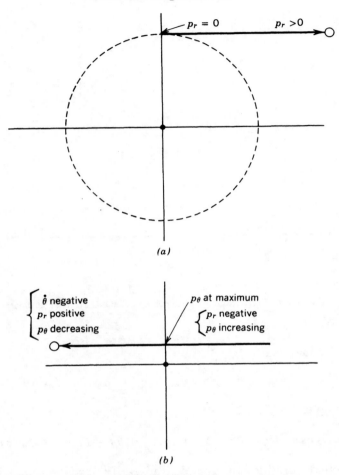

$p_r = 0$ $p_r > 0$

(a)

$\dot{\theta}$ negative
p_r positive
p_θ decreasing

p_θ at maximum
p_r negative
p_θ increasing

(b)

Figure 3.9 Virtual forces in a cylindrical coordinate system. (*a*) Centrifugal force: particle moves in straight line; p_r increases continually. (b) Coriolis force.

Coriolis force, and can be written

$$\text{Coriolis force} = -\gamma m_0 v_r v_\theta / r. \qquad \boxed{3.36}$$

Figure 3.9 illustrates the physical interpretation of the virtual forces. In the first example, a particle moves on a force-free, straight-line orbit. Viewed in the cylindrical coordinate system, the particle (with no initial v_r) appears to accelerate radially, propelled by the centrifugal force. At large radius, when $v_\theta \to 0$, the acceleration appears to stop, and the particle moves outward at constant velocity. The Coriolis force is demonstrated in the second example. A

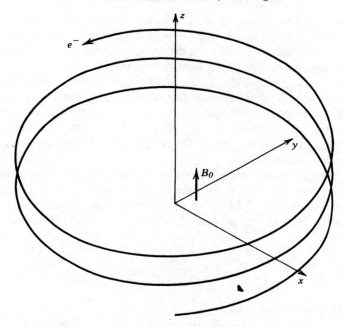

Figure 3.10   Helical orbit of a charged particle in a uniform magnetic field.

particle from large radius moves in a straight line past the origin with nonzero impact parameter. The azimuthal velocity, which was initially zero, increases as the particle moves inward with negative v_r and decreases as the particle moves out. The observer in the cylindrical coordinate system notes a negative and then positive azimuthal acceleration.

Cylindrical coordinates appear extensively in accelerator theory. Care must be exercised to identify properly the orientation of the coordinates. For example, the z axis is sometimes aligned with the beam axis, while in other cases, the z axis may be along a symmetry axes of the accelerator. In this book, to avoid excessive notation, (r, θ, z) will be used for all cylindrical coordinate systems. Illustrations will clarity the geometry of each case as it is introduced.

3.7   MOTION OF CHARGED PARTICLES IN A UNIFORM MAGNETIC FIELD

Motion of a charged particle in a uniform magnetic field directed along the z axis, $\mathbf{B} = B_0 \hat{z}$, is illustrated in Figure 3.10. Only the magnetic component of the Lorentz force is included. The equation of motion is

$$d\mathbf{p}/dt = d(\gamma m_0 \mathbf{v})/dt = q\mathbf{v} \times \mathbf{B}. \qquad (3.37)$$

By the nature of the cross product, the magnetic force is always perpendicular to the velocity of the particle. There is no force along a differential element of pathlength, **dx**. Thus, $\int \mathbf{F} \cdot \mathbf{dx} = 0$. According to Eq. (2.6), magnetic fields perform no work and do not change the kinetic energy of the particle. In Eq. (3.37), γ is constant and can be removed from the time derivative.

Since the force is perpendicular to **B**, there is no force along the z axis. Particles move in this direction with constant velocity. There is a force in the x–y plane. It is of constant magnitude (since the total particle velocity cannot change), and it is perpendicular to the particle motion. The projection of particle motion in the x–y plane is therefore a circle. The general three-dimensional particle orbit is a helix.

If we choose a cylindrical coordinate system with origin at the center of the circular orbit, then $dp_r/dt = 0$, and there is no azimuthal force. The azimuthal equation of motion [Eq. (3.34)] is satisfied trivially with these conditions. The radial equation [Eq. (3.33)] is satisfied when the magnetic force balances the centrifugal force, or

$$qv_\theta B_0 = \gamma m_0 v_\theta^2 / r.$$

The particle orbit radius is thus

$$r_g = \gamma m_0 v_\theta / |q| B_0. \qquad \boxed{3.38}$$

This quantity is called the *gyroradius*. It is large for high-momentum particles; the gyroradius is reduced by applying stronger magnetic field. The point about which the particle revolves is called the *gyrocenter*. Another important quantity is the angular frequency of revolution of the particle, the *gyrofrequency*. This is given by $\omega_g = v_\theta / r$, or

$$\omega_g = |q| B_0 / \gamma m_0. \qquad \boxed{3.39}$$

The particle orbits in Cartesian coordinates are harmonic,

$$x(t) - x_0 = r_g \cos \omega_g t,$$

$$y(t) - y_0 = r_g \sin \omega_g t,$$

where x_0 and y_0 are the coordinates of the gyrocenter. The gyroradius and gyrofrequency arise in all calculations involving particle motion in magnetic fields. Magnetic confinement of particles in circular orbits forms the basis for recirculating high-energy accelerators, such as the cyclotron, synchrotron, microtron, and betatron.

4

Steady-State Electric and Magnetic Fields

A knowledge of electric and magnetic field distributions is required to determine the orbits of charged particles in beams. In this chapter, methods are reviewed for the calculation of fields produced by static charge and current distributions on external conductors. Static field calculations appear extensively in accelerator theory. Applications include electric fields in beam extractors and electrostatic accelerators, magnetic fields in bending magnets and spectrometers, and focusing forces of most lenses used for beam transport.

Slowly varying fields can be approximated by static field calculations. A criterion for the static approximation is that the time for light to cross a characteristic dimension of the system in question is short compared to the time scale for field variations. This is equivalent to the condition that connected conducting surfaces in the system are at the same potential. Inductive accelerators (such as the betatron) appear to violate this rule, since the accelerating fields (which may rise over many milliseconds) depend on time-varying magnetic flux. The contradiction is removed by noting that the velocity of light may be reduced by a factor of 100 in the inductive media used in these accelerators. Inductive accelerators are treated in Chapters 10 and 11. The study of rapidly varying vacuum electromagnetic fields in geometries appropriate to particle acceleration is deferred to Chapters 14 and 15.

The static form of the Maxwell equations in regions without charges or currents is reviewed in Section 4.1. In this case, the electrostatic potential is determined by a second-order differential equation, the Laplace equation.

Magnetic fields can be determined from the same equation by defining a new quantity, the magnetic potential. Examples of numerical (Section 4.2) and analog (Section 4.3) methods for solving the Laplace equation are discussed. The numerical technique of successive overrelaxation is emphasized since it provides insight into the physical content of the Laplace equation. Static electric field calculations with field sources are treated in Section 4.4. The classification of charge is emphasized; a clear understanding of this classification is essential to avoid confusion when studying space charge and plasma effects in beams. The final sections treat the calculation of magnetic fields from specific current distributions through direct solution of the Maxwell equations (Section 4.5) and through the intermediary of the vector potential (Section 4.6).

4.1 STATIC FIELD EQUATIONS WITH NO SOURCES

When there are no charges or currents present, the Maxwell equations become

$$\nabla \cdot \mathbf{E} = 0, \tag{4.1}$$

$$\nabla \times \mathbf{E} = 0, \tag{4.2}$$

$$\nabla \cdot \mathbf{B} = 0, \tag{4.3}$$

$$\nabla \times \mathbf{B} = 0. \tag{4.4}$$

These equations resolve into two decoupled and parallel sets for electric fields [Eqs. (4.1) and (4.2)] and magnetic fields [Eqs. (4.3) and (4.4)]. Equations (4.1)–(4.4) hold in regions such as that shown in Figure 4.1. The charges or currents that produce the fields are external to the volume of interest. In electrostatic calculations, the most common calculation involves charge distributed on the surfaces of conductors at the boundaries of a vacuum region.

Equations (4.1)–(4.4) have straightforward physical interpretations. Similar conclusions hold for both sets, so we will concentrate on electric fields. The form for the divergence equation [Eq. (4.1)] in Cartesian coordinates is

$$\partial E_x/\partial x + \partial E_y/\partial y + \partial E_z/\partial z = 0. \tag{4.5}$$

An example is illustrated in Figure 4.2. The electric field is a function of x and y. The meaning of the divergence equation can be demonstrated by calculating the integral of the normal electric field over the surface of a volume with cross-sectional area A and thickness Δx. The integral over the left-hand side is $AE_x(x)$. If the electric field is visualized in terms of vector field lines, the integral is the *flux* of lines into the volume through the left-hand face. The electric field line flux out of the volume through the right-hand face is $AE_x(x + \Delta x)$.

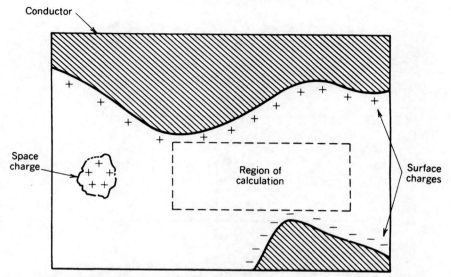

Figure 4.1 Calculation of static electric fields in a region with no source charge.

When the electric field is a smooth function of x, variations about a point can be approximated by a Taylor expansion. The right-hand integral is $A[E_x(x) + \Delta x \, \partial E_x / \partial x]$. The condition that $\partial E_x / \partial x = 0$ leads to a number of parallel conclusions.

1. The integrals of normal electric field over both faces of the volume are equal.
2. All field lines that enter the volume must exit.
3. The net flux of electric field lines into the volume is zero.
4. No field lines originate inside the volume.

Equation (4.5) is the three-dimensional equivalent of these statements.

The *divergence operator* applied to a vector quantity gives the effluence of the quantity away from a point in space. The *divergence theorem* can be written

$$\iint \mathbf{E} \cdot \hat{n} \, da = \iiint (\nabla \cdot \mathbf{E}) \, dV. \qquad \boxed{4.6}$$

Equation (4.6) states that the integral of the divergence of a vector quantity over all points of a volume is equal to the surface integral of the normal component of the vector over the surface of the volume. With no enclosed charges, field lines must flow through a volume as shown in Figure 4.3. The same holds true for magnetic fields. The main difference between electric and magnetic fields is that magnetic field lines have zero divergence under all

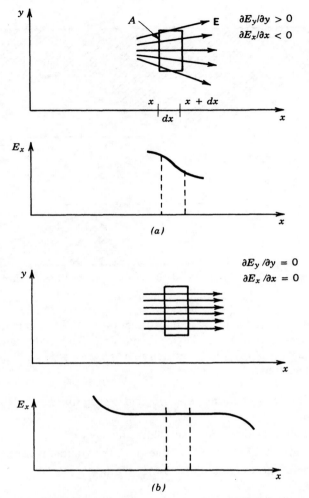

Figure 4.2 Relationship between gradients of electric field components and geometry of electric field lines; field with variations in x and y. a) Field with gradient at x. b) Field with zero gradient at x.

conditions, even in regions with currents. This means that magnetic field lines never emanate from a source point. They either extend indefinitely or are self-connected.

The curl equations determine another geometric property of field lines. This property proceeds from the Stokes theorem, which states that

$$\int \mathbf{E} \cdot \mathbf{dl} = \int\int (\nabla \times \mathbf{E}) \cdot \hat{n}\, da.$$

$$\boxed{4.7}$$

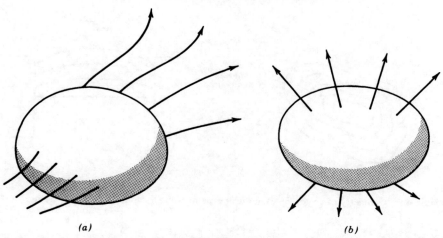

Figure 4.3 Geometry of vector field lines near a point with (*a*) zero divergence and (*b*) nonzero divergence.

The quantities in Eq. (4.7) are defined in Figure 4.4; S is a two-dimensional surface in space and **dl** is a length element oriented along the circumference. The integral on the left-hand side is taken around the periphery. The right-hand side is the surface integral of the component of the vector $\nabla \times \mathbf{E}$ normal to the surface. If the curl is nonzero at a point in space, then field lines form closed loops around the point. Figure 4.5 illustrates points in vector fields with zero and nonzero curl. The study of magnetic fields around current-carrying wires (Section 4.5) will illustrate a vector function with a nonzero curl.

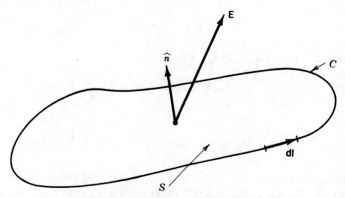

Figure 4.4 Definition of quantities used in Stokes theorem.

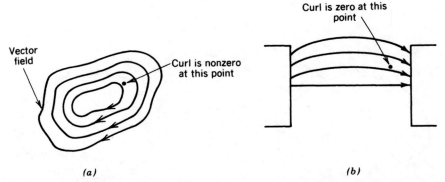

Figure 4.5   Geometry of vector field lines near a point with (*a*) nonzero curl and (*b*) zero curl.

For reference, the curl operator is written in Cartesian coordinates as

$$\nabla \times \mathbf{E} = \begin{vmatrix} \hat{x} & \hat{y} & \hat{z} \\ \partial/\partial x & \partial/\partial y & \partial/\partial z \\ E_x & E_y & E_z \end{vmatrix}. \tag{4.8}$$

The usual rule for evaluating a determinant is used. The expansion of the above expression is

$$\nabla \times \mathbf{E} = \hat{x}\left(\frac{\partial E_z}{\partial y} - \frac{\partial E_y}{\partial z}\right) + \hat{y}\left(\frac{\partial E_x}{\partial z} - \frac{\partial E_z}{\partial x}\right) + \hat{z}\left(\frac{\partial E_y}{\partial x} - \frac{\partial E_x}{\partial y}\right).$$

$$\tag{4.9}$$

The electrostatic potential function ϕ can be defined when electric fields are static. The electric field is the gradient of this function,

$$\mathbf{E} = -\nabla\phi. \tag{4.10}$$

Substituting for **E** in Eq. (4.1) gives

$$\nabla \cdot (\nabla\phi) = 0,$$

or

$$\nabla^2\phi = \partial^2\phi/\partial x^2 + \partial^2\phi/\partial y^2 + \partial^2\phi/\partial z^2 = 0. \qquad \boxed{4.11}$$

The operator symbolized by ∇^2 in Eq. (4.11) is called the *Laplacian operator*. Equation (4.11) is the *Laplace equation*. It determines the variation of ϕ (and hence **E**) in regions with no charge. The curl equation is automatically satisfied through the vector identity $\nabla \times (\nabla\phi) = 0$.

The main reason for using the Laplace equation rather than solving for electric fields directly is that boundary conditions can be satisfied more easily. The difficulty in solving the Maxwell equations directly lies in determining boundary conditions for vector fields on surrounding conducting surfaces. The electrostatic potential is a scalar function; we can show that the potential is a constant on a connected metal surface. Metals contain free electrons; an electric field parallel to the surface of a metal drives large currents. Electrons in the metal adjust their positions to produce a parallel component of field equal and opposite to the applied field. Thus, at a metal surface $E_{\parallel} = 0$ and $E_{\perp}$ is unspecified. Equation (4.10) implies that electric field lines are always normal to surfaces of constant ϕ. This comes about because the gradient of a function (which indicates the direction in which a function has maximum rate of variation) must always be perpendicular to surfaces on which the function is constant (Fig. 4.6). Since a metal surface is everywhere perpendicular to the

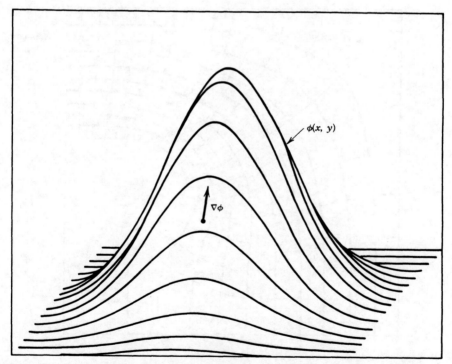

Figure 4.6   Three-dimensional plot of a scalar function $\phi(x, y)$, illustrating the orientation of the gradient vector $\nabla\phi$.

electric field, it must be an *equipotential surface* with the boundary condition ϕ = constant.

In summary, electric field lines have the following properties in source-free regions:

(a) Field lines are continuous. All lines that enter a volume eventually exit.

(b) Field lines do not kink, curl, or cross themselves.

(c) Field lines do not cross each other, since this would result in a point of infinite flux.

(d) Field lines are normal to surfaces of constant electrostatic potential.

(e) Electric fields are perpendicular to metal surfaces.

Fairly accurate electric field sketches can be made utilizing the laminar flow nature of electric field lines and the above properties. Even with the availability of digital computers, it is valuable to generate initial sketches of field patterns.

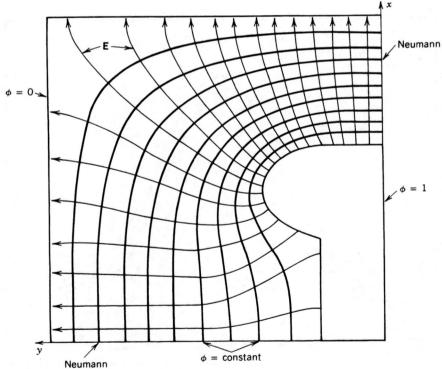

Figure 4.7 Equipotential lines and electric field lines around a high-voltage electrode in a grounded case. Electrode and case have infinite extent in z direction (out of page). Solution carried out only in one quadrant because of symmetry about x and y axes.

This saves time and gives insight into the nature of fields. An example of an electrostatic field pattern generated by the method of squares is shown in Figure 4.7. In this method, a number of equipotential lines between metal surfaces are sketched. Electric field lines normal to the equipotential lines and electrodes are added. Since the density of field lines is proportional to the distance between equipotentials, a valid final solution results when the elements between equipotential and field lines approach as close as possible to squares. The process is iterative and requires only some drawing ability and an eraser.

It is also possible to define formally a magnetic potential U_m such that $\mathbf{B} = \nabla U_m$. The magnetic potential satisfies the Laplace equation

$$\nabla^2 U_m = 0. \qquad \boxed{4.12}$$

The function U_m should not be confused with the vector potential. Methods used for electric field problems in source-free regions can also be applied to determine magnetic fields. We will defer use of Eq. (4.12) to Chapter 5. An understanding of magnetic materials is necessary to determine boundary conditions for U_m.

4.2 NUMERICAL SOLUTIONS TO THE LAPLACE EQUATION

The Laplace equation determines electrostatic potential as a function of position. Resulting electric fields can then be used to calculate particle orbits. Electrostatic problems may involve complex geometries with surfaces at many different potentials. In this case, numerical methods of analysis are essential.

Digital computers handle discrete quantities, so the Laplace equation must be converted from a continuous differential equation to a finite difference formulation. As shown in Figure 4.8, the quantity $\Phi(i, j, k)$ is defined at discrete points in space. These points constitute a three-dimensional mesh. For simplicity, the mesh spacing Δ between points in the three Cartesian directions is assumed uniform. The quantity Φ has the property that it equals $\phi(x, y, z)$ at the mesh points. If ϕ is a smoothly varying function, then a linear interpolation of Φ gives a good approximation for ϕ at any point in space. In summary, Φ is a mathematical construct used to estimate the physical quantity, ϕ.

The Laplace equation for ϕ implies an algebraic difference equation for Φ. The spatial position of a mesh point is denoted by (i, j, k), with $x = i\Delta$, $y = j\Delta$, and $z = k\Delta$. The x derivative of ϕ to the right of the point (x, y, z) is approximated by

$$\partial\phi(x + \Delta/2)/\partial x = [\Phi(i + 1, j, k) - \Phi(i, j, k)]/\Delta. \qquad (4.13)$$

A similar expression holds for the derivative at $x - \Delta/2$. The second deriva-

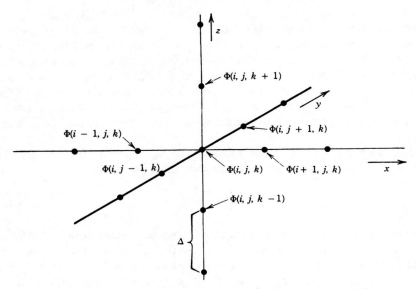

Figure 4.8 Finite difference approximation for electrostatic potential.

tive is the difference of derivatives divided by Δ, or

$$\frac{\partial}{\partial x}\left(\frac{\partial \phi(x)}{\partial x}\right) \cong \frac{1}{\Delta}\left(\frac{\partial \phi(x+\Delta/2)}{\partial x} - \frac{\partial \phi(x-\Delta/2)}{\partial x}\right). \quad (4.14)$$

Combining expressions,

$$\partial^2 \phi/\partial x^2 = [\Phi(i+1,j,k) - 2\Phi(i,j,k) + \Phi(i-1,j,k)]/\Delta^2.$$

Similar expressions can be found for the $\partial^2 \phi/\partial y^2$ and $\partial^2 \phi/\partial z^2$ terms. Setting $\nabla^2 \phi = 0$ implies

$$\Phi(i,j,k) = \tfrac{1}{6}[\Phi(i+1,j,k) + \Phi(i-1,j,k) + \Phi(i,j+1,k)$$

$$+ \Phi(i,j-1,k) + \Phi(i,j,k+1) + \Phi(i,j,k-1)].$$

$$\boxed{4.15}$$

In summary, (1) $\Phi(i,j,k)$ is a discrete function defined as mesh points, (2) the interpolation of $\Phi(i,j,k)$ approximates $\phi(x,y,z)$, and (3) if $\phi(x,y,z)$ satisfies the Laplace equation, then $\Phi(i,j,k)$ is determined by Eq. (4.15).

According to Eq. (4.15), individual values of $\Phi(i,j,k)$ are the average of their six neighboring points. Solving the Laplace equation is an averaging process; the solution gives the smoothest flow of field lines. The net length of all field lines is minimized consistent with the boundary conditions. Therefore, the solution represents the state with minimum field energy (Section 5.6).

There are many numerical methods to solve the finite difference form for the Laplace equation. We will concentrate on the *method of successive overrelaxation*. Although it is not the fastest method of solution, it has the closest relationship to the physical content of the Laplace equation. To illustrate the method, the problem will be formulated on a two-dimensional, square mesh. Successive overrelaxation is an iterative approach. A trial solution is corrected until it is close to a valid solution. Correction consists of sweeping through all values of an intermediate solution to calculate *residuals*, defined by

$$R(i, j) = \tfrac{1}{4}[\Phi(i + 1, j) + \Phi(i - 1, j)$$

$$+ \Phi(i, j + 1) + \Phi(i, j - 1)] - \Phi(i, j). \quad (4.16)$$

If $R(i, j)$ is zero at all points, then $\Phi(i, j)$ is the desired solution. An intermediate result can be improved by adding a correction factor proportional to $R(i, j)$,

$$\Phi(i, j)_{n+1} = \Phi(i, j)_n + \omega R_n. \quad (4.17)$$

The value $\omega = 1$ is the obvious choice, but in practice values of ω between 1 and 2 produce a faster convergence (hence the term overrelaxation). The succession of approximations resembles a time-dependent solution for a system with damping, relaxing to its lowest energy state. The elastic sheet analog (described in Section 4.3) is a good example of this interpretation. Figure 4.9 shows intermediate solutions for a one-dimensional mesh with 20 points and with $\omega = 1.00$. Information on the boundary with elevated potential propagates through the mesh.

The method of successive overrelaxation is quite slow for large numbers of points. The number of calculations on an $n \times n$ mesh is proportional to n^2. Furthermore, the number of iterations necessary to propagate errors out of the mesh is proportional to n. The calculational time increases as n^3. A BASIC algorithm to relax internal points in a 40×48 point array is listed in Table 4.1. Corrections are made continuously during the sweep. Sweeps are first carried out along the x direction and then along the y direction to allow propagation of errors in both directions. The electrostatic field distribution in Figure 4.10 was calculated by a relaxation program.

Advanced methods for solving the Laplace equation[†] generally use more efficient algorithms based on Fourier transforms. Most available codes to solve electrostatic problems utilize a more complex mesh. The mesh may have a rectangular or even triangular divisions to allow a close match to curved boundary surfaces.

[†]See D. Potter, *Computational Physics*, Wiley, New York, 1973, Chapter IV.

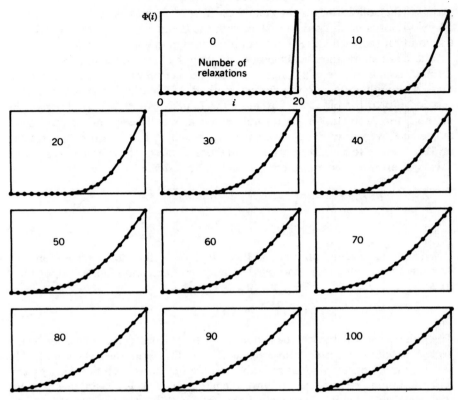

Figure 4.9 Relaxation method for solving the Laplace equation. Successive one-dimensional solutions for electrostatic potential between charged plates as a function of the number of mesh relaxations. Initial conditions: $\Phi(1) - \Phi(19) = 0$; $\Phi(20) = 1$.

Boundary conditions present special problems and must be handled differently from internal points representing the vacuum region. Boundary points may include those on the actual boundary of the calculational mesh, or points on internal electrodes maintained at a constant potential. The latter points are handled easily. They are marked by a flag to indicate locations of nonvariable potential. The relaxation calculation is not performed at such points. Locations on the mesh boundary have no neighbors outside the mesh, so that Eq. (4.16) can not be applied. If these points have constant potential, there is no problem since the residual need not be computed. Constant-potential points constitute a *Dirichlet boundary condition*.

The other commonly encountered boundary specification is the *Neumann condition* in which the normal derivative of the potential at the boundary is specified. In most cases where the Neumann condition is used, the derivative is zero, so that there is no component of the electric field normal to the

TABLE 4.1   Subroutines: Successive Over-Relaxation Program

```
1000 REM --- RELAXATION SUBROUTINES FOR POINTS
1010 REM --- INTERNAL TO BOUNDARY

1200 REM --- SWEEP ALONG I DIRECTION
1202 FOR J = 1 TO JMAX
1204 FOR I = 1 TO IMAX
1206 IF IB(I,J)<>0 THEN GOTO 1212:REM --- SPECIFIED POTENTIAL
1208 RES = (P(I+1,J) + P(I-1,J) + P(I,J+1) + P(I,J-1)) / 4 - P(I,J)
1210 P(I,J) = P(I,J) + OMEGA*RES
1212 NEXT I
1214 NEXT J
1216 RETURN

1300 REM --- SWEEP ALONG J DIRECTION
1302 FOR I = 1 TO IMAX
1304 FOR J = 1 TO JMAX
1306 IF IB(I,J)<>0 THEN GOTO 1312:REM --- SPECIFIED POTENTIAL
1308 RES = (P(I+1,J) + P(I-1,J) + P(I,J+1) + P(I,J-1)) / 4 - P(I,J)
1310 P(I,J) = P(I,J) + OMEGA*RES
1312 NEXT J
1314 NEXT I
1316 RETURN
```

boundary. This condition applies to boundaries with special symmetry, such as the axis in a cylindrical calculation or a symmetry plane of a periodic system. Residues can be calculated at Neumann boundaries since the potential outside the mesh is equal to the potential at the first point inside the mesh. For example, on the boundary $i = 0$, the condition $\Phi(-1, j) = \Phi(+1, j)$ holds. The residual is

$$R(0, j) = \tfrac{1}{4}[\Phi(0, j + 1) + 2\Phi(1, j) + \Phi(0, j - 1)] - \Phi(0, j).$$

$$(4.18)$$

Two-dimensional systems with cylindrical symmetry are often encountered in accelerator applications. Potential is a function of (r, z), with no azimuthal dependence. The Laplace equation for a cylindrical system is

$$\frac{1}{r} \frac{\partial}{\partial r}\left(r \frac{\partial \phi}{\partial r}\right) + \frac{\partial^2 \phi}{\partial z^2} = 0.$$

$$(4.19)$$

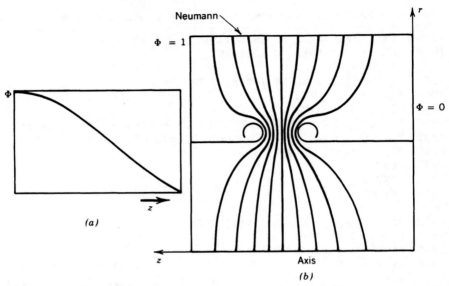

Figure 4.10 Two-dimensional solution for equipotentials between cylindrical electrodes of a drift tube linear accelerator calculated by the method of successive overrelaxation ($\omega = 1.75$). (*a*) Variation of potential on axis, $\Phi(0, z)$. (*b*) Boundary conditions, equipotential lines.

The finite difference form for the Laplace equation for this case is

$$\Phi(i, j) = \frac{1}{4}\left[\frac{\left(i + \frac{1}{2}\right)\Phi(i + 1, j)}{i} + \frac{\left(i - \frac{1}{2}\right)\Phi(i - 1, j)}{i}\right.$$
$$\left. + \Phi(i, j + 1) + \Phi(i, j - 1)\right]. \quad (4.20)$$

where $r = i\Delta$ and $z = j\Delta$.

Figure 4.10 shows results for a relaxation calculation of an electrostatic immersion lens. It consists of two cylinders at different potentials separated by a gap. Points of constant potential and Neumann boundary conditions are indicated. Also shown is the finite difference approximation for the potential variation along the axis, $\phi(0, z)$. This data can be used to determine the focal properties of the lens (Chapter 6).

4.3 ANALOG METHODS TO SOLVE THE LAPLACE EQUATION

Analog methods were used extensively to solve electrostatic field problems before the advent of digital computers. We will consider two analog techniques that clarify the nature of the Laplace equation. The approach relies on finding a physical system that obeys the Laplace equation but that allows easy measurements of a characteristic quantity (the analog of the potential).

One system, the tensioned elastic sheet, is suitable for two-dimensional problems (symmetry along the z axis). As shown in Figure 4.11, a latex sheet is stretched with uniform tension on a frame. If the sheet is displaced vertically a distance $H(x, y)$, there will be vertical restoring forces. In equilibrium, there is vertical force balance at each point. The equation of force balance can be determined from the finite difference approximation defined in Figure 4.11. In terms of the surface tension, the forces to the left and right of the point $(i\Delta, j\Delta)$ are

$$F\left[\left(i - \tfrac{1}{2}\right)\Delta\right] = T\left[H(i\Delta, j\Delta) - H([i - 1]\Delta, j\Delta)\right]/\Delta,$$

$$F\left[\left(i + \tfrac{1}{2}\right)\Delta\right] = T\left[H([i + 1]\Delta, j\Delta) - H(i\Delta, j\Delta)\right]/\Delta.$$

Similar expressions can be determined for the y direction. The height of the point $(i\Delta, j\Delta)$ is constant in time; therefore,

$$F\left[\left(i - \tfrac{1}{2}\right)\Delta\right] = -F\left[\left(i + \tfrac{1}{2}\right)\Delta\right],$$

and

$$F\left[\left(j + \tfrac{1}{2}\right)\Delta\right] = -F\left[\left(j - \tfrac{1}{2}\right)\Delta\right].$$

Substituting for the forces shows that the height of a point on a square mesh is the average of its four nearest neighbors. Thus, inverting the arguments of Section 4.2, $H(x, y)$ is described by the two-dimensional Laplace equation

$$\partial^2 H(x, y)/\partial x^2 + \partial^2 H(x, y)/\partial y^2 = 0.$$

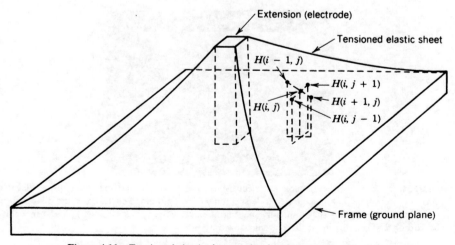

Figure 4.11   Tensioned elastic sheet analog for electrostatic potential.

Height is the analog of potential. To make an elastic potential solution, parts are cut to the shape of the electrodes. They are fastened to the frame to displace the elastic sheet up or down a distance proportional to the electrode potential. These pieces determine equipotential surfaces. The frame is the ground plane.

An interesting feature of the elastic sheet analog is that it can also be used to determine orbits of charged particles in applied electrostatic fields. Neglecting rotation, the total energy of a ball bearing on the elastic sheet is $E = T + mgH(x, y)$, where g is the gravitational constant. The transverse forces acting on a ball bearing on the elastic sheet are $F_x = \partial H/\partial x$ and $F_y = \partial H/\partial y$. Thus, ball bearings on the elastic sheet follow the same orbits as charged particles in the analogous electrostatic potential, although over a considerably longer time scale.

Figure 4.12 is a photograph of a model that demonstrates the potentials in a planar electron extraction gap with a coarse grid anode made of parallel wires. The source of the facet lens effect associated with extraction grids (Section 6.5) is apparent.

A second analog technique, the electrolytic tank, permits accurate measurements of potential distributions. The method is based on the flow of current in a liquid medium of constant-volume resistivity, ρ (measured in units of ohm meters). A dilute solution of copper sulfate in water is a common medium. A model of the electrode structure is constructed to scale from copper sheet and immersed in the solution. Alternating current voltages with magnitude proportional to those in the actual system are applied to the electrodes.

According to the definition of volume resistivity, the current density is proportional to the electric field

$$\mathbf{E} = \rho \mathbf{j}$$

Figure 4.12 Elastic sheet analog for electrostatic potential near an extraction grid. Elevated section represents a high voltage electrode surrounded by a grounded enclosure. Note distortion of potential near grid wires that results in focusing of extracted particles. (Photograph and model by the author. Latex courtesy of the Hygenic Corporation.)

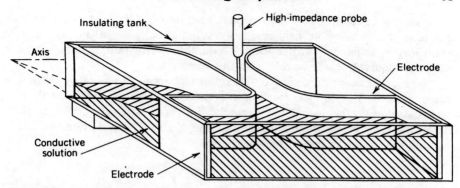

Figure 4.13 Electrolytic tank analog to solve the Laplace equation in a cylindrically symmetric system.

or

$$\mathbf{j} = -\nabla\phi/\rho. \qquad \boxed{4.21}$$

The steady-state condition that charge at any point in the liquid is a constant implies that all current that flows into a volume element must flow out. This condition can be written

$$\nabla \cdot \mathbf{j} = 0. \qquad (4.22)$$

Combining Eq. (4.21) with (4.22), we find that potential in the electrolytic solution obeys the Laplace equation.

In contrast to the potential in the real system, the potential in the electrolytic analog is maintained by a real current flow. Thus, energy is available for electrical measurements. A high-impedance probe can be inserted into the solution without seriously perturbing the fields. Although the electrolytic method could be applied to three-dimensional problems, in practice it is usually limited to two-dimensional simulations because of limitations on insertion of a probe. A typical setup is shown in Figure 4.13. Following the arguments given above, it is easy to show that a tipped tank can be used to solve for potentials in cylindrically symmetric systems.

4.4 ELECTROSTATIC QUADRUPOLE FIELD

Although numerical calculations are often necessary to determine electric and magnetic fields in accelerators, analytic calculations have advantages when they are tractable. Analytic solutions show general features and scaling relationships. The field expressions can be substituted into equations of motion to yield particle orbit expressions in closed form. Electrostatic solutions for a wide variety of electrode geometries have been derived. In this section, we will examine the quadrupole field, a field configuration used in all high-energy transport systems. We will concentrate on the electrostatic quadrupole; the magnetic equivalent will be discussed in Chapter 5.

The most effective procedure to determine electrodes to generate quadrupole fields is to work in reverse, starting with the desired electric field distribution and calculating the associated potential function. The equipotential lines determine a set of electrode surfaces and potentials that generate the field. We assume the following two-dimensional fields:

$$E_x = +kx = E_0 x/a, \qquad \boxed{4.23}$$

$$E_y = -ky = -E_0 y/a. \qquad \boxed{4.24}$$

It is straightforward to verify that both the divergence and curl of $\mathbf{E}$ are zero. The fields of Eqs. (4.23) and (4.24) represent a valid solution to the Maxwell equations in a vacuum region. The electric fields are zero at the axis and increase (or decrease) *linearly* with distance from the axis. The potential is related to the electric field by

$$\partial \phi / \partial x = -E_0 x/a, \qquad \partial \phi / \partial y = +E_0 y/a.$$

Integrating the partial differential equations

$$\phi = -E_0 x^2/2a + f(y) + C, \qquad \phi = +E_0 y^2/2a + g(x) + C'.$$

Taking $\phi(0,0) = 0$, both expressions are satisfied if

$$\phi(x, y) = (E_0/2a)(y^2 - x^2). \qquad (4.25)$$

This can be rewritten in a more convenient, dimensionless form:

$$\frac{\phi(x, y)}{E_0 a/2} = \left(\frac{y}{a}\right)^2 - \left(\frac{x}{a}\right)^2. \qquad (4.26)$$

Equipotential surfaces are hyperbolas in all four quadrants. There is an infinite set of electrodes that will generate the fields of Eqs. (4.23) and (4.24). The usual choice is symmetric electrodes on the equipotential lines $\phi_0 = \pm E_0 a/2$. Electrodes, field lines, and equipotential surfaces are plotted in Figure 4.14. The quantity a is the minimum distance from the axis to the electrode, and E_0 is the electric field on the electrode surface at the position closest to the origin. The equipotentials in Figure 4.14 extend to infinity. In practice, focusing fields are needed only near the axis. These fields are not greatly affected by terminating the electrodes at distances a few times a from the axis.

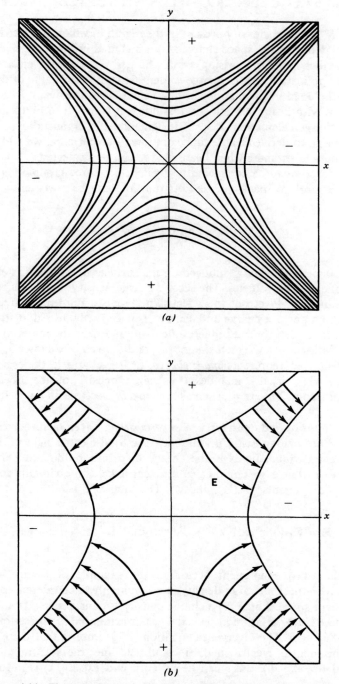

Figure 4.14 Electrostatic quadrupole field. (*a*) Equipotential lines. (*b*) Field lines.

4.5   STATIC ELECTRIC FIELDS WITH SPACE CHARGE

Space charge is charge density present in the region in which an electric field is to be calculated. Clearly, space charge is not included in the Laplace equation, which describes potential arising from charges on external electrodes. In accelerator applications, space charge is identified with the charge of the beam; it must be included in calculations of fields internal to the beam. Although we will not deal with beam self-fields in this book, it is useful to perform at least one space charge calculation. It gives insight into the organization of various types of charge to derive electrostatic solutions. Furthermore, we will derive a useful formula to estimate when beam charge can be neglected.

Charge density can be conveniently divided into three groups: (1) applied, (2) dielectric, and (3) space charge. Equation 3.13 can be rewritten

$$\nabla \cdot \mathbf{E} = (\rho_1 + \rho_2 + \rho_3)/\varepsilon_0. \tag{4.27}$$

The quantity ρ_1 is the charge induced on the surfaces of conducting electrodes by the application of voltages. The second charge density represents charges in *dielectric materials*. Electrons in dielectric materials cannot move freely. They are bound to a positive charge and can be displaced only a small distance. The dielectric charge density can influence fields in and near the material. Electrostatic calculations with the inclusion of ρ_2 are discussed in Chapter 5. The final charge density, ρ_3, represents *space charge*, or free charge in the region of the calculation. This usually includes the charge density of the beam. Other particles may contribute to ρ_3, such as low-energy electrons in a neutralized ion beam.

Electric fields have the property of *superposition*. Given fields corresponding to two or more charge distributions, then the total electric field is the vector sum of the individual fields if the charge distributions do not perturb one another. For instance, we could calculate electric fields individually for each of the charge components, $\mathbf{E}_1$, $\mathbf{E}_2$, and $\mathbf{E}_3$. The total field is

$$\mathbf{E} = \mathbf{E}_1 \text{ (applied)} + \mathbf{E}_2 \text{ (dielectric)} + \mathbf{E}_3 \text{ (space charge)}. \tag{4.28}$$

Only the third component occurs in the example of Figure 4.15. The cylinder with uniform charge density is a commonly encountered approximation for beam space charge. The charge density is constant, ρ_0, from $r = 0$ to $r = r_b$. There is no variation in the axial (z) or azimuthal (θ) directions so that $\partial/\partial z = \partial/\partial\theta = 0$. The divergence equation (3.13) implies that there is only a radial component of electric field. Since all field lines radiate straight outward (or inward for $\rho_0 < 0$), there can be no curl, and Eq. (3.11) is automatically satisfied.

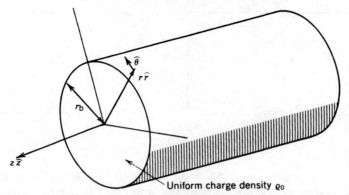

Figure 4.15 Quantities for calculating electric fields in and around a cylinder with uniform charge density (ρ_0).

Inside the charge cylinder, the electric field is determined by

$$\frac{1}{r}\frac{d(rE_r)}{dr} = \frac{\rho_0}{\varepsilon_0}. \tag{4.29}$$

Electric field lines are generated by the charge inside a volume. The size of the radial volume element goes to zero near the origin. Since no field lines can emerge from the axis, the condition $E_r\,(r=0) = 0$ must hold. The solution of Eq. (4.29) is

$$E_r\,(r < r_b) = \frac{\rho_0 r}{2\varepsilon_0}. \qquad \boxed{4.30}$$

Outside the cylinder, the field is the solution of Eq. (4.29) with the right-hand side equal to zero. The electric field must be a continuous function of radius in the absence of a charge layer. (A charge layer is a finite quantity of charge in a layer of zero thickness; this is approximately the condition on the surface of an electrode.) Thus, $E_r\,(r = r_b^+) = E_r\,(r = r_b^-)$, so that

$$E_r\,(r > r_b) = \frac{\rho_0 r_b^2}{2\varepsilon_0 r}. \tag{4.31}$$

The solution is plotted in Figure 4.16. The electric field increases linearly away from the axis in the charge region. It decreases as $1/r$ for $r > r_b$ because the field lines are distributed over a larger area.

The problem of the charge cylinder can also be solved through the electrostatic potential. The *Poisson equation* results when the gradient of the potential

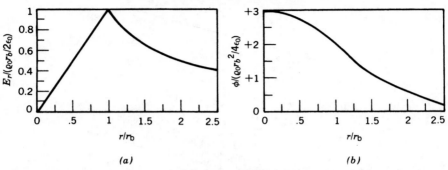

Figure 4.16 Solutions for the electrostatic fields of a cylinder with uniform charge density. (*a*) Normalized radial electric field $E_r/(\rho_0 r_b/2\varepsilon_0)$. (*b*) Normalized electrostatic potential, $\phi/(\rho_0 r_b^2/4\varepsilon_0)$.

is substituted in Eq. (3.13):

$$\nabla^2\phi = \frac{-\rho(x)}{\varepsilon_0}, \tag{4.32}$$

or

$$\frac{1}{r}\frac{d}{dr}\left(r\frac{d\phi}{dr}\right) = \frac{-\rho_0}{\varepsilon_0}. \tag{4.33}$$

The solution to the Poisson equation for the charge cylinder is

$$\phi\,(r < r_b) = -\frac{\rho_0 r^2}{4\varepsilon_0}, \tag{4.34}$$

$$\phi\,(r > r_b) = -\frac{\rho_0 r_b^2}{4\varepsilon_0}\left(2\ln\frac{r}{r_b} + 1\right). \tag{4.35}$$

The potential is also plotted in Figure 4.16.

The Poisson equation can be solved by numerical methods similar to those developed in Section 4.2. If the finite difference approximation to $\nabla^2\phi$ [Eq. (4.14)] is substituted in the Poisson equation in Cartesian coordinates (4.32) and both sides are multiplied by Δ^2, the following equation results

$$-6\Phi(i, j) + \Phi(i + 1, j) + \Phi(i - 1, j, k) + \Phi(i, j + 1, k)$$

$$+ \Phi(i, j - 1, k) + \Phi(i, j, k + 1) + \Phi(i, j, k - 1)$$

$$= -\rho(x, y, z)\Delta^3/\Delta\varepsilon_0. \tag{4.36}$$

The factor $\rho\Delta^3$ is approximately the total charge in a volume Δ^3 surrounding the mesh point (i, j, k) when (1) the charge density is a smooth function of position and (2) the distance Δ is small compared to the scale length for variations in ϕ. Equation (4.36) can be converted to a finite difference equation by defining $Q(i, j, k) = \rho(x, y, z)\Delta^3$. Equation (4.36) becomes

$$\Phi(i, j, k) = \tfrac{1}{6}[\Phi(i + 1, j) + \Phi(i - 1, j, k) + \Phi(i, j + 1, k)$$

$$+ \Phi(i, j - 1, k) + \Phi(i, j, k + 1) + \Phi(i, j, k - 1)]$$

$$+ Q(i, j, k)/6\,\Delta\varepsilon_0. \qquad \boxed{4.37}$$

Equation (4.37) states that the potential at a point is the average of its nearest neighbors elevated (or lowered) by a term proportional to the space charge surrounding the point.

The method of successive relaxation can easily be modified to treat problems with space charge. In this case, the residual [Eq. (4.16)] for a two-dimensional problem is

$$R(i, j) = \tfrac{1}{4}[\Phi(i + 1, j) + \Phi(i - 1, j) + \Phi(i, j + 1) + \Phi(i, j - 1)]$$

$$- \Phi(i, j) + Q(i, j)/4\,\Delta\varepsilon_0. \qquad (4.38)$$

4.6 MAGNETIC FIELDS IN SIMPLE GEOMETRIES

This section illustrates some methods to find static magnetic fields by direct use of the Maxwell equations [(4.3) and (4.4)]. The fields are produced by current-carrying wires. Two simple, but often encountered, geometries are included: the field outside a long straight wire and the field inside of solenoidal winding of infinite extent.

The wire (Fig. 4.17) has current I in the z direction. There are no radial magnetic field lines since $\nabla \cdot \mathbf{B} = 0$. There is no component B_z since the fields must be perpendicular to the current. Thus, magnetic field lines are azimuthal. By symmetry, the field lines are circles. The magnitude of the azimuthal field (or density of lines) can be determined by rewriting the static form of Eq. (3.12) in integral form according to the Stokes law [Eq. (4.7)],

$$\int \mathbf{B} \cdot \mathbf{dl} = \mu_0 \int \int j_z \, dA = \mu_0 I. \qquad (4.39)$$

Using the fact that field lines are circles, we find that

$$B_\theta = \mu_0 I/2\pi r. \qquad \boxed{4.40}$$

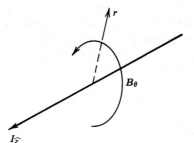

Figure 4.17 Magnetic field lines near a current-carrying wire.

The solenoidal coil is illustrated in Figure 4.18. It consists of a helical winding of insulated wire on a cylindrical mandrel. The wire carries current I. The quantity (N/L) is the number of turns per unit length. *Solenoid* derives from the Greek word for pipe; magnetic field lines are channeled through the windings. In a finite length winding, the field lines return around the outside. We will consider the case of an infinitely long structure with no axial variations. Furthermore, we assume there are many windings over a length comparable to the coil radius, or $(N/L)r_c \gg 1$. In this limit, we can replace the individual windings with a uniform azimuthal current sheet. The sheet has a current per unit length J (A/m) $= (N/L)I$.

The current that produces the field is azimuthal. By the law of Biot and Savart, there can be no component of azimuthal magnetic field. By symmetry, there can be no axial variation of field. The conditions of zero divergence and curl of the magnetic field inside the winding are written

$$\frac{1}{r}\frac{\partial(rB_r)}{\partial r} + \frac{\partial B_z}{\partial z} = 0, \qquad \frac{\partial B_r}{\partial z} - \frac{\partial B_z}{\partial r} = 0. \qquad (4.41)$$

Setting $\partial/\partial z$ equal to zero in Eqs. (4.41); we find that B_r is zero and that B_z has equal magnitude at all radii. The magnitude of the axial field can be determined by applying Eq. (4.39) to the loop illustrated in Figure 4.18. The field outside a long solenoid is negligible since return magnetic flux is spread over a large area. There are no contributions to the loop integral from the radial segments since fields are axial. The only component of the integral comes from the part of the path inside the solenoid, so that

$$B_0 = \mu_0 J = \mu_0(N/L)I. \qquad \boxed{4.42}$$

Many magnetic confinement systems for intense electron beams or for high-temperature plasmas are based on a solenoidal coil bent in a circle and connected, as shown in Figure 4.19. The geometry is that of a doughtnut or *torus* with circular cross section. The axial fields that circulate around the torus

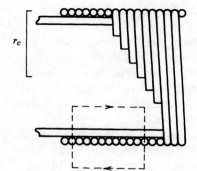

Figure 4.18   Solenoidal magnet coil.

are called *toroidal field lines*. Field lines are continuous and self-connected. All field lines are contained within the winding. The toroidal field magnitude inside the winding is not uniform. Modification of the loop construction of Figure 4.18 shows that the field varies as the inverse of the *major radius*. Toroidal field variation is small when the *minor radius* (the radius of the solenoidal windings) is much less than the major radius.

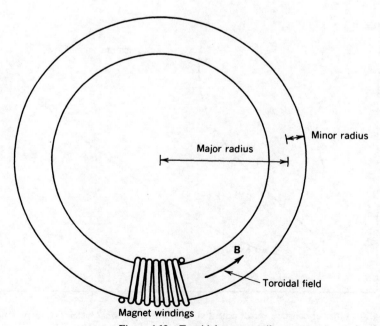

Figure 4.19   Toroidal magnet coil.

4.7 MAGNETIC POTENTIALS

The magnetic potential and the vector potential aid in the calculation of magnetic fields. In this section, we will consider how these functions are related and investigate the physical meaning of the vector potential in a two-dimensional geometry. The vector potential will be used to derive the magnetic field for a circular current loop. Assemblies of loop currents are used to generate magnetic fields in many particle beam transport devices.

In certain geometries, magnetic field lines and the vector potential are closely related. Figure 4.20 illustrates lines of constant vector potential in an axially uniform system in which fields are generated by currents in the z direction. Equation (3.24) implies that the vector potential has only an axial component, A_z. Equation (3.23) implies that

$$B_x = \partial A_z / \partial y, \qquad B_y = - \partial A_z / \partial x. \tag{4.43}$$

Figure 4.20 shows a surface of constant A_z in the geometry considered. This line is defined by

$$dA_z = 0 = (\partial A_z / \partial x)\, dx + (\partial A_z / \partial y)\, dy. \tag{4.44}$$

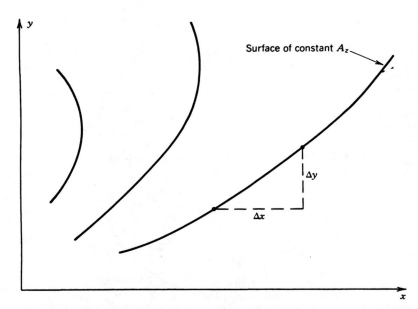

Figure 4.20 Lines of constant vector potential A_z. Two-dimensional system, symmetric along the z direction (out of page), axial currents only (j_z).

Substituting Eqs. (4.43) into Eq. (4.44), an alternate equation for a constant A_z line is

$$dy/dx = B_y/B_x. \tag{4.45}$$

Equation (4.45) is also the equation for a magnetic field line. To summarize, when magnetic fields are generated by axial currents uniform in z, *magnetic field lines are defined by lines of constant A_z.*

A similar construction shows that magnetic field lines are normal to surfaces of constant magnetic potential. In the geometry of Figure 4.20,

$$B_x = \partial U_m/\partial x, \qquad B_y = \partial U_m/\partial y, \tag{4.46}$$

by the definition of U_m. The equation for a line of constant U_m is

$$dU_m = (\partial U_m/\partial x)\, dx + (\partial U_m/\partial y)\, dy = B_x\, dx + B_y\, dy = 0.$$

Lines of constant magnetic potential are described by the equation

$$dy/dx = -B_x/B_y. \tag{4.47}$$

Analytic geometry shows that the line described by Eq. (4.47) is perpendicular to that of Eq. (4.45).

The correspondence of field lines and lines of constant A_z can be used to find magnetic fields of arrays of currents. As an example, consider the geometry illustrated in Figure 4.21. Two infinite length wires carrying opposed currents $\pm I$ are separated by a distance $2d$. It is not difficult to show that the vector potential for a single wire is

$$A_z = \pm \tfrac{1}{2}\mu_0 I \ln(x'^2 + y'^2),$$

where the origin of the coordinate system (x', y') is centered on the wire. The total vector potential is the sum of contributions from both wires. In terms of the coordinate system (x, y) defined in Figure 4.21, the total vector potential is

$$A_z = \frac{\mu_0 I}{2} \ln\left(\frac{(x - d)^2 + y^2}{(x + d)^2 + y^2} \right).$$

Lines of constant A_z (corresponding to magnetic field lines) are plotted in Figure 4.21.

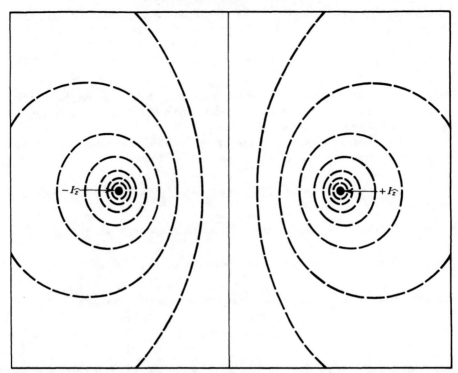

Figure 4.21 Lines of constant vector potential A_z from two wires of infinite extent in z (out of page) carrying equal and opposite currents.

There are many instances in accelerator applications in which magnetic fields are produced by azimuthal currents in cylindrically symmetric systems. For instance, the field of a solenoidal lens (Section 6.7) is generated by axicentered current loops of various radii. There is only one nonzero component of the vector potential, A_θ. It can be shown that magnetic field lines follow surfaces of constant $2\pi r A_\theta$. The function $2\pi r A_\theta$ is called the *stream function*. The contribution from many loops can be summed to find a net stream function.

The vector potential of a current loop of radius a (Fig. 4.22) can be found by application of Eq. (3.24). In terms of cylindrical coordinates centered at the loop axis, the current density is

$$j_\theta = I\delta(z')\delta(r' - a). \tag{4.48}$$

Care must be exercised in evaluating the integrals, since Eq. (3.24) holds only

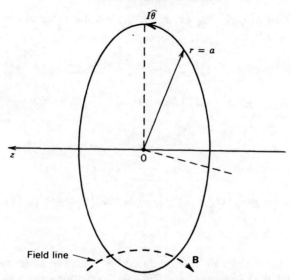

Figure 4.22 Quantities for calculating magnetic fields from a circular current loop.

for a Cartesian coordinate system. The result is

$$A_\theta(r, z) = \left(\frac{\mu_0 I a}{4\pi}\right) \int_0^{2\pi} \frac{\cos \theta' \, d\theta'}{(a^2 + r^2 + z^2 - 2ar\cos\theta')^{1/2}}. \quad (4.49)$$

Defining the quantity

$$M = 4ar/(a^2 + r^2 + z^2 + 2ar),$$

Eq. (4.49) can be written in terms of the complete elliptic integrals $E(M)$ and $K(M)$ as

$$A_\theta(r, z) = \frac{\mu_0 I a}{\pi(a^2 + r^2 + z^2 + 2ar)^{1/2}} \frac{(2 - M)K(M) - 2E(M)}{M}.$$

$$(4.50)$$

Although the expressions in Eq. (4.50) are relatively complex, the vector potential can be calculated quickly on a computer. Evaluating the elliptic integrals directly is usually ineffective and time consuming. A better approach is to utilize empirical series tabulated in many mathematical handbooks. These series give an accurate approximation in terms of power series and elementary transcendental functions. For example, the elliptic integrals are given to an accuracy of 4×10^{-5} by[†]

$$K(M) = 1.38629 + 0.111972(1 - M) + 0.0725296(1 - M)^2$$

$$+ \left[0.50000 + 0.121348(1 - M) + 0.0288729(1 - M^2)\right]$$

$$\times \ln[1/(1 - M)], \tag{4.51}$$

$$E(M) = 1 + 0.463015(1 - M) + 0.107781(1 - M)^2$$

$$+ \left[0.245273(1 - M) + 0.0412496(1 - M)^2\right]\ln[1/(1 - M)]. \tag{4.52}$$

The vector potential can be calculated for multiple coils by transforming coordinates and then summing A_θ. The transformations are $z \rightarrow (z - z_{cn})$ and $a \rightarrow r_{cn}$, where z_{cn} and r_{cn} are the coordinates of the nth coil. Given the net vector potential, the magnetic fields are

$$B_z = \frac{1}{r}\frac{\partial(rA_\theta)}{\partial r}, \qquad B_r = -\frac{\partial A_\theta}{\partial z}. \tag{4.53}$$

A quantity of particular interest for paraxial orbit calculations (Section 7.5) is the longitudinal field magnitude on the axis $B_z(0, z)$. The vector potential for a single coil [Eq. (4.49)] can be expanded for $r \ll a$ as

$$A_\theta(r, z) \simeq \left(\frac{\mu_0 Ia}{4\pi}\right)\int_0^{2\pi}\left(\frac{\cos\theta'\,d\theta'}{(a^2 + z^2)^{1/2}} + \frac{ar\cos^2\theta'\,d\theta'}{(a^2 + z^2)^{3/2}}\right). \tag{4.54}$$

The integral of the first term is zero, while the second term gives

$$A_\theta = \frac{\mu_0 Ia^2 r}{4(a^2 + z^2)^{3/2}}. \tag{4.55}$$

Applying Eq. (4.53), the axial field is

$$B_z(0, z) = \frac{\mu_0 Ia^2}{2(a^2 + z^2)^{3/2}}. \qquad \boxed{4.56}$$

[†]Adapted from M. Abramowitz and I. A. Stegun, Eds., *Handbook of Mathematical Functions*, Dover, New York, 1970, p. 591.

We can use Eq. (4.56) to derive the geometry of the *Helmholtz coil* configuration. Assume that two loops with equal current are separated by an axial distance d. A Taylor expansion of the axial field near the axis about the midpoint of the coils gives

$$B_z(0, z) = (B_1 + B_2) + (\partial B_1/\partial z + \partial B_2/\partial z)z$$

$$+ (\partial^2 B_2/\partial z^2 + \partial^2 B_2/\partial z^2)z^2 + \cdots .$$

The subscript 1 refers to the contribution from the coil at $z = -d/2$, while 2 is associated with the coil at $z = +d/2$. The derivatives can be determined from Eq. (4.56). The zero-order components from both coils add. The first derivatives cancel at all values of the coil spacing. At a spacing of $d = a$, the second derivatives also cancel. Thus, field variations near the symmetry point are only on the order of $(z/a)^3$. Two coils with $d = a$ are called Helmholtz coils. They are used when a weak but accurate axial field is required over a region that is small compared to the dimension of the coil. The field magnitude for Helmholtz coils is

$$B_z = \mu_0 I/(1.25)^{3/2} a. \qquad \boxed{4.57}$$

5

Modification of Electric and Magnetic Fields by Materials

Certain materials influence electric and magnetic fields through bound charges and currents. Their properties differ from those of metals where electrons are free to move. Dielectric materials contain polar molecules with spatially displaced positive and negative charge. Applied electric fields align the molecules. The resulting charge displacement reduces the electric field in the material and modifies fields in the vicinity of the dielectric. There are corresponding magnetic field effects in paramagnetic and ferromagnetic materials. These materials contribute to magnetic fields through orientation of atomic currents rather than a macroscopic flow of charge as in a metal.

Although the responses of materials to fields differ in scale, the general behavior is similar in form. This is the reason the contributions of dielectric and magnetic materials were singled out in Section 4.5 as ρ_2 and $\mathbf{j}_2$. It is often useful to define new field quantities that automatically incorporate the contributions of bound charges and currents. These quantities are **D** (the electric displacement vector) and **H** (the magnetic field intensity).

The study of the properties of dielectric and magnetic materials (including subsidiary field quantities and boundary conditions) is not conceptually exciting. This is especially true for ferromagnetic materials where there is considerable terminology. Nonetheless, it is essential to understand the properties of dielectric and ferromagnetic materials since they have extensive uses in all types of accelerators.

A partial list of applications of dielectric materials includes the following:

1.   Electric field distributions can be modified by adjustment of dielectric–vacuum boundaries. For example, dielectric boundary conditions must be applied to determine optimum shapes of high-voltage vacuum insulators.

2.   Dielectrics can store more electrostatic field energy than vacuum. The high-energy storage density of water (80 times that of vacuum) is the basis for much of modern pulsed power technology.

3.   Dielectrics reduce the velocity of propagation of electromagnetic waves (or photons). This helps to match the velocities of rf waves and high-energy particles for resonant acceleration. This effect is also important in designing energy storage transmission lines for pulse modulators.

All high-energy accelerators utilize ferromagnetic materials. The following are some important applications.

1.   Ferromagnetic materials shape magnetic fields. They play a role analogous to electrodes in electrostatics. Shaped iron surfaces (poles) are utilized to generate complex field distributions for focusing and bending magnets.

2.   Ferromagnetic materials amplify the flux change produced by a real current. The resulting increased inductance is essential to the operation of transformers. Inductive isolation is the basis of the betatron and linear induction accelerator.

3.   Ferromagnetic materials convey magnetic field lines in a manner analogous to the conduction of current by a low-resistivity wire. This effect leads to substantial reductions in power requirements for beam transport magnets.

4.   The nonlinear response of a ferromagnetic material to an applied field can be utilized for fast, high-power switching.

The physics of dielectric and ferromagnetic materials is reviewed in this chapter. Special emphasis is placed on the concept of the magnetic circuit. A section is included on permanent magnet circuits.

5.1 DIELECTRICS

Dielectric materials are composed of *polar molecules*. Such molecules have spatially separated positive and negative charge. The molecules may be either bound in one position (solids) or free to move (liquids and gases). Figure 5.1a shows a diagram of a water molecule. The electronegative oxygen atom attracts the valence electrons of the hydrogen atoms, leaving an excess of positive charge at the locations of the hydrogen atoms.

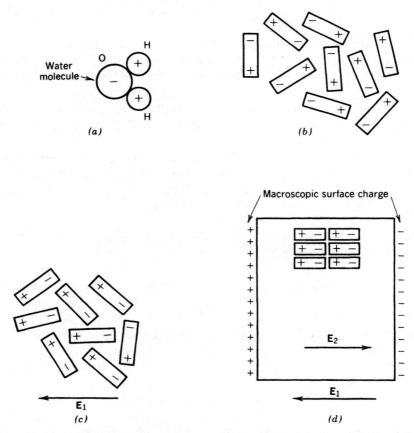

Figure 5.1 Behavior of dielectric materials. (*a*) Polar molecule (water). (*b*) Randomly oriented polar molecules. (*c*) Aligned polar molecules under the influence of an applied electric field. (*d*) Macroscopic charge on the surface of a dielectric material in an applied electric field from the alignment of polar molecules.

In the absence of an applied electric field, the molecules of a dielectric are randomly oriented (Fig. 5.1*b*). This results from the disordering effects of thermal molecular motion and collisions. Molecular ordering cannot occur spontaneously because a net electric field would result. With no external influence, there is no source of energy to generate such fields (Section 5.6). The randomized state is the state of lowest net energy (thermodynamic equilibrium).

Applied electric fields act on the charges of a polar molecule to align it as shown in Figure 5.1*c*. On the average, the molecular distribution becomes ordered as the change in electrostatic potential energy counteracts the randomizing thermal kinetic energy. The macroscopic effect of molecular alignment is shown in Figure 5.1*d*. Inside the material, a shifted positive charge in one molecule is balanced by a shifted negative charge of a nearby molecule. On the average, there is no net internal charge to contribute to fields. This balance does not occur on the surfaces. An applied electric field results in positive and negative charge layers at opposite surfaces. The field produced by these layers inside the dielectric is opposed to the applied field.

The simplest geometry to consider is a one-dimensional dielectric slab in a region of uniform applied electric field. The applied field is produced by a voltage difference between two parallel plates (Fig. 5.1d). The electric field resulting from the charges on the plates (Section 3.2) is denoted E_1. The surface charges induced on the dielectric produce a field, $-E_2$. The total field inside the dielectric is $E = E_1 - E_2$. Most dielectrics have the property that the degree of orientation of polar molecules is linearly proportional to the applied field at typical field strengths. Thus, the surface charge density is proportional to applied field, and $E_2 \sim E_1$.

The linear response of dielectrics comes about because the degree of alignment of molecules is small at normal temperatures and field strength. Increased applied field strength brings about a proportional increase in the orientation. Nonlinear effects are significant when the dielectric approaches *saturation*. In a saturated state, all molecules are aligned with the field so that an increase in applied field brings about no increase in surface charge. We can estimate the magnitude of the saturation electric field. At room temperature, molecules have about 0.025 eV of thermal kinetic energy. Saturation occurs when the electrostatic potential energy is comparable to the thermal energy. The decrease in potential energy associated with orientation of a polar molecule with charge separation d is $q\,dE_1$. Taking $q = e$ and $d = 1$ Å $(10^{-10}$ m), E_1 must be on the order of 250 MV/m. This is much higher than the strongest fields generated in rf accelerators, so that the linear approximation is well satisfied. In contrast, saturation effects occur in ferromagnetic materials at achievable values of applied magnetic field.

The net electric field inside a linear dielectric is proportional to the applied field. The constant of proportionality is defined by

$$\mathbf{E} = \frac{\mathbf{E}_1}{\varepsilon/\varepsilon_0} = \mathbf{E}_1 + \mathbf{E}_2. \qquad \boxed{5.1}$$

The quantity $\varepsilon/\varepsilon_0$ is the *relative dielectric constant*. Ordinary solid or liquid dielectrics reduce the magnitude of the electric field, so that $\varepsilon/\varepsilon_0 > 1$. Equation (5.1) is written in vector notation. This result can be derived from the above one-dimensional arguments by considering a differential cubic volume and treating each component of the field separately. This approach holds if the material is isotropic (liquids, glass). Equation (5.1) may not be valid for some solid materials. For instance, if polar molecules are bound in a crystal lattice, their response to an applied field may vary depending on the orientation of the field with respect to the crystal axes. The dielectric constant depends on the alignment of the field relative to the crystal. Such materials are called *bifringent*.

Water is a commonly encountered isotropic dielectric medium in electrical energy storage applications. The relative dielectric constant of liquid water is plotted in Figure 5.2 versus temperature and the frequency of an oscillating applied electric field. The low-frequency value is high since water molecules have large charge separation. The dielectric constant decreases with increasing temperature. This comes about because the molecules have higher thermal energy; therefore, they do not align as strongly in an applied electric field. At constant temperature, the relative dielectric constant decreases at high frequency (the microwave regime). This is because the inertia of the water molecules retards their response to the oscillating electric field. The alignment of the molecules lags in phase behind the electric field, so that the medium extracts energy from the field. Thus, water is not an ideal dielectric at high frequency. The loss process is usually denoted by an imaginary part of the dielectric constant, ε''. Higher temperatures randomize molecular motion and lessen the relative effect of the ordered phase lag. This explains the unusual result that the absorption of high-frequency electric fields in water is reduced at higher temperature.

It is useful to define the *displacement vector* **D** when the dielectric is linear. The displacement vector is proportional to the sum of field components excluding the contribution of dielectrics, or (in the notation of Chapter 4)

$$\mathbf{D} = \varepsilon_0(\mathbf{E}_1 + \mathbf{E}_3). \tag{5.2}$$

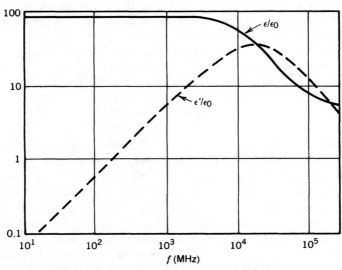

Figure 5.2 Real and imaginary parts of the relative dielectric constant of water as a function of the applied electric field frequency. (Adapted from J. B. Halstead, "Liquid Water—Dielectric Properties" in *Water—a Comprehensive Treatise*, F. Franks Ed., Plenum, New York, 1972.)

Thus, **D** arises from *free charges* (either on electrodes or in the volume). Combining Eqs. (5.1) and (5.2) the electric displacement is related to the net field inside a dielectric region by

$$\mathbf{D} = \varepsilon\mathbf{E}.$$

$\boxed{5.3}$

If a dielectric is inserted into a vacuum field region (Section 4.1), the following equations hold:

$$\nabla \cdot \mathbf{D} = 0, \qquad \nabla \cdot \mathbf{E} \neq 0. \tag{5.4}$$

The meaning of these equations is illustrated in Figure 5.3. A thin differential volume element that includes a vacuum–dielectric boundary is illustrated. There is no flux of **D** lines out of the volume since there are no free charges to act as sources. The divergence of **E** is nonzero because the magnitude is different on both faces. The volume includes a net positive charge in the form of the dielectric surface charge.

When dielectrics are included in a vacuum region, the Laplace equation can be written

$$\nabla \cdot \left[\left(\varepsilon(\mathbf{x})/\varepsilon_0 \right) \nabla \phi \right] = 0.$$

$\boxed{5.5}$

Equation (5.5) proceeds from Eq. (5.4), which implies that $\nabla \cdot \varepsilon\mathbf{E} = 0$. The potential is still given by $-\nabla\phi = \mathbf{E}$ since the force on a particle depends on the net electric field, independent of the sources of the field. Numerical methods to solve Eq. (5.5) are similar to those of Section 4.2. A value of the

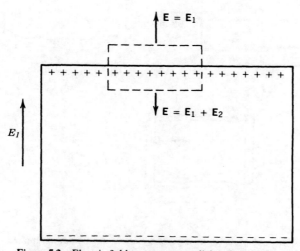

Figure 5.3 Electric fields at a vacuum–dielectric boundary.

relative dielectric constant is associated with each point and must be included in the finite difference formulation of the Laplacian operator.

The concept of the dielectric constant often leads to confusion in treating plasmas. A plasma is a relatively dense region of equal positive and negative free charges (Fig. 5.4). The clearest approach to describe the interactions of plasmas and electric fields is to include the electron and ion space charge as contributions to ρ_3 (free space charge). Nonetheless, it is a common practice to introduce the concept of a plasma dielectric constant to describe phenomena such as the refraction of optical radiation. This permits utilization of familiar optical definitions and equations. Referring to the plasma slab illustrated in Figure 5.4*a*, the plasma dielectric constant is clearly undefined for a steady-state applied field since positive and negative charges are free to move in opposite directions. At very low frequency, plasmas support real currents, as in a metal conductor. When a medium-frequency ac electric field is applied, the heavy ions are relatively immobile. The electrons try to move with the field, but displacements lead to charge separation. The space charge field acts to cancel the applied field. The electrons are thus bound to the ions. The result is that at medium frequency, electric fields are excluded from plasmas. Alternatively, the plasma can be described by a relative dielectric constant much greater than unity. Note the geometric similarity between Figure 5.4*b* and 5.1*d*. At high frequencies, electron inertia becomes an important factor. At high frequencies (such as the optical regime), the electron motion is 180° out of phase with

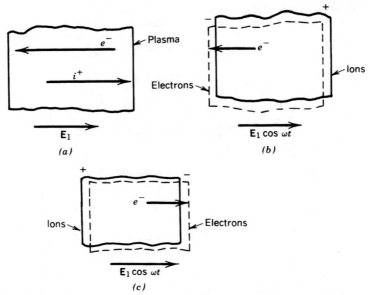

Figure 5.4 Response of particles in a plasma slab to an oscillating applied electric field. (*a*) Direct-current field (zero frequency). (*b*) Low-frequency ac field. (*c*) High-frequency ac field.

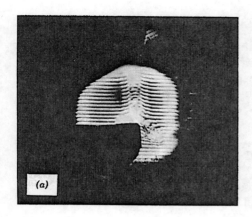

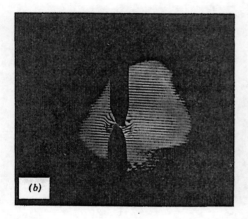

Figure 5.5 Infrared laser interferogram of plasma in dense neutral backgrounds. (a) Exploding wire: note that the dense expanding neutrals cause an upward shift of fringes ($\varepsilon/\varepsilon_0 > 1$), while the electrode plasma causes a downward fringe shift ($\varepsilon/\varepsilon_0 < 1$). ($b$) Spark in atmospheric air: dense plasma causes downward fringe shift. (Photographs by the author.)

applied electric fields. In this case, the electron space charge (oscillating about the immobile positive charge) *adds* to the applied field, so that $\varepsilon/\varepsilon_0 < 1$. Plasmas are a very unusual dielectric material at high frequencies. This effect is important in laser interferometry of plasmas. Figure 5.5 shows far-infrared holographic interferograms of an exploding wire and a plasma spark in atmospheric air. The direction of displacement of the fringes shows the dielectric constant relative to vacuum. Note that there are displacements in both directions in Figure 5.5a because of the presence of a dense shock wave of neutrals ($\varepsilon/\varepsilon_0 > 1$) and an electrode plasma ($\varepsilon/\varepsilon_0 < 1$).

5.2 BOUNDARY CONDITIONS AT DIELECTRIC SURFACES

Methods for the numerical calculation for vacuum electric fields in the presence of dielectrics were mentioned in Section 5.1. There are also numerous analytic methods. Many problems involve uniform regions with different values of $\varepsilon/\varepsilon_0$. It is often possible to find general forms of the solution in each region by the Laplace equation, and then to determine a general solution by matching field components at the interfaces. In this section, we shall consider how electric fields vary passing from a region with $\varepsilon/\varepsilon_0 \neq 1$ to a vacuum. Extensions to interfaces between two dielectrics is straightforward.

The electric fields at a dielectric–vacuum interface are divided into components parallel and perpendicular to the surface (Fig. 5.6). The magnitude of the electric field is different in each region (Section 5.1); the direction may also change. The relationship between field components normal to the interface is demonstrated by the construction of Figure 5.6b. A surface integral is taken over a thin volume that encloses the surface. The main contributions come from integration over the faces parallel to the surface. Using Eq. (5.3) and the divergence theorem,

$$\iint dA \left(\varepsilon_0 E_{\perp \alpha} - \varepsilon E_{\perp \beta} \right) = 0.$$

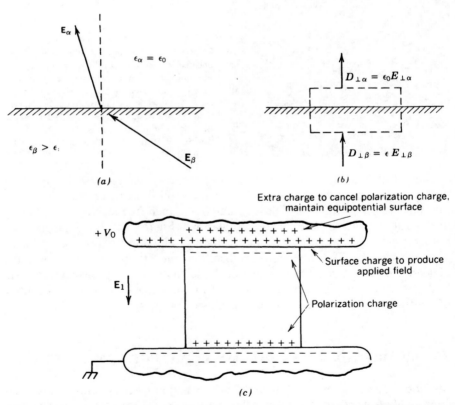

Figure 5.6 Boundary conditions for electric field components at a vacuum–dielectric boundary. (*a*) Electric field lines at boundary. (*b*) Geometry to find the relationship of normal field components. (*c*) Geometry to find the relationship of parallel field components.

This gives the matching condition for perpendicular field components,

$$E_{\perp\beta}/E_{\perp\alpha} = \varepsilon_0/\varepsilon \qquad \boxed{5.6}$$

or $D_{\perp\beta} = D_{\perp\alpha}$.

Matching conditions for the parallel field components can be determined from the construction of Figure 5.6c. A slab of dielectric extends between two parallel metal plates at different voltages. The dielectric–vacuum interface is normal to the plates. The geometry of Figure 5.6c is the simplest form of *capacitor*. The charges that produce the electric field must be moved against the potential difference in order to charge the plates. During this process, work is performed on the system; the energy can be recovered by reversing the process. Thus, the capacitor is a storage device for electrostatic energy.

In the absence of the dielectric, electric field lines normal to the plates are produced by positive and negative surface charge layers on the plates. When the dielectric is introduced, polarization charge layers are set up that try to reduce the electric field inside the dielectric. Since there is no net charge between the plates or inside the dielectric, the condition $\nabla \cdot \mathbf{E} = 0$ holds everywhere between the plates. Field lines are thus straight lines parallel to the dielectric–vacuum interface. The integral $-\int \mathbf{E} \cdot \mathbf{dx}$ has the constant value V_0 on any path between the equipotential plates. In particular, the integral can be taken just inside and outside the dielectric interface. This implies that the parallel electric field is the same inside and outside the dielectric surface. This fact can be reconciled with the presence of the polarization charge by noting that additional surface charge is distributed on the plates. The extra charge cancels the effect of polarization charge on the electric field, as shown in Figure 5.6c.

The matching condition for parallel components of electric field at a dielectric field at a dielectric–vacuum surface is

$$E_{\|\alpha} = E_{\|\beta}. \qquad \boxed{5.7}$$

The construction also shows that a dielectric fill allows a capacitor to store more plate charge at the same voltage. Since the electrostatic energy is proportional to the charge, the energy density is proportional to $\varepsilon/\varepsilon_0$. This explains the predominance of water as a medium for high-power density-pulsed voltage systems. Compact high-voltage capacitors are produced using barium titanate, which has a relative dielectric constant which may exceed 10^4.

The combined conditions of Eqs. (5.6) and (5.7) imply that the normal components of electric field lines entering a medium of high $\varepsilon/\varepsilon_0$ from vacuum are small. Inside such a medium, electric field lines are thus bent almost parallel to the interface. Figure 5.7 shows an example of applied dielectric boundary conditions. Equipotential lines are plotted at the output of a high-power, water-filled pulser. The region contains water ($\varepsilon/\varepsilon_0 = 80$), a lucite

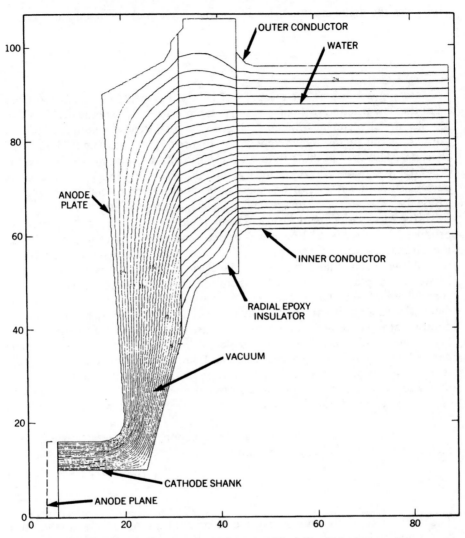

Figure 5.7 Equipotential lines near the vacuum insulator of a high-power pulse generator. (Courtesy J. Benford, Physics International Company.)

insulator ($\varepsilon/\varepsilon_0 \cong 3$), and a vacuum region ($\varepsilon/\varepsilon_0 = 1$) for electron beam acceleration. The aim of the designer was to distribute equipotentials evenly across the vacuum side of the insulator for uniform field stress and to shape boundaries so that field lines enter the surface at a 45° angle for optimum hold-off (see Section 9.5). Note the sharp bending of equipotential lines at the lucite–water boundary. The equipotentials in the water are evenly spaced straight lines normal to the boundary. They are relatively unaffected by the field distribution in the low dielectric constant region.

5.3 FERROMAGNETIC MATERIALS

Some materials modify applied magnetic fields by alignment of bound atomic currents. Depending on the arrangement of electrons, atoms may have a *magnetic moment*. This means that the circulating electrons produce magnetic fields outside the atom. The fields, illustrated in Figure 5.8, have the same form as those outside a circular current loop (Section 4.7); therefore, the circular loop is often used to visualize magnetic interactions of atoms.

The magnetic moment p_m of a loop of radius a carrying a current I is

$$p_m = I(\pi a^2).\tag{5.8}$$

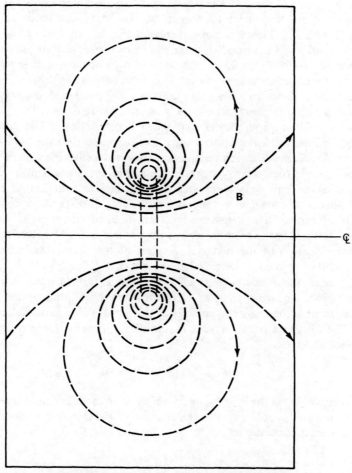

Figure 5.8 Magnetic field lines near an atom with a magnetic moment.

In classical physics, the atomic p_m is visualized to originate from the circular current loop of a valence rotating about the atom. The current I is about $qev/2\pi a$ (where v is the orbital velocity) and a is about 1 Å. Although this gives a rough estimate of typical atomic p_m, the microscopic problem must be approached by quantum mechanics. The correct result is that the magnetic moment is quantified, and can have values

$$p_m = \pm meh/4\pi m_e, \tag{5.9}$$

where h is the Planck constant,

$$h = 6.63 \times 10^{-34} \text{ J-s},$$

and m is an integer which depends on the arrangement of electrons in the atom.

On a macroscopic scale, when two fixed adjacent current loops have the same orientation, the magnetic forces act to rotate the loops to opposite polarity (Fig. 5.9*a*). This is a consequence of the fact that when the magnetic moments are aligned antiparallel, magnetic fields cancel so that the field energy is minimized. With no applied field, atomic currents are oriented randomly, and there is no macroscopic field. The situation is analogous to that of molecules in a dielectric. Material that can be described by this classical viewpoint is called *paramagnetic*, which means "along or parallel to the field." Reference to Figure 5.9*b* shows that when a magnetic field is applied to a paramagnetic material, the atomic currents line up so that the field inside the loop is in the same direction as the applied field while the return flux is in opposition. As in dielectrics, the fractional alignment is small since the change in magnetostatic energy is much less than the average thermal energy of an atom. Figure 5.9*c* shows what the net magnetic field looks like when magnetic moments of atoms in a dense medium are aligned; there is an increase of magnetic flux inside the material above the applied field. Negative flux returns around the outside of the material. A view of the atomic and real currents normal to the applied field clarifies the process (Fig. 5.9*d*). Alignment of magnetic moments does not produce a net atomic current inside the material, but results in a surface current in the same direction as the applied field. The surface current is the magnetic analogy of the surface charge of dielectrics.

The field inside a paramagnetic material is approximately proportional to the applied field, or

$$\mathbf{B} = (\mu/\mu_0)\mathbf{B}_1 = \mathbf{B}_1 + \mathbf{B}_2. \tag{5.10}$$

The quantity μ/μ_0 is the *relative permeability*. The magnetic field intensity $\mathbf{H}$ is a vector quantity proportional to the magnetic field minus the contribution of atomic material currents, or

$$\mathbf{H} = \mathbf{B}_1/\mu_0.$$

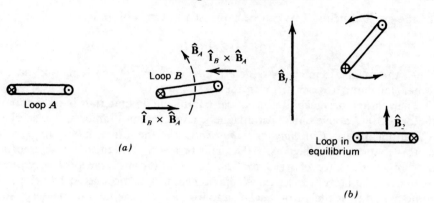

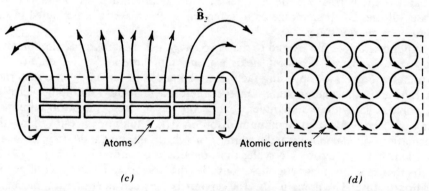

Figure 5.9 Behavior of magnetic materials. (*a*) Interactions between atoms with magnetic moments; force exerted on one current loop (*B*) by another (*A*). (*b*) Response of atoms in a paramagnetic material to an applied magnetic field. (*c*) Macroscopic magnetic fields produced by alignment of atomic currents in a material. (*d*) Macroscopic surface current in a material resulting from alignment of atoms with magnetic moments.

The field intensity is related to the magnetic field inside a magnetic material by

$$\mathbf{H} = \mathbf{B}/\mu. \qquad \boxed{5.11}$$

Magnetic fields obey the principle of superposition. Equation (4.39) can be written

$$\int \mathbf{B}_1 \cdot d\mathbf{l} = \mu_0 I_1,$$

for the applied fields. This can be expressed in terms of H by

$$\int \mathbf{H} \cdot \mathbf{dl} = I_1. \qquad (5.12)$$

Thus, the magnetic intensity is determined only by free currents and has the dimensions amperes per meter in the mks system.

The relative permeability in typical paramagnetic materials is only about a factor of 10^{-6} above unity. Paramagnetic effects are not important in accelerator applications. *Ferromagnetic materials*, on the other hand, have μ/μ_0 factors that can be as high as 10,000. This property gives them many important uses in magnets for charged particle acceleration and transport. Ferromagnetism is a quantum mechanical phenomenon with no classical analogy. In some materials (chiefly iron, nickel, and iron alloys), the minimum energy state consistent with the exclusion principle has atomic magnetic moments aligned parallel rather than antiparallel. The energy involved in this alignment is greater than thermal kinetic energies at room temperature. On a microscopic scale, all the magnetic moments in a ferromagnetic material are aligned in the minimum energy state.

Alignment does not extend to macroscopic scales. Macroscopic alignment of magnetic moments produces fields *outside* the material which require additional energy. Two opposing factors are balanced in ferromagnetic materials in the minimum energy state. On the microscopic scale, minimum energy is associated with atomic alignment, while on the macroscopic scale minimum energy is equivalent to maximum disorder. The situation is resolved by the formation of *domains*, small regions in which all magnetic moments are aligned. On a macroscopic scale, the domains are randomized (Fig. 5.10) so that there is no magnetic field outside the ferromagnetic material in its ordinary state. The domain size (the separatrix between the quantum mechanical and classical regimes) is about 10^{-5} cm, or 1000 atoms wide.

Ferromagnetic materials respond to applied magnetic fields by shifting domain boundaries to favor domains aligned with the field. In contrast to paramagnetic materials, the resulting high degree of atomic orientation produces large magnetic effects. Saturation (total alignment) can occur at attainable applied field strengths (~ 2 T). Although the magnetic field is a

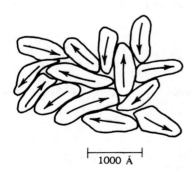

$\vdash\!\!-\!\!-\!\!-\!\!\dashv$
1000 Å

Figure 5.10 Random orientation of domains in unmagnetized material.

monotonic function of the applied field, we cannot expect the response to be linear or reversible. Equation (5.11) is no longer valid. We can preserve the concept of the permeability by considering the response of ferromagnetic materials to small excursions in the applied field about an equilibrium value. The *small signal* μ is defined by

$$\Delta \mathbf{B} = \mu(H)\, \Delta \mathbf{H},$$

or $\hspace{11cm}$ (5.13)

$$\mu(H) = (dB/dH)_H.$$

5.4 STATIC HYSTERESIS CURVE FOR FERROMAGNETIC MATERIALS

In this section we shall look in more detail at the response of ferromagnetic materials to an applied field. In unmagnetized material, the directions of domains are randomized because energy is required to generate magnetic fields outside the material. If the external magnetic field energy is supplied by an outside source, magnetic moments may become orientated, resulting in large amplified flux *inside the material.* In other words, an applied field tips the energy balance in favor of macroscopic magnetic moment alignment.

A primary use of ferromagnetic materials in accelerators is to conduct magnetic flux between vacuum regions in which particles are transported. We shall discuss relationships between fields inside and outside ferromagnetic materials when we treat magnetic circuits in Section 5.7. In this section we limit the discussion to fields confined inside ferromagnetic materials. Figure 5.11 illustrates such a case; a ferromagnetic torus is enclosed in a tight uniform magnet wire winding. We want to measure the net toroidal magnetic field inside the material, B, as a function of the applied field B_1 or the field intensity H. The current in the winding is varied slowly so that applied field permeates the material uniformly. The current in the winding is related to B_1 through Eq. (4.42). By Eq. (5.11), $H = NI/L$, hence the designation of H in ampere-turns per meter. The magnetic field inside the material could be measured by a probe inserted in a thin gap. A more practical method is illustrated in Figure 5.11. The voltage from a loop around the torus in integrated electronically. According to Section 3.5, the magnetic field enclosed by the loop of area A can be determined from $B = \int V\, dt/A$.

With zero current in the windings, a previously unmagnetized core has randomly orientated domains and has no macroscopic magnetization ($H = 0$, $B = 0$). Domains become aligned as the applied field is raised. Both H and B increase, as shown in Figure 5.12. The field in the material (the sum of the applied and atomic contributions) may be over 1000 times that of the applied field alone; thus, the small signal μ is high. At some value of applied field, all the domains are aligned. This is called *saturation.* Beyond this point, there is no increase in the contribution of material currents to the field with increasing

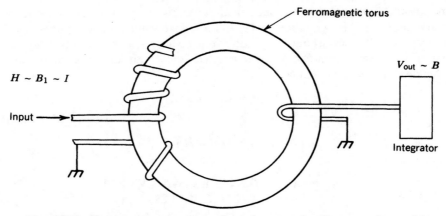

Figure 5.11   Circuit to measure response of atomic currents in a ferromagnetic material.

applied field; therefore, the small signal μ drops to μ_0. The portion of the $B-H$ *curve* from ($H = 0$, $B = 0$) to saturation (the dashed line in Fig. 5.12) is called the *virgin magnetization curve*. Unless the material is completely demagnetized, it will not be repeated again.

The next step in the hypothetical measurement is to reduce the applied field. If the magnetization process were reversible, the $B-H$ curve would follow the

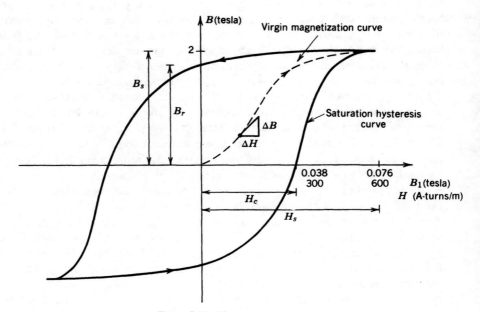

Figure 5.12   Hysteresis curve.

virgin magnetization curve back to the origin. This does not happen since it takes energy to shift domain boundaries. When H is returned to zero, the domains are energetically able to retain much of their alignment so that the torus remains magnetized. This occurs because all field lines are contained in the torus and no energy need be expended to produce external fields. The magnetization would be reduced if there were an air gap in the core (see Section 5.8). If a reverse current is applied to the driving circuit, the magnetic field will return to zero and eventually be driven to reverse saturation. The $B-H$ curve of a magnetic material exhibits *hysteresis* or nonreversibility. The term derives from a Greek word meaning shortcoming or lagging. The $B-H$ curve described is a particular case, the *saturation hysteresis curve*. There is a nested family of hysteresis curves converging to the origin, depending on the magnitude of current in the driving circuit.

Magnet engineering is based on some straightforward concepts and a large body of terminology. A clear understanding of the definitions of magnetic properties is essential to utilize data on magnetic materials. To facilitate this, important terms will be singled out in this section and in Sections 5.7 and 5.8. Terms related to the hysteresis curve are illustrated in Figure 5.12. Most data on magnetic materials and permanent magnets is given in cgs units, so it is important to know the transformation of quantities between mks and cgs.

H, Magnetic Intensity. Also called magnetizing force. The cgs unit is oersteds (Oe), where $1 \text{ A} \cdot \text{turn}/\text{m} = 0.01256 \text{ Oe}$.

B, Magnetic Induction Field. Also called magnetic field, magnetic induction, magnetic flux density. The cgs unit is the gauss (G), where 1 tesla (T) = 1 weber (Wb)$/\text{m}^2 = 10^4$ gauss.

B_s, Saturation Induction. The magnetic field in a ferromagnetic material when all domains are aligned.

H_s. The magnetizing force necessary to drive the material to saturation.

H_c, Coercive Force. The magnetic intensity necessary to reduce the magnetization of a previously saturated material to zero.

B_r, Remanence Flux. Also called residual induction. The value of the magnetic field on the saturation hysteresis curve when the driving current is zero. It is assumed that all magnetic flux is contained in the material.

Soft Magnetic Material. Ferromagnetic materials which require a small magnetizing force to be driven to saturation, typically 10 Oe. The area enclosed by the saturation hysteresis curve is relatively small, and we shall see that this is equivalent to small energy input to magnetize or demagnetize the material. Soft magnetic materials are used to conduct field lines and as isolators in induction accelerators.

Hard Magnetic Materials. Ferromagnetic materials which require considerable energy to reorient the domains. The coercive force can be as high as 8000 Oe. The large amount of energy stored in hard magnetic materials during

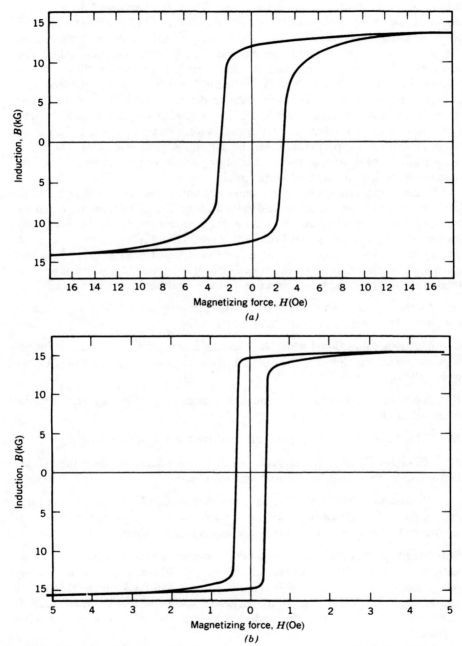

Figure 5.13 Saturation hysteresis curves. (*a*) Low-carbon steel. (*b*) Steel with 3.25% silicon. (Courtesy M. Wilson, National Bureau of Standards.)

magnetization means that more energy is available to produce fields external to the material. Hard magnetic materials are used for permanent magnets.

Figure 5.13 shows hysteresis curves for carbon steel (a material used for magnet poles and return flux yokes) and silicon steel (used for pulsed transformer cores).

5.5 MAGNETIC POLES

Figure 5.14 shows a boundary between a magnetic material with permeability μ and vacuum with μ_0. A thin volume element encloses the boundary. The equation $\nabla \cdot \mathbf{B} = 0$ implies that the integral of the normal component of $\mathbf{B}$ on the surfaces of the volume is zero (Fig. 5.14a). The main contributions to the integral are from the upper and lower faces, so that

$$B_{\perp\alpha} = B_{\perp\beta}. \qquad \boxed{5.14}$$

Noting that there is no free current enclosed, Eq. (5.12) can be applied around the periphery of the volume (Fig. 5.14b). The main contributions to the circuital integral are on the faces.

$$\int (\mathbf{H}_\alpha - \mathbf{H}_\beta) \cdot \mathbf{dl} = 0 \qquad (5.15)$$

or

$$B_{\|\alpha}/\mu_0 = B_{\|\beta}/\mu. \qquad \boxed{5.16}$$

For ferromagnetic materials ($\mu \gg \mu_0$), the parallel component of magnetic field outside the ferromagnetic material is much smaller than the parallel component inside the material. Thus, magnetic field lines just outside a ferromagnetic material are almost normal to the surface. This simple boundary

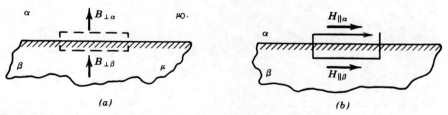

(a)   *(b)*

Figure 5.14 Boundary conditions for magnetic field components at a boundary between vacuum and a ferromagnetic material. (*a*) Geometry to find the relationship of normal field components. (*b*) Geometry to find the relationship of parallel field components.

condition means that ferromagnetic materials define surfaces of constant magnetic potential U_m. Ferromagnetic surfaces can be used to generate magnetostatic field distributions in the same way that electrodes are used for electrostatic fields. Since both the electrostatic and magnetic potentials satisfy the Laplace equation, all electrostatic solutions can be applied to magnetic fields. Ferromagnetic surfaces used to shape magnetic field lines in a vacuum (or air) region are called *pole pieces*. By convention, magnetic field vectors point from the North to the South pole. Figure 5.15 shows an example of a numeric calculation of magnetic fields for a synchrotron magnet. Shaping of the pole piece near the gap provides the proper field gradient for beam focusing.

The boundary condition is not valid when the pole material becomes saturated. In high-field magnets, regions of high flux may become saturated before rest of the pole piece. This distorts the magnetic field pattern. The fields of partially saturated magnets are difficult to predict. Note that local saturation is avoided in the design of Figure 5.15 by proper shaping and avoidance of sharp edges. This allows the maximum magnetic field without distortion. In the limit of fields well above the saturation limit ($\gg 2$ T), the effective relative permeability decreases to unity and the field pattern approaches that of the exciting coil only.

The analogy between electrostatic and magnetostatic solutions leads to the *magnetic quadrupole*, illustrated in Figure 5.16. The pole pieces follow hyperbolic surfaces of constant magnetic potential [Eq. (4.26)]. In contrast to the electrostatic quadrupole, the x–y forces on a beam moving along the z axis are perpendicular to the magnetic field lines. It is usually more convenient to analyze a magnetic quadrupole in terms of x–y axes rotated 45° from those used for the electric field version. In the coordinate system of Figure 5.16, the

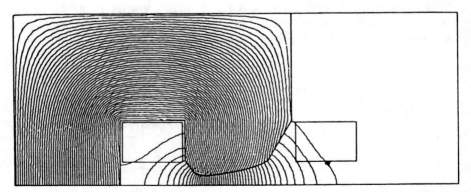

Figure 5.15 Numerical calculation of magnetic field lines in a synchrotron magnet. (Adapted from J. S. Colonias, *Particle Accelerator Design; Computer Programs*, used by permission, Academic Press.)

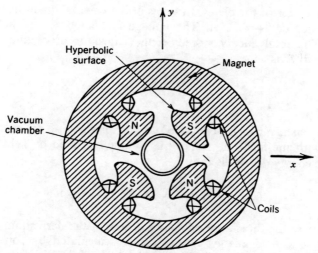

Figure 5.16   Magnetic quadrupole lens (cross section).

magnetic field components are

$$B_x = B_0 y/a, \qquad B_y = B_0 x/a.$$

$$\boxed{5.17}$$

5.6   ENERGY DENSITY OF ELECTRIC AND MAGNETIC FIELDS

As mentioned in Section 3.2, the field description summarizes the electromagnetic interactions between charged particles. Although exchange of energy in a system takes place between charged particles, it is often more convenient to imagine that energy resides in the fields themselves. In this section, we shall use two examples to demonstrate the correspondence between the electromagnetic energy of particles and the concept of *field energy density*.

The electric field energy density can be determined by considering the parallel plate capacitor (Fig. 5.6c). Initially, there is no voltage difference between the plates, and hence no stored charge or energy. Charge is moved slowly from one plate to another by a power supply. The supply must perform work to move charge against the increasing voltage. If the plates have a voltage V', the differential energy transfer from the power supply to the capacitor to move an amount of charge dQ' is $\Delta U = V' \, dQ'$. Modifying Eq. (3.9) for the presence of the dielectric, the electric field is related to the total charge moved between the plates by

$$E = \frac{Q/A}{\varepsilon},$$

where A is the area of the plates. In terms of incremental quantities during the charging cycle, $dQ' = \varepsilon\, dE'\, A$. The voltage is related to the field through $V' = E'd$. The total energy stored in the capacitor proceeding from zero electric field to E is

$$U = \int_0^E (E'd)(\varepsilon A\, dE') = (\varepsilon E^2/2)(Ad). \qquad (5.18)$$

The total energy can be expressed as the product of an energy density associated with the field lines times the volume occupied by field (Ad). The electrostatic energy density is thus

$$U(E) = \varepsilon E^2/2 = \mathbf{D} \cdot \mathbf{E}/2. \ (\mathrm{J/m^3}). \qquad \boxed{5.19}$$

The last form is the three-dimensional extension of the derivation.

The magnetic field energy density can be calculated by considering the toroidal core circuit of Figure 5.11. Since the core is ferromagnetic, we will not assume a linear variation and well-defined μ. The coil has N turns around a circumference C. A power supply slowly increases the current in the coil. To do this, it must counteract the inductive voltage V'. The energy transferred from the power supply to the circuit is $\int V' I\, dt$. The inductive voltage is $V' = NA(dB'/dt)$, where B' is the sum of contributions to the field from I and the atomic current of the material. The applied field is related to the circuit current by $B_1 = \mu_0 NI/C$. Summarizing the above considerations,

$$U = \int_0^B NA \frac{dB'}{dt} \frac{B_1 C}{N\mu_0}\, dt = AC \int_0^B H\, dB'. \qquad (5.20)$$

Recognizing that the volume occupied by fields is AC, the magnetic field energy density is

$$U(B) = \int_0^B H\, dB' \ (\mathrm{J/m^3}). \qquad \boxed{5.21}$$

If the relationship between B and H is known, the energy density of the final state can be evaluated. For instance, with a linear variation (constant μ)

$$U(B) = B^2/2\mu = \mathbf{H} \cdot \mathbf{B}/2 \ (\mathrm{J/m^3}). \qquad \boxed{5.22}$$

The magnetic field energy density in vacuum is $B^2/2\mu_0$.

The magnetic field energy density can be determined for nonlinear materials given the appropriate $B–H$ curve. For instance, consider magnetizing a ferromagnetic toroidal core following the saturation hysteresis curve (Fig. 5.17a). Assume initially that the material is biased to $-B_r$, and a positive driving current is applied to bring it to $+B_s$ (Fig. 5.17b). Both H and dB are positive,

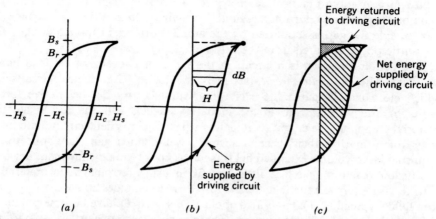

Figure 5.17 Energy required to magnetize and demagnetize a ferromagnetic material. (*a*) Saturation hysteresis curve. (*b*) Quantities to calculate energy changes moving along saturation hysteresis curve. (*c*) Energy supplied by circuit (cross-hatched area) and energy returned to circuit (shaded portion).

so work must be performed by the power source. The energy transfer is given by the shaded area of the graph (Fig. 5.17*c*) multiplied by the volume of the core. If the power supply is turned off, the core returns to $+B_r$. During this part of the cycle, H is positive but dB is negative, so that some energy is returned to the supply as an induced voltage. This energy is denoted by the darkly shaded portion of the graph; the net energy transfer in a half-cycle is the lightly shaded remainder. A similar process occurs for negative H. In most induction accelerators, a full cycle is tranversed around the saturation hysteresis curve for each beam pulse. An amount of energy equal to the area circumscribed by the hysteresis curve is lost to the core in each full cycle. This energy is expended in the irreversible process of domain reorientation. It ultimately appears in the core in the form of heat. In applications with continued and rapid recycling, it is clearly advantageous to use a ferromagnetic material with the smallest hysteresis curve area (soft material).

5.7 MAGNETIC CIRCUITS

Magnetic fields used for charged particle transport are usually localized to small regions. This is the case for a bending magnet where the beam traverses a narrow vacuum region of parallel field lines. The condition of zero divergence implies that the field lines must curve around outside the transport region to return to the gap. Thus, most of the volume occupied by magnetic field serves no purpose for the application. If the surrounding region is vacuum or air ($\mu/\mu_0 = 1$), most of the power input to the magnet is consumed to support the return flux. A more practical geometry for a bending magnet uses ferromagnetic material in the return flux region. We shall see that in this case magnetic flux outside the gap is supported by the atomic currents in the material, with very little power required from the external supply.

The bending magnet is a circuit in the sense that the magnetic field lines circulate. The *magnetic circuit* has many analogies with electric circuits in which electrons circulate. The excitation windings provide the motive force (voltage), the vacuum gap is the load (resistance), and the ferromagnetic material completes the circuit (conducting wire). The magnetic circuit with iron is useful primarily when there is a small, well-defined gap. In applications requiring large-volume extended fields (such as hot plasma confinement devices for fusion reactors), there is little benefit in using ferromagnetic materials. Furthermore, ferromagnetic materials lose their advantages above their saturation field (typically 2 T). High-field applications ($\sim$ 6 T) have become practical through the use of superconductors. There is no energy penalty in supporting return flux lines in vacuum since the excitation windings draw no power.

Figure 5.18 illustrates the advantage of including ferromagnetic material in ordinary magnetic circuits. Assume that both the air core and iron core geometries produce the same field, B_g, in equal gaps. In order to compare the circuits directly, windings are included in the air core circuit so that the return flux is contained in the same toroidal volume. The magnetic flux in any cross section is a constant and is the same for both circuits. The gap has cross-sectional area A_g, and the core (or return flux coil) has area A_c. The length of the gap is g, while the length of the core (coil) is l. The excitation coils have an ampere turn product given by the number of windings multiplied by the current input to the windings, NI. The wires that carry the current have resistivity in an ordinary magnet; the power necessary to support the field is proportional to N (the length of the wire) and to I^2. It is desirable to make NI as small as possible.

The ampere turn products for the two circuits of Figure 5.18 can be related to the magnetic field in the circuit through Eq. (5.12):

$$\int (\mathbf{B}/\mu) \cdot \mathbf{dl} = NI. \tag{5.23}$$

The constant circuit flux is given by

$$\Psi = B_g A_g = B_c A_c. \tag{5.24}$$

For the air core circuit, Eq. (5.23) becomes

$$B_g(g/\mu_0) + B_c(l/\mu_0) = NI,$$

or

$$\Psi(g/A_g\mu_0 + l/A_c\mu_0) = NI. \tag{5.25}$$

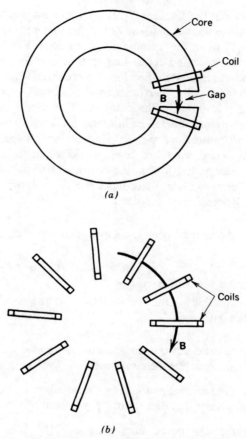

Figure 5.18 Comparison of energy required to generate a gap field by (*a*) an iron core magnet and (*b*) an air core magnet.

Similarly, the following equation describes the ferromagnetic circuit:

$$\Psi\left(g/A_g\mu_0 + l/A_c\mu\right) = NI. \tag{5.26}$$

Comparing Eqs. (5.25) and (5.26), the ampere turn product for equal flux is much smaller for the case with the ferromagnetic core when $\mu \gg \mu_0$ and $g \ll l$ (a small gap).

An iron core substantially reduces power requirements for a dc magnet. An alternate view of the situation is that the excitation coil need support only the magnetic field in the gap since the second term in Eq. (5.26) is negligible. The return flux is supported by the atomic currents of the material. The excitation coils are located at the gap in Figure 5.18*a* to clarify this statement. In practice, the coils can be located anywhere in the circuit with about the same result. This follows from the Laplace equation which implies that the field line configuration minimizes the net field energy of the system. The field energy is a minimum if the flux flows in the ferromagnetic material. The field lines are thus conducted through the core and cross the gap in a manner consistent with the boundary condition discussed in Section 5.5, relatively independent of the location of the exciting coils. Ferromagnetic materials also help in pulsed magnet circuits, such as those in the betatron. The *duty cycle* (time on/time off) of such magnets is usually low. Thus, the total field energy is of greater concern than the instantaneous power. Energy for pulsed fields is usually supplied from a switched capacitor bank. A ferromagnetic return flux core reduces the circuit energy and lowers the cost and size of the capacitor bank.

To summarize, ferromagnetic materials have the following applications in nonsuperconducting accelerator magnets:

1. The iron can be shaped near the gap to provide accurate magnetic field gradients.
2. The exciting coil power (or net field energy) is reduced significantly compared to an air core circuit.
3. The iron conducts flux lines so that the exciting windings need not be located at the gap.

Equation (5.26) has the same form as that for an electric circuit consisting of a power source and resistive elements with the following substitutions.

Magnetic Flux Ψ.   The analogy of current in an electric circuit. Although the magnitude of B may vary, the flux in any cross section is constant.

Magnetomotive Force.   (Ampere turn product, NI.) The driving force for magnetic flux. Magnetomotive force corresponds to the voltage in an electric circuit.

Reluctance.   Corresponds to the resistance. Equation 5.26 contains two reluctances in series, $R_g = gA_g/\mu_0$ and $R_l = lA_c/\mu$. The higher the reluctance, the lower the flux for a given magnetomotive force. The reluctance of the iron return flux core is much smaller than that of the gap (the load), so it acts in the same way as a low-resistivity wire in an electric circuit.

Permanence.   The inverse of reluctance and the analogue of conductance.

The circuit analogy is useful for estimating operating parameters for complex magnets such as those found in electric motors. To illustrate a magnetic field calculation, consider the spectrometer magnet illustrated in Figure 5.19*a*. The components of reluctance already mentioned can be supplemented by additional paths representing *fringing flux* (magnetic field lines bulging out near the gap) and *leakage flux* (magnetic flux returning across the magnetic yoke without traversing the gap). These reluctances can be determined by a

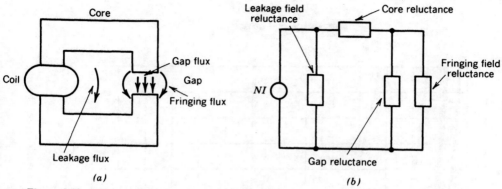

Figure 5.19 Magnet circuit. (*a*) Geometry of spectrometer magnet. (*b*) Equivalent magnetic circuit.

solution of the Laplace equation for the magnet. Reluctances are combined in series and in parallel, just as resistances. The equivalent circuit is illustrated in Figure 5.19*b*. The effect of leakage flux on the field in the gap can be minimized by placing the excitation coils as close to the gap as possible.

A first-order estimate of driving coil parameters can be derived by neglecting the leakage and fringing contributions and assuming that the circuit reluctance resides predominantly in the gap. In this case it is sufficient to use Eq. (5.23), so that $NI \cong B_g g / \mu_0$. For example, production of a field of 1 T in a gap with a 0.02 m spacing requires 16-kA turns (160 turns of wire if a 100-A supply is available). For a given supply and excitation coil winding, the field magnitude is inversely proportional to the spacing of the magnet poles.

5.8 PERMANENT MAGNET CIRCUITS

Permanent magnet circuits have the advantage that a dc magnetic field can be maintained with no power input. There are two drawbacks of permanent magnet circuits: (1) it is difficult to vary the field magnitude in the gap and (2) bulky magnets are needed to supply high fields over large areas. The latter problem has been alleviated by the development of rare-earth samarium cobalt magnets which have a maximum energy product three times that of conventional Alnico alloys. In other words, the same field configuration can be produced with a magnet of one-third the volume. Permanent magnet quadrupole lenses (Section 6.10) are an interesting option for focusing in accelerators.[†] In this section, we shall review some of the properties of permanent magnetic materials and first-order principles for designing magnetic circuits.

[†]See K. Halbach, "Physical and Optical Properties of Rare Earth Cobalt Magnets," *Nucl. Instrum. Methods*, **187**, 109 (1981).

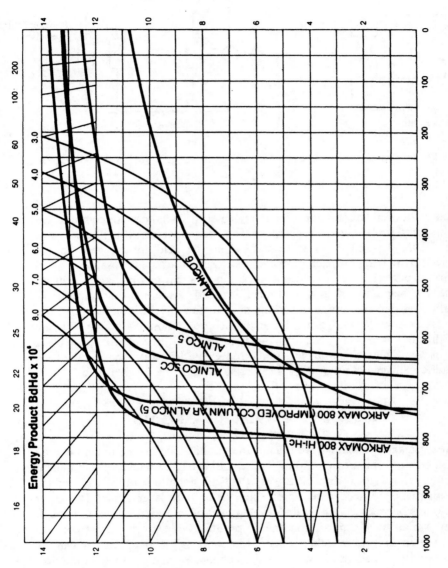

Figure 5.20 Demagnetization curve for some common permanent magnet materials. (Used by permission of Arnold Engineering Company.)

The second quadrant of a hysteresis curve for some common permanent magnet materials is shown in Figure 5.20. The plot is usually called the *demagnetization curve*. The most striking difference between Figure 5.20 and the hysteresis curve for soft iron (Fig. 5.13) is that the coercive force is about 100 times larger for the permanent magnets. In other words, it takes considerably more energy to align the domains and to demagnetize the material. Generation of magnetic fields in vacuum requires energy; a permanent magnet can produce fields because of the stored energy received during magnetization.

Figure 5.20 can be used to calculate the field produced in the gap of a magnetic circuit. The method used to find the operating point on the demagnetization curve is illustrated in Figure 5.21. In the first part of the figure, the

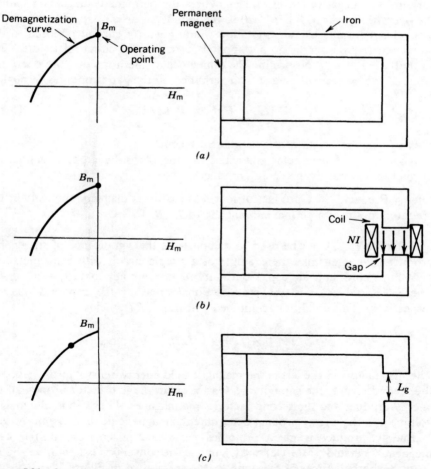

Figure 5.21 Operating point of a permanent magnet. (*a*) Permanent magnet with a continuous iron flux conductor, zero magnetizing force. (*b*) Addition of an air gap with a coil to supply field energy. (*c*) Deactivation of the gap coil.

permanent magnet is included in a zero reluctance circuit containing an ideal iron core ($H_s = 0$) with no gap. There is no free current. The circuital integral of H is zero, so that H_m (the magnetic intensity in the permanent magnet) is zero. The magnetic field in the loop is equal to B_{rm}, the remanence field of the permanent magnet. Next, assume that an air gap is introduced into the circuit, but excitation windings are placed around the gap with the proper current to produce a field equal to B_{rm} (Fig. 5.21b). The current in the windings is in the same direction as the atomic currents in the ferromagnetic materials. Since the energy for the vacuum fields is supplied by an external source, the circuit still appears to have zero reluctance. The operating point of the permanent magnet remains at $H_m = 0$, $B_m = B_{rm}$).

In the final state (Fig. 5.21c), the current in the excitation coils drops to zero. This is equivalent to the addition of a negative current to the existing current. The negative current demagnetizes the permanent magnet, or moves the operating point in Figure 5.20 to the left. Thus, an air gap in a permanent magnet circuit acts like an excitation winding with a current opposed to the atomic currents. There is no net applied current in the circuit of Figure 5.21c so that $\int H \cdot dl = 0$. Neglecting the reluctance of the iron core, the operating point of the permanent magnet is determined by the gap properties through

$$H_m L_m = H_g L_g = B_g L_g / \mu_0, \qquad (5.27)$$

where L_m is the length of the permanent magnet.

An important parameter characterizing the performance of a permanent magnet in a circuit is the energy product.

Energy Product.   The product of magnet field times magnetic intensity at the operating point of a permanent magnet, or $H_m B_m$.

Equation (5.27) can be used to demonstrate the significance of the energy product. We again take the example of a simple circuit with a magnet, zero reluctance core, and air gap. Continuity of flux implies that $B_g A_g = B_m A_m$, where A_g and A_m are the cross-sectional areas of the gap and magnet, respectively. This condition, combined with Eq. (5.27), yields

$$\left(B_g^2 / 2\mu_0 \right)\left(A_g L_g \right) = \left(H_m B_m / 2 \right)\left(A_m L_m \right). \qquad (5.28)$$

The first factor on the left is the magnetic field energy density in the gap, and the second term is the gap volume. On the right, the first factor is one-half the energy product and the second factor is the magnet volume. Thus, the magnet volume and the energy product determine the magnetic field energy in the gap.

Energy product is given in joules per cubic meter (mks units) or in megagauss oersteds (MG-Oe) in CGS units. The conversion is 1 MG-Oe = 3980 J/m^3. Hyperbolic lines of constant $B_m H_m$ are plotted in Figure 5.20. This is a graphic aid to help determine the energy product at different points of the demagnetization curve. A goal in designing a permanent magnet circuit is to produce the required gap field with the minimum volume magnet. This occurs at the point on the demagnetization curve where $B_m H_m$ is maximum. In Figure 5.20, parameters of the circuit should be chosen so that H_m is about 550 Oe.

Two examples will serve to illustrate methods of choosing permanent magnets. Both involve first-order design of a simple circuit with no leakage or fringing flux; the second-order design must invoke field calculations, tabulated gap properties or modeling experiments for an accurate prediction. To begin, suppose that we constrain the gap parameters ($B_g = 8$ kG, $A_g = 10$ cm^2, and $L_g = 1$) and the type of magnetic material (Alnico 5). We must now determine the dimensions of the magnet that will produce the gap field. Using Eq. (5.27) with $H_g = 8$ kOe and $H_m = 600$ Oe, the length of the magnet must be 13.3 cm. The magnetic field is $B_m = 8$ kG, so that the minimum magnet cross-sectional area is 10 cm^2.

In the second example, assume that the dimensions of the gap and magnet are constrained by the application. The goal is to determine what magnetic material will produce the highest gap flux and the value of this flux. For the simple circuit, the condition of constant flux can be combined with Eq. (5.27) to give

$$H_g L_g / B_g A_g = H_m L_m / B_m A_m = L_g / \mu_0 A_g.$$

The expression on the right-hand side is the reluctance of the gap, R_g. We can then write

$$B_m / H_m = L_m / R_g A_m. \tag{5.29}$$

With the stated conditions, the quantity B_m / H_m must be a constant ratio. This motivates the definition of the permanence coefficient.

Permanence Coefficient, or Load Line.   Equal to (B_m / H_m). The permanence coefficient is a function only of the geometries of the magnetic load (system reluctance) and the magnet.

Fiducial points are usually included in demagnetization curves to lay out load lines. Given the load line, operating points on various permanent magnet materials can be determined. The highest gap flux will be produced by the material with the highest energy product at the intersection. The gap flux can then be determined from the magnet operating point.

6

Electric and Magnetic
Field Lenses

The subject of charged particle optics is introduced in this chapter. The concern is the control of the transverse motion of particles by shaped electric and magnetic fields. These fields bend charged particle orbits in a manner analogous to the bending of light rays by shaped glass lenses. Similar equations can be used to describe both processes. Charged particle lenses have extensive applications in such areas as electron microscopy, cathode ray tubes, and accelerator transport.

In many practical cases, beam particles have small velocity perpendicular to the main direction of motion. Also, it is often permissible to use special forms for the electric and magnetic fields near the beam axis. With these approximations, the transverse forces acting on particles are linear; they increase proportional to distance from the axis. The treatment in this chapter assumes such forces. This area is called *linear* or *Gaussian* charged particle optics.

Sections 6.2 and 6.3 derive electric and magnetic field expressions close to the axis and prove that any region of linear transverse forces acts as a lens. Quantities that characterize thick lenses are reviewed in Section 6.4 along with the equations that describe image formation. The bulk of the chapter treats a variety of static electric and magnetic field focusing devices that are commonly used for accelerator applications.

6.1 TRANSVERSE BEAM CONTROL

Particles in beams always have components of velocity perpendicular to the main direction of motion. These components can arise in the injector; charged particle sources usually operate at high temperature so that extracted particles have random thermal motions. In addition, the fields in injectors may have imperfections of shape. After extraction, space charge repulsion can accelerate particles away from the axis. These effects contribute to expansion of the beam. Accelerators and transport systems have limited transverse dimensions. Forces must be applied to deflect particles back to the axis. In this chapter, the problem of confining beams about the axis will be treated. When accelerating fields have a time dependence, it is also necessary to consider longitudinal confinement of particles to regions along the axis. This problem will be treated in Chapter 13.

Charged particle lenses perform three types of operations. One purpose of lenses is to *confine* a beam, or maintain a constant or slowly varying radius (see Fig. 6.1a). This is important in high-energy accelerators where particles must travel long distances through a small bore. Velocity spreads and space

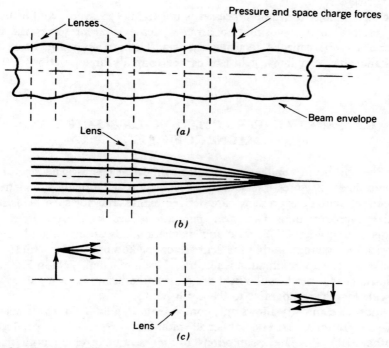

Figure 6.1 Functions of charged particle lenses. (*a*) Beam confinement. (*b*) Focusing to a spot. (*c*) Imaging.

charge repulsion act to increase the beam radius. Expansion can be countered by continuous confining forces which balance the outward forces or through a periodic array of lenses which deflect the particles toward the axis. In the latter case, the beam outer radius (or *envelope*) oscillates about a constant value.

A second function of lenses is to *focus* beams or compress them to the smallest possible radius (Fig. 6.1*b*). If the particles are initially parallel to the axis, a linear field lens aims them at a common point. Focusing leads to high particle flux or a highly localized beam spot. Focusing is important for applications such as scanning electron microscopy, ion microprobes, and ion-beam-induced inertial fusion.

A third use of charged particle lenses is forming an *image*. (Fig. 6.1*c*). When there is a spatial distribution of beam intensity in a plane, a lens can make a modified copy of the distribution in another plane along the direction of propagation. An image is formed if all particles that leave a point in one plane are mapped into another, regardless of their direction. An example of charged particle image formation is an image intensifier. The initial plane is a photocathode, where electrons are produced proportional to a light image. The electrons are accelerated and deflected by an electrostatic lens. The energetic electrons produce an enhanced copy of the light image when they strike a phosphor screen.

The terminology for these processes is not rigid. Transverse confinement is often referred to as focusing. An array of lenses that preserves the beam envelope may be called a focusing channel. The processes are, in a sense, interchangeable. Any linear field lens can perform all three functions.

6.2 PARAXIAL APPROXIMATION FOR ELECTRIC AND MAGNETIC FIELDS

Many particle beam applications require cylindrical beams. The electric and magnetic fields of lenses for cylindrical beams are azimuthally symmetric. In this section, analytic expressions are derived expressions for such fields in the paraxial approximation. The term *paraxial* comes from the Greek *para* meaning "alongside of." Electric and magnetic fields are calculated at small radii with the assumption that the field vectors make small angles with the axis. The basis for the approximation is illustrated for a magnetic field in Figure 6.2. The currents that produce the field are outside the beam and vary slowly in z over scale lengths comparable to the beam radius.

Cylindrical symmetry allows only components B_r and B_z for static magnetic fields. Longitudinal currents at small radius are required to produce an azimuthal field B_θ. The assumptions of this section exclude both particle currents and displacement currents. Similarly, only the electric field components E_r and E_z are included. In the paraxial approximation, **B** and **E** make small angles with the axis so that $E_r \ll E_z$ and $B_r \ll B_z$.

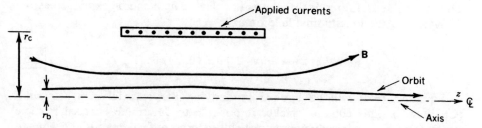

Figure 6.2 Validity conditions for the paraxial field approximation in a magnetic field lens. Paraxial approximation: $r_b \ll r_c$; $v_r \ll v_z$; $B_r \ll B_z$.

The following form for electrostatic potential is useful to derive approximations for paraxial electric fields:

$$\phi(r, z) \cong \phi(0, z) + Ar(\partial\phi/\partial z)|_0 + Br^2(\partial^2\phi/\partial z^2)|_0$$

$$+ Cr^3(\partial^3\phi/\partial z^3)|_0 + Dr^4(\partial^4\phi/\partial z^4)|_0 + \cdots . \qquad (6.1)$$

The z derivatives of potential are evaluated on the axis. Note that Eq. (6.1) is an assumed form, not a Taylor expansion. The form is valid if there is a choice of the coefficients $A, B, C, \ldots$, such that $\phi(r, z)$ satisfies the Laplace equation in the paraxial approximation. The magnitude of terms decreases with increasing power of r. A term of order n has the magnitude $\phi_0(\Delta r/\Delta z)^n$, where Δr and Δz are the radial and axial scale lengths over which the potential varies significantly. In the paraxial approximation, the quantity $\Delta r/\Delta z$ is small.

The electric field must go to zero at the axis since there is no included charge. This implies that $A = 0$ in Eq. (6.1). Substituting Eq. (6.1) into (4.19), we find that the coefficients of all odd powers of r must be zero. The coefficients of the even power terms are related by

$$4B(\partial^2\phi/\partial z^2) + 16Dr^2(\partial^4\phi/\partial z^4)$$

$$+ \cdots + (\partial^2\phi/\partial z^2) + Br^2(\partial\phi^4/\partial z^4) + \cdots = 0. \qquad (6.2)$$

This is consistent if $B = -1/4$ and $D = -B/16 = 1/64$. To second order in $\Delta r/\Delta z$, $\phi(r, z)$ can be expressed in terms of derivatives evaluated on axis by

$$\phi(r, z) \cong \phi(0, z) - (r^2/4)(\partial^2\phi/\partial z^2)|_0. \qquad (6.3)$$

The axial and radial fields are

$$E_z(0, z) \cong -(\partial\phi/\partial z)|_0, \qquad E_r(r, z) \cong (r/2)(\partial^2\phi/\partial z^2)|_0. \qquad (6.4)$$

This gives the useful result that the radial electric field can be expressed as the derivative of the longitudinal field on axis:

$$E_r(r, z) \cong -(r/2)[\partial E_z(0, z)/\partial z].$$

6.5

Equation (6.5) will be applied in deriving the paraxial orbit equation (Chapter 7). This equation makes it possible to determine charged particle trajectories in cylindrically symmetric fields in terms of field quantities evaluated on the axis. A major implication of Eq. (6.5) is that all transverse forces are linear in the paraxial approximation. Finally, Eq. (6.5) can be used to determine the radial variation of E_z. Combining Eq. (6.5) with the azimuthal curl equation $(\partial E_z/\partial r - \partial E_r/\partial z = 0)$ gives

$$E_z(r, z) \cong E_z(0, z) - (r^2/4)[\partial^2 E_z(0, z)/\partial z^2].$$

6.6

The variation of E_z is second order with radius. In the paraxial approximation, the longitudinal field and electrostatic potential are taken as constant in a plane perpendicular to the axis.

A parallel treatment using the magnetic potential shows that

$$B_r(r, z) \cong -(r/2)[\partial B_z(0, z)/\partial z].$$

6.7

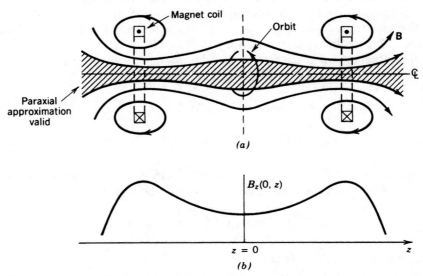

Figure 6.3 Magnetic mirror field. (*a*) Geometry, field components, and forces on a particle with an axi-centered orbit. (*b*) Variation of axial field on-axis, $B_z(0, z)$.

Figure 6.3 is an example of a paraxial magnetic field distribution. The fields are produced by two axicentered circular coils with currents in the same direction. In plasma research, the field distribution is called a *magnetic mirror*. It is related to the fields used in cyclotrons and betatrons. The magnitude of $B_z(0, z)$ is maximum at the coils. The derivative of B_z is positive for $z > 0$ and negative for $z < 0$. Consider a positively charged particle with an axicentered circular orbit that has a positive azimuthal velocity. If the particle is not midway between the coils, there will be an axial force $qv_\theta B_r$. Equation (6.7) implies that this force is in the negative z direction for $z > 0$ and the converse when $z < 0$. A magnetic mirror can provide radial and axial confinement of rotating charged particles. An equivalent form of Eq. (6.6) holds for magnetic fields. Since $\partial^2 B_z/\partial z^2$ is positive in the mirror, the magnitude of B_z decreases with radius.

6.3 FOCUSING PROPERTIES OF LINEAR FIELDS

In this section, we shall derive the fact that all transverse forces that vary linearly away from an axis can focus a parallel beam of particles to a common point on the axis. The parallel beam, shown in Figure 6.4, is a special case of *laminar* motion. Laminar flow (from *lamina*, or layer) implies that particle orbits at different radii follow streamlines and do not cross. The ideal laminar beam has no spread of transverse velocities. Such beams cannot be produced, but in many cases laminar motion is a valid first approximation. The derivation in this section also shows that linear forces preserve laminar flow.

The radial force on particles is taken as $F_r(r) = -A(z, v_z)r$. Section 6.2 showed that paraxial electric forces obey this equation. It is not evident that magnetic forces are linear with radius since particles can gain azimuthal velocity passing through radial magnetic fields. The proof that the combination

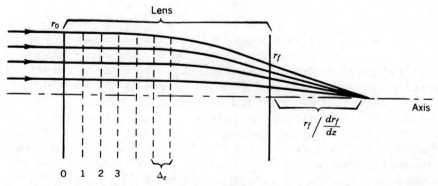

Figure 6.4 Finite difference calculation of the effect of linear focusing fields on particle orbits.

of magnetic and centrifugal forces gives a linear radial force variation is deferred to Section 6.7.

Particle orbits are assumed paraxial; they make small angles with the axis. This means that $v_r \ll v_z$. The total velocity of a particle is $v_0^2 = v_r^2 + v_z^2$. If v_0 is constant, changes of axial velocity are related to changes of radial velocity by

$$\Delta v_z / v_z \cong (v_r / v_z)(\Delta v_z / v_z).$$

Relative changes of axial velocity are proportional to the product of two small quantities in the paraxial approximation. Therefore, the quantity v_z is almost constant in planes normal to z. The average axial velocity may vary with z because of acceleration in E_z fields. If v_z is independent of r, time derivatives can be converted to spatial derivatives according to

$$d/dt \rightarrow v_z(d/dz). \qquad \boxed{6.8}$$

The interpretation of Eq. (6.8) is that the transverse impulse on a particle in a time Δt is the same as that received passing through a length element $\Delta z = v_z \, \Delta t$. This replacement gives differential equations expressing radius as a function of z rather than t. In treatments of steady-state beams, the orbits $r(z)$ are usually of more interest than the instantaneous position of any particle.

Consider, for example, the nonrelativistic transverse equation of motion for a particle moving in a plane passing through the axis in the presence of azimuthally symmetric radial forces

$$d^2 r / dt^2 = F_r(r, z, v_z) / m_0.$$

Converting time derivatives to axial derivatives according to Eq. 6.8 yields

$$d(v_z r') / dt = v_z r' v_z' + v_z^2 r'' = F_r / m_0$$

or

$$dr'/dz = F(r, z, v_z) / m_0 v_z^2 - v_z' r' / m_0 v_z, \qquad r' = dr/dz. \qquad (6.9)$$

A primed quantity denotes an axial derivative. The quantity r' is the angle between the particle orbit and the axis. The motion of a charged particle through a lens can be determined by a numerical solution of Eqs. (6.9). Assume that the particle has $r = r_0$ and $r' = 0$ at the lens entrance. Calculation of the final position, r_f, and angle, r_f', determines the focal properties of the fields. Further, assume that F_r is linear and that $v_z(0, z)$ is a known function calculated from $E_z(0, z)$. The region over which lens forces extend is divided into a number of elements of length Δz. The following numerical algorithm (the *Eulerian difference method*) can be used to find a particle orbit.

$$r(z + \Delta z) = r(z) + r'(z)\,\Delta z,$$

$$r'(z + \Delta z) = r'(z) - \left[A(z, v_z)r/m_0 v_z^2 + v_z' r'/m_0 v_z \right]\Delta z \quad (6.10)$$

$$= r' - \left[a_1(z)r + a_2(z)r' \right]\Delta z.$$

More accurate difference methods will be discussed in Section 7.8.

Applying Eq. (6.10), position and velocity at the first three positions in the lens are

$$r_0 = r_0, \qquad r_0' = 0,$$

$$r_1 = r_0, \qquad r_1' = -a_1(0)r_0\,\Delta z,$$

$$r_2 = r_0 - a_1(0)r_0, \qquad\qquad\qquad (6.11)$$

$$r_2' = -a_1(0)r_0\,\Delta z - a_1(\Delta z)r_0\,\Delta z + a_2(\Delta z)a_1(0)r_0\,\Delta z.$$

Note that the quantity r_0 appears in all terms; therefore, the position and angle are proportional to r_0 at the three axial locations. By induction, this conclusion holds for the final position and angle, r_f and r_f'. Although the final orbit parameters are the sum of a large number of terms (becoming infinite as Δz approaches zero), each term involves a factor of r_0. There are two major results of this observation.

1. The final radius is proportional to the initial radius for all particles. Therefore, orbits do not cross. A linear force preserves laminar motion.

2. The final angle is proportional to r_0; therefore, r_f' is proportional to r_f. In the paraxial limit, the orbits of initially parallel particles exiting the lens form similar triangles. All particles pass through the axis at a point a distance r_f/r_f' from the lens exit (Fig. 6.4).

The conclusion is that any region of static, azimuthally symmetric electric fields acts as a lens in the paraxial approximation. If the radial force has the form $+A(z, v_z)r$, the final radial velocity is positive. In this case, particle orbits form similar triangles that emanate from a point upstream. The lens, in this case, is said to have a *negative focal length*.

6.4 LENS PROPERTIES

The lenses used in light optics can often be approximated as *thin lenses*. In the thin-lens approximation, rays are deflected but have little change in radius passing through the lens. This approximation is often invalid for charged

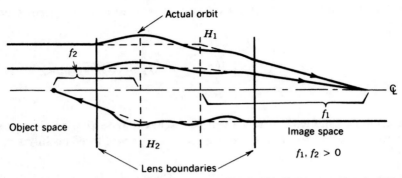

Figure 6.5 Charged particle lens. Definition of quantities used in linear approximation (Gaussian optics): principal planes (H_1, H_2) and focal lengths (f_1, f_2).

particle optics; the Laplace equation implies that electric and magnetic fields extend axial distances comparable to the diameter of the lens. The particle orbits of Figure 6.4 undergo a significant radius change in the field region. Lenses in which this occurs are called *thick lenses*. This section reviews the parameters and equations describing thick lenses.

A general charged particle lens is illustrated in Figure 6.5. It is an axial region of electric and magnetic fields that deflects (and may accelerate) particles. Particles drift in *ballistic orbits* (no acceleration) in field-free regions outside the lens. The upstream field-free region is called the *object space* and the downstream region is called the *image space*. Lenses function with particle motion in either direction, so that image and object spaces are interchangeable.

Orbits of initially parallel particles exiting the lens form similar triangles (Fig. 6.5). If the exit orbits are projected backward in a straight line, they intersect the forward projection of entrance orbits in a plane perpendicular to the axis. This is called the *principal plane*. The location of the principal plane is denoted H_1. The distance from H_1 to the point where orbits intersect is called the *focal length*, f_1. When H_1 and f_1 are known, the exit orbit of any particle that enters the lens parallel to the axis can be specified. The exit orbit is the straight line connecting the focal point to the intersection of the initial orbit with the principal plane. This construction also holds for negative focal lengths, as shown in Figure 6.6.

There is also a principal plane (H_2) and focal length (f_2) for particles with negative v_z. The focal lengths need not be equal. This is often the case with electrostatic lenses where the direction determines if particles are accelerated or decelerated. Two examples of $f_1 \neq f_2$ are the aperture lens (Section 6.5) and the immersion lens (Section 6.6). A *thin lens* is defined as one where the axial length is small compared to the focal length. Since the principal planes are contained in the field region, $H_1 = H_2$. A thin lens has only one principal plane. Particles emerge at the same radius they entered but with a change in direction.

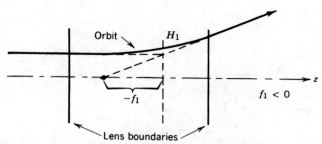

Figure 6.6 Construction of a particle orbit in a lens with a negative focal length.

There are two other common terms applied to lenses, the *lens power* and the *f-number*. The strength of a lens is determined by how much it bends orbits. Shorter focal lengths mean stronger lenses. The lens power P is the inverse of the focal length, $P = f^{-1}$. If the focal length is measured in meters, the power is given in m^{-1} or *diopters*. The f-number is the ratio of focal length to the lens diameter: f-number $= f/D$. The f-number is important for describing focusing of nonlaminar beams. It characterizes different optical systems in terms of the minimum focal spot size and maximum achievable particle flux.

If the principal planes and focal lengths of a lens are known, the transformation of an orbit between the lenses entrance and exit can be determined. This holds even for nonparallel entrance orbits. The conclusion follows from the fact that particle orbits are described by a linear, second-order differential equation. The relationship between initial and final orbits ($r_1, r_1' \to r_f, r_f'$) can be expressed as two algebraic equations with four constant coefficients. Given the two initial conditions and the coefficients (equivalent to the two principal planes and focal lengths), the final orbit is determined. This statement will be proved in Chapter 8 using the ray transfer matrix formalism.

Chapter 8 also contains a proof that a linear lens can produce an *image*. The definition of an image is indicated in Figure 6.7. Two special planes are defined on either side of the lens: the *object plane* and the *image plane* (which depends on the lens properties and the location of the object). An image is produced if all particles that leave a point in the object plane meet at a point in the image plane, independent of their initial direction. There is a mapping of spatial points from one plane to another. The image space and object space are interchangeable, depending on the direction of the particles. The proof of the existence of an image is most easily performed with matrix algebra. Nonetheless, assuming this property, the principal plane construction gives the locations of image and object planes relative to the lens and the magnification passing from one to another.

Figure 6.7 shows image formation by a lens. Orbits in the image and object space are related by the principal plane construction; the exit orbits are

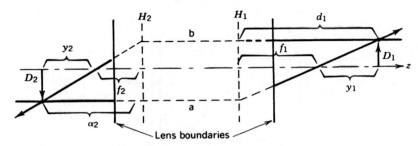

(Actual orbits unspecified in this region)

Figure 6.7   Quantities for calculating the imaging properties of a thick lens.

determined by the principal plane and focal length. These quantities give no detailed information on orbits inside the lens. The arrows represent an intensity distribution of particles in the transverse direction. Assume that each point on the source arrow (of length D_2) is mapped to a point in the image plane. The mapping produces an image arrow of length D_1. Parallel orbits are laminar, and the distance from the axis to a point on the image is proportional to its position on the source. We want to find the locations of the image and object planes (d_1 and d_2) relative to the principal planes, as well as the magnification, $M_{21} = D_1/D_2$.

The image properties can be found by following two particle orbits leaving the object. Their intersection in the image space determines the location of the image plane. The orbit with known properties is the one that enters the lens parallel to the axis. If a parallel particle leaves the tip of the object arrow, it exits the lens following a path that passes through the intersection with the principal plane at $r = D_2$ and the point f_1. This orbit is marked a in Figure 6.7. In order to determine a second orbit, we can interchange the roles of image and object and follow a parallel particle that leaves the right-hand arrow in the $-z$ direction. This orbit, marked b, is determined by the points at H_2 and f_2. A property of particle dynamics under electric and magnetic forces is time reversibility. Particles move backward along the same trajectories if $-t$ is substituted for t. Thus, a particle traveling from the original object to the image plane may also follow orbit b. If the two arrows are in object and image planes, the orbits must connect as shown in Figure 6.7.

The image magnification for particles traveling from left to right is $M_{21} = D_1/D_2$. For motion in the opposite direction, the magnification is $M_{12} = D_2/D_1$. Therefore,

$$M_{21}M_{12} = 1. \tag{6.12}$$

Referring to Figure 6.7, the following equations follow from similar triangles:

$$D_1/y_1 = D_2/f_1, \qquad D_1/f_2 = D_2/y_2. \tag{6.13}$$

These are combined to give $(f_1)(f_2) = y_1 y_2$. This equation can be rewritten in terms of the distances d_1 and d_2 from the principal planes as $(f_1)(f_2) = (d_2 - f_2)(d_1 - f_1)$. The result is the *thick-lens equation*

$$f_1/d_1 + f_2/d_2 = 1. \qquad \boxed{6.14}$$

In light optics, the focal length of a simple lens does not depend on direction. In charged particle optics, this holds for magnetic lenses or *unipotential* electrostatic lenses where the particle energy does not change in the lens. In this case, Eq. (6.14) can be written in the familiar form

$$1/d_1 + 1/d_2 = 1/f, \qquad \boxed{6.15}$$

where $f_1 = f_2 = f$.

In summary, the following procedure is followed to characterize a linear lens. Measured data or analytic expressions for the fields of the lens are used to calculate two special particle orbits. The orbits enter the lens parallel to the axis from opposite axial directions. The orbit calculations are performed analytically or numerically. They yield the principal planes and focal lengths. Alternately, if the fields are unknown, lens properties may be determined experimentally. Parallel particle beams are directed into the lens from opposite directions to determine the lens parameters. If the lens is linear, all other orbits and the imaging properties are found from two measurements.

In principle, the derivations of this section can be extended to more complex optical systems. The equivalent of Eq. (6.14) could be derived for combinations of lenses. On the other hand, it is much easier to treat optical systems using ray transfer matrices (Chapter 8). Suceeding sections of this chapter are devoted to the calculation of optical parameters for a variety of discrete electrostatic and magnetostatic charged particle lenses.

6.5 ELECTROSTATIC APERTURE LENS

The *electrostatic aperture lens* is an axicentered hole in an electrode separating two regions of axial electric field. The lens is illustrated in Figure 6.8. The fields may be produced by grids with applied voltage relative to the aperture plate. If the upstream and downstream electric fields differ, there will be radial components of electric field near the hole which focus or defocus particles.

In Figure 6.8, axial electric fields on both sides of the plate are positive, and the field at the left is stronger. In this case, the radial fields point outward and the focal length is negative for positively charged particles traveling in either direction. With reversed field direction (keeping the same relative field strength)

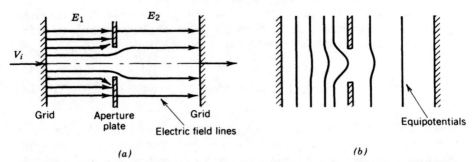

Figure 6.8 Electrostatic aperture lens. (*a*) Geometry, showing electric field lines. (*b*) Equipotential lines.

or with stronger field on the right side (keeping the same field polarity), the lens has positive focal length. The transverse impulse on a particle passing through the hole is proportional to the time spent in the radial electric fields. This is inversely proportional to the particle velocity which is determined, in part, by the longitudinal fields. Furthermore, the final axial velocity will depend on the particle direction. These factors contribute to the fact that the focal length of the aperture lens depends on the transit direction of the particle, or $f_1 \neq f_2$.

Radial electric fields are localized at the aperture. Two assumptions allow a simple estimate of the focal length: (1) the relative change in radius passing through the aperture is small (or, the aperture is treated in the thin lens approximation) and (2) the relative change in axial velocity is small in the vicinity of the aperture. Consider a particle moving in the $+z$ direction with $v_r = 0$. The change in v_r for nonrelativistic motion is given by the equation

$$dv_r/dz = qE_r/m_0 v_z \cong -(q/2m_0 v_z)r\left[dE_z(0, z)/dz\right]. \qquad (6.16)$$

In Eq. (6.16), the time derivative was converted to a spatial derivative and E_r was replaced according to Eq. (6.5).

With the assumption of constant r and v_z in the region of nonzero radial field, Eq. (6.16) can be integrated directly to yield

$$v_{rf}/v_{zf} = r_f' \cong -qr(E_{z2} - E_{z1})/2m_0 v_{za} v_{zf}, \qquad (6.17)$$

where v_{za} is the particle velocity at the aperture and v_{rf} is the radial velocity after exiting. The quantity v_{zf} is the final axial velocity; it depends on the final location of the particle and the field E_2. The focal length is related to the final radial position (r) and the ratio of the radial velocity to the final axial velocity

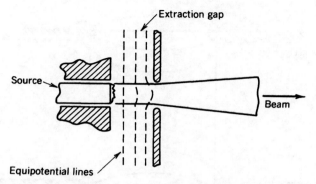

Figure 6.9 Extraction gap of an electron gun showing negative-lens effect.

by $v_{\mathrm{rf}}/v_{\mathrm{zf}} \cong r/f$. The focal length is

$$f \cong 2m_0 v_{\mathrm{za}} v_{\mathrm{zf}} q (E_{z2} - E_{z1}).\qquad(6.18)$$

When the particle kinetic energy is large compared to the energy change passing through the lens, $v_{\mathrm{za}} \cong v_{\mathrm{zf}}$, and we find the usual approximation for the aperture lens focal length

$$f \cong 2m_0 v_{\mathrm{zf}}^2/q(E_{z2} - E_{z1}) = 4T/q(E_{z2} - E_{z1}).\quad\boxed{6.19}$$

The *charged particle extractor* (illustrated in Fig. 6.9) is a frequently encountered application of Eq. (6.19). The extractor gap pulls charged particles from a source and accelerates them. The goal is to form a well-directed low-energy beam. When there is high average beam flux, grids cannot be used at the downstream electrode and the particles must pass through a hole. The hole acts as an aperture lens, with $E_1 > 0$ and $E_2 = 0$. The focal length is negative; the beam emerging will diverge. This is called the *negative lens effect* in extractor design. If a parallel or focused beam is required, a focusing lens can be added downstream or the source can be constructed with a concave shape so that particle orbits converge approaching the aperture.

6.6 ELECTROSTATIC IMMERSION LENS

The geometry of the *electrostatic immersion lens* is shown in Figure 6.10. It consists of two tubes at different potential separated by a gap. Acceleration gaps between drift tubes of a standing-wave linear accelerator (Chapter 14) have this geometry. The one-dimensional version of this lens occurs in the gap between the Dees of a cyclotron (Chapter 15). Electric field distributions for a cylindrical lens are plotted in Figure 6.10.

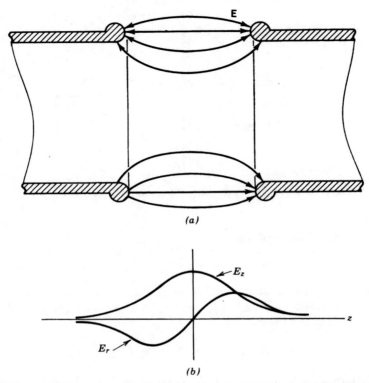

Figure 6.10 Electrostatic immersion lens. (*a*) Geometry and electric field lines. (*b*) Relative variations of longitudinal electric field, $E_z(0, z)$ (on-axis), and transverse field, $E_r(r, z)$ (off-axis), moving through the lens.

Following the treatment used for the aperture lens, the change in radial velocity of a particle passing through the gap is

$$\Delta v_r = v_{\text{rf}} = \int dz \, \left[qE_r(r, z)/m_0 v_z \right]. \tag{6.20}$$

The radial electric field is symmetric. There is no deflection if the particle radius and axial velocity are constant. In contrast to the aperture lens, the focusing action of the immersion lens arises from changes in r and v_z in the gap region. Typical particle orbits are illustrated in Figure 6.11. When the longitudinal gap field accelerates particles, they are deflected inward on the entrance side of the lens and outward on the exit side. The outward impulse is smaller because (1) the particles are closer to the axis and (2) they move faster on the exit side. The converse holds for a decelerating gap. Particles are deflected to larger radii on the entrance side and are therefore more strongly influenced by the radial fields on the exit side. Furthermore, v_z is lower at the

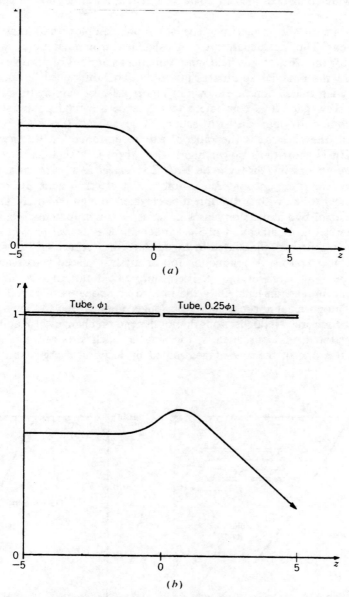

Figure 6.11 Particle trajectories in an immersion lens. (*a*) Accelerating lens (final kinetic energy equals 4 times initial energy). (*b*) Decelerating lens (final kinetic energy equals 0.25 times initial energy). (Note expanded radial scale).

exit side which enhances focusing. The focal length for either polarity or charge sign is positive.

The orbits in the immersion lens are more complex than those in the aperture lens. The focal length must be calculated from analytic or numerical solutions for the electrostatic fields and numerical solutions of particle orbits in the gap. In the paraxial approximation, only two orbits need be found. The results of such calculations are shown in Figure 6.12 for varying tube diameter with a narrow gap. It is convenient to reference the tube potentials to the particle source so that the exit energy is given by $T_f = qV_2$. With this convention, the abscissa is the ratio of exit to entrance kinetic energy. The focal length is short (lens power high) when there is a significant change in kinetic energy passing through the lens. The *einzel lens* is a variant of the immersion lens often encountered in low-energy electron guns. It consists of three colinear tubes, with the middle tube elevated to high potential. The einzel lens consists of two immersion lenses in series; it is a unipotential lens.

An interesting modification of the immersion lens is *foil or grid focusing*. This focusing method, illustrated in Figure 6.13, has been used in low-energy linear ion accelerators. A conducting foil or mesh is placed across the downstream tube of an accelerating gap. The resulting field pattern looks like half of that for the immersion lens. Only the inward components of radial field are present. The paraxial approximation no longer applies; the foil geometry has first-order focusing. Net deflections do not depend on changes of r and v_z as in the immersion lens. Consequently, focusing is much stronger. Foil focusing demonstrates one of the advantages gained by locating charges and currents

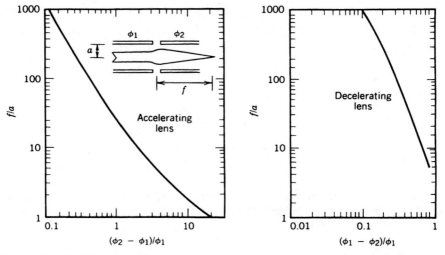

Figure 6.12 Focal lengths of immersion lenses in terms of relative kinetic energy change in lens. (*a*) Accelerating lenses. (*b*) Decelerating lenses.

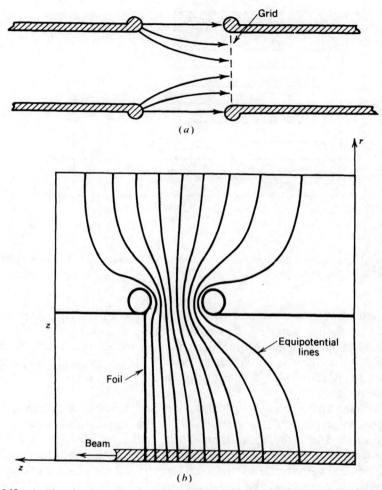

Figure 6.13 Accelerating gap of a drift tube linear accelerator with a grid for enhanced electrostatic focusing. (*a*) Geometry. (*b*) Equipotential lines.

within the beam volume, rather than relying on external electrodes or coils. The charges, in this case, are image charges on the foil. An example of internal currents, the toroidal field magnetic sector lens, is discussed in Section 6.8.

6.7 SOLENOIDAL MAGNETIC LENS

The *solenoidal magnetic lens* is illustrated in Figure 6.14. It consists of a region of cylindrically symmetric radial and axial magnetic fields produced by axicentered coils carrying azimuthal current. This lens is the only possible

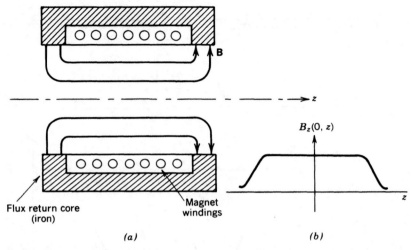

Figure 6.14 Solenoidal magnetic lens. (*a*) Geometry and field lines. (*b*) Variation of longitudinal magnetic field on axis, $B_z(0, z)$.

magnetic lens geometry consistent with cylindrical paraxial beams. It is best suited to electron focusing. It is used extensively in cathode ray tubes, image intensifiers, electron microscopes, and electron accelerators. Since the magnetic field is static, there is no change of particle energy passing through the lens; therefore, it is possible to perform relativistic derivations without complex mathematics.

Particles enter the lens through a region of radial magnetic fields. The Lorentz force ($ev_z \times B_r$) is azimuthal. The resulting v_θ leads to a radial force when crossed into the B_z fields inside the lens. The net effect is a deflection toward the axis, independent of charge state or transit direction. Since there is an azimuthal velocity, radial and axial force equations must be solved with the inclusion of centrifugal and coriolis forces.

The equations of motion (assuming constant γ) are

$$\gamma m_0 \dot{v}_r = -q v_\theta B_z + \gamma m_0 v_\theta^2 / r, \tag{6.21}$$

$$\gamma m_0 \dot{v}_\theta = -q v_z B_r - \gamma m_0 v_r v_\theta / r. \tag{6.22}$$

The axial equation of motion is simply that v_z is constant. We assume that r is approximately constant and that the particle orbit has a small net rotation in the lens. With the latter condition, the Coriolis force can be neglected in Eq. (6.22). If the substitution $dv_\theta/dt \cong v_z(dv_\theta/dz)$ is made and Eq. (6.7) is used to express B_r in terms of $dB_z(0, z)/dz$, Eq. (6.22) can be integrated to give

$$v_\theta - qrB_z/2\gamma m_0 = \text{const.} = 0. \qquad \boxed{6.23}$$

Equation (6.23) is an expression of conservation of canonical angular momentum (see Section 7.4). It holds even when the assumptions of this calculation are not valid. Equation (6.23) implies that particles gain no net azimuthal velocity passing completely through the lens. This comes about because they must cross negatively directed radial magnetic field lines at the exit that cancel out the azimuthal velocity gained at the entrance. Recognizing that $d\theta/dt = v_\theta/r$ and assuming that B_z is approximately constant in r, the angular rotation of an orbit passing through the lens is

$$(\theta - \theta_0) = [qB_z(0, z)/2\gamma m_0 v_z](z - z_1). \qquad \boxed{6.24}$$

Rotation is the same for all particles, independent of radius. Substituting Eq. (6.23) and converting the time derivative to a longitudinal derivative, Eq. (6.21) can be integrated to give

$$r_f' = \frac{v_{\text{rf}}}{v_z} \cong \frac{-\int dz \, [qB_z(0, z)/\gamma m_0 v_z]^2 r}{4}. \qquad (6.25)$$

The focal length for a solenoidal magnetic lens is

$$f = \frac{-r_f}{r_f'} = \frac{4}{\int dz \, [qB_z(0, z)/\gamma m_0 v_z]^2}. \qquad \boxed{6.26}$$

The quantity in brackets is the reciprocal of a gyroradius [Eq. (3.38)]. Focusing in the solenoidal lens (as in the immersion lens) is second order; the inward force results from a small azimuthal velocity crossed into the main component of magnetic field. Focusing power is inversely proportional to the square of the particle momentum. The magnetic field must increase proportional to the relativistic mass to maintain a constant lens power. Thus, solenoidal lenses are effective for focusing low-energy electron beams at moderate field levels but are seldom used for beams of ions or high-energy electrons.

6.8 MAGNETIC SECTOR LENS

The lenses of Sections 6.5–6.7 exert cylindrically symmetric forces via paraxial electric and magnetic fields. We now turn attention to devices in which focusing is one dimensional. In other words, if the plane perpendicular to the axis is resolved into appropriate Cartesian coordinates (x, y), the action of focusing forces is different and independent in each direction. The three examples we shall consider are (1) horizontal focusing in a sector magnet (Section 6.8), (2) vertical focusing at the edge of a sector magnet with an inclined boundary (Section 6.9), and (3) quadrupole field lenses (Section 6.10).

A sector magnet (Fig. 6.15) consists of a gap with uniform magnetic field extending over a bounded region. Focusing about an axis results from the location and shape of the field boundaries rather than variations of the field properties. To first approximation, the field is uniform $[\mathbf{B} = B_x(x, y, z) = B_0]$ inside the magnet and falls off sharply at the boundary. The x direction

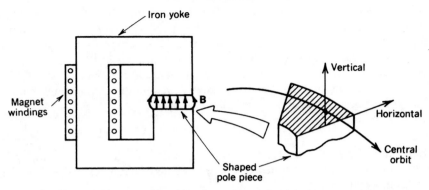

Figure 6.15 Sector field magnet for deflecting a beam showing definition of horizontal and vertical directions with respect to the main beam orbit.

(parallel to the field lines) is usually referred to as the *vertical* direction. The *y* direction (perpendicular to the field lines) is the *horizontal* direction. The beam axis is curved. The axis corresponds to one possible particle orbit called the *central orbit*. The purpose of focusing devices is to confine nonideal orbits about this line. Sector field magnets are used to bend beams in transport lines and circular accelerators and to separate particles according to momentum in charged particle spectrometers.

The 180° spectrograph (Fig. 6.16) is an easily visualized example of horizontal focusing in a sector field. Particles of different momentum enter the field through a slit and follow circular orbits with gyroradii proportional to momentum. Particles entering at an angle have a displaced orbit center. Circular orbits have the property that they reconverge after one half revolution with only a second-order error in position. The sector magnet focuses all particles of the same momentum to a line, independent of entrance angle.

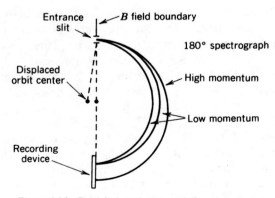

Figure 6.16 Particle focusing by a 180° sector magnet.

Focusing increases the *acceptance* of the spectrometer. A variety of entrance angles can be accepted without degrading the momentum resolution. The input beam need not be highly collimated so that the flux available for the measurement is maximized. There is no focusing in the vertical direction; a method for achieving simultaneous horizontal and vertical focusing is discussed in Section 6.9.

A sector field with angular extent less than 180° can act as a thick lens to produce a horizontal convergence of particle orbits after exiting the field. This effect is illustrated in Figure 6.17. Focusing occurs because off-axis particles travel different distances in the field and are bent a different amount. If the field boundaries are perpendicular to the central orbit, we can show, for initially parallel orbits, that the difference in bending is linearly proportional to the distance from the axis.

The orbit of a particle (initially parallel to the axis) displaced a distance Δr_1 from the axis is shown in Figure 6.17. The final displacement is related by $\Delta y_f = \Delta y_1 \cos \alpha$, where α is the angular extent of the sector. The particle emerges from the lens at an angle $\Delta \theta = -\Delta y_1 \sin \alpha / r_g$, where r_g is the gyroradius in the field B_0. Given the final position and angle, the distance from the field boundary to the focal point is

$$f' = r_g / \tan \alpha. \qquad \boxed{6.27}$$

The focal distance is positive for $\alpha < 90°$; emerging particle orbits are convergent. It is zero at $\alpha = 90°$; initially parallel particles are focused to a point at the exit of a 90° sector. At 180° the focusing distance approaches infinity.

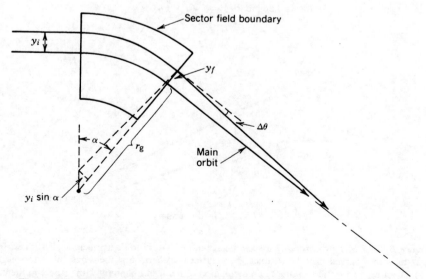

Figure 6.17   Geometry for calculating horizontal focal properties of a sector magnet.

The sector field magnet must usually be treated as a thick lens. This gives us an opportunity to reconsider the definition of the principal planes, which must be clarified when the beam axis is curved. The plane H_1 is the surface that gives the correct particle orbits *in the image space*. The appropriate construction is illustrated in Figure 6.18a. A line parallel to the beam axis in the image plane is projected backward. The principal plane is perpendicular to this line. The exit orbit intersects the plane at a distance equal to the entrance distance. If a parallel particle enters the sector field a distance y_1 from the beam axis, its exit orbit is given by the line joining the focal point with a point on the principal plane y_1 from the axis. The focal length is the distance from the principal plane to the focal point. The plane H_2 is defined with respect to orbits in the $-z$ direction.

The focal length of a sector can be varied by inclining the boundary with respect to the beam axis. Figure 18b shows a boundary with a positive inclination angle, β. When the inclination angle is negative, particles at a larger

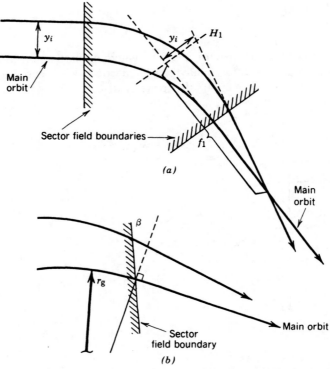

Figure 6.18 Focal properties of a sector magnet. (*a*) Definition of principal planes and focal lengths with a curved main orbit; thick-lens treatment of horizontal focusing in a sector magnet. (*b*) Effect on horizontal focusing of an inclination of the field boundary with respect to the main orbit.

distance from the central orbit gyrocenter travel longer distances in the field and are bent more. The focusing power of the lens in the horizontal direction is increased. Conversely, for $\beta > 0$, horizontal focusing is decreased. We will see in Section 6.9 that in this case there is vertical focusing by the fringing fields of the inclined boundary.

A geometric variant of the sector field is the *toroidal field sector lens*. This is shown in Figure 6.19. A number of magnet coils are arrayed about an axis to produce an azimuthal magnetic field. The fields in the spaces between coils are similar to sector fields. The field boundary is determined by the coils. It is assumed that there are enough coils so that the fields are almost symmetric in azimuth. The field is not radially uniform but varies as $B_\theta(R, Z) = B_0 R_0/R$, where R is the distance from the lens axis. Nonetheless, boundaries can still be determined to focus particles to the lens axis; the boundaries are no longer straight lines. The figure shows a toroidal field sector lens designed to focus a parallel, annular beam of particles to a point.

The location of the focal point for a toroidal sector lens depends on the particle momentum. Spectrometers based on the toroidal fields are called *orange spectrometers* because of the resemblance of the coils to the sections of an orange when viewed from the axis. They have the advantage of an

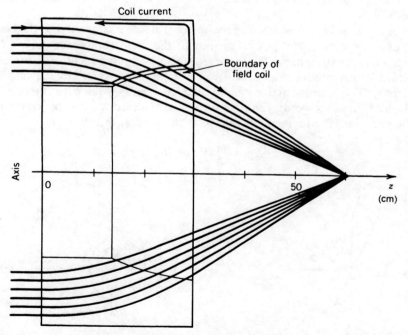

Figure 6.19 Particle orbits in a toroidal field lens with an exit boundary optimized for focusing an annular beam to a spot.

extremely large solid angle of acceptance and can be used for measurements at low flux levels. The large acceptance outweighs the disadvantage of particle loss on the coils.

The toroidal field sector lens illustrates the advantages gained by locating applied currents within the volume of the beam. The lens provides first-order focusing with cylindrical symmetry about an axis, as contrasted to the solenoidal field lens, which has second-order focusing. The ability to fine tune applied fields in the beam volume allows correction of focusing errors (aberrations) that are unavoidable in lenses with only external currents.

6.9 EDGE FOCUSING

The term *edge focusing* refers to the vertical forces exerted on charged particles at a sector magnet boundary that is inclined with respect to the main orbit. Figure 6.20 shows the fringing field pattern at the edge of a sector field. The vertical field magnitude decreases away from the magnet over a scale length comparable to the gap width. Fringing fields were neglected in treating perpendicular boundaries in Section 6.8. This is justified if the gap width is small compared to the particle gyroradius. In this case, the net horizontal deflection is about the same whether or not some field lines bulge out of the gap. With perpendicular boundaries, there is no force in the vertical direction since $B_y = 0$.

In analyzing the inclined boundary, the coordinate z is parallel to the beam axis, and the coordinate ξ is referenced to the sector boundary (Fig. 6.20). When the inclination angle β is nonzero, there is a component of $\mathbf{B}$ in the y direction which produces a vertical force at the edge when crossed into the particle v_z. The focusing action can be calculated easily if the edge is treated as a thin lens; in other words, the edge forces are assumed to provide an impulse to the particles. The momentum change in the vertical direction is

$$\Delta p_x = \int dt \left(q v_z B_y \right). \tag{6.28}$$

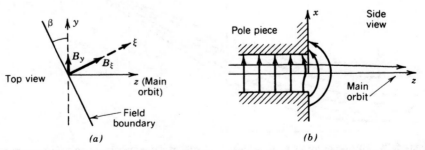

Figure 6.20 Geometry for calculating vertical focusing by a sector magnet boundary inclined with respect to the main orbit (edge focusing). (*a*) Field components at exit boundary, viewed along vertical direction (top view). (*b*) View of exit boundary along horizontal direction (side view).

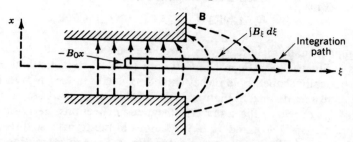

Figure 6.21 Geometry for evaluating field integral relevant to edge focusing.

The integral is taken over the time the particle is in the fringing field. The y component of magnetic field is related to the ξ component of the fringing field by $B_y = B_\xi \sin \beta$. The integral of Eq. (6.28) can be converted to an integral over the particle path noting that $v_z\, dt = dz$. Finally, the differential path element can be related to the incremental quantity $d\xi$ by $dz = d\xi/\cos \beta$. Equation (6.28) becomes

$$\Delta v_x = (e/\gamma m_0) \int d\xi\, B_\xi \tan \beta. \qquad (6.29)$$

This integral can be evaluated by applying the equation $\int \mathbf{B} \cdot \mathbf{ds} = 0$ to the geometry of Figure 6.21. The circuital integral extends from the uniform field region inside the sector magnet to the zero field region outside. This implies that $\int B_\xi\, d\xi = B_0 x$. Substituting into Eq. (6.29), the vertical momentum change can be calculated. It is proportional to x, and the focal length can be determined in the usual manner as

$$f_x = \frac{(\gamma m_0 v_z/qB_0)}{\tan \beta} = \frac{r_{g0}}{\tan \beta}. \qquad \boxed{6.30}$$

The quantity r_{g0} is the particle gyroradius inside the constant-field sector magnet. When $\beta = 0$ (perpendicular boundary), there is no vertical focusing, as expected. When $\beta > 0$, there is vertical focusing, and the horizontal focusing is decreased. If β is positive and not too large, there can still be a positive horizontal focal length. In this case, the sector magnet can focus in both directions. This is the principal of the dual-focusing magnetic spectrometer, illustrated in Figure 6.22. Conditions for producing an image of a point source can be calculated using geometric arguments similar to those already used. Combined edge and sector focusing has also been used in a high-energy accelerator, the zero-gradient synchrotron[†] (Section 15.5).

[†]A. V. Crewe, *Proc. Intern. Conf. High Energy Accelerators*, CERN, Geneva, 1959, p. 359.

6.10 MAGNETIC QUADRUPOLE LENS

The magnetic quadrupole field was introduced in Section 5.8. A quadrupole field lens is illustrated in Figure 5.16. It consists of a magnetic field produced by hyperbolically shaped pole pieces extending axially a length l. In terms of the transverse axes defined in Figure 5.16, the field components are $B_x = B_0 y/a$ and $B_y = B_0 x/a$. Since the transverse magnetic deflections are normal to the field components, $F_x \sim x$ and $F_y \sim y$. Motions in the transverse directions are independent, and the forces are linear. We can analyze motion in each direction separately, and we know (from Section 6.3) that the linear fields will act as one-dimensional focusing (or defocusing) lenses.

The orbit equations are

$$d^2 y/dz^2 = (qB_0/\gamma m_0 a v_z) y, \qquad (6.31)$$

$$d^2 x/dz^2 = -(qB_0/\gamma m_0 a v_z) x. \qquad (6.32)$$

The time derivatives were converted to axial derivatives. The solutions for the

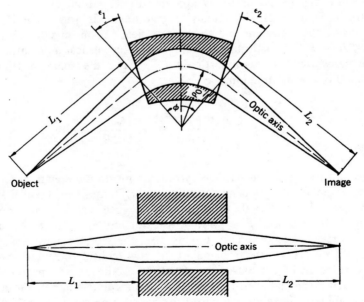

Figure 6.22 A homogeneous sector field in which the particle beam crosses the boundary obliquely. Proper choice of the angles ϵ_1 and ϵ_2 gives stigmatic focusing in both the radial and vertical directions. (Courtesy, H. Wollnik.)

particle orbits are

$$x(z) = x_1\cos\sqrt{\kappa_m}\,z + x_1'\sin\sqrt{\kappa_m}\,z/\sqrt{\kappa_m}, \tag{6.33}$$

$$x'(z) = -x_1\sqrt{\kappa_m}\,\sin\sqrt{\kappa_m}\,z + x_1'\cos\sqrt{\kappa_m}\,z, \tag{6.34}$$

$$y(z) = y_1\cosh\sqrt{\kappa_m}\,z + y_1'\sinh\sqrt{\kappa_m}\,z/\sqrt{\kappa_m}, \tag{6.35}$$

$$y'(z) = y_1\sqrt{\kappa_m}\,\sinh\sqrt{\kappa_m}\,z + y_1'\cosh\sqrt{\kappa_m}\,z, \tag{6.36}$$

where x_1, y_1, x_1', and y_1' are the initial positions and angles. The parameter κ_m

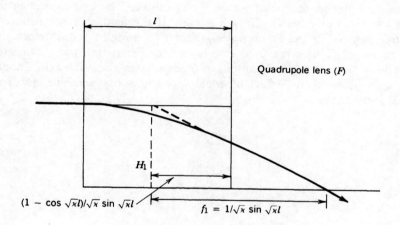

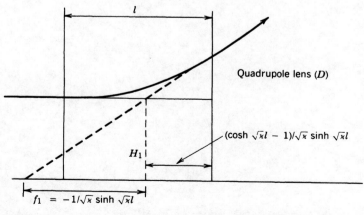

Figure 6.23 Principal planes and focal lengths for a magnetic quadrupole lens.

is

$$\kappa_m = qB_0/\gamma m_0 a v_z \quad (\text{m}^{-2})$$

$$\boxed{6.37}$$

where B_0 is the magnetic field magnitude at the surface of the pole piece closest to the axis and a is the minimum distance from the axis to the pole surface. A similar expression applies to the electrostatic quadrupole lens

$$\kappa_e = qE_0/\gamma m_0 a v_z^2 \quad (\text{m}^{-2})$$

$$\boxed{6.38}$$

In the electrostatic case, the x and y axes are defined as in Figure 4.14.

The principal plane and focal length for a magnetic quadrupole lens are shown in Figure 6.23 for the x and y directions. They are determined from the orbit expressions of Eqs. (6.33) and (6.34). The lens acts symmetrically for particle motion in either the $+z$ or $-z$ directions. The lens focuses in the x direction but defocuses in the y direction. If the field coils are rotated 90° (exchanging North and South poles), there is focusing in y but defocusing in x. Quadrupole lenses are used extensively for beam transport applications. They must be used in combinations to focus a beam about the axis. Chapter 8 will show that the net effect of equal focusing and defocusing quadrupole lenses is focusing.

7

Calculation of Particle Orbits in Focusing Fields

In this chapter, we shall study methods for calculating particle orbits in extended regions of static electromagnetic forces. One important application of such calculations is the determination of electric or magnetic lens properties. Once the fields have been calculated by the Laplace equation, orbit solutions lead to the four discrete quantities that characterize a linear lens (focal lenths and principal points). Properties of orbits in complex optical systems with many focusing elements or in long periodic lens arrays can then be predicted by the combination of the discrete lens parameters through matrix algebra; this is the subject of Chapter 8. The matrix approach gives information on orbits only at the lens boundaries. If detailed information on orbits inside the lenses is required, then the numerical methods of this chapter must be extended through the entire optical system. The matrix approach is not applicable to some accelerators, such as the betatron, with focusing forces that are invariant along the direction of beam propagation. In this case, the best approach is the direct solution of orbits in the extended fields.

Section 7.1 discusses general features of particle orbits in constant transverse forces and introduces betatron oscillations. Some concepts and parameters fundamental to charged particle accelerators are introduced in Section 7.2, including beam distribution functions, focusing channel acceptance, the ν value in a circular accelerator, and orbital resonance instabilities. An example of focusing in continuous fields, particle motion in a magnetic gradient field, is treated in Section 7.3. This transport method is used in betatrons and cyclotrons.

137

Methods to solve particle orbits in cylindrically symmetric electric and magnetic fields are reviewed in the second part of the chapter. The paraxial ray equation is derived in Section 7.5. This equation has extensive applications in all regimes of particle acceleration, from cathode ray tubes to multi-GeV linacs. As a preliminary to the paraxial ray equation, the Busch theorem is derived in Section 7.4. This theorem demonstrates the conservation of canonical angular momentum in cylindrically symmetric electric and magnetic fields. Section 7.6 introduces methods for the numerical solution of the paraxial ray equation. Two examples are included: determination of the beam envelope in an electrostatic acceleration column and calculation of the focal length of an electrostatic immersion lens. The numerical methods can also be applied to the direct solution of three-dimensional equations of particle motion.

7.1   TRANSVERSE ORBITS IN A CONTINUOUS LINEAR FOCUSING FORCE

In many instances, particle motion transverse to a beam axis is separable along two Cartesian coordinates. This applies to motion in a magnetic gradient field (Section 7.3) and in an array of quadrupole lenses (Section 8.7). Consider one-dimensional transverse paraxial particle motion along the z axis in the presence of a linear force, $F_x = F_0(x/x_b)$. The quantity F_0 is the force at the edge of the beam, x_b. The y motion can be treated separately. The axial velocity is approximately constant (Section 6.3). The equation of motion is (Section 2.10)

$$\gamma m_0 (d^2x/dt^2) \cong F_0(x/x_b). \tag{7.1}$$

In the absence of acceleration, Eq. (7.1) can be expressed as

$$(d^2x/dz^2) = (F_0/\gamma m_0 v_z^2)(x/x_b) \tag{7.2}$$

in the paraxial approximation. Equation (7.2) has the solution

$$x(z) = x_0 \cos(2\pi z/\lambda_x + \phi), \tag{7.3}$$

where

$$\lambda_x = 2\pi (\gamma m_0 v_z^2 x_b/F_0)^{1/2}.$$

Particle motion is harmonic. All particle orbits have the same wavelength; they differ only in amplitude and phase. Transverse particle motions of this type in accelerators are usually referred to as *betatron oscillations* since they were first described during the development of the betatron.[†] The quantity λ_x is called the *betatron wavelength*.

[†] D. W. Kerst and R. Serber, *Phys. Rev.*, **60**, 53 (1941).

It is often useful to replace the action of an array of discrete lenses by an equivalent continuous transverse focusing force rather than apply the formalisms developed in Chapter 8. This approximation is used, for example, to compare the defocusing effect of continuous-beam space charge forces with confinement forces to derive the maximum allowed current in a quadrupole channel. Orbits in an array of identical, discrete lenses approach the harmonic solution of Eq. (7.3) when the distance between lenses is small compared to the betatron wavelength, as shown in Figure 7.1. Consider lenses of focal length, f, and axial spacing, d, in the limit that $d \ll f$ (thin-lens approximation). We want to calculate the change in x and v_x passing through one drift space and one lens. If v_x is the transverse velocity in the drift region, then (following Fig. 7.1)

$$\Delta x = (v_x/v_z)d. \tag{7.4}$$

Similarly, by the definition of the focal length,

$$\Delta v_x/v_z = -x/f, \tag{7.5}$$

where x is the particle position entering the thin lens. Equations (7.4) and (7.5) can be converted to differential equations by associating Δz with d and letting $\Delta z \to 0$,

$$dx/dz = v_x/v_z \quad \text{and} \quad dv_x/dz = -(v_z/fd)x.$$

The solution to this equation is again harmonic, with $\lambda_x = (fd)^{1/2}$. The condition for the validity of the continuous approximation, $\lambda_x \gg d$, is equivalent to $d \ll f$. Averaging the transverse force over many lenses gives

$$\langle F_x \rangle = (\gamma m_0 v_z^2/fd)x. \tag{7.6}$$

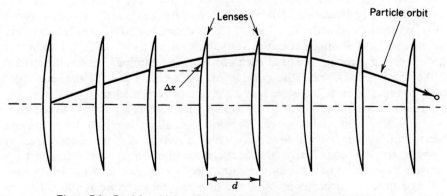

Figure 7.1 Particle orbit in an array of uniform, equally spaced, thin lenses.

Equation (7.6) is derived by expressing Eq. (7.5) in terms of time derivatives and multiplying both sides by γm_0. The conclusion is that particle orbits in any periodic linear focusing system approach harmonic orbits when the betatron wavelength is long compared to the period of the system.

7.2 ACCEPTANCE AND v OF A FOCUSING CHANNEL

Perfectly laminar beams cannot be achieved in practice. Imperfections have their origin, in part, in the particle source. For example, thermonic sources produce spatially extended beams at finite current density with a random distribution of transverse velocity components resulting from thermal effects. Laminarity can also be degraded by field errors in beam transport elements. The result is that the particles that constitute a beam have a spread in position and a spread in velocity at each position. The capability of a focusing system is parametrized by the range of particle positions and transverse velocities that can be transported without beam loss. This parameter is called the *acceptance*.

In order to understand the significance of acceptance, we must be familiar with the concept of a *particle distribution*. The transverse particle distribution of a beam at an axial location is the set of the positions and velocities of all particles $(x_i(z), v_{xi}(z), y_i(z), v_{yi}(z))$. In cylindrical beams, particles are parametrized by $(r_i(z), v_{ri}(z), \theta_i, v_{\theta i}(z))$. Methods of collective physics have been developed to organize and manipulate the large amount of information contained in the distribution. When motions in two dimensions are independent (as in the betatron or a quadrupole focusing channel), distributions in $x - v_x$ and $y - v_y$ can be handled independently.

One-dimensional distributions are visualized graphically by a *phase space plot*. For motion in the x direction, the axes are v_x (ordinate) and x (abscissa). At a particular axial location, each particle is represented by a point on the plot $(x_i(z), v_{xi}(z))$. A beam distribution is illustrated in Figure 7.2. Normally, there are too many particles in a beam to plot each point; instead, contour lines have been drawn to indicate regions of different densities of particle points. In the specific case shown, particles in the beam are symmetric about the x axis but asymmetric in velocity. This corresponds to a beam launched on the axis with an aiming error. The vertical extent of the distribution indicates that there is a velocity spread at each position. In a laminar beam, all particles at the same position have the same transverse velocity. The phase space plot for a laminar beam is therefore a straight line of zero thickness.

Acceptance is the set of all particle orbit parameters (x_i, v_{xi}) at the entrance to a focusing system or optical element that allow particles to propagate through the system without loss. Acceptance is indicated as a bounded area on a phase space plot. Clearly, the particle distribution at the entrance to a focusing channel must be enclosed within the acceptance to avoid beam loss. For example, the acceptance of a one-dimensional aperture is illustrated in Figure 7.3a. All particles with displacement $x_i < x_0$ pass through, indepen-

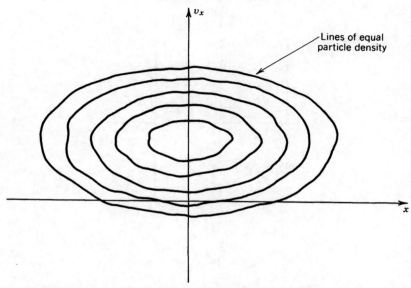

Figure 7.2 Phase space distribution of particles in a beam. Density of particles ((number of particles)/($\Delta x \, \Delta v_x$)) versus x and v_x. Beam is centered on axis, but aimed in $+x$ direction.

dent of v_{xi}. An acceptable particle distribution boundary is designated as a dashed line. *Matching* a beam to a focusing system consists of operating on the beam distribution so that it is enclosed in the acceptance. For example, matching to the aperture is simple. The beam is focused by an upstream lens so that the displacement of the peripheral particles is less than x_0.

As a second example, consider the acceptance of the continuous linear focusing system discussed in Section 7.1. We assume that there is some maximum dimension beyond which particles are lost, $x_{max} = x_0$. This may be the width of the vacuum chamber. The focusing force and the maximum width define the maximum transverse velocity for confined particle orbits,

$$v_x \leq v_z(2\pi x_0/\lambda_x). \tag{7.7}$$

The acceptance of a linear focusing system has boundaries on both axes of an x–v_x plot (Fig. 7.3*b*). A more detailed analysis shows that the allowed orbit parameters fill an ellipse with major (or minor) radii x_0 and $(2\pi x_0/\lambda_x)v_z$. The maximum velocity for the discrete lens system (in the continuous approximation) is

$$v_x \leq v_z\left(x_0^2/fd\right)^{1/2}. \tag{7.8}$$

A large acceptance means that the system can transport particles with a wide spread in entrance position and angle. A transport system with small

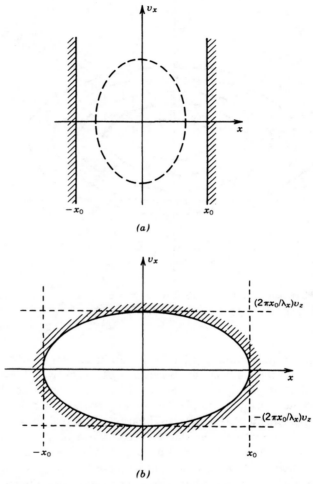

Figure 7.3 Acceptance of transport elements. (*a*) A one-dimensional aperture; slot of width $2x_0$. Dashed line indicates a beam distribution that can pass without attenuation. (*b*) One-dimensional continuous focusing system with linear forces.

acceptance places constraints on the size of and current density from a particle source. Generally, low beam flux is associated with small acceptance. For a given maximum channel dimension, the acceptance grows with increased focusing force (F_0). This results in shorter betatron wavelength.

The quantity ν is defined for circular transport systems, such as betatrons, synchrotrons, or storage rings. It is the ratio of the circumference of the system (C) to the betatron wavelength. The betatron wavelength may differ in the two transverse directions, so that

$$\nu_x = C/\lambda_x, \qquad \nu_y = C/\lambda_y.$$

$$\boxed{7.9}$$

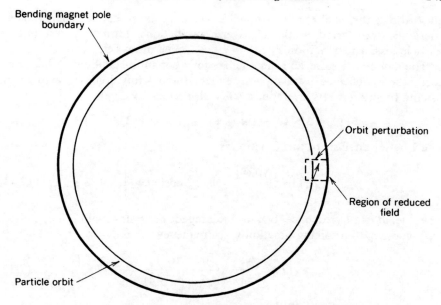

Figure 7.4 Bending magnetic field of a circular accelerator with a field error.

The number ν is also given by

$$\nu = \frac{\text{frequency of transverse oscillations}}{\text{frequency of rotation}}. \qquad \boxed{7.10}$$

A short betatron wavelength is equivalent to a high value of ν. Thus, increased ν is associated with stronger focusing and increased acceptance.

When the quantity ν has an integer value, there is the possibility of a *resonance instability* in a circular system.[†] The physical basis of a resonance instability at $\nu_x = 1$ is illustrated in Figure 7.4. Assume that a circular accelerator has an imperfection at an azimuthal position, such as a dip in field magnitude of a bending magnet. The reduced field represents a localized outward force that gives particles an impulse each time they pass. If $\nu_x \neq 1$, then the impulse acts during different phases of the betatron oscillation. The driving force is nonresonant, and the resulting oscillation amplitude is finite. When $\nu_x = 1$, the driving force is in resonance with the particle oscillation. The impulse is correlated with the betatron oscillation so that the oscillation amplitude increases on each revolution.

Another view of the situation is that the field perturbation acts to couple periodic motion in the axial direction (rotation) to periodic motion in the transverse direction (betatron oscillation). Coupling is strong when the harmonic modes have the same frequency. The distribution of beam energy is

[†] E. D. Courant, *J. Appl. Phys.*, **20**, 611 (1949).

anisotropic; the axial energy is much larger than the transverse energy. Thus, energy is transferred to the transverse oscillations, ultimately resulting in particle loss. The condition $\nu_x = 1$ or $\nu_y = 1$ must always be avoided.

The model of Figure 7.5 is an analogy for resonant instabilities. A harmonic oscillator is deflected by a time-dependent forcing function. We assume the forcing function exerts an impulse Δp with periodicity $2\pi/\omega$:

$$F(t) = \Delta p \delta(t - 2m\pi/\omega) \qquad (m = 1, 2, 3, \dots).$$

The Fourier analysis of this function is

$$F(t) = \frac{\Delta p \, \omega}{\pi}\left[\frac{1}{2} + \sum_{n=1}^{\infty} \cos(n\omega t)\right]. \tag{7.11}$$

The steady-state response of an undamped oscillator with characteristic frequency ω_0 to a forcing function with frequency $n\omega$ is

$$x(t) = \frac{\Delta p \, \omega}{m\pi} \frac{\cos(n\omega t)}{\left(\omega_0^2 - (n\omega)^2\right)}. \tag{7.12}$$

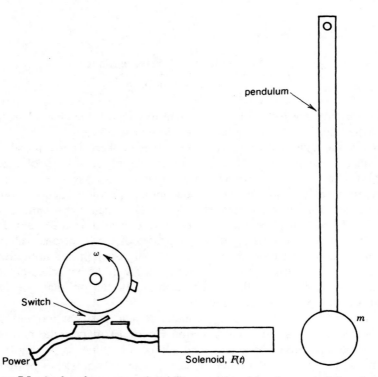

Figure 7.5 Analogy for resonance instability; pendulum driven by a periodic impulse.

The response of the linear system to the force of Eq. (7.11) is (by superposition)

$$x(t) = \frac{\Delta p\,\omega}{m\pi}\left(\frac{\cos\omega t}{\left(\omega_0^2 - \omega^2\right)} + \frac{\cos 2\omega t}{\left(\omega_0^2 - 4\omega^2\right)} + \frac{\cos 3\omega t}{\left(\omega_0^2 - 9\omega^2\right)} + \cdots\right).$$

$$(7.13)$$

The pendulum mass represents a beam particle deflected from the primary orbit. The betatron oscillation frequency is ω_0. If we assume that $F(t)$ represents a system perturbation at a particular azimuth, then ω is the rotation frequency of the particle. By Eq. (7.10), $\nu = \omega_0/\omega$. Equation (7.13) can be rewritten

$$x(t) = \frac{\Delta p}{m\pi\omega}\left(\frac{\cos\omega t}{\nu^2 - 1} + \frac{\cos 2\omega t}{\nu^2 - 4} + \frac{\cos 3\omega t}{\nu^2 - 9} + \cdots\right). \qquad (7.14)$$

Thus, the steady-state amplitude of particle oscillations diverges when ν has an integer value. A complete analysis of the time-dependent response of the particles shows that the maximum growth rate of oscillations occurs for the $\nu = 1$ resonance and decreases with higher-order numbers. Systems with high ν and strong focusing are susceptible to resonance instabilities. Resonance instabilities do not occur in weak focusing systems with $\nu < 1$.

7.3 BETATRON OSCILLATIONS

The most familiar example of focusing in an accelerator by continuous fields occurs in the betatron. In this device, electrons are confined for relatively long times as they are inductively accelerated to high energy. We shall concentrate on transport problems for constant energy particles in this section. The treatment is generalized to particles with changing energy in Chapter 11.

Section 3.7 showed how a uniform magnetic field confines an energetic charged particle in a circular orbit normal to the field. Particles can drift freely along the field direction. In a betatron, particles are enclosed in a toroidal vacuum chamber centered on the main circular orbit. Beams always have spreads in angle and position of particle orbits about the main orbit. It is necessary to supplement the uniform field with additional field components so that nonideal particle orbits oscillate about the main axis rather than drift away. Coordinates to analyze particle motion are shown in Figure 7.6. One set (x, z, s) is referenced to the primary orbit: s points along the beam axis, x is normal to z and the field, while z is directed along the field. We shall also use polar coordinates (r, θ, z) centered at the main orbit gyrocenter, which is also the axis of symmetry of the field.

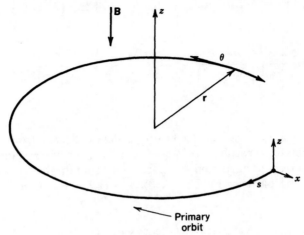

Figure 7.6 Coordinate system for analyzing betatron oscillations in an axisymmetric magnetic field with a radial gradient.

To begin, consider motion in a uniform magnetic field, B_0. Orbits viewed normal to the field are circles, including the orbits of perturbed particles (Fig. 7.7). The particle path shown as a solid line Figure 7.7 has nonzero radial velocity at the primary orbit. This is equivalent to a displacement of its gyrocenter from the system axis. Note that excursions away from the primary orbit are bounded and approximately harmonic. In this sense, there is *radial focusing* in a uniform magnetic field. A particle distribution with a spread in v_x occupies a bounded region in x. This is not true in the z direction. A spread in v_z causes the boundary of the particle distribution to expand indefinitely. In order to confine a beam, we must either add additional focusing lenses around the torus to supplement the bending field or modify the bending field. We shall

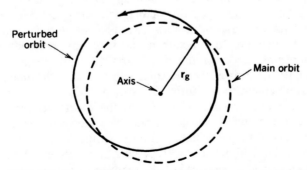

Figure 7.7 Radial focusing in a uniform bending field. Transverse velocity with respect to the main orbit corresponds to a displaced gyrocenter.

consider the latter option and confine attention to fields with azimuthal symmetry.

The zero-order particle velocity is in the θ direction relative to the field axis. The magnetic field must have a radial component in order to exert a force in the z direction. Consider, first, the field distribution from an air core magnet. Figure 7.8a shows a uniform solenoidal field. If extra windings are added to the ends of the solenoid (Fig. 7.8b), the field magnitude increases away from the midplane. The bent field lines have a component B_r. In the paraxial approximation

$$B_r \sim -(r/2)\left[dB_z(0, z)/dz\right]. \tag{7.15}$$

The variation of B_r and B_z about the axis of symmetry of the field is sketched in Figure 7.8c. The axial field is minimum at $z = 0$; therefore, B_r varies approximately linearly with z about this plane at a constant radius. Directions of the cross products are indicated in Figure 7.8b. The axial force is in the $-z$ direction for $z > 0$ and in the $+z$ direction for $z < 0$. Particles are confined about the plane of minimum B_z. A field with magnitude decreasing away from a plane is defocusing.

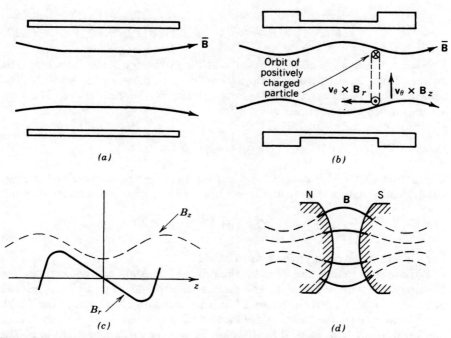

Figure 7.8 Axisymmetric magnetic field with radial gradient. (*a*) Field lines of a solenoidal coil. (*b*) Addition of end coils to generate magnetic mirror; forces on particle with axicentered orbit. (*c*) Axial variations of B_z and B_r in a mirror field. (*d*) Generation of mirror field in a gap by shaping magnet poles to conform to magnetic equipotential lines.

The geometry of Figure 7.8b is used for the confinement of hot fusion plasmas. It is called the *magnetic mirror* since particles are reflected from the region of increased B_z. In accelerator applications, particles are confined to a small region about $z = 0$. Following the discussion of Section 5.7, iron is used to carry the magnetic flux in all other regions to minimize power requirements. We can easily determine the shape of nonsaturated pole pieces that will generate the same field patterns as Figures 7.8a and b by following the method of Section 5.5. The pole faces must lie on surfaces of constant magnetic potential normal to the field lines. After choosing the gap width, the surfaces are determined by connecting a curve everywhere normal to the field lines (Fig. 7.8d). Most vacuum field patterns generated by distributed coils can be duplicated by shaped pole pieces.

The equation $\nabla \times \mathbf{B} = 0$ implies that

$$\partial B_z / \partial r = \partial B_r / \partial z. \tag{7.16}$$

Taking the partial derivative of Eq. (7.15) leads to

$$\partial B_z / \partial r = -(r/2)\left[\partial^2 B_z(0, z)/\partial z^2\right]$$

or

$$B_z(r, z) \cong B_0 - (r^2/4)\left[\partial^2 B_z(r, 0)/\partial z^2\right]. \tag{7.17}$$

The second derivative of B_z is positive at the symmetry plane for a magnetic mirror. Thus, $B_z(r, 0)$ is not uniform, but has a negative radial gradient. The bending field gradient is usually parametrized by the *field index*. The field index is a function of radius given by

$$n(r) = -\left[r/B_z(r, 0)\right]\left[\partial B_z(r, 0)/\partial r\right]. \qquad \boxed{7.18}$$

Comparing Eqs. (7.17) and (7.18), the magnitude of the field index increases quadratically with radius in the symmetry plane of a paraxial magnetic field:

$$n(r) = \left[r^2/2B_z(r)\right]\left[\partial^2 B_z(r, 0)/\partial z^2\right].$$

The field index is positive for a mirror field.

We can now derive equations for the orbits of constant energy particles in a bending field with a gradient. Assume that the spread of particle angles and position is small so that the beam is localized about an ideal circular orbit with $r = r_g$ and $z = 0$. The bending field and field index values at the position of the ideal orbit are $B_0 = B_z(r_g, 0)$ and $n_0 = -(r_g/B_0)[\partial B_z(r_g, 0)/\partial r]$. The bending field and radius of the ideal orbit are related by $r_g = \gamma_0 m_0 v_\theta / qB_0$. Small orbit perturbations imply that $x = r - r_g \ll r_g$, $z \ll r_g$, $v_x \ll v_\theta$, $v_z \ll v_\theta$, and $v_\theta \cong v_0$ (a constant). The magnetic field components are approximated by a Taylor expansion about the point $(r_g, 0)$.

$$B_r \cong B_r(r_g, 0) + (\partial B_r/\partial r)x + (\partial B_r/\partial z)z, \tag{7.19}$$

$$B_z \cong B_z(r_g, 0) + (\partial B_z/\partial r)x + (\partial B_z/\partial z)z. \tag{7.20}$$

The derivatives are taken at $r = r_g$, $z = 0$. A number of terms can be eliminated. In the symmetry plane, B_z is a minimum so that $\partial B_z/\partial z = 0$. Inspection of Figure 7.8b shows that $B_r(r_g, 0) = 0$. Since B_r is zero everywhere in the symmetry plane, $\partial B_r/\partial r = 0$. Combining this with the condition $\nabla \cdot \mathbf{B} = 0$ implies that $\partial B_z/\partial z = 0$. Finally, the equation $\nabla \times \mathbf{B} = 0$ sets the constraint that $\partial B_r/\partial z = \partial B_z/\partial r$. The magnetic field expressions become

$$B_r \cong (\partial B_z/\partial r)z, \tag{7.21}$$

$$B_z \cong B_0 + (\partial B_z/\partial r)x. \tag{7.22}$$

Replacing the bending field gradient with the field index, we find that

$$B_r \cong -(n_0 B_0/r_g)z, \tag{7.23}$$

$$B_z \cong B_0 - (n_0 B_0/r_g)x. \tag{7.24}$$

We can verify by direct substitution that these fields satisfy the paraxial relation of Eq. (6.7).

The next step is to determine the particle equations of motion for small x and z. The particle energy is a constant (parametrized by γ_0) since only static magnetic fields are present. The relativistic equations of motion are

$$\gamma_0 m_0 (d^2 r/dt^2) = \gamma_0 m_0 v_\theta^2/r - q v_\theta B_z, \tag{7.25}$$

$$\gamma_0 m_0 (d^2 z/dt^2) = + q v_\theta B_r. \tag{7.26}$$

The radial equation can be simplified by substituting $x = r - r_g$ and the field expression of Eq. (7.24),

$$d^2 x/dt^2 = v_\theta^2/(r_g + x) - q v_\theta (B_0 - n_0 B_0 x/r_g)/\gamma_0 m_0$$

$$\cong (v_\theta^2/r_g) - (v_\theta/r_g)^2 x - q v_\theta B_0/\gamma_0 m_0 + (n_0 q v_\theta B_0/\gamma_0 m_0 r_g)x. \tag{7.27}$$

The second expression was derived using the binomial expansion and retaining only first-order terms in x/r_g. The zero-order force terms on the right-hand side cancel. The first-order equation of radial motion is

$$(d^2x/dt^2) \cong -(v_\theta/r_g)^2(1 - n_0)x = -\omega_g^2(1 - n_0)x. \qquad (7.28)$$

The last form is written in terms of the gyrofrequency of the ideal particle orbit. The equation for motion along the bending field lines can be rewritten as

$$(d^2z/dt^2) \cong -\omega_g^2 n_0 z. \qquad (7.29)$$

Equations (7.28) and (7.29) describe uncoupled harmonic motions and have the solutions

$$x = x_0\cos\left[(1 - n_0)^{1/2}\omega_g t + \phi_1\right], \qquad \boxed{7.30}$$

$$z = z_0\cos\left[(n_0)^{1/2}\omega_g t + \phi_2\right]. \qquad \boxed{7.31}$$

The equations for betatron oscillations have some interesting features. In a uniform field ($n_0 = 0$), first-order radial particle motion is harmonic about the ideal orbit with frequency ω_g. This is equivalent to a displaced circle, as shown in Figure 7.7. When a field gradient is added, the transverse oscillation frequency is no longer equal to the rotation frequency. We can write

$$\omega_x/\omega_g = (1 - n_0)^{1/2} = \nu_x. \qquad (7.32)$$

When n_0 is positive (a negative field gradient, or mirror field), ν_x is less than unity and the restoring force decreases with radius. Particles moving outward are not reflected to the primary axis as soon; conversely, particles moving inward are focused back to the axis more strongly. The result is that the orbits precess, as shown in Figure 7.9. When $n_0 > 1$, the field drops off too rapidly to restore outward moving particles to the axis. In this case, $(1 - n_0)^{1/2}$ is imaginary so that x grows exponentially. This is an example of an *orbital instability*.

In the y direction, particle orbits are stable only when $n_0 > 0$. Relating the oscillation frequency to ω_g gives

$$\nu_y = (n_0)^{1/2}. \qquad (7.33)$$

We conclude that particles can be focused around the primary circular orbit in an azimuthally symmetric bending magnet if the field has a negative radial gradient such that

$$0 < n_0 < 1. \qquad \boxed{7.34}$$

Focusing of this type in a circular accelerator is called *weak focusing*. In the

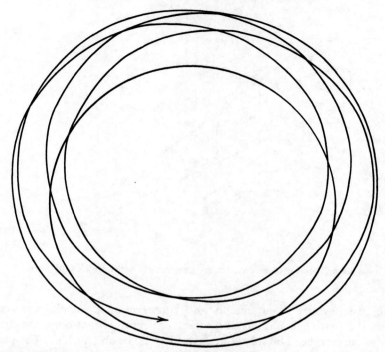

Figure 7.9 Precessing particle orbit in an axisymmetric magnetic field with field index $n = 0.5$; z axis out of page.

stable regime, both ν_x and ν_y are less than unity. Therefore, resonant instability is not a critical problem in weak focusing machines.

7.4 AZIMUTHAL MOTION OF PARTICLES IN CYLINDRICAL BEAMS

In principle, we have collected sufficient techniques to calculate particle orbits in any steady-state electric or magnetic field. Three equations of motion must be solved simultaneously for the three dimensions. Three components of electric and magnetic fields must be known throughout the region of interest. In practice, this is a difficult procedure even with digital computers. It is costly in either calculational time or memory space to specify field components at all positions.

The paraxial ray equation affords a considerable simplification for the special case of cylindrical beams where particle orbits have small inclination angles with respect to the axis. Use of paraxial forms (Section 6.2) means that electric and magnetic fields can be calculated quickly from specified quantities

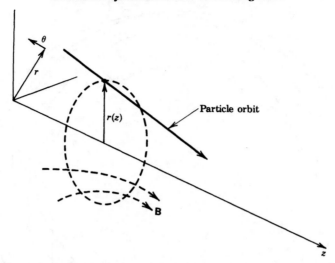

Figure 7.10   Coordinate system and quantities to derive Busch's theorem.

on the axis. Symmetry arguments and conservation principles are used to simplify the particle equations of motion. The result is that only one equation for radial motion need be solved to determine the beam envelope. The paraxial ray equation is derived in Section 7.5. In this section, we shall determine a result used in the derivation, the conservation of canonical angular momentum. The method used, the Busch theorem,[†] provides more physical insight than the usual approach from the Hamiltonian formulation.

Consider a region of azimuthally symmetric, paraxial electric and magnetic fields where particles move almost parallel to the z axis. A cylindrical coordinate system (r, θ, z) is referenced to the beam axis (Fig. 7.10). (Note that this coordinate system differs from the one used in Section 7.3.) We seek an equation that describes the azimuthal motion of charged particles in terms of on-axis quantities. The Lorentz force law for azimuthal motion [Eq. (3.34)] is

$$\dot{p}_\theta + p_r \dot{\theta} = q(\dot{r}B_z - \dot{z}B_r) + qE_\theta, \tag{7.35}$$

where time derivatives are indicated by the dot notation. Inspection of Eq. (7.35) shows that changes in v_θ arise from either interactions of particle velocity with static magnetic field components or an azimuthal electric field. In the paraxial approximation, we will see that such an electric field can arise only from time variations of axial magnetic flux. We will treat the two terms on the right-hand side of Eq. (7.35) separately and then combine them to a modified equation for azimuthal motion. In the modified equation, v_θ depends only on the quantity $B_z(0, z)$.

[†] H. Busch, *Z. Phys.*, **81**, 974 (1926).

To begin, consider a particular moving through a region of static field components B_r and B_z (Fig. 7.10). The *flux function* ψ is defined as the flux of axial magnetic field enclosed in a circle of radius r at z, where r and z are particle coordinates. (Note that this is a calculational definition; the particle does not necessarily move in a circle.) The flux function changes as the particle moves along its orbit. First, assume that the particle moves radially a distance Δr. The change in ψ is

$$\Delta \psi \cong B_z(r, z) 2\pi r \Delta r. \tag{7.36}$$

Equation (7.36) reflects the fact that B_z is almost constant in the paraxial approximation and that the area of the circle increases by about $2\pi r \Delta r$. Similarly, axial motion produces a change in ψ of

$$\Delta \psi \cong (\pi r^2)(\partial B_z / \partial z) \Delta z = -2\pi r B_r \Delta z. \tag{7.37}$$

The area of the circle is constant, but the magnitude of B_z may change if there is an axial gradient. The last form is derived by substitution from Eq. (6.7).

The change in ψ from a particle motion is obtained by adding Eqs. (7.36) and (7.37). Dividing both sides of the sum by Δt and taking the limit $\Delta t \to 0$, we find that

$$d\psi / dt = 2\pi r \left[B_z(dr/dt) - B_r(dz/dt) \right]. \tag{7.38}$$

The derivative indicates the change in flux function arising from particle motion when the magnetic fields are static. The right-hand side of Eq. (7.38) is identical to the right side of Eq. (7.35) multiplied by $2\pi r$. Substituting, Eq. (7.35) becomes

$$\dot{p}_\theta + p_r \dot{\theta} = \dot{\psi} / 2\pi r. \tag{7.39}$$

The left-hand side of Eq. (7.39) can be modified by noting that $p_\theta = \gamma m_0 r(d\theta/dt)$ and $p_r = \gamma m_0(dr/dt)$:

$$\dot{p}_\theta + p_r \dot{\theta} = \gamma m_0(r\ddot{\theta} + \dot{r}\dot{\theta} + \dot{r}\dot{\theta}) = (1/r)d(\gamma m_0 r^2 \dot{\theta})/dt.$$

The final result is that

$$d(\gamma m_0 r^2 \dot{\theta})/dt = q\dot{\psi}/2\pi$$

or

$$\gamma m_0 r v_\theta + q\psi/2\pi = q\psi_0/2\pi. \tag{7.40}$$

The quantity ψ_0 is a constant equal to the value of ψ when $v_\theta = 0$. Solving for v_θ,

$$v_\theta = -(q/2\pi\gamma m_0 r)(\psi - \psi_0). \tag{7.41}$$

The azimuthal velocity is a function only of r, ψ, and ψ_0. Since particles usually have $v_\theta = 0$ when they are generated, ψ_0 is taken as the flux in the source plane enclosed in a circle of radius r_0, where r_0 is the extraction radius. Variation of B_z is second order with radius in the paraxial approximation, so that $\psi \cong (\pi r^2) B_z(0, z)$.

Equation (7.40) can be cast in the form familiar from Hamiltonian dynamics by substituting Eq. (4.53) into Eq. (3.25):

$$\psi = \int_0^r (2\pi r' \ dr')[(1/r') \partial (r'A_\theta)/\partial r'] = 2\pi r A_\theta. \tag{7.42}$$

A_θ is the only allowed component of the vector potential in the paraxial approximation. In this circumstance, the vector potential has a straightforward physical interpretation. It is proportional to the flux enclosed within a radius divided by $2\pi r$. The quantity $\psi/2\pi$ is thus synonymous with the stream function (Section 4.7). Lines of constant enclosed flux must lie along field lines since field lines do not cross one another.

To complete the derivation, we must include the effects on azimuthal motion when the magnetic field changes in time. An azimuthal electrostatic field is not consistent with symmetry in θ since the existence of such a field implies variation of ϕ along θ. A symmetric azimuthal electric field can be generated inductively by a changing axial magnetic field. We will derive a modified form for the term qE_θ in Eq. (7.35) by neglecting spatial gradients of B_z. By Faraday's law [Eq. (3.26)], the azimuthal electric field acting on a particle at radius r is $E_\theta = \partial \psi/\partial t$, where the derivative implies variation of ψ in time neglecting spatial variations. If we add contributions from spatial and temporal variations of ψ, we arrive at Eq. (7.41) as the general modified equation of azimuthal motion. The time derivative is interpreted as the change in ψ arising from all causes. In the Hamiltonian formulation of particle dynamics, the canonical angular momentum in the presence of a magnetic field is constant and is defined as

$$P_\theta = \gamma m_0 r v_\theta + q r A_\theta = q\psi_0/2\pi. \qquad \boxed{7.43}$$

7.5 THE PARAXIAL RAY EQUATION

The paraxial ray equation is derived by combining the properties of paraxial fields, the conservation of canonical angular momentum, and the conservation of energy. It can be used to determine the *envelope* (outer radius) of a beam as a function of position along the axis in terms of the electrostatic potential and longitudinal magnetic field on the axis, $\phi(0, z)$ and $B_z(0, z)$.

The equation is based on the following assumptions.

1. The beam is cylindrically symmetric.
2. Beam properties vary in space but not in time.
3. The fields are cylindrically symmetric, with components E_r, E_z, B_r, and B_z. This encompasses all axisymmetric electrostatic lenses and the solenoidal magnetic lens.
4. The fields are static.
5. Particle motion is paraxial.
6. Fields are paraxial and transverse forces are linear.

In the following derivation, it is also assumed that particle orbits are laminar and that there are no self-fields. Terms can be added to the paraxial ray equation to represent spreads in transverse velocity and space charge forces.

The laminarity of orbits means that the radial projections of all particle orbits are similar. They differ only in amplitude. It is thus sufficient to treat only the boundary orbit. The axial velocity is approximately constant in any plane normal to the axis. Time derivatives are replaced by $v_z(d/dz)$ since we are interested in the steady-state beam envelope as a function of axial position. The azimuthal equation of motion is (Section 7.4)

$$\dot{\theta} = -(q/2\pi\gamma m_0 r^2)(\psi - \psi_0). \tag{7.44}$$

The quantity γ may vary with axial position since zero-order longitudinal electric fields are included. The only nontrivial equation of motion is in the radial direction:

$$d(\gamma m_0 \dot{r})/dt - \gamma m_0 v_\theta^2/r = q(E_r + v_\theta B_z) = m_0 \dot{r}\dot{\gamma} + \gamma m_0 \ddot{r} - \gamma m_0 v_\theta^2/r. \tag{7.45}$$

Conservation of energy can be expressed as

$$(\gamma - 1)m_0 c^2 = \int_{z_0}^{z} dz' \, qE_z = q\phi. \tag{7.46}$$

The quantity z_0 is the location of the source where particles have zero kinetic energy. The absolute electrostatic potential is used. A change in position leads to a change in γ:

$$\Delta\gamma = qE_z \, \Delta z/m_0 c^2.$$

Dividing both sides by Δt and taking the limit of zero interval, we find that

$$\dot{\gamma} = qE_z v_z/m_0 c^2 = qE_z \beta/m_0 c. \tag{7.47}$$

Equations (7.44) and (7.47) are used to replace γ and θ in Eq. (7.45).

$$\ddot{r} + \frac{qE_z\beta\dot{r}}{\gamma m_0 c} + \frac{q^2 B_z^2 r}{4\gamma^2 m_0^2} - \frac{q^2\psi_0^2}{4\pi^2\gamma^2 m_0^2}\bigg/ r^3 - \frac{qE_r}{\gamma m_0} = 0. \qquad (7.48)$$

The radial electric field is replaced according to $E_r = (-r/2)(dE_z/dz)$. The derivative is a total derivative since the fields are assumed static and E_z has no first-order radial variation. The final step is to replace all time derivatives with derivatives in z. For instance, it can be easily shown from Eq. (7.47) that $\partial E_z/\partial z = (m_0 c^2/q)(d^2\gamma/dz^2)$. The second time derivative of r becomes $v_z^2(d^2 r/dz^2) + v_z(dr/dz)(dv_z/dz)$. When the substitutions are carried out, the paraxial ray equation is obtained.

$$r'' + \frac{\gamma' r'}{\beta^2\gamma} + \left[\frac{\gamma''}{2\beta^2\gamma} + \left(\frac{qB_z}{2\beta\gamma m_0 c}\right)^2\right]r - \left(\frac{q\psi_0}{2\pi\beta\gamma m_0 c}\right)^2\bigg/ r^3 = 0.$$

$$\boxed{7.49}$$

The prime symbol denotes a differentiation with respect to z. The quantities γ, γ', γ'', and B_z are evaluated on the axis. The quantity $\gamma(0, z)$ and its derivatives are related to the electrostatic potential [Eq. (7.46)] which is determined via the Laplace equation.

Consider the final term in Eq. (7.49). The quantity ψ_0 is the magnetic flux enclosed by the beam at the source. This is elucidated in Figure 7.11 for an electron beam. The figure shows an *immersed cathode* located within the magnetic field. When ψ_0 is nonzero, the final term in Eq. (7.49) has a strong radial dependence through the $1/r^3$ factor. The term has a dominant defocusing effect when the beam is compressed to a small spot. This is a result of the fact that particles produced in a magnetic field have a nonzero canonical

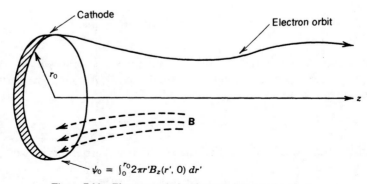

Figure 7.11 Electron emission from an immersed cathode.

angular momentum and are unable to pass through the axis. Thus, care should be taken to exclude magnetic fields from the cathode of the electron source in applications calling for fine focusing. When $\psi_0 = 0$, the paraxial ray equation is linear since the remaining terms contain only first powers of r'', r', and r.

The nonrelativistic approximation can be used for beams of ions or low-energy electrons. Substituting $\gamma = 1 + q\phi/m_0 c^2$ in the limit $q\phi \ll m_0 c^2$, Eq. (7.49) becomes

$$2\phi r'' + \phi' r' + \left(\frac{\phi''}{2} + \frac{qB_z^2}{4m_0} \right) r - \frac{q\psi_0^2/4\pi^2 m_0}{r^3} = 0. \qquad \boxed{7.50}$$

If there are only electric fields present, Eq. (7.50) reduces to the familiar form

$$r'' + (\phi'/2\phi)r' + (\phi''/4\phi)r = 0, \qquad \boxed{7.51}$$

where, again, the prime symbol indicates an axial derivative. An alternate form for Eq. (7.51) is

$$\frac{d(r'\phi)}{dz} + \frac{r^3}{4} \frac{d(\phi'/r^2)}{dz} = 0. \qquad (7.52)$$

7.6 NUMERICAL SOLUTIONS OF PARTICLE ORBITS

The paraxial ray equation is a second-order differential equation that describes a beam envelope or the radial orbit of a particle. It is possible to solve the equation analytically in special field geometries, such as the narrow-diameter aperture lens or the solenoidal lens with sharp field boundaries. In most realistic situations, it is rare to find closed forms for field variations on axis that permit analytic solutions. Thus, numerical methods are used in almost all final designs of charged particle optical systems. In this section, we shall briefly consider a computational method to solve the paraxial ray equation. The method is also applicable to all second-order differential equations, such as the general particle equations of motion.

A second-order differential equation can be written as two first-order equations by treating r' as a variable:

$$(dr/dz) - r' = 0, \qquad (7.53)$$

$$(dr'/dz) - f(r', r, z) = 0, \qquad (7.54)$$

where $f(r', r, z)$ is a general function. For instance, $f = -(\phi' r'/2\phi + \phi'' r/4\phi)$ in Eq. (7.51). Equations (7.53) and (7.54) must be solved simultaneously. In order to illustrate the method of solution, consider the simpler example of the equation

$$(dr/dz) - f(r, z) = 0. \tag{7.55}$$

An obvious first approach to solving Eq. (7.55) by finite difference methods would be to approximate the derivative by $(\Delta r / \Delta z)$ to arrive at the algorithm,

$$r(z + \Delta z) \cong r(z) + f(r, z)\,\Delta z. \quad \boxed{7.56}$$

Given an initial condition, $r(z_0)$, the solution $r(z)$ can be determined in steps.

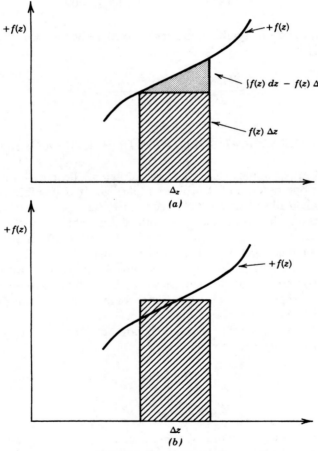

Figure 7.12 Approximations to the integral for the numerical solution of Eq. (7.56). (*a*) Eulerian finite difference method. (*b*) Two-step finite difference method.

The algorithm of Eq. (7.56) is called the *Eulerian difference method*. The drawback of this approach is apparent if we express Eq. (7.55) in integral form.

$$r(z + \Delta z) - r(z) = \int_{z}^{z+\Delta z} f(r', z') \, dz'. \tag{7.57}$$

Since we do not know $r(z)$ over the interval Δz, the essence of a finite difference method is to find an estimate of the integral that leads to a good approximation for the continuous function. The estimate of the Eulerian method is $\int f(r', z') \, dz' \cong f(r, z) \Delta z$, as shown in Figure 7.12*a*. This introduces a first-order error, so that the final result will be accurate to only order Δz.

We could get a much better approximation for the integral if we knew the value of $f(r, z)$ at $z + \Delta z/2$. The integral is then approximated as $f(r, z + \Delta z/2) \Delta z$. There is cancellation of first-order errors as shown in Figure 7.12*b*. The basis of the *two-step method* is to make an initial estimate of the value of $f(r, z + \Delta z/2)$ by the Eulerian method and then to use the value of advance r. The two-step algorithm is

Step 1: $\qquad r(z + \Delta z/2) \cong r(z) + f(r, z) \Delta z/2,$

$$\boxed{7.58}$$

Step 2: $\qquad r(z + \Delta z) \cong r(z) + f(r + \Delta z/2, z + \Delta z/2) \Delta z.$

Although this involves two operations to advance r, it saves computation time in the long run since the error between the finite difference approximation and the actual value is of order Δz^2. The two-step method is an example of a *space-centered* (or *time-centered*) difference method since the integral is estimated symmetrically in the interval. The extension of the algorithm of Eqs. (7.58) to the case of two coupled first-order differential equations [Eqs. (7.53) and (7.54)] is

Step 1: $\quad r(z + \Delta z/2) = r(z) + r' \Delta z/2,$

$\qquad\qquad r'(z + \Delta z/2) = r'(z) - f[r(z), r'(z), z] \Delta z/2.$

$$\boxed{7.59}$$

Step 2: $\quad r(z + \Delta z) = r(z) + r'(z + \Delta z/2) \Delta z,$

$\qquad\qquad r'(z + \Delta z) = r'(z) + f[r(z + \Delta z/2), r'(z + \Delta z/2), z + \Delta z/2] \Delta z.$

The extra work involved in programming a higher-order finite difference scheme is worthwhile even for simple orbit problems. Figure 7.13 shows results for the calculation of a circular orbit described by the equations $(d^2x/dt^2) + y = 0$, $(d^2y/dt^2) - x = 0$. The initial coordinates are $(x = 1, y = 0)$. The relative error in final position $(\Delta x/x)$ after one revolution is plotted as a

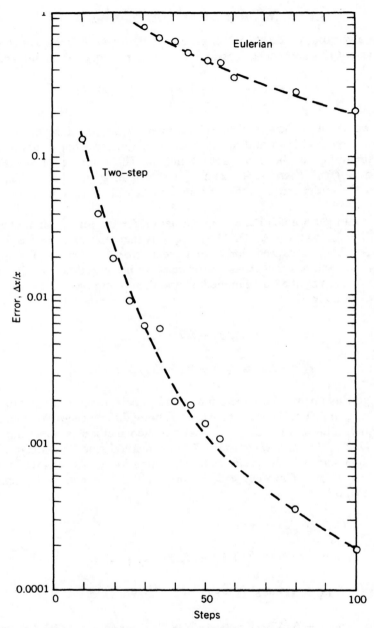

Figure 7.13 Convergence of solutions; Two-step versus Eulerian method. Solutions of equations to generate a circle of unit radius. Difference in radius after one revolution plotted versus number of computational steps per revolution.

function of the number of time steps ($\Delta t = 2\pi/n$). The two-step method achieves 1% accuracy with only 25 steps, while the accuracy of the Eulerian method is no better than 20% after 100 steps.

As an example of the solution of the paraxial ray equation, consider ion acceleration in an electrostatic acceleration column. The nonrelativistic equation with electric fields only [Eq. (7.51)] can be used. The column, illustrated in Figure 7.14, is found on Van de Graaff accelerators and on electrostatic injectors for high-energy accelerators. A static potential in the range 0.1 to 10 MeV exists between the entrance and the extraction plates. The inside of the column is at high vacuum while the outside is usually immersed in a high-voltage insulating medium such as sulfur hexafluoride gas or transformer oil. A solid vacuum insulator separates the two regions. The insulator has greater immunity to vacuum breakdown if it is separated into a number of short sections by metal *grading rings* (see Section 9.5). The rings are connected to a high-resistance voltage divider so that the voltage is evenly distributed between the insulator sections. Thick grading rings can also play a role in focusing a beam on the axis of the column.

Consider, first, the field distribution when the rings are thin. If the thin rings extend to a radius that is large compared to the aperture, then the electric field

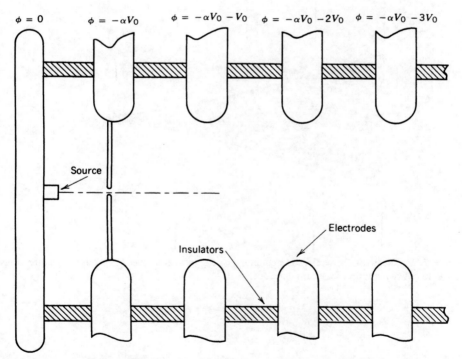

Figure 7.14 Definition of quantities to determine particle orbits in an electrostatic acceleration column.

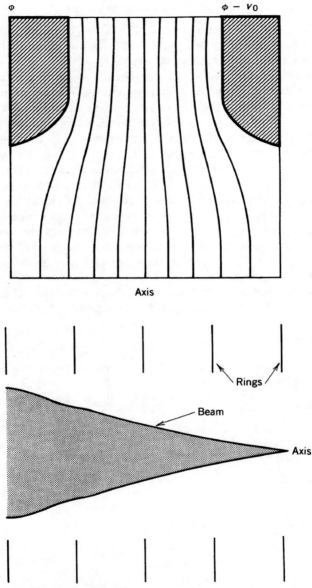

Figure 7.15 Electrostatic acceleration column. (*a*) Equipotential lines between two grading rings. (*b*) Paraxial ray solution for beam envelope.

on the axis is uniform with z. The quantity ϕ'' is zero; hence the third term in Eq. (7.51) vanishes and there is no radial focusing. The second term in Eq. (7.51) is nonzero, but it does not lead to beam focusing. This term corresponds to the decrease in dr/dz from the increase in v_z with acceleration. In contrast, there are focusing forces when the grading rings are thick. In this case, equipotential lines are periodically compressed and expanded along the axis (Fig. 7.15a) leading to radial electric fields.

We express potential relative to the particle source (absolute potential). Assume that ions are injected into the column with energy $q\alpha V_0$, where $-\alpha V_0$ is the absolute potential of the entrance plate. The potential decreases moving down the column (Fig. 7.14). Radial electric field is taken in the form

$$E_r(r, z) \cong -(\beta r/2)\sin(2\pi z/d), \qquad (7.60)$$

where d is the ring spacing and the origin ($z = 0$) corresponds to the entrance plate. The form of Eq. (7.60) is confirmed by a numerical solution of the Laplace equation (Fig. 7.15a), which also gives the parameter β. The longitudinal electric field is calculated from Eqs. (6.5) and (7.60) as

$$E_z(r, z) \sim V_0/d - V_0(\beta d/2\pi V_0)\cos(2\pi z/d), \qquad (7.61)$$

where V_0 is the voltage between rings. This leads to the following expression

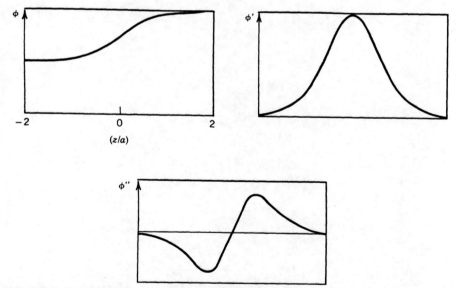

Figure 7.16 Solutions for axial variation of electrostatic potential (ϕ) and its first two z derivatives (ϕ', ϕ'') crossing an immersion lens with tube radius a.

for the absolute potential:

$$\phi(0, z) \sim V_0\left[-\alpha - z/d + (\beta d^2/4\pi^2)\sin(2\pi z/d)\right]. \qquad (7.62)$$

The derivatives ϕ' and ϕ'' can be determined from Eq. (7.62).

Substitution of Eq. (7.62) and its derivatives into Eq. (7.51) leads to the results illustrated in Figure 7.15b for a choice of $\beta = 0.27$ and $\alpha = 1.5$. Note that in the region between the symmetry planes of rings there is a focusing section followed by a defocusing section. The cumulative effect is focusing since the particle gains axial energy and is less affected by positive radial fields on the downstream side.

As a final application, consider the immersion lens of Section 6.6. The variation of axial potential corresponding to Figure 6.10 is plotted in Figure 7.16. An empirical fit to the potential between two accelerating tubes with a small gap and radius a is given[†] by

$$\phi(0, z) \cong [\kappa + 1][1 + (1 - \kappa)\tanh(1.318z/a)/(1 + \kappa)]/2, \qquad (7.63)$$

where ϕ_1 is the upstream absolute potential, ϕ_2 is the downstream potential, $\kappa = \phi_1/\phi_2$, and a is the tube radius. Equation (7.63) was used as input to a two-step paraxial ray equation program to obtain the orbits of Figure 6.11. The orbit solutions lead to the thin-lens focal length of Figure 6.12.

[†]See P. Grivet, *Electron Optics*, *Vol. 1* (transl. by P. W. Hawkes), Pergamon, Oxford, 1972, p. 208.

8

Transfer Matrices and Periodic Focusing Systems

Periodic focusing channels are used to confine high-energy beams in linear and circular accelerators. Periodic channels consist of a sequence of regions called focusing cells containing one or more charged particle optical elements. A focusing cell is the smallest unit of periodicity in the channel. The main application of periodic channels is in high-energy accelerators that utilize strong focusing. For example, the focusing channel of a linear ion accelerator consists typically of a series of magnetic quadrupole lenses with alternating north–south pole orientation. Thus, along either transverse axis, the lenses are alternately focusing and defocusing. We shall see that such a combination has a net focusing effect that is stronger than a series of solenoid lenses at the same field strength. A quadrupole focusing channel can therefore be constructed with a much smaller bore diameter than a solenoid channel of the same acceptance. The associated reduction in the size and power consumption of focusing magnets has been a key factor in the development of modern high-energy accelerators. Periodic focusing channels also have application at low beam energy. Configurations include the electrostatic accelerator column, the electrostatic Einzel lens array and periodic permanent magnet (PPM) channels used in high-power microwave tubes.

The transfer matrix description of beam transport in linear optical elements facilitates the study of periodic focusing channels. The matrix description is a mathematical method to organize information about the transverse motions of particles about the main beam axis. Matrices are particularly helpful when dealing with systems with a large number of different elements. The effects on particles of a single element and combinations of elements are described by the familiar rules of matrix algebra. All the lenses and beam bending devices described in Chapter 6 have associated transfer matrices.

The transfer matrices for the focusing and defocusing axes of a quadrupole lens are derived in Section 8.1. Section 8.2 lists transfer matrices for a variety of common optical elements. The rules for combining matrices to describe complex optical systems are reviewed in Section 8.3. The rules are applied in Section 8.4 to the quadrupole doublet and triplet lenses. These lenses combine quadrupole fields to provide focusing along both transverse axes. Periodic systems are introduced by the example of an array of thin one-dimensional lenses separated by drift spaces (Section 8.5). The discussion illustrates the concepts of phase advance and orbital stability. Matrix algebra is used to extend the treatment to general linear focusing systems. Given the transverse matrix for a focusing cell, the stability limits on beam transport can be derived by studying the mathematical properties of the matrix power operation (Section 8.6). The chapter concludes with a detailed look at orbit stability in a long quadrupole channel (Section 8.7).

8.1 TRANSFER MATRIX OF THE QUADRUPOLE LENS

Transfer matrices describe changes in the transverse position and angle of a particle relative to the main beam axis. We assume paraxial motion and linear fields. The axial velocity v_z and the location of the main axis are assumed known by a previous equilibrium calculation. If x and y are the coordinates normal to z, then a particle orbit at some axial position can be represented by the four-dimensional vector (x, x', y, y'). In other words, four quantities specify the particle orbit. The quantities x' and y' are angles with respect to the axis; they are equivalent to transverse velocities if v_z is known. We further assume that the charged particle optical system consists of a number of separable focusing elements. Separable means that boundary planes between the elements can be identified. We seek information on orbit vectors at the boundary planes and do not inquire about details of the orbits within the elements. In this sense, an optical element operates on an entrance orbit vector to generate an output orbit vector. The transfer matrix represents this operation.

Orbits of particles in a magnetic quadrupole lens were discussed in Section 6.10. The same equations describe the electric quadrupole with a correct choice of transverse axes and the replacement $\kappa_m \to \kappa_e$. In the following discussion, κ can represent either type of lens. According to Eqs. (6.31) and (6.32), motions in the x and y directions are separable. Orbits can therefore be represented by two independent two-dimensional vectors, $\mathbf{u} = (x, x')$ and $\mathbf{v} = (y, y')$. This separation holds for other useful optical elements, such as the magnetic sector field (Section 6.8) and the focusing edge (Section 6.9). We shall concentrate initially on analyses of orbits along one coordinate. Orbit vectors have two components and transfer matrices are 2×2.

Consider motion in the x direction in a quadrupole lens oriented as shown in Figure 5.16. The lens is focusing in the x direction. If the lens has a length l, the exit parameters are related to the entrance parameters by

$$x_f = x_i \cos(\sqrt{\kappa}\, l) + x_i' \sin(\sqrt{\kappa}\, l)/\sqrt{\kappa}\,, \tag{8.1}$$

$$x_f' = -x_i \sqrt{\kappa} \sin(\sqrt{\kappa}\, l) + x_i' \cos(\sqrt{\kappa}\, l). \tag{8.2}$$

The lens converts the orbit vector $\mathbf{u}_i = (x_i, x_i')$ into the vector $\mathbf{u}_f = (x_f, x_f')$. The components of $\mathbf{u}_f$ are linear combinations of the components of $\mathbf{u}_i$. The operation can be written in matrix notation as

$$\mathbf{u}_f = \mathbf{A}_F \mathbf{u}_i, \tag{8.3}$$

if $\mathbf{A}_F$ is taken as

$$\mathbf{A}_F = \begin{bmatrix} \cos(\sqrt{\kappa}\, l) & \sin(\sqrt{\kappa}\, l)/\sqrt{\kappa} \\ -\sqrt{\kappa} \sin(\sqrt{\kappa}\, l) & \cos(\sqrt{\kappa}\, l) \end{bmatrix}, \tag{8.4}$$

where the subscript F denotes the focusing direction. For review, the rule for multiplication of a 2×2 matrix times a vector is

$$\begin{bmatrix} a_{11} & a_{12} \\ a_{21} & a_{22} \end{bmatrix} \begin{pmatrix} x \\ x' \end{pmatrix} = \begin{pmatrix} a_{11}x + a_{12}x' \\ a_{21}x + a_{22}x' \end{pmatrix}. \tag{8.5}$$

If the poles in Figure 5.16 are rotated 90°, the lens defocuses in the x direction. The transfer matrix in this case is

$$\mathbf{A}_D = \begin{bmatrix} \cosh(\sqrt{\kappa}\, l) & \sinh(\sqrt{\kappa}\, l)/\sqrt{\kappa} \\ \sqrt{\kappa} \sinh(\sqrt{\kappa}\, l) & \cosh(\sqrt{\kappa}\, l) \end{bmatrix}. \tag{8.6}$$

Quadrupole lenses are usually used in the limit $\sqrt{\kappa}\, l \leq 1$. In this case, the trigonometric and hyperbolic functions of Eqs. (8.4) and (8.6) can be expanded in a power series. For reference, the power series forms for the transfer matrices are

$$\mathbf{A}_F = \begin{bmatrix} 1 - \Gamma^2/2 + \Gamma^4/24 + \cdots & (\Gamma - \Gamma^3/6 + \cdots)/\sqrt{\kappa} \\ -\sqrt{\kappa}\,(\Gamma - \Gamma^3/6 + \cdots) & 1 - \Gamma^2/2 + \Gamma^4/24 + \cdots \end{bmatrix}, \tag{8.7}$$

and

$$\mathbf{A}_D = \begin{bmatrix} 1 + \Gamma^2/2 + \Gamma^4/24 + \cdots & (\Gamma - \Gamma^3/6 + \cdots)/\sqrt{\kappa} \\ \sqrt{\kappa}\,(\Gamma + \Gamma^3/6 + \cdots) & 1 + \Gamma^2/2 + \Gamma^4/24 + \cdots \end{bmatrix},$$

$$(8.8)$$

where $\Gamma = \sqrt{\kappa}\,l$. The example of the quadrupole illustrates the method for finding the transfer matrix for a linear optical element. Numerical or analytic orbit calculations lead to the identification of the four matrix components. The transfer matrix contains complete information on the properties of the lens as an orbit operator.

When the action of a focusing system is not decoupled in x and y, the full four-dimensional vector must be used and the transfer matrices have the form

$$\begin{bmatrix} \blacksquare & \blacksquare & \square & \square \\ \blacksquare & \blacksquare & \square & \square \\ \square & \square & \blacksquare & \blacksquare \\ \square & \square & \blacksquare & \blacksquare \end{bmatrix}.$$

A focusing system consisting of quadrupole lenses mixed with axisymmetric elements (such as solenoid lens) has coupling of x and y motions. The transfer matrix for this system has coupling components represented by the open boxes above. Sometimes, in the design of particle spectrometers (where beam energy spread is of prime concern), an extra dimension is added to the orbit vector to represent *chromaticity*, or the variations of orbit parameters with energy.[†] In this case, the orbit vector is represented as $\mathbf{u} = (x, x', y, y', T)$.

8.2 TRANSFER MATRICES FOR COMMON OPTICAL ELEMENTS

The following examples illustrate the concept of ray transfer matrices and indicate how they are derived. The simplest case is the thin one-dimensional lens, illustrated in Figure 8.1. Only the angle of the orbit changes when a particle passes through the lens. Following Section 6.4, the transformation of orbit variables is

$$x_f = x_i,$$
$$x'_f = x'_i - x_i/f,$$

$$(8.9)$$

where f is the focal length. This can be written in the form of Eq. (8.3) with the transfer matrix,

$$\mathbf{A} = \begin{bmatrix} 1 & 0 \\ -1/f & 1 \end{bmatrix}. \qquad \boxed{8.10}$$

[†]See P. Dahl, *Introduction to Electron and Ion Optics*, Academic, New York, 1973, Chapter 2.

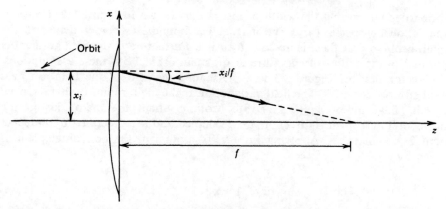

Figure 8.1 Effect of a thin lens (focal length f) on a particle orbit initially parallel to the axis.

The matrix for a diverging lens is the same except for the term a_{21}, which equals $+1/f$. In general, the sign of a_{21} indicates whether the optical element (or combination of elements) is focusing or defocusing.

An optical element is defined as any region in a focusing system that operates on an orbit vector to change the orbit parameters. Thus, there is a transfer matrix associated with translation in field-free space along the z axis (Fig. 8.2). In this case, the distance from the axis changes according to $x_f = x_i + x_i'd$, where d is the length of the drift space. The angle is unchanged. The transfer matrix is

$$\mathbf{A} = \begin{bmatrix} 1 & d \\ 0 & 1 \end{bmatrix}.$$

$$\boxed{8.11}$$

We have already studied the magnetic sector lens with uniform field (Section 6.8). A gradient can be added to the sector field to change the focal

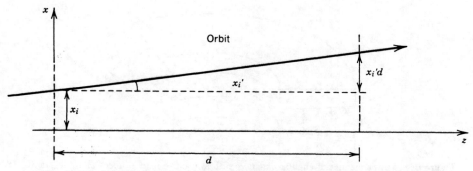

Figure 8.2 Modification of a particle orbit passing through a drift region of length d.

properties by varying the width of the gap, as shown in Figure 8.3. Consider the following special case. The magnet has boundaries perpendicular to the main orbit so that there is no edge focusing. Furthermore, the field gradient is parallel to the radius of curvature of the main orbit. With these assumptions, the sector field of Figure 8.3 is a pie-shaped segment of the betatron field studied in Section 7.3. The field variation near the main radius is characterized by the field index n_0 [Eq. (7.18)]. Motions about the main axis in the horizontal and vertical direction are decoupled and are described by independent 2×2 matrices. Applying Eq. (7.30), motion in the horizontal plane is given by

$$x = A \cos\left[\sqrt{1 - n_0}\,(z/r_\phi) + \phi\right], \tag{8.12}$$

$$x' = dx/dz = -\left[\sqrt{1 - n_0}/r_g\right] A \sin\left[\sqrt{1 - n_0}\,(z/r_g) + \phi\right]. \tag{8.13}$$

The initial position and angle are related to the amplitude and phase by $x_i = A\cos\phi$ and $x_i' = -\sqrt{1 - n_0}\, A \sin\phi/r_g$. In order to determine the net effect of a sector (with transit distance $-d = \alpha r_g$) on horizontal motion, we consider two special input vectors, $(x_i, 0)$ and $(0, x_i')$. In the first case $\phi = 0$ and in the second $\phi = \pi/2$. According to Eqs. (8.12) and (8.13), the final orbit parameters for a particle entering parallel to the main axis are

$$x_f = x_i \cos\left(\sqrt{1 - n_0}\,d/r_g\right), \tag{8.14}$$

and

$$x_f' = -\left(\sqrt{1 - n_0}/r_g\right) x_i \sin\left(d/r_g\right). \tag{8.15}$$

Similarly, if the particle enters on the main axis at an angle, the final orbit

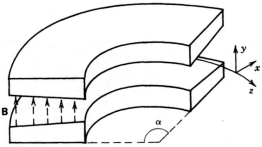

Figure 8.3 Sector magnet of angular extent α with a negative field gradient along the radius of curvature of the main particle orbit.

parameters are

$$x_f = x'_i r_g \sin\left(\sqrt{1 - n_0}\, d/r_g\right)/\sqrt{1 - n_0}, \qquad (8.16)$$

and

$$x'_f = x'_i \cos\left(\sqrt{1 - n_0}\, d/r_g\right). \qquad (8.17)$$

The factor d/r_g is equal to α, the angle subtended by the sector. Combining the results of Eqs. (8.14)–(8.17), the transfer matrix is

$$\mathbf{A}_H = \begin{bmatrix} \cos\left(\sqrt{1-n_0}\,\alpha\right) & r_g \sin\left(\sqrt{1-n_0}\,\alpha\right)/\sqrt{1-n_0} \\ -\sqrt{1-n_0}\,\sin\left(\sqrt{1-n_0}\,\alpha\right)/r_g & \cos\left(\sqrt{1-n_0}\,\alpha\right) \end{bmatrix}.$$

$$\boxed{8.18}$$

Similarly, for the vertical direction,

$$\mathbf{A}_V = \begin{bmatrix} \cos\left(\sqrt{n_0}\,\alpha\right) & r_g \sin\left(\sqrt{n_0}\,\alpha\right)/\sqrt{n_0} \\ -\sqrt{n_0}\,\sin\left(\sqrt{n_0}\,\alpha\right)/r_g & \cos\left(\sqrt{n_0}\,\alpha\right) \end{bmatrix}. \qquad \boxed{8.19}$$

Following the development of Section 6.8, initially parallel beams are focused in the horizontal direction. The focal point is located a distance

$$f' = r_g/\tan\sqrt{1-n_0}\,\alpha \qquad (8.20)$$

beyond the sector exit. This should be compared to Eq. (6.27). When n_0 is negative (a positive field gradient moving out along the radius of curvature), horizontal focusing is strengthened. Conversely, a positive field index decreases the horizontal focusing. There is also vertical focusing when the field index is positive. The distance to the vertical focal point is

$$f' = r_g/\tan\sqrt{n_0}\,\alpha. \qquad (8.21)$$

If $n_0 = +0.5$, horizontal and vertical focal lengths are equal and the sector lens can produce a two-dimensional image. The dual-focusing property of the gradient field is used in charged particle spectrometers.

Particles may travel in either direction through a sector magnet field, so there are two transfer matrices for the device. The matrix for negatively directed particles can be calculated directly. The transfer matrix for particles moving backward in the sector field is the *inverse* of the matrix for forward motion. The inverse of a 2×2 matrix

$$\mathbf{A} = \begin{bmatrix} a_{11} & a_{12} \\ a_{21} & a_{22} \end{bmatrix}$$

is

$$\mathbf{A}^{-1} = \begin{bmatrix} a_{22} & -a_{12} \\ -a_{21} & a_{11} \end{bmatrix} \bigg/ \det \mathbf{A}. \qquad \boxed{8.22}$$

The quantity $\det \mathbf{A}$ is the determinant of the matrix $\mathbf{A}$, defined by

$$\det \mathbf{A} = a_{11} a_{22} - a_{12} a_{21}. \qquad \boxed{8.23}$$

The determinant of the sector field transfer matrix in the horizontal direction is equal to 1. The inverse is

$$\mathbf{A}^{-1} = \begin{bmatrix} \cos\left(\sqrt{1 - n_0}\, \alpha\right) & -r_g \sin\left(\sqrt{1 - n_0}\, \alpha\right)/\sqrt{1 - n_0} \\ \sqrt{1 - n_0}\, \sin\left(\sqrt{1 - n_0}\, \alpha\right)/r_g & \cos\left(\sqrt{1 - n_0}\, \alpha\right) \end{bmatrix}.$$

$$(8.24)$$

Equation (8.24) is equal to Eq. (8.18) with the replacement $\alpha \to -\alpha$. The negative angle corresponds to motion in the $-z$ direction. The effect of the element is independent of the direction. The same holds true for any optical element in which the energy of the charged particle is unchanged. We can verify that in this case $\det \mathbf{A} = 1$.

Acceleration gaps in linear accelerators have the geometry of the immersion lens (Figure 6.10). This lens does not have the same focal properties for particle motion in different directions. Assume the focal length for motion of nonrelativistic particles in the accelerating direction, f_a, is known. This is a function of the lens geometry as well as the absolute potentials of each of the tubes. The upstream potential is ϕ_1 while the downstream potential is ϕ_2. The quantity ξ is defined as the ratio of the exit velocity to the entrance velocity and is equal to $\xi = (\phi_2/\phi_1)^{1/2}$. In the thin-lens approximation, a particle's position is constant but the transverse angle is changed. If the particle entered parallel to the axis in the accelerating direction, it would emerge at an angle $-x_i/f_a$. Similarly, a particle with an entrance vector $(0, x_i')$ emerges at an angle x_i'/ξ. The traverse velocity is the same, but the longitudinal velocity increases. The general form for the transfer matrix of a thin electrostatic lens with acceleration is

$$\mathbf{A}_a = \begin{bmatrix} 1 & 0 \\ -1/f_a & 1/\xi \end{bmatrix}. \qquad (8.25)$$

The determinant has the value $\det \mathbf{A} = 1/\xi \neq 1$. The transfer matrix for a

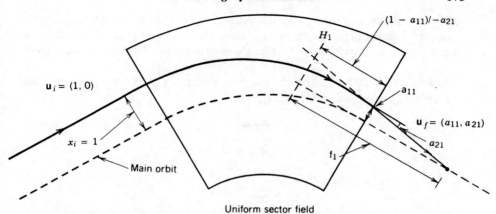

Figure 8.4 Horizontal particle motion in a uniform-field sector magnet. Relationship between the elements of the transfer matrix and the principal planes and focal lengths of Gaussian optics.

decelerating lens is the inverse of **A**. Applying Eq. (8.22) and inverting signs so that the particle travels in the $+z$ direction,

$$\mathbf{A}_d = \begin{bmatrix} 1 & 0 \\ -\xi/f_a & \xi \end{bmatrix}. \tag{8.26}$$

In the thin-lens limit, the accelerating and decelerating focal lengths are related by $f_d = f_a/\xi$.

To conclude the discussion of transfer matrices, we consider how the four components of transfer matrices are related to the focal lengths and principal planes of Gaussian optics (Chapter 6). Consider the uniform sector field of Figure 8.4. This acts as a thick lens with a curved main axis. An orbit vector $(1, 0)$ is incident from the left. The relationship between the emerging orbit and the matrix components as well as the focal length and principal plane H_1 are indicated on the figure. Applying the law of similar triangles, the focal length is given by $f_1 = -1/a_{11}$. The principal plane is located a distance $z_1 = (1 - a_{11})/a_{21}$ from the boundary. Thus, the components a_{11} and a_{21} are related to f_1 and H_1. When the matrix is inverted, the components a_{12} and a_{22} move to the first column. They are related to f_2 and H_2 for particles traveling from right to left. The matrix and Gaussian descriptions of linear lenses are equivalent. Lens properties are completely determined by four quantities.

8.3 COMBINING OPTICAL ELEMENTS

Matrix algebra makes it relatively easy to find the cumulative effect of a series of transport devices. A single optical element operates on an entrance orbit vector, $\mathbf{u}_0$, changing it to an exit vector, $\mathbf{u}_1$. This vector may be the entrance vector to another element, which subsequently changes it to $\mathbf{u}_2$. By the superposition property of linear systems, the combined action of the two elements can be represented by a single matrix that transforms $\mathbf{u}_0$ directly to $\mathbf{u}_2$.

If $\mathbf{A}$ is the transfer matrix for the first element and $\mathbf{B}$ for the second, the process can be written symbolically,

$$\mathbf{u}_1 = \mathbf{A}\mathbf{u}_0,$$

$$\mathbf{u}_2 = \mathbf{B}\mathbf{u}_1 = \mathbf{B}(\mathbf{A}\mathbf{u}_0),$$

or

$$\mathbf{u}_2 = \mathbf{C}\mathbf{u}_0.$$

The matrix $\mathbf{C}$ is a function of $\mathbf{A}$ and $\mathbf{B}$. The functional dependence is called *matrix multiplication* and is denoted $\mathbf{C} = \mathbf{B}\mathbf{A}$. The rule for multiplication of two 2×2 matrices is

$$\mathbf{C} = \begin{bmatrix} c_{11} & c_{12} \\ c_{21} & c_{22} \end{bmatrix} = \begin{bmatrix} b_{11} & b_{12} \\ b_{21} & b_{22} \end{bmatrix} \begin{bmatrix} a_{11} & a_{12} \\ a_{21} & a_{22} \end{bmatrix}, \tag{8.27}$$

where

$$c_{11} = b_{11}a_{11} + b_{12}a_{21}, \qquad c_{12} = b_{11}a_{12} + b_{12}a_{22},$$

$$c_{21} = b_{21}a_{11} + b_{22}a_{21}, \qquad \text{and } c_{22} = b_{21}a_{21} + b_{22}a_{22}.$$

We shall verify the validity of Eq. (8.27) for the example illustrated in Figure 8.5. The optical system consists of two one-dimensional elements, a thin

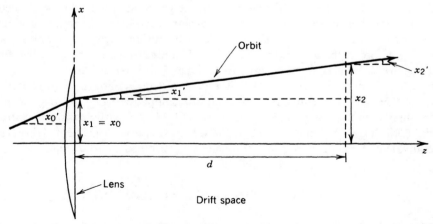

Figure 8.5 Orbit of a particle passing through a thin lens of focal length f and along a drift distance d.

lens with focal length f followed by a drift space d. The particle entrance orbit is (x_0, x_0'). The position and angle emerging from the lens are $x_1 = x_0$ and $x_1' = x_0' - x_0/f$. Traveling through the drift space, the orbit angle remains constant but the displacement changes by an amount $\Delta x = x_1'd$. The total transformation is

$$x_2 = x_0 + (x_0' - x_0/f)d = x_0(1 - d/f) + x_0'd, \quad (8.28)$$

$$x_2' = x_0' - x_0/f. \quad (8.29)$$

Inspection of Eqs. (8.28) and (8.29) yields the 2×2 transformation matrix,

$$\mathbf{C} = \begin{bmatrix} 1 - d/f & d \\ -1/f & 1 \end{bmatrix} = \begin{bmatrix} 1 & d \\ 0 & 1 \end{bmatrix}\begin{bmatrix} 1 & 0 \\ -1/f & 1 \end{bmatrix}. \quad (8.30)$$

We can easily verify that $\mathbf{C}$ is the matrix product of Eq. (8.11) by Eq. (8.10).

It is important to note that the mathematic order of matrix multiplication must replicate the geometric order in which the elements are encountered. Matrix multiplication is not commutative, so that $\mathbf{AB} \neq \mathbf{BA}$. The inequality can be demonstrated by calculating the transfer matrix for a drift space followed by a lens. The effect of this combination is not the same as a lens followed by a drift space. Consider a parallel orbit entering the two systems. In the drift–lens geometry, the particle emerges at the same position it entered. In the second combination, the final position will be different. Multiplying transfer matrices in the improper order is a frequent source of error. To reiterate, if a particle travels in sequence through elements represented by $\mathbf{A}_1, \mathbf{A}_2, \ldots, \mathbf{A}_n$, then the combination of these elements is a matrix given by

$$\mathbf{C} = \mathbf{A}_n\mathbf{A}_{n-1} \cdots \mathbf{A}_2\mathbf{A}_1. \quad (8.31)$$

The astigmatic focusing property of quadrupole doublets (Section 8.4) is an important consequence of the noncommutative property of matrix multiplication.

We can use matrix algebra to investigate the imaging property of a one-dimensional thin lens. The proof that a thin lens can form an image has been deferred from Section 6.4. The optical system consists of a drift space of length d_2, a lens with focal length f, and another drift space (d_1) (see Fig. 6.7). The vectors (x_0, x_0') and (x_3, x_3') represent the orbits in the planes σ_1 and σ_2. The planes are object and image planes if all rays that leave a point in σ_1 pass through a corresponding point in σ_2, regardless of the orbit angle. An equivalent statement is that x_3 is a function of x_0 with no dependence on x_0'. The transfer matrix for the system is

$$\mathbf{C} = \begin{bmatrix} 1 & d_1 \\ 0 & 1 \end{bmatrix}\begin{bmatrix} 1 & 0 \\ -1/f & 1 \end{bmatrix}\begin{bmatrix} 1 & d_2 \\ 0 & 1 \end{bmatrix}$$

$$= \begin{bmatrix} 1 - d_1/f & d_1 + d_2 - d_1d_2/f \\ -1/f & 1 - d_2/f \end{bmatrix}. \quad (8.32)$$

The position of the output vector in component form is $x_3 = c_{11}x_0 + c_{12}x_0'$. An image is formed if $c_{12} = 0$. This is equivalent to $1/f = (1/d_1) + (1/d_2)$. This is the thin-lens formula of Eq. (6.15). When the image condition holds, $M = x_2/x_1 = c_{11}$.

8.4   QUADRUPOLE DOUBLET AND TRIPLET LENSES

A quadrupole lens focuses in one coordinate direction and defocuses in the other. A single lens cannot be used to focus a beam to a point or to produce a two-dimensional image. Two-dimensional focusing can be accomplished with combinations of quadrupole lenses. We will study the focal properties of two (doublets) and three quadrupole lenses (triplets). Quadrupole lens combinations form the basis for most high-energy particle transport systems. They occur as extended arrays or as discrete lenses for final focus to a target. Quadrupole lens combinations are convenient to describe since transverse motions are separable in x and y if the poles (electrodes) are aligned with the axes as shown in Figures 4.14 (for the electrostatic lens) and 5.16 (for the magnetic lens). A 2×2 matrix analysis can be applied to each direction.

The magnetic quadrupole doublet is illustrated in Figure 8.6. We shall consider identical lenses in close proximity, neglecting the effects of gaps and edge fields. It is not difficult to extend the treatment to a geometry with a drift space between the quadrupoles. Relative to the x direction, the first element is

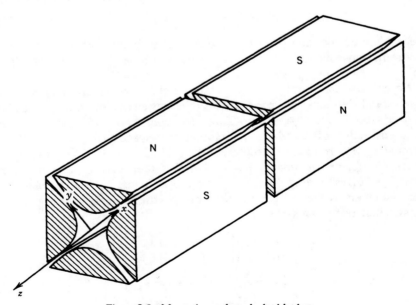

Figure 8.6   Magnetic quadrupole doublet lens.

focusing and the second is defocusing. This is represented symbolically as

$$\boxed{F \;\big|\; D}.$$

where particles move from left to right. Conversely, in the y direction the doublet is denoted

$$\boxed{D \;\big|\; F}.$$

The transfer matrices for the combination of the two elements can be found by matrix multiplication of Eqs. (8.4) and (8.6). The multiplication must be performed in the proper order. The result for an FD channel is

$$\mathbf{C}_{FD} = \begin{bmatrix} \cos\Gamma\cosh\Gamma - \sin\Gamma\sinh\Gamma & (\cosh\Gamma\sin\Gamma + \cos\Gamma\sin h\Gamma)/\sqrt{\kappa} \\ (\cos\Gamma\sinh\Gamma - \cosh\Gamma\sin\Gamma)\sqrt{\kappa} & \cos\Gamma\cosh\Gamma + \sin\Gamma\sinh\Gamma \end{bmatrix},$$

$$(8.33)$$

where $\Gamma = \sqrt{\kappa}\,l$. Similarly, for a DF channel,

$$\mathbf{C}_{DF} = \begin{bmatrix} \cos\Gamma\cosh\Gamma + \sin\Gamma\sinh\Gamma & (\cosh\Gamma\sin\Gamma + \cos\Gamma\sinh\Gamma)/\sqrt{\kappa} \\ (\cos\Gamma\sinh\Gamma - \cosh\Gamma\sin\Gamma)\sqrt{\kappa} & \cos\Gamma\cosh\Gamma - \sin\Gamma\sinh \end{bmatrix}.$$

$$(8.34)$$

Equations (8.33) and (8.34) have two main implications. First, the combination of equal defocusing and focusing elements leads to net focusing, and, second, focusing is different in the x and y directions. As we found in the previous section, the term c_{21} of the transfer matrix determines whether the lens is focusing or defocusing. In this case, $c_{21} = \cos\sqrt{\kappa}\,l\,\sinh\sqrt{\kappa}\,l - \cosh\sqrt{\kappa}\,l\,\sin\sqrt{\kappa}\,l$. We can verify by direct computation that $c_{21} = 0$ at $\sqrt{\kappa}\,l = 0$, and it is a monotonically decreasing function for all positive values of $\sqrt{\kappa}\,l$. The reason for this is illustrated in Figure 8.7, which shows orbits in the quadrupoles for the FD and DF directions. In both cases, the orbit displacement is larger in the focusing section than in the defocusing section; therefore, the focusing action is stronger. Figure 8.7 also shows that the focal points in the x and y directions are not equal. An initially parallel beam is compressed to a line rather than a point in the image planes. A lens with this property is called *astigmatic*. The term comes from the Latin word *stigma*, meaning a small mark. A lens that focuses equally in both directions can focus to a point or produce a two-dimensional image. Such a lens is called *stigmatic*. The term *anastigmatic* is also used. Astigmatism in the doublet arises from the displacement term c_{11}. Although initially parallel orbits emerge from FD and DF

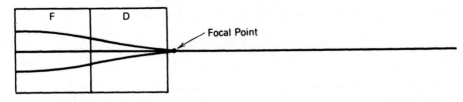

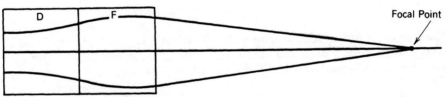

Figure 8.7 Astigmatism in a doublet lens. Orbits of particles initially parallel to the axis projected in the x and y planes.

doublets with the same angle, the displacement is increased in the DF combination, and decreased in the FD.

The transfer matrix for a three-element optical system consisting of a drift space of length $\frac{1}{2}l$, a thin lens with focal length f, and another drift space is

$$\mathbf{A} = \begin{bmatrix} 1 - l/2f & l - l^2/4f \\ -1/f & 1 - l/2f \end{bmatrix}. \tag{8.35}$$

Comparison of Eq. (8.35) with Eqs. (8.7) and (8.8) shows a correspondence if we take $f = 1/\pm \kappa l$. Thus, to order $(\sqrt{\kappa}\, l)^2$, quadrupole elements can be replaced by a drift space of length l with a thin lens at the center. This construction often helps to visualize the effect of a series of quadrupole lenses. A similar power series approximation can be found for the total ray transfer matrix of a doublet. Combining Eqs. (8.7) and (8.8) by matrix multiplication

$$\mathbf{C}_{DF} = \begin{bmatrix} 1 + \kappa l^2 & 2l \\ -2\kappa^2 l^3/3 & 1 - \kappa l^2 \end{bmatrix}. \tag{8.36}$$

$$\mathbf{C}_{FD} = \begin{bmatrix} 1 - \kappa l^2 & 2l \\ -2\kappa^2 l^3/3 & 1 + \kappa l^2 \end{bmatrix}. \tag{8.37}$$

Equations (8.36) and (8.37) are correct to order $(\sqrt{\kappa}\, l)^4$.

Stigmatism can be achieved with quadrupoles in a configuration called the triplet. This consists of three quadrupole sections. The entrance and exit sections have the same length ($l/2$) and pole orientation, while the middle

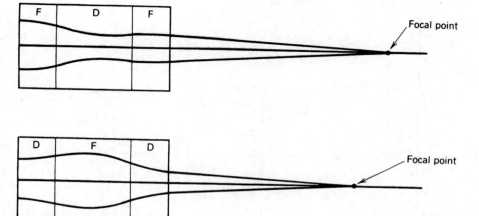

Figure 8.8 Improved stigmatic properties of a quadrupole triplet lens. Orbits of particles initially parallel to the axis projected in the x and y planes.

section is rotated 90° and has length l. Orbits in the x and y planes of the triplet are illustrated in Figure 8.8. An exact treatment (using the trigonometric-hyperbolic forms of the transfer matrices) shows that the exit displacements are identical in both planes for equal entrance displacements. The power series expansions [Eqs. (8.7) and (8.8)] can be used to show that the exit angles are approximately equal. When the calculation is carried out, it is found that all terms of order $(\sqrt{\kappa}\, l)^2$ mutually cancel from the total matrix. The following result holds for both the FDF and DFD combinations:

$$\mathbf{C}_{\text{triplet}} = \begin{bmatrix} 1 & 2l \\ -\kappa^2 l^3/6 & 1 \end{bmatrix}. \tag{8.38}$$

Equation (8.38) is accurate to order $(\sqrt{\kappa}\, l)^4$.

8.5 FOCUSING IN A THIN-LENS ARRAY

As an introduction to periodic focusing, we shall study the thin-lens array illustrated in Figure 8.9. Orbits in this geometry can be determined easily. The focusing cell boundaries can have any location as long as they define a periodic collection of identical elements. We will take the boundary at the exit of a lens. A focusing cell consists of a drift space followed by a lens, as shown in Figure 8.9.

The goal is to determine the positions and angles of particle orbits at cell boundaries. The following equations describe the evolution of the orbit parameters traveling through the $(n + 1)$th focusing cell in the series [see Eqs. (8.10)

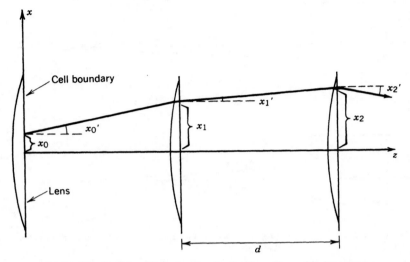

Figure 8.9   Particle orbit in first three cells of a uniform thin-lens array.

and (8.11)]. The subscript n denotes the orbit parameter at the exit of the nth focusing cell:

$$x_{n+1} = x_n + dx'_n, \qquad (8.39)$$

$$x'_{n+1} = x'_n - x_{n+1}/f. \qquad (8.40)$$

Equation (8.39) can be solved for x'_n,

$$x'_n = (x_{n+1} - x_n)/d. \qquad (8.41)$$

Equation (8.41) can be substituted in Eq. (8.40) to yield

$$x'_{n+1} = [(1 - d/f)x_{n+1} - x_n]/d. \qquad (8.42)$$

Finally, an equation similar to Eq. (8.41) can be written for the transition through the $(n + 2)$th focusing cell,

$$x'_{n+1} = (x_{n+2} - x_{n+1})/d. \qquad (8.43)$$

Setting Eqs. (8.42) and (8.43) equal gives the difference equation

$$x_{n+2} - 2(1 - d/2f)x_{n+1} + x_n = 0. \qquad (8.44)$$

This is the finite difference equivalent of a second-order differential equation. We found in Section 4.2 that the finite difference approximation to the second derivative of a function involves the values of the function at three adjacent mesh points.

We seek a mathematical solution of Eq. (8.44) in the form

$$x_n = x_0 \exp(jn\mu).\tag{8.45}$$

Defining $b = 1 - d/2f$ and substituting in Eq. (8.44),

$$\exp[j(n+2)\mu] - 2b \exp[j(n+1)\mu] + \exp(jn\mu) = 0$$

or

$$\exp(2j\mu) - 2b \exp(j\mu) + 1 = 0.\tag{8.46}$$

Applying the quadratic formula, the solution of Eq. (8.46) is

$$\exp(j\mu) = b \pm j(1 - b^2)^{1/2}.\tag{8.47}$$

The complex exponential can be rewritten as

$$\exp(j\mu) = \cos\mu + j \sin\mu = \cos\mu + j(1 - \cos^2\mu)^{1/2}.\tag{8.48}$$

Comparing Eqs. (8.47) and (8.48), we find that

$$\mu = \pm\cos^{-1}b = \pm\cos^{-1}(1 - d/2f).\tag{8.49}$$

The solution of Eq. (8.47) is harmonic when $|b| \le 1$. The particle displacement at the cell boundaries is given by the real part of Equation 8.45,

$$x_n = x_0 \cos(n\mu + \phi).\tag{8.50}$$

Equation (8.50) gives the displacement measured at a cell boundary. It does not imply that particle orbits between boundaries are harmonic. In the thin-lens array, it is easy to see that orbits are straight lines that connect the points of Eq. (8.45). The quantity μ is called the *phase advance* in a focusing cell. The meaning of the phase advance is illustrated in Figure 8.10. The case shown has $\mu = 2\pi/7 = 51.4°$. Particle orbits between cell boundaries in other focusing systems are generally not straight lines. Orbits in quadrupole lens arrays are more complex, as we will see in Section 8.7.

The orbit angle at cell boundaries can be determined by substituting Eq. (8.45) in Eq. (8.41). The result is

$$x_n' = (x_0/d)\{\exp[j(n+1)\mu] - \exp(jn\mu)\}$$

$$= (x_0/d)\{\exp(jn\mu)[\exp(j\mu) - 1]\}.\tag{8.51}$$

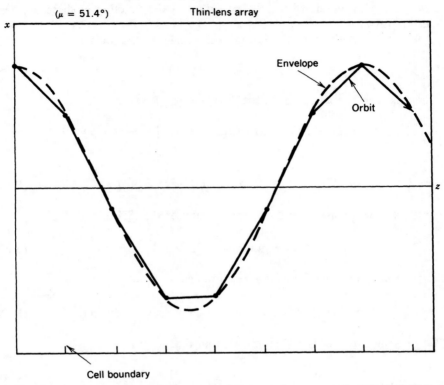

Figure 8.10 Particle orbit in a thin-lens array with phase advance per cell of 51.4°. Solid line: Actual particle orbit. Dotted line: Envelope function to calculate particle displacement at cell boundaries.

Note that when $\mu \to 0$, particle orbits approach the continuous envelope function $x(z) = x_0 \cos(\mu z/d + \phi)$. In this limit, the last factor in Eq. (8.51) is approximately $j\mu$ so that the orbit angle becomes

$$x'_n \cong \mathrm{Re}(jx_0\mu/d)[\exp(jn\mu)] \cong -(x_0\mu/d)\sin(n\mu z/d). \qquad (8.52)$$

An important result of the thin-lens array derivation is that there are parameters for which all particle orbits are unstable. Orbits are no longer harmonic but have an exponentially growing displacment when

$$|b| = |1 - d/2f| > 1. \qquad (8.53)$$

Setting $1 - d/2f$ equal to -1 gives the following stability condition for transport in a thin-lens array,

$$f \geq \tfrac{1}{4}d \quad \text{(stability)}. \qquad \boxed{8.54}$$

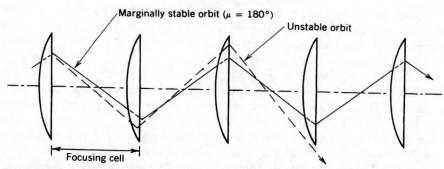

Figure 8.11 Orbital instability in a thin-lens array. Solid-line: Marginally stable orbit ($\mu \to 180°$). Dashed line: Unstable orbit resulting from a slight decrease of lens focal length.

There is a maximum usable lens strength for a given cell length. The physical interpretation of this instability can be visualized by reference to Figure 8.11. The lens system has $f = \frac{1}{4}d$ so that the particle orbit is marginally stable. In this case, $b = -1$ and $\mu = 180°$. The orbit crosses the boundary with a displacement of equal magnitude but opposite sign. If the focusing strength of the lens is increased slightly ($f < \frac{1}{4}d$), then the particle has an increased magnitude of displacement at each cell boundary as shown by the dotted line. The amplitude of displacement increases without limit.

8.6 RAISING A MATRIX TO A POWER

We want to generalize the treatment of the previous section to investigate orbits and stability properties for any linear focusing system. Focusing cells may be complex in high-energy accelerators. They may include quadrupole lenses, bending magnets, gradient fields, and edge focusing. Nonetheless, the net effect of a focusing cell can be represented by a single 4×4 transfer matrix no matter how many sequential elements it contains. When transverse motions along Cartesian coordinates are decoupled, each direction is separately characterized by a 2×2 matrix.

A periodic focusing system consists of a series of identical cells. We shall restrict consideration to transport in the absence of acceleration. This applies directly to storage rings, beamlines, and electron confinement in a microwave tube. It is also a good approximation for accelerators if the betatron wavelength is short compared to the distance over which a particle's energy is doubled.

If a particle enters a periodic channel with an orbit vector $\mathbf{u}_0$, then the vector at the boundary between the first and second focusing cells is $\mathbf{u}_1 = \mathbf{C}\mathbf{u}_0$. The quantity $\mathbf{C}$ is the transfer matrix for a cell. After traversing n cells, the orbit vector is

$$\mathbf{u}_n = \mathbf{C}^n\mathbf{u}_0. \tag{8.55}$$

The quantity $\mathbf{C}^n$ denotes the matrix multiplication of $\mathbf{C}$ by itself n times. The behavior of particle orbits in periodic focusing systems is determined by the matrix power operation. In particular, if all components of $\mathbf{C}^n$ are bounded as $n \to \infty$, then particle orbits are stable.

Analytic expressions for every small powers of a matrix can rapidly become unwieldy. The involved terms encountered in matrix multiplication are evident in Eqs. (8.33) and (8.34). We must use new methods of analysis to investigate the matrix power operation. We will concentrate on 2×2 matrices; the extension to higher-order matrices involves more algebra but is conceptually straightforward. We have already encountered the determinant of a matrix in Section 8.2. The determinant of a transfer matrix is always equal to unity when there is no acceleration. Another useful quantity is the *trace* of a matrix, defined as the sum of diagonal elements. To summarize, if

$$\mathbf{C} = \begin{bmatrix} c_{11} & c_{12} \\ c_{21} & c_{22} \end{bmatrix},$$

then

$$\det \mathbf{C} = c_{11}c_{22} - c_{12}c_{21} \qquad \boxed{8.56}$$

and

$$\mathrm{Tr}\, \mathbf{C} = c_{11} + c_{22}. \qquad \boxed{8.57}$$

Transfer matrices have *eigenvalues* and *eigenvectors*. These quantities are defined in the following way. For most square matrices, there are orbit vectors and numerical constants that satisfy the equation

$$\mathbf{C}v_i = \lambda_i v_i. \qquad (8.58)$$

The vectors for which Eq. (8.58) is true are called eigenvectors (characteristic vectors) of the matrix. The numerical constants (which may be complex numbers) associated with the vectors are called eigenvalues.

The following results are quoted without proof from the theory of linear algebra. The order of a square matrix is the number of rows (or columns). A square matrix of order m with nonzero determinant has m eigenvectors and m different eigenvalues. The eigenvectors have the property of *orthogonality*. Any m-dimensional orbit vector can be represented as a linear combination of eigenvectors. In the case of a 2×2 transfer matrix, there are two eigenvectors. Any orbit vector at the entrance to a periodic focusing system can be written in the form

$$\mathbf{u}_0 = a_1 v_1 + a_2 v_2. \qquad (8.59)$$

If the orbit given by the input vector of Eq. (8.59) passes through n focusing cells of a periodic system, it is transformed to

$$\mathbf{u}_n = \mathbf{C}^n \mathbf{u}_0 = a_1 \lambda_1^n \mathbf{v}_1 + a_1 \lambda_2^n \mathbf{v}_2. \tag{8.60}$$

Equation (8.60) demonstrates the significance of the eigenvector expansion for determining the power of a matrix. The problem of determining the orbit after a large number of focusing cells is reduced to finding the power of two numbers rather than the power of a matrix. If λ_1^n and λ_2^n are bounded quantities for $n \gg 1$, then orbits are stable in the focusing system characterized by the transfer matrix $\mathbf{C}$.

The eigenvalues for a 2×2 matrix can be calculated directly. Writing Eq. (8.58) in component form for a general eigenvector (ν, ν'),

$$(c_{11} - \lambda)\nu + c_{12}\nu' = 0, \tag{8.61}$$

$$c_{21}\nu + (c_{22} - \lambda)\nu' = 0. \tag{8.62}$$

Multiplying Eq. (8.62) by $c_{12}/(c_{22} - \lambda)$ and subtracting from Eq. (8.61), we find that

$$\frac{(c_{11} - \lambda)(c_{22} - \lambda) - c_{12}c_{21}}{c_{22} - \lambda}\nu = 0. \tag{8.63}$$

Equation (8.63) has a nonzero solution when

$$(c_{11} - \lambda)(c_{22} - \lambda) - c_{12}c_{21} = 0. \tag{8.64}$$

This is a quadratic equation that yields two values of λ. The values can be substituted into Eq. (8.61) or (8.62) to give ν_1' in terms of ν_1 and ν_2' in terms of ν_2. Equation (8.64) can be rewritten

$$\lambda^2 - \lambda \operatorname{Tr} \mathbf{C} + \det \mathbf{C} = 0. \tag{8.65}$$

The solution to Eq. (8.65) can be found from the quadratic formula using the fact that $\det \mathbf{C} = 1$.

$$\lambda_1, \lambda_2 = (\operatorname{Tr} \mathbf{C}/2) \pm \left[(\operatorname{Tr} \mathbf{C}/2)^2 - 1 \right]^{1/2}. \tag{8.66}$$

The product of the two eigenvalues of Eq. (8.66) is

$$\lambda_1 \lambda_2 = (\operatorname{Tr} \mathbf{C}/2)^2 - \left[(\operatorname{Tr} \mathbf{C}/2)^2 - 1 \right] \equiv 1. \tag{8.67}$$

The fact that the product of the eigenvalues of a transfer matrix is identically equal to unity leads to a general condition for orbital stability. We know that the eigenvalues are different if $\det \mathbf{C}$ and $\operatorname{Tr} \mathbf{C}$ are nonzero. If both eigenvalues are real numbers, then one of the eigenvalues must have a magnitude greater than unity if the product of eigenvalues equals 1. Assume, for instance, that $\lambda_1 > 1$. The term λ_1 will dominate in Eq. (8.60). The magnitude of the orbit displacement will diverge for a large number of cells so that orbits are unstable. Inspecting Eq. (8.65), the condition for real eigenvalues and instability is $|\text{Tr}\,\mathbf{C}/2| > 1$.

When $|\text{Tr}\,\mathbf{C}/2| \le 1$, the square root quantity is negative in Eq. (8.65) so that the eigenvalues are complex. In this case, Eq. 8.66 can be rewritten

$$\lambda_1, \lambda_2 = (\text{Tr}\,\mathbf{C}/2) \pm j\left[1 - (\text{Tr}\,\mathbf{C}/2)^2\right]^{1/2} \tag{8.68}$$

If we make the formal substitution

$$\text{Tr}\,\mathbf{C}/2 = \cos\mu, \qquad \boxed{8.69}$$

then Eq. (8.68) becomes

$$\lambda_1, \lambda_2 = \cos\mu \pm j\sin\mu = \exp(\pm j\mu). \tag{8.70}$$

Euler's formula was applied to derive the final form. The eigenvalues to the nth power are $\exp(\pm nj\mu)$. This is a periodic trigonometric function. The magnitude of both eigenvalues is bounded for all powers of n. Thus, the orbit displacement remains finite for $n \to \infty$ and the orbits are stable. To reiterate, if the action of a cell of a periodic focusing system is represented by a transfer matrix $\mathbf{C}$, then orbits are stable if

$$|\text{Tr}\,\mathbf{C}/2| \le 1 \quad \text{(stability)}. \qquad \boxed{8.71}$$

This simple rule holds for any linear focusing cell.

We have concentrated mainly on mathematics in this section. We can now go back and consider the implications of the results. The example of the drift space and thin lens will be used to illustrate application of the results. The transfer matrix is

$$\mathbf{C} = \begin{bmatrix} 1 & d \\ -1/f & 1 - d/f \end{bmatrix}. \tag{8.72}$$

The determinant is equal to unity. The trace of the matrix is $\text{Tr}\,\mathbf{C} = 2 - d/f$. Applying Eq. (8.71), the condition for stable orbits is

$$|1 - d/2f| \le 1, \tag{8.73}$$

as we found before. Similarly, $\cos\mu = 1 - d/2f$. Thus, the parameter μ has the same value as Eq. (8.49) and is associated with the phase advance per cell. This holds in general for linear systems. When the orbits are stable and the eigenvalues are given by Eq. (8.70), the orbit vector at the cell boundaries of any linear focusing cell is

$$\mathbf{u}_n = a_1 \mathbf{v}_1 \exp(jn\mu) + a_2 \mathbf{v}_2 \exp(-jn\mu). \tag{8.74}$$

The solution is periodic; orbits repeat after passing through $N = 2\pi/\mu$ cells.

The eigenvectors for the transfer matrix [Eq. (8.72)] can be found by substituting the eigenvalues in Eq. (8.61). The choice of v is arbitrary. This stands to reason since all orbits in a linear focusing system are similar, independent of magnitude, and the eigenvectors can be multiplied by a constant without changing the results. Given a choice of the displacement, the angle is

$$v' = (\lambda - c_{11})v/c_{12}. \tag{8.75}$$

The eigenvectors for the thin-lens array are

$$\mathbf{v}_1 = (1, [\exp(j\mu) - 1]/d), \tag{8.76}$$

$$\mathbf{v}_2 = (1, [\exp(-j\mu) - 1]/d). \tag{8.77}$$

Suppose we wanted to treat the same orbit considered in Section 8.5 [Eq. (8.45)]. The particle enters the first cell parallel to the axis with a displacement equal to x_0. The following linear combination of eigenvectors is used.

$$\mathbf{u}_0 = x_0 \mathbf{v}_1. \tag{8.78}$$

The displacement after passing through n cells is

$$x_n = x_0 \text{Re}[\exp(jn\mu)] = x_0 \cos(n\mu), \tag{8.79}$$

as we found from the finite difference solution. The angle after n cells is

$$x'_n = x_0 \exp(jn\mu)[\exp(j\mu) - 1]/d. \tag{8.80}$$

This is identical to Eq. (8.51). Both models lead to the same conclusion. We can now proceed to consider the more complex but practical case of FD quadrupole focusing cells.

8.7 QUADRUPOLE FOCUSING CHANNELS

Consider the focusing channel illustrated in Figure 8.12. It consists of a series of identical, adjacent quadrupole lenses with an alternating 90° rotation. The cell boundary is chosen so that the cell consists of a defocusing section followed by focusing section (DF) for motion in the x direction. The cell is therefore represented as FD in the y direction. Note that individual lenses do not comprise a focusing cell. The smallest element of periodicity contains both a focusing and a defocusing lens. The choice of cell boundary is arbitrary. We could have equally well chosen the boundary at the entrance to an F lens so that the cell was an FD combination in the x direction. Another valid choice is to locate the boundary in the middle of the F lens so that focusing cells are quadrupole triplets, FDF. Conclusions related to orbital stability do not depend on the choice of cell boundary.

The transfer matrix for a cell is the product of matrices $\mathbf{C} = \mathbf{A}_F\mathbf{A}_D$. The individual matrices are given by Eqs. (8.4) and (8.6). Carrying out the multiplication,

$$
\mathbf{C}_{DF} = \begin{bmatrix} \cos\Gamma\cosh\Gamma + \sin\Gamma\sinh\Gamma & (\cos\Gamma\sinh\Gamma + \sin\Gamma\cosh\Gamma)/\sqrt{\kappa} \\ \sqrt{\kappa}\,(\cos\Gamma\sinh\Gamma - \sin\Gamma\cosh\Gamma) & \cos\Gamma\cosh\Gamma - \sin\Gamma\sinh\Gamma \end{bmatrix},
$$

$$(8.81)$$

where $\Gamma = \sqrt{\kappa}\,l$. Taking $|\mathrm{Tr}\,\mathbf{C}/2| < 1$, the condition for stable orbits is

$$
-1 \le \cos\Gamma\cosh\Gamma \le +1. \qquad \boxed{8.82}
$$

Figure 8.13 shows a plot of $f(\Gamma) = \cos\Gamma\cosh\Gamma$ versus Γ. Only positive values

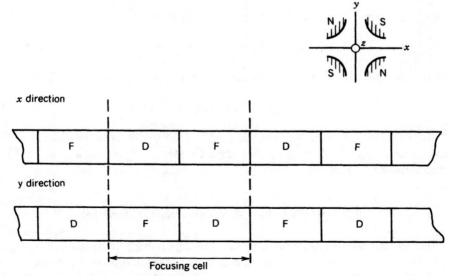

Figure 8.12 Array of quadrupole lenses in an FD configuration. Action of lenses indicated for motion in x and y directions.

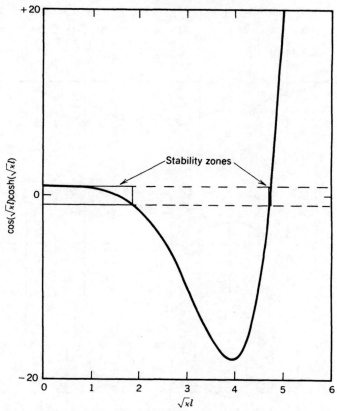

Figure 8.13 Orbital stability in an FD quadrupole array. Plot of $\cos(\sqrt{\kappa}\,l)\cosh(\sqrt{\kappa}\,l)$ versus $\sqrt{\kappa}\,l$.

of Γ have physical meaning. Orbits are stable for Γ in the range

$$0 \leq \Gamma \leq 1.86. \qquad \boxed{8.83}$$

A stable orbit (calculated by direct solution of the equation of motion in the lens fields) is plotted in Figure 8.14a. The orbit has $\Gamma = 1$, so that $\mu = 0.584 = 33.5°$. Higher values of Γ correspond to increased lens strength for the same cell length. Orbits are subject to the same type of overshoot instability found for the thin-lens array [Eq. (8.53)]. An unstable orbit with $\Gamma = 1.9$ is plotted in Figure 8.14b. At higher values of Γ, there are regions of Γ in which stable propagation can occur. Figure 8.14c illustrates a stable orbit with $\Gamma = 4.7$ ($\mu = 270°$). Orbits such as those of Figure 8.14c strike a fragile balance between focusing and defocusing in a narrow range of Γ. Higher-order stability bands are never used in real transport systems. For practical purposes, Γ must be in the range indicated by Eq. (8.83). The same result applies to

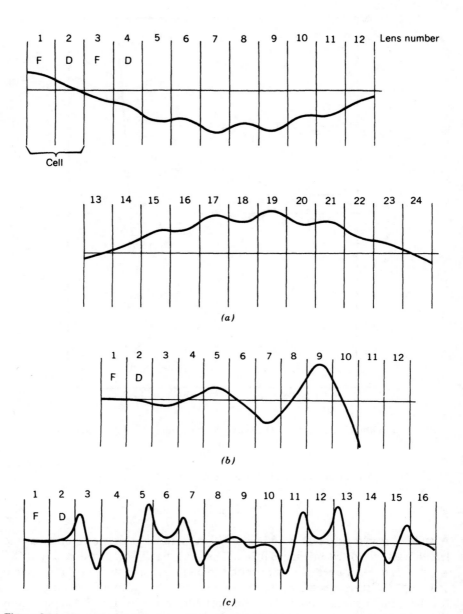

Figure 8.14 Numerical calculations of actual particle orbits in an FD quadrupole array. (*a*) $\sqrt{\kappa}\, l = 1$ (stable orbit, first passband). (*b*) $\sqrt{\kappa}\, l = 1.9$ (unstable orbit). (*c*) $\sqrt{\kappa}\, l = 4.7$ (stable orbit, second passband).

motion in the y direction. The matrix coefficients are different from those of Eq. (8.81), but the quantity Tr $\mathbf{C}/2$ is the same.

Given a range of Γ for stable operation, it remains to determine an optimum value. Small phase advance ($\mu \ll 1$) has advantages. The effect of any one lens is small so that orbits approach the continuous focusing limit (Section 7.1). Beam envelope oscillations in a cell are small, and particle orbits are less affected by errors and misalignments of the lenses. Small μ is an effective choice for an array of lenses with positive focal length, such as the thin-lens array discussed in Section 8.5. The thin-lens array is a good model for a series of solenoidal magnetic lenses or unipotential electrostatic lenses. In such lens arrays, we can show that to first order the focusing strength is independent of μ for given applied fields. The focusing strength is proportional to the average transverse force that the lenses exert on the particles [Eq. (7.6)]. Neglecting edge fields, the focal length of a solenoidal lens is inversely proportional to the length of the lens, d. Thus, the product fd in Eq. (7.6) does not depend on the number of individual lenses per unit axial length. In other words, given a series of solenoids with phase advance μ, focal length f, and lens length d, the acceptance would be unchanged if the lens length and phase advance were halved and the focal length were doubled. Thus, it is generally best to use small phase advance and a fine division of cells in channels with only focusing lenses. The minimum practical cell length is determined by ease of mechanical construction and the onset of nonlinearities associated with the edge fields between lenses.

This conclusions does not apply to FD-type focusing channels. In order to investigate scaling in this case, consider the quadrupole doublet treated using the expansions for small $\sqrt{\kappa}\, l$ [Eqs. (8.7), (8.8), and (8.35)]. In this approximation, the doublet consists of two lenses with focal lengths

$$f_F = +1/\kappa l, \qquad f_D = -1/\kappa l,$$

separated by a distance l. As with the solenoidal lens, the focal lengths of the individual lenses are inversely proportional to the lengths of the lens. The net positive focal length of the combination from Eqs. (8.36) and (8.37) is

$$f_{FD} = 3/2\kappa^2 l^3 = 3|f_F f_D|/2l \sim 1/d^3, \qquad (8.84)$$

where $d = 2l$ is the cell length.

If we divide the quadrupole doublet system into smaller units with the same applied field, the scaling behavior is different from that of the solenoid channel. The average focusing force decreases, since the product $f_{FD}d$ is proportional to d^{-2}. This scaling reflects the fact that the action of an FD combination arises from the difference in average radius of the orbits in the F and D sections. Dividing the lenses into smaller units not only decreases the focusing strength of the F and D sections but also reduces the relative difference in transverse displacement. The conclusion is that FD focusing channels should be designed for the highest acceptable value of phase advance. The value used in practice is well below the stability limit. Orbits in channels with high μ are sensitive to misalignments and field errors. A phase advance of $\mu = 60°$ is usually used.

In many strong focusing systems, alternate cells may not have the same length or focusing strength. This is often true in circular accelerators. This case is not difficult to analyze. Defining $\Gamma_1 = \sqrt{\kappa_1}\, l_1$ and $\Gamma_2 = \sqrt{\kappa_2}\, l_2$, the transfer matrix for motion in the DF direction is

$$
\mathbf{C} = \begin{bmatrix} \cos \Gamma_1 & \sin \Gamma_1/\sqrt{\kappa_1} \\ -\sqrt{\kappa_1}\sin \Gamma_1 & \cos \Gamma_1 \end{bmatrix} \begin{bmatrix} \cosh \Gamma_2 & \sinh \Gamma_2/\sqrt{\kappa_2} \\ \sqrt{\kappa_2}\sinh \Gamma_2 & \cosh \Gamma_2 \end{bmatrix}.
$$

(8.85)

Performing the matrix multiplication and taking $\mathrm{Tr}\,\mathbf{C}/2$ gives the following phase advance:

$$
\cos \mu_{\mathrm{DF}} = \cos \Gamma_1 \cosh \Gamma_2 + \left(\frac{\sin \Gamma_1 \sinh \Gamma_2}{2} \right)\left(\frac{\Gamma_2 l_1}{\Gamma_1 l_2} - \frac{\Gamma_1 l_2}{\Gamma_2 l_1} \right). \quad (8.86)
$$

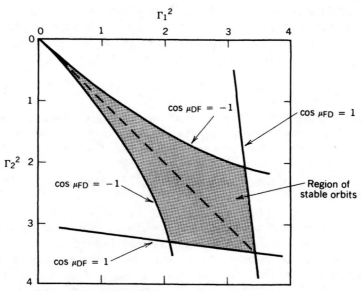

Figure 8.15 Necktie diagram. Orbital stability in an FD (DF) quadrupole array; the two lenses of a cell have unequal focusing strength. $\Gamma_1 = \sqrt{\kappa_1}\, l_1$, $\Gamma_2 = \sqrt{\kappa_2}\, l_2$. Region of parameter space with orbital stability in both x and y directions shaded. FD (lens 1 focusing, lens 2 defocusing), DF (lens 1 defocusing, lens 2 focusing).

Since the two lenses of the cell have unequal effects, there is a different phase advance for the FD direction, given by

$$\cos \mu_{\text{FD}} = \cosh \Gamma_1 \cos \Gamma_2 + \left(\frac{\sinh \Gamma_1 \sin \Gamma_2}{2} \right) \left(\frac{\Gamma_1 l_2}{\Gamma_2 l_1} - \frac{\Gamma_2 l_1}{\Gamma_1 l_2} \right). \qquad (8.87)$$

There is little difficulty deriving formulas such as Eqs. (8.86) and (8.87). Most

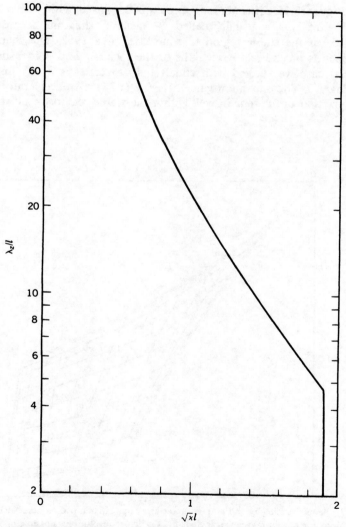

Figure 8.16 Betatron wavelength in a quadrupole focusing channel normalized to the length of a lens as a function of $\sqrt{\kappa_2}\, l$. FD channel with uniform lenses.

of the problem is centered on plotting and interpreting the results. There are stability conditions in two directions that must be satisfied simultaneously for stable orbits:

$$-1 \leq \cos \mu_{FD} \leq +1, \tag{8.88}$$

$$-1 \leq \cos \mu_{DF} \leq +1. \tag{8.89}$$

The stability results are usually plotted in terms of Γ_1^2 and Γ_2^2 in a diagram such as Figure 8.15. This region of parameter space with stable orbits is shaded. Figure 8.15 is usually called a "necktie" diagram because of the resemblance of the stable region to a necktie, circa 1952. The shape of the region depends on the relative lengths of the focusing and defocusing lenses. The special case we studied with equal lens properties is a 45° line on the $l_1 = l_2$ diagram. The maximum value of Γ^2 is $(1.86)^2$. An accelerator designer must ensure that orbits remain well within the stable region for all regimes of

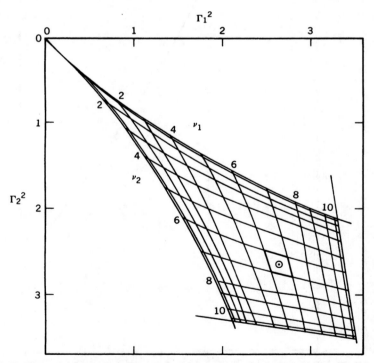

Figure 8.17 Beam focusing by an FD quadrupole array in a circular accelerator. Necktie diagram with conditions for orbital resonances included (24 cells per revolution). Possible operating point indicated at $\nu = 6.4$. (M. S. Livingston and J. P. Blewett, *Particle Accelerators*, used by permission, McGraw-Hill Book Co.).

operation. This is a particular concern in a synchrotron where the energy of particles and strength of the focusing field varies during the acceleration cycle.

The betatron wavelength for orbits in a quadrupole channel is

$$\lambda_z = (2\pi d/\mu), \tag{8.90}$$

where d is the cell length and $d = 2l$. When F and D cells have the same parameters, Eq. (8.90) can be written

$$\lambda_z = 2\pi d/\cos^{-1}(\cos\Gamma\cosh\Gamma). \tag{8.91}$$

The quantity λ_z/l is plotted in Figure 8.16 as a function of Γ. The betatron wavelength is important in circular machines. Particles may suffer resonance instabilities when the circumference of the accelerator or storage ring is an integral multiple of the betatron wavelength (Section 7.3). If we include the possibility of different focusing properties in the horizontal and verticle directions, resonance instabilities can occur when

$$\mu_{DF} = 2\pi n(d/C) \tag{8.92}$$

or

$$\mu_{FD} = 2\pi n(d/C), \tag{8.93}$$

with $n = 1, 2, 3, 4, \ldots$. Equations (8.92) and (8.93) define lines of constant $\cos\mu_{DF}$ or $\cos\mu_{FD}$, some of which are inside the region of general orbital stability. These lines, are included in the necktie diagram of Figure 8.17. They have the same general orientation as the lines $\cos\mu_{DF} = -1$ and $\cos\mu_{DF} = +1$. The main result is that particle orbits in a circular machine with linear FD forces must remain within a quadrilateral region of $\Gamma_1-\Gamma_2$ space. If this is not true over the entire acceleration cycle, orbital resonance instabilities may occur with large attendant particle losses.

9

Electrostatic Accelerators and Pulsed High Voltage

In this chapter we begin the study of charged particle acceleration. Subsequent chapters describe methods for generating high-energy charged particle beams. The kinetic energy of a charged particle is increased by electric fields according to $\Delta T = \int \mathbf{E}(\mathbf{x}, t) \cdot d\mathbf{x}$ [Eq. (3.16)], where the integral is taken along the particle orbit. This equation applies to all accelerators. The types of accelerators discussed in the next six chapters differ by the origin and characteristics of the electric field. In the most general case, $\mathbf{E}$ varies in time and position. In this chapter, we limit consideration to the special case of static electric fields.

Static electric fields can be derived from a potential function with no contribution from time-varying magnetic flux. An electrostatic accelerator consists basically of two conducting surfaces with a large voltage difference V_0. A particle with charge q gains a kinetic energy qV_0. In this chapter, we shall concentrate mainly on methods for generating high voltage. The description of voltage sources, particularly pulsed voltage generators, relies heavily on passive circuit analysis. Section 9.1 reviews the properties of resistors, capacitors, and inductors. A discussion of circuits to generate dc voltage follows in Section 9.2. The equations to describe ideal transformers are emphasized. A thorough knowledge of the transformer is essential to understand linear induction accelerators (Chapter 10) and betatrons (Chapter 11).

Section 9.4 introduces the Van de Graaff voltage generator. This device is used extensively in low-energy nuclear physics. It also provides preacceleration for beams on many high-energy accelerators. The principles of dc accelerators are easily understood; the main difficulties associated with these machines are related to insulation technology, which is largely an empirical field. Properties of insulators are discussed in Section 9.3 and Paschen's law for spark breakdown of an insulating gas is derived. The important subject of vacuum breakdown is reviewed in Section 9.5.

The remainder of the chapter is devoted to techniques of pulsed voltage generation. Acceleration by pulsed voltage accelerators is well described by the electrostatic approximation since the transit time of particles and the propagation time for electromagnetic waves in acceleration gaps are small compared to typical voltage pulselengths ($\Delta t_p \geq 50$ ns). Although the static approximation describes the acceleration gap, the operation of many pulsed power circuits, such as the transmission line, involves electromagnetic wave propagation. Pulsed voltage generators have widespread use in accelerators characterized by cyclic operation. In some instances they provide the primary accelerating power, such as in high-current relativistic electron beam generators and the linear induction accelerator. In other cases, they are used for power conditioning, such as klystron drivers in high-gradient rf electron linear accelerators. Finally, pulsed voltage modulators are necessary to drive pulsed extraction and injection fields for synchrotrons and storage rings.

Pulsed voltage circuits must not only produce a high voltage but must also shape the voltage in time. Sections 9.5–9.14 introduce many of the circuits and techniques used. Material is included on critically damped circuits (Section 9.6), impulse generators (Section 9.7), transmission line modulators (Section 9.9), the Blumlein transmission line (Section 9.10), pulse forming networks (Section 9.11), power compression systems (Section 9.12), and saturable core magnetic switching (Section 9.13). The equations for electromagnetic wave propagation in a transmission line are derived in Section 9.8. This section introduces the principles of interacting electric and magnetic fields that vary in time. The results will be useful when we study electromagnetic wave phenomena in rf accelerators. Material is also included on basic methods to measure fast-pulsed voltage and current (Section 9.14).

9.1 RESISTORS, CAPACITORS, AND INDUCTORS

High-voltage circuits can usually be analyzed in terms of five elements: resistors, capacitors, inductors, ideal diodes, and ideal switches. These elements are two-terminal passive components. Two conductors enter the device, and two quantities characterize it. These are the current through the device and the voltage between the terminals. The action of the circuit element is described by the relationship between voltage and current. Standard symbols and polarity conventions are shown in Figure 9.1.

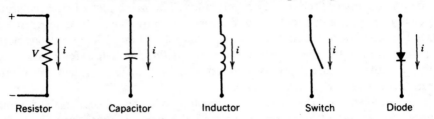

Resistor   Capacitor   Inductor   Switch   Diode

Figure 9.1   Symbols and polarity conventions for common circuit elements.

A. Voltage–Current Relationships

The ideal diode has zero voltage across the terminals when the current is positive and passes zero current when the voltage is negative. Diodes are used for rectification, the conversion of a bipolar voltage waveform into a dc voltage. A switch is either an open circuit (zero current at any voltage) or a closed circuit (zero voltage at any current). In pulsed voltage circuits, a *closing switch* is an open circuit for times $t < 0$ and a short circuit for $t \geq 0$. An *opening switch* has the inverse properties. The spontaneous breakdown gap used on some pulse power generators and pulse sharpening circuits is an example of a two-terminal closing switch. Triggered switches such as the thyratron and silicon-controlled rectifier are more common; they are three terminal, two-state devices.

In this section, we consider the properties of resistors, capacitors, and inductors in the *time-domain*. This means that time variations of the total voltage are related to time variations of the total current. Relationships are expressed as differential equations. It is also possible to treat circuits in the *frequency-domain*. This approach is useful for oscillating harmonic circuits, and will be applied to rf cavities in Chapter 12. In the frequency-domain analysis, time variations are Fourier analyzed in terms of the angular frequency ω. Each harmonic component is treated separately. In the frequency-domain analysis, voltage and current in individual elements and multielement circuits are related simply by $V = a(\omega)I$, where $a(\omega)$ may be a complex number.

A resistor contains material that impedes the flow of electrons via collisions. The flow of current is proportional to the driving voltage

$$I = V/R, \qquad \boxed{9.1}$$

where I is in amperes, V in volts, and R is the resistance in ohms (Ω). Energy is transferred from flowing electrons to the resistive material. With the polarity shown in Figure 9.1, electrons flow into the bottom of the resistor. Each electron absorbs an energy eV_0 from the driving circuit during its transit through the resistor. This energy acts to accelerate the electrons between collisions. They emerge from the top of the resistor with low velocity since most of the energy gained was transferred to the material as heat. The number of electrons passing through the resistor per second is I/e. The power deposited is

$$P \text{ (watts)} = VI = V^2/R = I^2R. \qquad \boxed{9.2}$$

As we saw in Section 5.2, the basic capacitor geometry has two conducting plates separated by a dielectric (Figure 5.6c). The voltage between the plates is proportional to the stored charge on the plates and the geometry of the capacitor:

$$V = Q/C. \qquad (9.3)$$

The quantity V is in volts, Q is in coulombs, and C is in farads (F). Neglecting fringing fields, the capacitance of the parallel plate geometry can be determined from Eq. (3.9):

$$C = \frac{\varepsilon_0(\varepsilon/\varepsilon_0)A}{\delta} \text{ (F)}, \qquad \boxed{9.4}$$

where $\varepsilon_0 = 8.85 \times 10^{-12}$, $\varepsilon/\varepsilon_0$ is the relative dielectric constant of the material between the plates, A is the plate area in square meters, and δ is the plate spacing in meters. Small capacitors in the pF range (10^{-12} F) look much like Figure 5.6c with a dielectric such as Mylar ($\varepsilon/\varepsilon_0 \sim 2$-3). High values of capacitance are achieved by combining convoluted reentrant geometries (for large A) with high dielectric constant materials. The current through a capacitor is the time rate of change of the stored charge. The derivative of Eq. (9.3) gives

$$I = C\,dV/dt. \qquad \boxed{9.5}$$

The capacitor contains a region of electric field. The inductor is configured to produce magnetic field. The most common geometry is the solenoidal winding (Fig. 4.18). The magnetic flux linking the windings is proportional to the current in the winding. The voltage across the terminals is proportional to the time-rate of change of magnetic flux. Therefore,

$$V = L(di/dt), \qquad \boxed{9.6}$$

where L is a constant dependent on inductor geometry. Inductance is measured in Henries (H) in the mks system.

B. Electrical Energy Storage

A resistor converts electrical to thermal energy. There is no stored electrical energy that remains in a resistor when the voltage supply is turned off. Capacitors and inductors, on the other hand, store electrical energy in the form of electric and magnetic fields. Electrical energy can be extracted at a latter time to perform work. Capacitors and inductors are called *reactive* elements (i.e., the energy can act again).

We can prove that there is no average energy lost to a reactive element from any periodic voltage waveform input. Voltage and current through the element can be resolved into Fourier components. Equations (9.5) and (9.6) imply that the voltage and current of any harmonic component are 90° out of phase. In other words, if the voltage varies as $V_0\cos(\omega t)$, the current varies as $\pm I_0\sin(\omega t)$. Extending the arguments leading to Eq. (9.2) to reactive elements, the total energy change in an element is

$$\Delta U = \int P\,dt = I_0 V_0 \int \cos(\omega t)\sin(\omega t).$$

Although the energy content of a reactive element may change over an oscillation period, the average over many periods is zero.

Energy is stored in a capacitor in the form of electric fields. Multiplying Eq. (5.19) by the volume of a parallel-plate capacitor, the stored energy is

$$U_e = \tfrac{1}{2}\left[\varepsilon_0(\varepsilon/\varepsilon_0)(V/\delta)^2\right]A\delta. \tag{9.7}$$

Comparing Eq. (9.7) to (9.4),

$$U_e = \tfrac{1}{2}CV^2. \boxed{9.8}$$

The magnetic energy stored in an inductor is

$$U_m = \tfrac{1}{2}LI^2. \boxed{9.9}$$

C. Common Capacitor and Inductor Geometries

The coaxial capacitor (Fig. 9.2*a*) is often used as an energy storage device for high-voltage pulsed power generators. The electrodes are cylinders of length d with radii R_0 and R_i. The cylinders have a voltage difference V_0, and there is a medium with relative dielectric constant $(\varepsilon/\varepsilon_0)$ between them. Neglecting fringing fields at the ends, the electric field in the dielectric region is

$$E_r = Q/2\pi\varepsilon r d. \tag{9.10}$$

The integral of Eq. (9.10) from R_i to R_0 equals V_0, or

$$V_0 = (Q/2\pi\varepsilon d)\ln(R_0/R_i). \tag{9.11}$$

Comparing Eq. (9.11) with (9.3) shows the capacitance is

$$C = 2\pi\varepsilon d/\ln(R_0/R_i). \boxed{9.12}$$

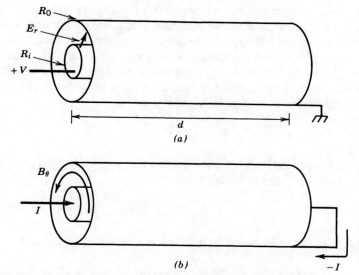

Figure 9.2 Lumped circuit elements. (*a*) Coaxial capacitor. (*b*) Coaxial inductor.

All current-carrying elements produce magnetic field and thus have an inductance. There is a magnetic field between the cylinders of Figure 9.2*b* if current flows along the center conductor. An equal and opposite current must return along the outer conductor to complete the circuit; therefore, there is no magnetic field for $r > R_0$. Equation (4.40) specifies that the field between the cylinders is

$$B_\theta(r) = \mu_0 I/2\pi r \tag{9.13}$$

if there is no ferromagnetic material in the intercylinder volume. The total magnetic field energy is

$$U_m = d \int dr (2\pi r)(B_\theta^2/2\mu_0)$$

$$= (\mu_0/2\pi)(I^2/2)(d \ln(R_0/R_i)). \tag{9.14}$$

Setting U_m equal to $\frac{1}{2}LI^2$ implies that

$$L = (\mu_0/2\pi)d \ln(R_0/R_i). \boxed{9.15}$$

Equation (9.15) should be multiplied by μ/μ_0 if the coaxial region contains a ferromagnetic material with an approximately linear response.

Coaxial inductors generally have between 0.1 and 1 μH per meter. Higher inductances are produced with the solenoidal geometry discussed in Section 4.6. We shall make the following assumptions to calculate the inductance of a solenoid: edge fields and curvature effects are neglected, the winding is completely filled with a linear ferromagnetic material, and N series windings of cross-sectional area A are uniformly spaced along a distance d. The magnetic field inside the winding for a series current I is

$$B_z = \mu I (N/d) \tag{9.16}$$

according to Eq. (4.42). Faraday's law implies that a time variation of magnetic field linking the windings produces a voltage

$$V = NA(dB_z/dt). \tag{9.17}$$

We can identify the inductance by combining Eqs. (9.16) and (9.17):

$$L = \mu_0(\mu/\mu_0)N^2A/d. \qquad \boxed{9.18}$$

Compact solenoids can assume a wide range of inductance values because of the strong scaling with N.

D.   Introductory Circuits

Figure 9.3 shows a familiar circuit combining resistance, capacitance, and a switch. This is the simplest model for a pulsed voltage circuit; electrical energy is stored in a capacitor and them dumped into a load resistor via a switch. Continuity of current around the circuit combined with Eqs. (9.1) and (9.5) implies the following differential equation for the load voltage after switching:

$$C(dV/dt) + V/R = 0. \tag{9.19}$$

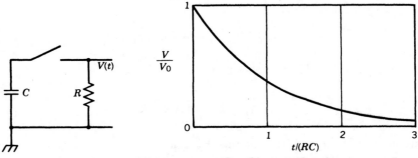

Figure 9.3   Time-dependent load voltage in a switched RC circuit.

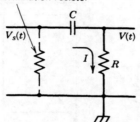

Figure 9.4 Passive integrator.

The solution (plotted in Fig. 9.3) for switching at $t = 0$ is $V(t) = V_0\exp(-t/RC)$ with V_0 the initial charge voltage. The *RC time* $(\Delta t = RC)$ is the characteristic time for the transfer of energy from a capacitor to a resistor in the absence of inductance.

The passive integrator (Fig. 9.4) is a useful variant of the *RC* circuit. The circuit can integrate fast signals in the nanosecond range. It is often used with the fast diagnostics described in Section 9.14. Assume that voltage from a diagnostic, $V_s(t)$, is incident from a terminated transmission line (Section 9.10). The signal has duration Δt. When $\Delta t \ll RC$, the voltage across the capacitor is small compared to the voltage across the resistor. Thus, current flowing into the circuit is limited mainly by the resistor and is given by $I \cong V_s(t)/R$. Applying Eq. (9.5), the circuit output voltage is

$$V_{out} \cong \int V_s(t)\, dt/RC. \qquad \boxed{9.20}$$

The passive integrator is fast, resistant to noise, and simple to build compared to a corresponding circuit with an operational amplifier. The main disadvantage is that there is a droop of the signal. For instance, if V_s is a square pulse, the signal at the end of the pulse is low by a factor $1 - \Delta t/RC$. An accurate signal integration requires that $RC \gg \Delta t$. This condition means that the output signal is reduced greatly, but this is usually not a concern for the large signals available from fast diagnostics.

A circuit with an inductor, resistor, voltage source, and switch is shown in Figure 9.5. This circuit models the output region of a pulsed voltage generator when the inductance of the leads and the load is significant. Usually, we want a rapid risetime for power into the load. The time for initiation of current flow to the load is limited by the undesirable (or parasitic) inductance. Continuity of current and the *V–I* relations of the components [Eqs. (9.1) and (9.6)] give the following differential equation for the load voltage:

$$\int V\, dt/L = V/R = 0. \qquad (9.21)$$

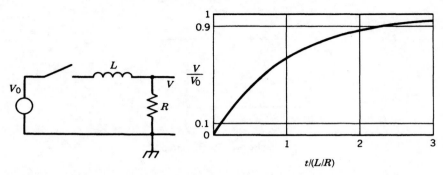

Figure 9.5   Time-dependent load voltage from a pulse generator with a series inductance.

The time variation for load voltage, plotted in Figure 9.5, is

$$V(t) = V_0\left\{1 - \exp\left[-\frac{t}{L/R}\right]\right\}. \qquad (9.22)$$

The *L/R time* determines how fast current and voltage can be induced in the load. The 10–90% risetime for the voltage pulse is $2.2(L/R)$. The load power varies as

$$P(t) = \frac{V^2(t)}{R} = \frac{V_0^2}{R}\left\{1 - \exp\left[-\frac{t}{L/R}\right]\right\}^2. \qquad (9.23)$$

The 10–90 time for the power pulse is $2.6(L/R)$. As an example, if we had a 50-ns pulse generator to drive a 25-Ω load, the total inductance of the load circuit must be less than 0.12 μH if the risetime is to be less than 25% of the pulsewidth.

9.2   HIGH-VOLTAGE SUPPLIES

The transformer is a prime component in all high-voltage supplies. It utilizes magnetic coupling to convert a low-voltage ac input to a high-voltage ac output at reduced current. The transformer does not produce energy. The product of voltage times current at the output is equal to or less than that at the input. The output can be rectified for dc voltage.

We will first consider the air core transformer illustrated in Figure 9.6. Insulated wire is wound uniformly on a toroidal insulating mandrel with $\mu \cong \mu_0$. There are two overlapped windings, the primary and the secondary. Power is introduced in the primary and extracted from the secondary. There is no direct connection between them; coupling is inductive via the shared

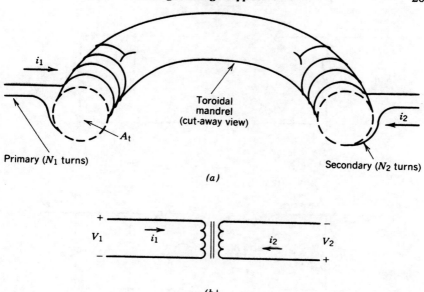

Figure 9.6   Air core transformer. (*a*) Geometry. (*b*) Circuit symbol.

magnetic flux in the torus. Assume that the windings have cross-sectional area A_t and average major radius r_t. The primary winding has N_1 turns, and secondary winding has N_2 turns. The symbol for the transformer and polarity conventions is indicated in Figure 9.6. The windings are oriented so that positive i_1 and positive i_2 produce magnetic fields in opposite directions.

We will determine V_2 and i_2 in terms of V_1 and i_1 and find a simple model for a transformer in terms of the circuit elements of Section 9.1. First, note that energy entering the transformer on the primary can have two destinations. It can be transferred to the secondary or it can produce magnetic fields in the torus. The ideal transformer transfers all input energy to a load connected to the secondary; therefore, the second process is undesirable. We will consider each of the destinations separately and then make a combined circuit model.

To begin, assume that the secondary is connected to an open circuit so that $i_2 = 0$. Thus, no energy is transferred to the secondary. At the primary, the transformer appears to be an inductor with $L_1 = \mu_0 N_1^2 A_t / 2\pi r_t$. All the input energy is converted to magnetic fields; the load is reactive. The equivalent circuit is shown in Figure 9.7*a*.

Next, suppose the secondary is connected to a resistive load and that there is a way to make L_1 infinitely large. In this case, there is no magnetic field energy and all energy is transferred to the load. Infinite inductance means there are no magnetic fields in the torus with finite driving voltage. The field produced by current flow in the primary is exactly canceled by the secondary

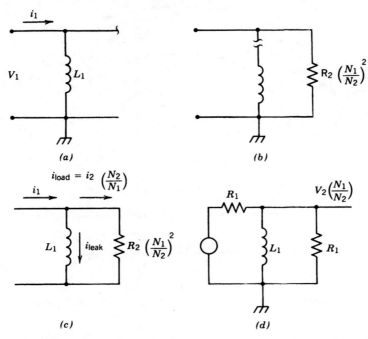

Figure 9.7 Equivalent circuit models for a transformer with ideal coupling. (*a*) Open circuit load on secondary. (*b*) Infinite primary shunt inductance. (*c*) Basic circuit model for an ideal transformer with a resistive load on the secondary. (*d*) Driving a resistive load on the secondary by a pulse modulator with matched impedance at the primary.

field. This means that

$$N_1 i_1 \cong N_2 i_2. \qquad \boxed{9.24}$$

Equation (9.24) holds if the transformer is perfectly wound so that the primary and secondary windings enclose the same area. This situation is called *ideal coupling*. With ideal coupling, the windings enclose equal magnetic flux so that

$$N_1(d\Phi/dt) = V_1, \qquad (9.25a)$$

$$N_2(d\Phi/dt) = V_2, \qquad (9.25b)$$

and hence,

$$V_2 = V_1(N_2/N_1). \qquad \boxed{9.26}$$

Equations (9.24)–(9.26) indicate that energy exchange between the windings is through a changing magnetic flux and that there is a voltage step-up when

$N_2 > N_1$. In the case of infinite L_1 the combination of Equations 9.24 and 9.26 gives the condition for conservation of energy,

$$V_1 i_1 = V_2 i_2. \tag{9.27}$$

When the secondary is connected to a load resistor R_2, the primary voltage is proportional to current via $V_1 = i_1 R_2 (N_1/N_2)^2$. Thus, when viewed from the primary, the circuit is that illustrated in Figure 9.7b with a transformed load resistance.

We have found voltage–current relationships for the two energy paths with the alternate path assumed to be an open circuit. The total circuit (Fig. 9.7c) is the parallel combination of the two. The model shown determines the primary current in terms of the input voltage when the secondary is connected to a resistive load. The input current is expended partly to produce magnetic field in the transformer (i_{leak}) with the remainder coupled to the secondary (i_{load}). The secondary voltage is given by Eq. (9.26). The secondary current is given by Eq. (9.27) with i_{load} substituted for i_1. The conclusion is that if the primary inductance is low, a significant fraction of the primary current flows in the reactance; therefore, $i_1 > (N_2/N_1) i_2$.

The reactive current component is generally undesirable in a power circuit. The extra current increases resistive losses in the transformer windings and the ac voltage source. If the transformer is used to amplify the voltage of a square pulse from a pulsed voltage generator (a common accelerator application), then leakage currents contribute to droop of the output voltage waveform. Consider applying a square voltage pulse of duration Δt from a voltage generator with an output impedance $R_1 = R_2 (N_1/N_2)^2$. The equivalent circuit has resistance R_1 in series with the primary (Fig. 9.7d). We will see in Section 9.10 that this is a good model for the output of a charged transmission line. The output pulse shape is plotted in Figure 9.8 as a function of the ratio of the circuit time $L/[R_2(N_1/N_2)^2]$ to the pulselength, Δt. The output pulse is a square pulse when $L/R_1 \gg \Delta t$. If this condition is not met, the pulse droops. Energy remains in transformer magnetic fields at the end of the main pulse; this energy appears as a negative post-pulse. Although no energy is lost in the ideal transformer, the negative post-pulse is generally useless. Therefore, pulse transformers with low primary inductance have poor energy transfer efficiency and a variable voltage output waveform. Drooping waveforms are often unacceptable for accelerator applications.

Leakage current is reduced by increasing the primary inductance. This is accomplished by constructing the toroidal mandrel from ferromagnetic material. The primary inductance is increased by a factor of μ/μ_0. Depending on the operating regime of the transformer, this factor may be as high as 10,000. Another way to understand the role of iron in a transformer is to note that a certain flux change is necessary to couple the primary voltage to the secondary. In an air core transformer, the flux change arises from the difference in the ampere turns between the primary and secondary; the flux

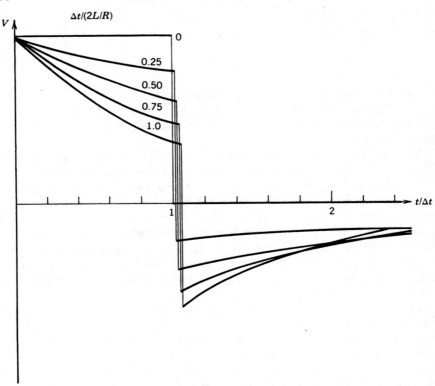

Figure 9.8 Pulse waveforms; resistive load driven by a matched pulse modulator through a transformer: Δt is the duration of the voltage pulse, R is the output impedance of the modulator and the load resistance viewed at the primary, and L is the primary shunt inductance.

change is generated by the leakage current. When a ferromagnetic material is added, atomic currents contribute to the flux change so that the leakage current can be much smaller.

The energy transfer efficiency of real transformers falls short of that for the ideal transformer of Figure 9.7c. One reason for power loss is that ferromagnetic materials are not ideal inductors. Eddy current losses and hysteresis losses in iron transformer cores for fast pulses are discussed in Section 10.2. Another problem is that ideal coupling between the primary and secondary cannot be achieved. There is a region between the windings in which there is magnetic field that is not canceled when both windings have the same ampere turns; the effect of this region is represented as a series inductance in the primary or secondary. Minimizing the effects of nonideal coupling is one of the motivations for using an iron core to increase the primary inductance of the transformer rather than simply increasing the number of turns in the primary and secondary windings of an air core transformer.

There are limitations on the primary voltage waveforms that can be handled by pulse transformers with ferromagnetic cores. The rate of change of flux enclosed by the primary is given by Eq. (9.25a). The flux enclosed in a transformer with a toroidal core area A_t is $A_t B(t)$. Inserting this into Eq. (9.25a) and integrating, we find that

$$[B(t) - B(0)] N_1 A_t = \int V(t) \, dt \tag{9.28}$$

Equation (9.28) constrains the input voltage in terms of the core geometry and magnetic properties of the core material. An input signal at high voltage or low frequency may drive the core to saturation. If saturation occurs, the primary inductance drops to the air core value. In this case, the primary impedance drops, terminating energy transfer to the load. Referring to the hysteresis curve of Figure 5.12, the maximum change in magnetic field is $2B_s$. If the primary input is an ac signal, $V_1(t) = V_0 \sin(2\pi f t)$, then Eq. (9.28) implies that

$$V_0/f \text{ (volt-s)} \le 2\pi N_1 A_t B_s. \qquad \boxed{9.29}$$

Transformers are usually run well below the limit of Eq. (9.29) to minimize hysteresis losses.

The basic circuit for a high-voltage dc supply is shown in Figure 9.9. The configuration is a *half-wave rectifier*; the diode is oriented to pass current only on the positive cycle of the transformer output. A capacitor is included to reduce ripple in the voltage. The fractional drop in voltage during the negative half-cycle is on the order of $(1/2f)/RC$, where R is the load resistance. Output voltage is controlled by a variable autotransformer in the primary.

Because of the core volume and insulation required, transformers are inconvenient to use at voltages above 100 kV. The ladder network illustrated in

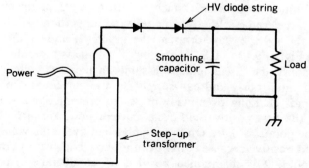

Figure 9.9 Half-wave-rectifier circuit to generate high-voltage dc power from a transformed ac waveform.

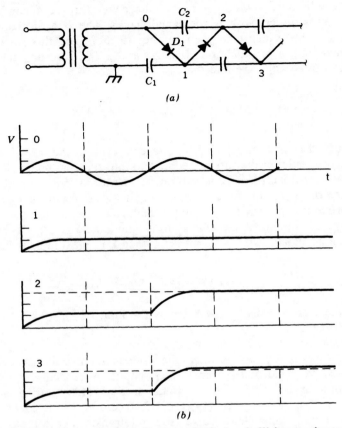

Figure 9.10 (*a*) Voltage multiplication ladder network (Cockcroft–Walton accelerator). (*b*) Ideal voltage waveforms (neglecting the ac generator impedance) shown at various points of the network.

Figure 9.10*a* can supply voltages in the 1-MV range. It is the basis of the Cockcroft–Walton accelerator,[†] which is utilized as a preinjector for many high-energy accelerators. We can understand operation of the circuit by considering the voltage waveforms at the input (0) and at the three points indicated in Figure 9.10*b* (1, 2, 3). The input voltage is a bipolar ac signal. In order to simplify the discussion, we assume an ideal ac voltage source that turns on instantaneously and can supply infinite current; in reality, a series resistor extends charging over many cycles to prevent damage to the transformer. On the first positive half-cycle, current flows through the diodes to charge all the points to $+V_0$. On the negative half-cycle, the voltage at point 1 is maintained positive because current cannot flow backward in diode D_1. The voltage at point 2 is maintained positive by conductance from C_1 to C

[†]J. D. Cockcroft and E. T. S. Walton, *Proc. Roy. Soc. (London)*, **A136**, 619 (1932).

through diode D_2. At the time of maximum negative voltage at the input, there is a voltage difference greater than V_0 across capacitor C_2. In the steady state, the voltage difference approaches $2V_0$. On the second positive cycle, the voltage at point 2 is boosted to $3V_0$. The voltage at point 3 also approaches $3V_0$ because of charging through D_3. The reasoning can be extended to higher points on the ladder, leading to the unloaded steady-state voltages indicated in Figure 9.10.

At the same voltage, a ladder network requires about the same number of diodes and capacitors as a power supply based on a transformer and rectifier stack. The main advantages of a ladder network is that it utilizes a smaller transformer core and it is easier to insulate. Insulation of the secondary of a transformer is difficult at high voltage because the secondary winding must encircle the transformer core. A large core must be used with oil-impregnated insulation. In contrast, the ladder network is extended in space with natural voltage grading along the column. It is possible to operate Cockcroft–Walton-type accelerators at megavolt levels with air insulation at atmospheric pressure by locating the capacitor–diode stack in a large shielded room.

9.3 INSULATION

Insulation is the prevention of current flow. It presents the major technological problem of high-voltage electrostatic acceleration. At low values of electric field stress, current flow through materials such as glass, polyethylene, transformer oil, or dry air is negligible. Problems arise at high voltage because there is sufficient energy to induce ionization in materials. Portions of the material can be converted from a good insulator to a conducting ionized gas (plasma). When this happens, the high-voltage supply is shorted. Plasma breakdowns can occur in the solid, liquid, or gaseous insulation of high-voltage supplies and cables. Breakdown may also occur in vacuum along the surface of solid insulators. Vacuum breakdown is discussed in the Section 9.5.

There is no simple theory of breakdown in solids and liquids. Knowledge of insulating properties is mainly empirical. These properties vary considerably with the chemical purity and geometry of the insulating material. The *dielectric strength* and relative dielectric constant of insulators commonly used in high-voltage circuits are given in Table 9.1. The dielectric strength is the maximum electric field stress before breakdown. The actual breakdown level can differ considerably from those given; therefore, it is best to leave a wide safety factor in designing high-voltage components. The dielectric strength values of Table 9.1 hold for voltage pulses of submicrosecond duration. Steady-state values are tabulated in most physical handbooks.

Measurements show that for voltage pulses in the submicrosecond range the dielectric strength of liquids can be considerably higher than the dc value. The

TABLE 9.1 Properties of Some Common Insulators[a]

Material	Relative Dielectric Constant ($\varepsilon/\varepsilon_0$)	Dielectric Strength (MV/cm)
Transformer oil	3.4	1
Mylar	3	1.8
Polyethylene	2.25	1.8
Teflon	2.1	4.3
Polycarbonate	2.96	5.5

[a]Adapted from J. F. Francis, *High Voltage Pulse Techniques*, Air Force Office of Scientific Research, AFOSR-74-2639-5, 1974.

following empirical formulas describe breakdown levels in transformer oil and purified water,[†] two media used extensively in high-power pulse modulators. The pulsed voltage dielectric strength of transformer oil is approximately

$$E_{max} \cong 0.5/t_p^{0.33}A^{0.1}. \qquad \boxed{9.30}$$

The quantity E_{max} (in megavolts per centimeter) is the highest value of electric field in the insulator at the peak of the voltage pulse. The quantity t_p is the time during which the voltage is above 63% of the maximum value (in microseconds), and A is the surface area of the high voltage electrode in centimeters squared. The breakdown level of water has the same scaling as Eq. (9.30) but is polarity dependent. The dielectric strength of water for a negative high-voltage electrode is given by

$$E_{max} \cong 0.6/t_p^{0.33}A^{0.1} \quad \text{(negative)} \qquad \boxed{9.31}$$

in the same units as Eq. (9.30). With a positive polarity, the breakdown level is approximately

$$E_{max} \cong 0.3/t_p^{0.33}A^{0.1} \quad \text{(positive)}. \qquad \boxed{9.32}$$

The polarity dependence of Eqs. (9.31) and (9.32) is important for designing optimum pulsed power systems. For example, a charged coaxial transmission line filled with water can store four times the electrostatic energy density if the center conductor is negative with respect to the outer conductor rather than positive.

The enhanced dielectric strength of liquids for fast pulsed voltages can be attributed to the finite time for a breakdown streamer to propagate through the

[†]See R. B. Miller, *Intense Charged Particle Beams*, Plenum, New York, 1982, p. 16.

Gas (pressure p)

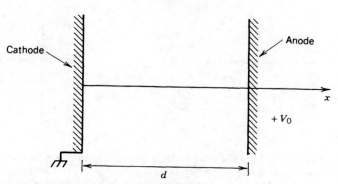

Figure 9.11 Gas-filled high-voltage gap, planar electrodes.

medium to short the electrodes. Section 9.12 will show how pulse-charge overvolting is used to increase the electrostatic energy density of a pulse modulator for high output power. This process can be applied in liquid insulators because they are *self-healing*. On the other hand, the damage caused by streamers is cumulative in solids. Solid insulators cannot be used above the steady-state levels.

Gas insulation is used in most steady-state high-voltage electrostatic accelerators. Gases have $\varepsilon/\varepsilon_0 \sim 1$; therefore, gas insulators do not store a high density of electrostatic energy. Gases, like liquids, are self-healing after spark breakdowns. The major advantage of gas insulation is cleanliness. A fault or leak in an accelerator column with oil insulation usually leads to a major cleanup operation. Although the gas itself is relatively inexpensive, the total gas insulation system is costly if the gas must be pressurized. Pressurization requires a large sealed vessel.

We shall study the theory of spark breakdown in gas in some detail. The topic is essential for the description of most types of fast high-voltage switches and it is relevant to insulation in most high-voltage electrostatic accelerators. Consider the one-dimensional gas-filled voltage gap of Figure 9.11. The electrodes have separation d; the applied voltage is V_0. There is negligible transfer of charge between the electrodes when V_0 is small. We will determine how a large interelectrode current can be initiated at high voltage.

Assume that a few electrons are produced on the negative electrode. They may be generated by photoemission accompanying cosmic ray or ultraviolet bombardment. The small source current density leaving the electrode is represented by j_0. The electrons are accelerated by the applied field and move between the widely spaced gas molecules, as shown in Figure 9.12a. In a collision with a molecule, an electron is strongly deflected and much of its kinetic energy is absorbed. To construct a simple model, we assume that electrons are accelerated between molecules and lose all their directed energy

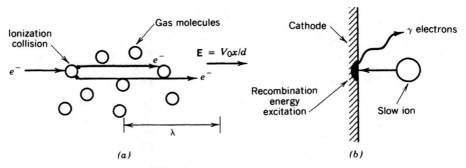

Figure 9.12 Processes in a gas-filled, high-voltage gap. (*a*) Electron migration in presence of a strong electric field showing electron acceleration, ionizing collision, and the mean free path. (*b*) Liberation of secondary electrons from a negative polarity electrode (cathode) by recombination of slow ions.

in a collision. The parameter λ is the *mean free path*, the average distance traversed between collisions. The mean free path is inversely proportional to the density of gas molecules. The density is, in turn, proportional to the pressure, p, so that $\lambda \sim 1/p$.

The average energy gained by an electron between collisions is

$$\Delta T = \lambda e E = \lambda e V_0/d. \qquad (9.33)$$

An electron may ionize a molecule in a collision if ΔT is high enough ($\Delta T \geq 30$ eV). In the ionization process, the excess kinetic energy of the electron drives an electron out of the molecule, leaving a positive ion. The two electrons move forward under the influence of the field, producing further electrons. The current density increases geometrically along the electron drift direction. The ions drift in the opposite direction and do not contribute to ionization in the gas. The motion of drifting ions is dominated by collisions; generally, they cannot reach high enough velocity to eject electrons in a collision with a molecule.

Electron multiplication in a gas with an applied electric field is characterized by α, the first Townsend coefficient.[†] This parameter is defined by

$$\alpha = \frac{\text{number of ionizations induced by an electron}}{\text{cm of pathlength}}. \qquad (9.34)$$

Consider an element of length Δx at the position x. The quantity $n(x)$ is the density of electrons at x. According to the definition of α, the total number of additional electrons produced in a volume with cross-sectional area dA and length Δx is

$$dN \cong \alpha n(x) \, dA \, \Delta x.$$

[†]J. S. Townsend, *Electricity in Gases*, *Phil. Trans.* **A193**, 129 (1900).

Dividing both sides of the equation by $dA \, \Delta x$, and taking the limit of small Δx, we find that

$$dn\,(x)/dx = \alpha n(x). \tag{9.35}$$

The solution of Eq. (9.35) implies an exponential electron density variation. Expressed in terms of the current density of electrons (assuming that the average electron drift velocity is independent of x), Eq. (9.35) implies that

$$j(x) = j_0 \exp(\alpha x). \tag{9.36}$$

if the negative electrode is located at $x = 0$. The current density arriving at the positive electrode is

$$j_a = j_0 \exp(\alpha d). \tag{9.37}$$

Although the amplification factor may be high, Eq. (9.37) does not imply that there is an insulation breakdown. The current terminates if the source term is removed. A breakdown occurs when current flow is *self-sustaining*, or independent of the assumed properties of the source. In the breakdown mode, the current density rapidly multiplies until the voltage supply is shorted (discharged). Ion interactions at the negative electrode introduce a mechanism by which a self-sustained discharge can be maintained. Although the positive ions do not gain enough energy to ionize gas molecules during their transit, they may generate electrons at the negative electrode through secondary emission (Fig. 9.12b). This process is parametrized by the secondary emission coefficient, γ. The coefficient is high at high ion energy (> 100 keV). The large value of γ reflects the fact that ions have a short stopping range in solid matter and deposit their energy near the surface. Ions in a gas-filled gap have low average energy. The secondary emission coefficient has a nonzero value for slow ions since there is available energy from recombination of the ion with an electron at the surface. The secondary emission coefficient for a zero-velocity ion is in the range $\gamma \approx 0.02$.

With ion interactions, the total electron current density leaving the negative electrode consists of the source term plus a contribution from secondary emission. We define the following current densities: j_0 is the source, j_c is the net current density leaving the negative electrode, j_i is the ion current density arriving at the negative electrode, and j_a is the total electron flux arriving at the positive electrode. Each ionizing collision creates one electron and one ion. By conservation of charge, the ion current is given by

$$j_i = j_a - j_c. \tag{9.38}$$

Equation (9.38) states that ions and electrons leave the gap at the same rate in the steady state. The portion of the electron current density from the negative electrode associated with secondary emission is thus,

$$j_c - j_0 = \gamma(j_a - j_c). \tag{9.39}$$

Equation (9.39) implies that

$$j_c = (j_0 + \gamma j_a)/(1 + \gamma). \tag{9.40}$$

By the definition of α,

$$j_a = j_c \exp(\alpha d). \tag{9.41}$$

Combining Eqs. (9.40) and (9.41) gives

$$\frac{j_a}{j_0} = \frac{\exp(\alpha d)}{\{1 - \gamma[\exp(\alpha d) - 1]\}}. \tag{9.42}$$

Comparing Eq. (9.42) with (9.37), the effects of ion feedback appear in the denominator. The current density amplification becomes infinite when the denominator equals zero. This means that there is a finite current for zero source current. When the conditions are such that the denominator of Eq. (9.42) is zero, a small charge at the negative electrode can initiate a self-sustained discharge. The discharge current grows rapidly over a time scale equal to the ion drift time, heating and ionizing the gas. The end result is a *spark*. A spark is a high-current density plasma channel. The condition for spark formation is

$$\exp(\alpha d) \le (1/\gamma) + 1 \cong 1/\gamma. \tag{9.43}$$

The discharge continues until the electrode voltage is depleted.

Calculation of the absolute value of α as a function of the gap parameters involves complex atomic physics. We will, instead, develop a simple scaling relationship for α as a function of gas pressure and electric field. Such a relationship helps to organize experimental data and predict breakdown properties in parameter regimes where data is unavailable. The ionization coefficient is proportional to the number of collisions between an electron and gas atoms per centimeter traveled, or $\alpha \sim 1/\lambda$. The ionization coefficient also depends on the average electron drift energy, $E\lambda$. The ionization rate is proportional to the fraction of collisions in which an electron enters with kinetic energy greater than the ionization energy of the atom, I. The problem of a particle traveling through a random distribution of collision centers is treated in texts on nuclear physics. The familiar result is that the distance traveled between collisions is a random variable that follows a Poisson distribution, or

$$P(x) = \exp(-x/\lambda), \tag{9.44}$$

where λ is the mean free path. The energy of a colliding electron is, by definition, $T = xeV_0/d$. The fraction of collisions in which an electron has $T > I$ is given by the integral of Eq. (9.44) from $x = Id/eV_0$ to ∞.

The result of these considerations is that α is well described by the scaling law

$$\alpha = (A'/\lambda)\exp(-B'd/V_0\lambda) = (Ap)\exp(-Bpd/V_0). \tag{9.45}$$

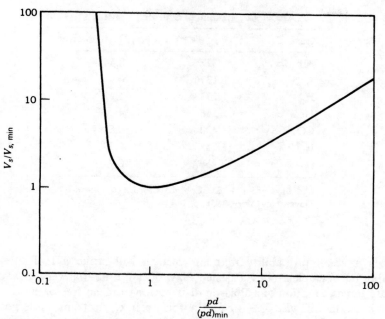

Figure 9.13 Normalized Paschen curve; sparking voltage versus the product of pressure and gap width.

The quantities A and B are determined from experiments or detailed collision theory. The gap voltage for sparking, V_s, is found by substituting Eq. (9.45) into (9.43):

$$\ln(1/\gamma) = (Apd)\exp(-Bpd/V_s).$$

Solving for V_s,

$$V_s = Bpd/\ln \frac{Apd}{\ln(1/\gamma)}. \qquad \boxed{9.46}$$

Equation (9.46) is Paschen's law[†] for gas breakdown. The sparking voltage is a function of the product pd and constants which depend on the gas properties. The values of A and B are relatively constant over a wide voltage range when the average electron energy is less than I. Furthermore, V_s is insensitive to γ; therefore, the results are almost independent of the electrode material.

Figure 9.13 is a normalized plot of V_s versus pd. The sparking voltage reaches a minimum value, $V_{s,\min}$ at a value $(pd)_{\min}$. Voltage hold-off increases at both low and high values of pd. Voltage hold-off is high at low pd since

[†] F. Paschen, *Wied. Ann.*, **37**, 69 (1889).

TABLE 9.2   Minimum Sparking Constants[a]

Gas	$V_{s,\,min}$ (V)	$(pd)_{min}$ (torr-cm)
Air (dry)	327	0.567
A	137	0.9
H_2	273	1.15
He	156	4.0
CO_2	420	0.51
N_2	251	0.67
O_2	450	0.7

[a]Adapted from J. D. Cobine, *Gaseous Conductors*, Dover, New York, 1958, p. 164.

there is a small probability that an electron will strike a molecule while traveling between electrodes. Paschen's law does not hold at very low vacuum because the assumption of a Poisson distribution of mean free paths is invalid. At high values of pd, $V_s \sim (pd)$. In this regime, electrons undergo many collisions, but the mean free path is short. Few electrons gain enough energy to produce an ionization.

Equation (9.46) can be rewritten,

$$V_s = \frac{V_{s,\,min}\left[pd/(pd)_{min}\right]}{\ln\left[2.72\ pd/(pd)_{min}\right]}. \qquad \boxed{9.47}$$

Table 9.2 gives a table of spark parameters for some common gases. Note that the minimum sparking voltage is high for electronegative gases like oxygen and low for gases with little probability of electron capture like helium and argon. In electronegative gases, there is a high probability that electrons are captured to form negative ions. The heavy negative ions cannot produce further ionization, so that the electron is removed from the current multiplication process. The result is that gaps with electronegative gases can sustain high voltage without breakdown. Sulfur hexafluoride, an extremely electronegative gas, is often mixed with air or used alone to provide strong gas insulation at high pressure. The expense of the gas is offset by cost savings in the pressure vessel surrounding the high-voltage system. The addition of 8% (by volume) SF_6 to air increases V_s by a factor of 1.7. Pure SF_6 has V_s 2.2 times that of air for the same pd. Vacuum insulation is seldom used in power supplies and high-voltage generators because of technological difficulties in maintaining high vacuum and the possibility of long breakdown paths. Vacuum insulation in accelerator columns is discussed in Section 9.5.

Our discussion of gas breakdown has been limited to a one-dimensional geometry. Two-dimensional geometries are difficult to treat analytically since a

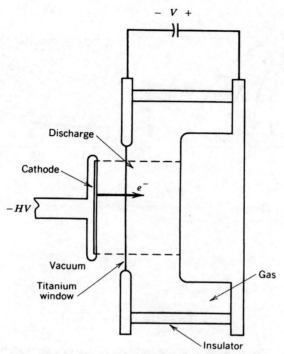

Figure 9.14 Electron-beam-controlled gas discharge laser.

variety of electron trajectories are possible and the electric field varies along the electron paths. Breakdown voltages are tabulated for special electrode geometries such as two spheres. In some circumstances (such as gas lasers), a stable, uniform discharge must be sustained over a large area, as shown in Figure 9.14. The interelectrode voltage must be less than V_s to prevent a localized spark; therefore, a source current is required to sustain the discharge. The source is often provided by injection of a high-energy electron beam through a foil on the negative electrode. Demands on the source are minimized if there is a large multiplication factor, $\exp(\alpha d)$; therefore, the main gap is operated as close to V_s as possible. In this case, care must be taken with shaping of the electrodes. If the electrode has a simple radius (Fig. 9.15a), then the field stress is higher at the edges than in the body of the discharge, leading to sparks. The problem is solved by special shaping of the edges. One possibility is the *Rogowski profile*,[†] illustrated in Figure 9.15b. This shape, derived by conformal mapping, has the property that the field stress on the electrodes decreases monotonically moving away from the axis of symmetry. The shape for two symmetric profiled electrodes is described by the parametric

[†] W. Rogowski and H. Rengier, *Arch. Elekt.*, **16**, 73 (1926).

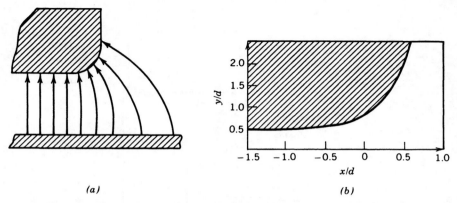

Figure 9.15 Electrode shaping. (*a*) Field enhancement at the edge of an electrode with a simple radius. (*b*) Electrode with a Rogowski profile.

equations

$$x = d\phi/\pi,$$

$$y = (d/\pi)[\pi/2 + \exp(\phi)].$$

$$\boxed{9.48}$$

where d is the minimum separation between electrodes. As indicated in Figure 9.15b, x indicates the transverse distance from the center of the electrodes, and $\pm y$ is the distance from the midplane (between electrodes) to the electrode surfaces.

Corona discharges appear in gases when electrodes have strong two-dimensional variations (Figure 9.16). Corona (crown in Latin) is a pattern of bright sparks near a pointed electrode. In such a region, the electric field is enhanced above the breakdown limit so that spark discharges occur. Taking the dimension of the corona region as d_c, the approximate condition for breakdown is

$$\langle E \rangle d_c \approx V_{s,\min}(pd_c/(pd)_{\min})/\ln[2.72(pd_c/(pd)_{\min})] \qquad (9.49)$$

where $\langle E \rangle$ is the average electric field in the corona region. A self-sustained breakdown cannot be maintained by the low electric field in the bulk of the gap. Current in the low-field region is conducted by a Townsend (or dark) discharge, with the corona providing the source current j_0. The system is stabilized by the high resistivity of the dark discharge. A relatively constant current flows, even though the sparks of the corona fluctuate, extinguish, and reform rapidly. Inspection of Eq. (9.49) shows that for constant geometry, the size of the corona region grows with increasing electrode voltage. The voltage drop in the highly ionized corona is low; therefore, the voltage drop is concentrated in the dark discharge region. At some voltage, bulk breakdown occurs and the electrodes are shorted.

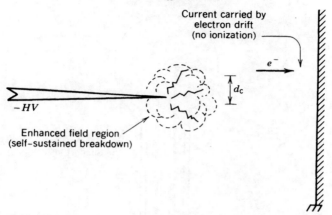

Figure 9.16 Physical basis for corona discharge near a pointed electrode.

9.4   VAN DE GRAAFF ACCELERATOR

Two types of voltage generators are used for low current electrostatic accelerators in the megavolt range, the Cockcroft–Walton and Van de Graaff generators. We have already discussed the principle of the ladder network voltage multiplier used in the Cockcroft–Walton accelerator in Section 9.2. Cockcroft–Walton accelerators are used mainly for injectors with voltage of approximately 1 MV. In this section, we shall discuss the Van de Graaff generator, which can sustain steady-state voltages up to 15 MV.

A Van de Graaff accelerator[†] for electrons is illustrated in Figure 9.17. The principle of operation is easily understood. A corona discharge from an array of needles in gas is used as a convenient source of electrons. The electrons drift toward the positive electrode and are deposited on a moving belt. The belt, composed of an insulating material with high dielectric strength, is immersed in insulating gas at high pressure. The attached charge is carried mechanically against the potential gradient into a high-voltage metal terminal. The terminal acts as a Faraday cage; there is no electric field inside the terminal other than that from the charge on the belt. The charge flows off the belt when it comes in contact with a metal brush and is deposited on the terminal.

The energy to charge the high-voltage terminal is supplied by the belt motor. The current available to drive a load (such as an accelerated beam) is controlled by either the corona discharge current or the belt speed. Typical currents are in the range of 10 μA. Power for ion sources or thermonic electron sources at high voltage can be supplied by a generator attached to the belt pulley inside the high-voltage terminal. The horizontal support of long belts

[†] R. J. Van de Graaff, *Phys. Rev.*, **38**, 1919 (1931).

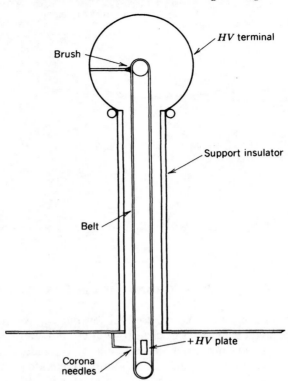

(a)

Figure 9.17 Van de Graaff accelerator. (*a*) Principle of operation. (*b*) Los Alamos National Laboratory, 7 MeV Van de Graaff accelerator, utilized as injector for 24.5 MeV tandem Van de Graaff accelerator facility. (Courtesy, J. R. Tesmer, Los Alamos National Laboratory.)

and accelerator columns is difficult; therefore, many high-voltage Van de Graaff accelerators are constructed vertically, as in Figure 9.17.

Van de Graaff accelerators are excellent research tools since they provide a steady-state beam with good energy regulation. Continuous low-current beams are well suited to standard nuclear diagnostics that detect individual reaction products. Although the primary use of Van de Graaff accelerators has been in low-energy nuclear physics, they are finding increased use for high-energy electron microscopes and ion microprobes. The output beam energy of a Van de Graaff accelerator can be extended a factor of 2 (up to the 30 MeV range) through the tandem configuration illustrated in Figure 9.18. Negative ions produced by a source at ground potential are accelerated to a positive high-voltage terminal and pass through a stripping cell. Collisions in the cell remove electrons, and some of the initial negative ions are converted to

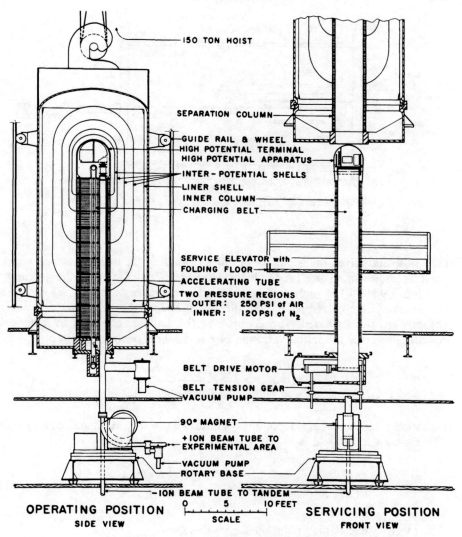

150 TON HOIST

SEPARATION COLUMN

GUIDE RAIL & WHEEL
HIGH POTENTIAL TERMINAL
HIGH POTENTIAL APPARATUS

INTER - POTENTIAL SHELLS
LINER SHELL
INNER COLUMN
CHARGING BELT

SERVICE ELEVATOR with
FOLDING FLOOR
ACCELERATING TUBE
TWO PRESSURE REGIONS
    OUTER: 250 PSI of AIR
    INNER: 120 PSI of N$_2$

BELT DRIVE MOTOR

BELT TENSION GEAR
VACUUM PUMP

90° MAGNET

+ION BEAM TUBE TO
EXPERIMENTAL AREA

VACUUM PUMP
ROTARY BASE

-ION BEAM TUBE TO TANDEM

OPERATING POSITION
SIDE VIEW

0   5   10 FEET
SCALE

SERVICING POSITION
FRONT VIEW

LOS ALAMOS VERTICAL VAN de GRAAFF

Figure 9.17 (*Continued*).

positive ions. They are further accelerated traveling from the high-voltage terminal back to ground.

Voltage hold-off is maximized when there are no regions of enhanced electric field. In other words, the electric field stress should be made as uniform as possible in the region of gas insulation. High-voltage terminals are usually constructed as large, smooth spheres to minimize peak electric field stress. We

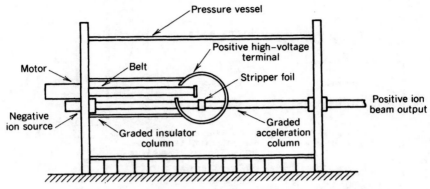

Figure 9.18 Schematic diagram of a tandem Van de Graaff accelerator.

can show that, given the geometry of the surrounding grounded pressure vessel, there is an optimum size for the high-voltage terminal. Consider the geometry of Figure 9.19a. The high-voltage terminal is a sphere of radius R_0 inside a grounded sphere of radius R_2. Perturbing effects of the belt and accelerator column are not included. The solution to the Laplace equation in spherical coordinates gives the radial variation of potential between the spheres

$$\phi(r) = \frac{V_0 R_0 (R_2/r - 1)}{R_2 - R_0}. \qquad (9.50)$$

Equation (9.50) satisfies the boundary conditions $\phi(R_0) = V_0$ and $\phi(R_2) = 0$. The radial electric field is

$$E_r(r) = -\partial\phi/\partial r = \frac{V_0 R_0 (R_2/r^2)}{R_2 - R_0}. \qquad (9.51)$$

The electric field is maximum on the inner sphere,

$$E_{r,\,\text{max}} = \frac{(V_0/R_2)(R_2/R_0)}{1 - R_0/R_2}. \qquad (9.52)$$

We assume R_2 is a fixed quantity, and determine the value of R_0/R_2 that will minimize the maximum electric field. Setting $\partial E_{r,\,\text{max}}/\partial(R_0/R_2) = 0$, we find that

$$R_0/R_2 = \tfrac{1}{2}. \qquad \boxed{9.53}$$

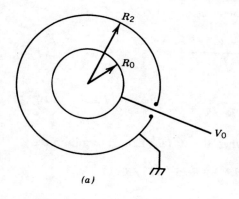

(a)

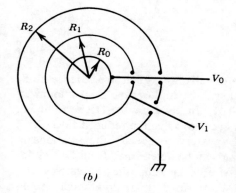

(b)

Figure 9.19 Geometry for calculating electric fields on spherical electrodes. (*a*) Two concentric spheres. (*b*) Concentric spheres with an equipotential shield.

A similar calculation can be carried out for concentric cylinders. In this case, the optimum ratio of inner to outer cylinder radii is

$$R_0/R_2 = 1/e = 0.368.$$

$$\boxed{9.54}$$

A better electric field distribution can be obtained through the use of *equipotential shields*. Equipotential shields are biased electrodes located between the high-voltage terminal and ground, as in Figure 9.17. Voltage on the shields is maintained at specific intermediate values by a high-voltage resistive divider circuit. The simplified geometry of three nested spheres (Fig. 9.19*b*) will help in understanding the principle of equipotential shields. The quantities R_2 and V_0 are constrained by the space available and the desired operating voltage. We are free to choose R_0, R_1, and V_1 (the potential of the shield) for optimum voltage hold-off. The electric fields on outer surfaces of the nested spheres, E_1 $(r = R_1^+)$ and E_0 $(r = R_0^+)$, are simultaneously minimized.

The fields on the outer surfaces of the electrodes are

$$E_0 = \frac{(V_0 - V_1)/R_0}{1 - R_0/R_1},$$ (9.55)

$$E_1 = \frac{V_1/R_1}{1 - R_1/R_2}.$$ (9.56)

The following equations are satisfied when the surface fields are minimized:

$$\Delta E_0 = (\partial E_0/\partial V_1)\,\Delta V_1 + (\partial E_0/\partial R_1)\,\Delta R_1 + (\partial E_0/\partial R_0)\,\Delta R_0 = 0,$$

          (9.57)

$$\Delta E_1 = (\partial E_1/\partial V_1)\,\Delta V_1 + (\partial E_1/\partial R_1)\,\Delta R_1 + (\partial E_1/\partial R_0)\,\Delta R_0 = 0.$$

          (9.58)

The final term in Eq. (9.58) is zero since there is no explicit dependence of E_1 on R_0. The previous study of two nested spheres implies that E_0 is minimized with the choice $R_0 = R_1/2$ given V_1 and R_1. In this case, the last term in Eq. (9.57) is independently equal to zero. In order to solve the reduced Eqs. (9.57) and (9.58), an additional equation is necessary to specify the relationship between ΔR_1 and ΔV_1. Such a relationship can be determined from Eqs. (9.55) and (9.56) with the choice $E_0 = E_1$. After evaluating the derivatives and substituting into Eqs. (9.57) and (9.58), the result is two simultaneous equations for R_1 and V_1. After considerable algebra, the optimum parameters are found to be

$$R_0 = \tfrac{1}{2}R_1, \qquad R_1 = \tfrac{5}{8}R_2, \qquad V_1 = \tfrac{3}{5}V_0.$$ (9.59)

Addition of a third electrode lowers the maximum electric field stress in the system. To compare, the optimized two electrode solution (with $R_0 = R_2/2$) has

$$E_0 = 4V_0/R_2.$$ (9.60)

The peak field for the three electrode case is

$$E_0 = E_1 = (192/75)V_0/R_2 = 2.56 V_0/R_2.$$ (9.61)

Larger numbers of nested shells reduce the peak field further. The intent is to counter the geometric variation of fields in spherical and cylindrical geometries to provide more uniform voltage grading. For a given terminal and pressure vessel size, multiple shells have approximately equal voltage increments if they are uniformly spaced.

9.5 VACUUM BREAKDOWN

A column for electron beam acceleration in an electrostatic accelerator is shown in Figure 7.14. It consists of insulating disks separated by metal grading rings. The disks and rings are sealed to hold high vacuum, either by low vapor pressure epoxy or by a direct metal to ceramic bond. We have already discussed the role grading rings play in focusing particle beams through long columns (Section 7.7). We shall now consider how rings also improve the voltage hold-off capability of solid insulators in vacuum. High-vacuum sparking is determined by complex phenomena occurring on solid surfaces under high-field stress. Effects on both conductors and insulators can contribute to breakdown.

When a metal is exposed to a strong electric field, electrons may be produced by *field emission*. Field emission is a quantum mechanical tunneling process. Strong electric fields lower the energy barrier at the metal surface, resulting in electron emission. Field emission is described by the Fowler–Nordheim equation,[†] which has the scaling

$$j_{ef} \sim E^2 \exp(-B/E), \tag{9.62}$$

where E is the electric field. Typical current density predicted by Eq. (9.62) in a gap with an electric field of 25 MV/m is on the order of 10 nA/cm^2.

Equation (9.62) implies that a plot of $\ln(j_{ef}/E^2)$ versus $1/E$ is a straight line. The scaling is observed in experiments on clean metal surfaces, but the slope of the plot indicates that the field magnitude on the surface is 10–100 times the macroscopic field (voltage divided by gap spacing). The discrepancy is explained by the presence of *whiskers* on the metal surface.[†] Whiskers are small-scale protrusions found on all ordinary metal surfaces. The electric field is enhanced near whiskers, as shown in Figure 9.20a. Although the high-field regions account for a small fraction of the surface area, there is a significant enhancement of emitted current because of the strong scaling of Eq. (9.62) with E.

The area-averaged whisker-enhanced current density from a metal surface at 10 MV/m is only 1–100 μA/cm^2. The leakage current by this process is not a significant concern in high-voltage systems. The main problem is that whiskers may be vaporized by the high-current density at the tips, ejecting bursts of material into highly stressed vacuum regions. Vaporization of even a small amount of solid material (at a density of 10^{22} cm^{-3}) on the tip of the whisker can eject a significant gas pulse into the vacuum. A conventional gas spark can then occur. Vaporization occurs at field stress exceeding 10 MV/m (100

[†] See J. Thewlis (Ed.), *Encyclopaedic Dictionary of Physics*, Vol. 3, Macmillan, New York, 1962, p. 120.
[†] See P. A. Chatterton, "Vacuum Breakdown," in *Electrical Breakdown in Gases*, edited by J. M. Meek and J. D. Craggs, Ed., Wiley, New York, 1978.

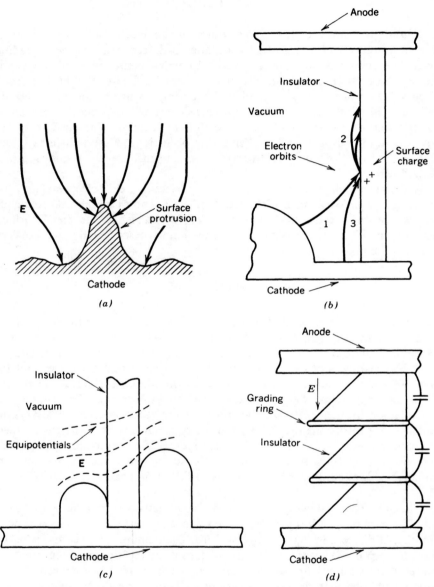

Figure 9.20 Processes influencing electric breakdowns in vacuum. (*a*) Microscopic view of surface of negative electrode showing field enhancement on tips of whiskers. (*b*) High-voltage electrodes separated by insulator; processes leading to insulator breakdown: (1) Electrons emitted from cathode (by field emission or whisker vaporization) collide with insulator surface. (2) If secondary emission coefficient exceeds unity, electrons are liberated. (3) Excess positive charge attracts other electrodes, increasing growth of surface charge. (*c*) Shielding triple point to prevent deposition of electrons emitted from negative electrode on insulator. (*d*) Common high-voltage insulator design, with insulator shaping (to minimize charge imbalance on surfaces) and grading rings.

kV/cm) for machined metal surfaces. The level can be raised by careful chemical preparation of the surface or by conditioning. In conditioning, a metal surface is slowly raised to the final electric field operating level. Controlled vaporization of whiskers often smooths the surface, allowing higher operating field.

Whisker vaporization probably accounts for *microbursts* in steady-state electrostatic accelerators. Microbursts are random pulses of current observed in highly stressed accelerator columns. Whisker vaporizaton is used to advantage in pulsed electron accelerators. In this application, a rapidly pulsed electric field is applied to the cathode surface. Many whiskers vaporize simultaneously, covering the cathode with a dense, cold plasma. The plasma has zero work function; electrons can be extracted at current density exceeding 1 kA/cm². The main problem with *cold cathodes* is that the plasma expands into the acceleration gap (*plasma closure*), causing deterioration of the beam optics and ultimately shorting the gap. The expansion rate of the plasma, the *closure velocity*, is enhanced by electron flow through the plasma. Beam plasma instabilities rapidly heat the plasma. Closure velocities as high as 10 cm/μs have been observed. Depending on the acceleration gap width, cold cathodes are useful only for pulses in the 1-μs range.

Breakdowns occur on insulator surfaces in a vacuum at lower field levels than those that cause whisker vaporization on metals. Steady-state operating levels for insulators are in the 2 MV/m (20 kV/cm) range. Therefore, vacuum insulators are the weak point in any high-voltage electrostatic accelerator. The mechanisms of breakdowns on insulator surfaces are not well understood. The following qualitative observations lead to useful procedures for optimizing insulator hold-off.

1. There are two main differences between insulators and metals that affect vacuum breakdown. First, electric fields parallel to surfaces can exist near insulators. Second, regions of space charge can be built up on insulators. These charges produce local distortions of electric fields.

2. A full-scale breakdown on an insulator is a complex process. Fortunately, it is usually sufficient to understand the low-current initiation processes that precede breakdown to predict failure levels.

3. Discharges on insulators are not initiated by whisker explosion since current cannot flow through the material. If the electric field is below the bulk dielectric strength of the insulator, discharge initiation is probably due to processes on nearby metal surfaces.

4. When voltage pulses are fast (< 50 ns), current flow from field emission on nearby electrodes is too small to cause serious charge accumulation on an insulator. In this case, the onset of insulator breakdown is probably associated with charge from whisker explosions on the metal electrodes. This is consistent with experiments; short-pulse breakdown levels on insulators between metal plates are in the 10–15 MV/m range.

5.   Electrons produced in a whisker explosion move away from a metal electrode because the electric field is normal to the surface. In contrast, electrons can be accelerated parallel to the surface of an insulator. High-energy electrons striking the insulator can lead to ejection of surface material. A microburst between metal electrodes in vacuum has a slow growth of discharge current (over many ion transit times) and may actually quench. In contrast, a discharge initiated along an insulator has a rapid growth of current and usually leads to a complete system short.

6.   For long voltage pulses, field emission from nearby metal electrodes leads to accumulated space charge on insulator surfaces and field distortion (Fig. 9.20b). Secondary emission coefficients on insulators are generally above unity in the energy range from 100 eV to a few keV. The impact of field-emitted electrons results in a net positive surface charge. This charge attracts more electrons, so that the electric field distortion increases. The nonuniform surface electric field fosters breakdown at levels well below that for whisker explosion on surrounding electrodes ($\sim$ 2 MV/m).

There are some steps that can be taken to maximize voltage hold-off along an insulator surface:

1.   The *triple point* is the location where metal and insulator surfaces meet in a vacuum. In high-voltage pulsed systems, whisker explosions near the triple points at both ends of the insulator can spray charge on the insulator surface. Electrostatic shielding of the triple point to prevent whisker explosions, as shown in Figure 9.20c, improves voltage hold-off.

2.   Electrons from whisker explosions are the most dangerous particles since they travel rapidly and have a secondary emission coefficient about two orders of magnitude higher than ions in the keV range. Voltage hold-off is high if the insulator is shaped so that electrons emitted from the negative electrode travel directly to the positive electrode without striking the insulator surface. A common insulator configuration is illustrated in Figure 9.20d. Insulator sections with angled surfaces are separated by grading rings. With the angle shown, all electrons emitted from either metal or insulator surfaces are collected on the rings. The hold-off level in both pulsed and dc columns may be doubled with the proper choice of insulator angle. A graph of breakdown level for pulsed voltages as a function of insulator angle is shown in Figure 9.21.

3.   Space-charge-induced field distortions in dc columns are minimized if the insulator is divided into a number of short sections separated by metal grading rings. The potential drop between rings is evenly distributed by divider network outside the vacuum. Current flowing through the divider relieves charge accumulation on the rings from field emission currents.

4.   Many pulsed voltage insulating columns are constructed with grading rings in the configuration shown in Figure 9.20d, although the utility of rings

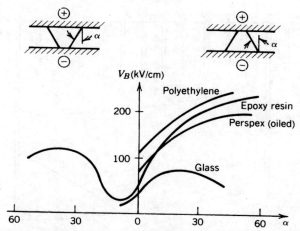

Figure 9.21 Insulator breakdown levels as function of insulator inclination angle for materials subject to fast voltage pulse (~ 30 ns). (Adapted from I. D. Smith, *Proc. First Int. Symp. Discharges and Elec. Insul. in Vac.*, MIT, Cambridge MA, 1964, p. 261.)

uncertain. Capacitive grading is usually sufficient to prevent field distortion for fast voltage pulses, especially if the medium outside the vacuum insulator has a high dielectric constant. Vacuum insulators in pulsed voltage systems usually have a breakdown level proportional to their length along the field.

Insulator hold-off in dc accelerator columns may be reduced with the introduction of a beam. Peripheral beam particles may strike the insulator directly or the beam may produce secondary particles. Beam induced ions are dangerous if they reach energy in the range > 100 keV. These ions have a high coefficient of secondary emission (> 10) and deposit large energy in a narrow range near the surface when they collide with an insulator. Figure 9.22 illustrates a section of a Van de Graaff accelerator column designed to minimize insulator bombardment. Grading rings are closely spaced and extend inward a good distance for shielding. Furthermore, the gradient along the rings is purposely varied. As we saw in studying the paraxial ray equation (Section 7.5), variation of E_z has an electrostatic lens effect on low-energy particle orbits. Secondary electrons, ions, and negative ions are overfocused and collected on the rings before they are able to accelerate to high energy along the column.

9.6 *LRC* CIRCUITS

This section initiates our study of pulsed voltage generators. All pulsed voltage circuits have an energy storage element where electrical energy is contained in the form of electric or magnetic fields. The energy is transferred by a fast

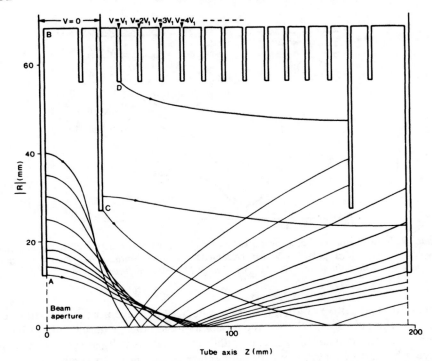

Figure 9.22 Insulator column for a Van de Graaff accelerator showing orbits of secondary ions liberated from electrodes. Variation of the grading ring spacing results in electrostatic overfocusing and radial loss of ions. (Courtesy T. Joy, Daresbury Laboratory.)

switch to a load. The speed of transfer (or the maximum power attainable) is limited by parasitic inductance or capacitance in the circuit. The voltage pulse waveform is determined by the configuration of the energy storage element and the nature of the load. In the next sections, we shall concentrate on simple resistive loads. The combination of energy storage element and switch is usually called a *voltage modulator*. The circuit produces a modulation, or variation in time, of the voltage.

There are four main accelerator applications of pulsed voltage circuits.

1.   High electric field can often be sustained in small systems for short times because of time-dependent processes controlling breakdowns. Sparking on vacuum insulators is a good example of this fact. Therefore, pulsed voltage modulators can be used to generate rapidly pulsed high-energy beams in compact systems.

2.   Pulsed accelerators are often required for the study of fast physical processes. Applications include pulsed X-ray radiography, material response at high pressure and temperatures, and inertial fusion.

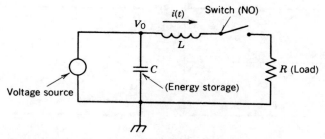

Figure 9.23 Pulse modulator with capacitive energy storage and closing switch.

3. Some beam applications require power at the multimegawatt level. Such systems are usually run at low-duty cycle. A system capable of supplying such power on a steady-state basis would require extensive equipment and cooling capability. In contrast, a pulsed power modulator stores energy over a long time and releases it in a short pulse. This process is called *power compression*. High-energy electron linacs (Section 14.1) illustrate this process. Strong accelerating gradient is obtained by injecting high pulsed electromagnetic energy flux into a slow-wave structure. Klystron tubes powered by modulated dc voltage from a capacitor bank generate the rf power.

4. Accelerators that utilize inductive isolation by ferromagnetic cores, such as the induction linac and the betatron, must operate in a pulsed mode.

The simplest electrical energy storage device is a single capacitor. The voltage modulator of Figure 9.23 consists of a capacitor (charged to voltage V_0) and a shorting switch to transfer the energy. The energy is deposited in a load resistor, R. The flow of current involved in the transfer generates magnetic fields, so we must include the effect of a series inductance L in the circuit.

The time-dependent voltage across the circuit elements is related to the current by

$$V_c = -V_0 + \int i\,dt/C,$$

$$V_L = L\,di/dt,$$

$$V_R = iR,$$

Setting the loop voltage equal to zero gives

$$L(d^2i/dt^2) + R(di/dt) + i/C = 0. \tag{9.63}$$

Equation (9.63) is solved with the boundary conditions $i(0+) = 0$ and $di/dt(0+) = V_0/L$, where $t = 0$ is the switching time. The second condition

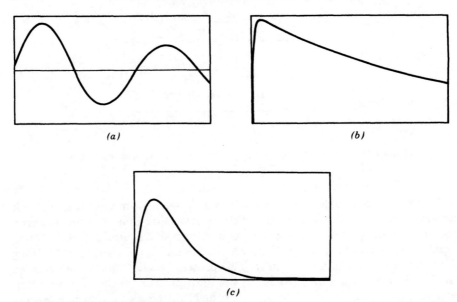

Figure 9.24 Voltage on a load driven by a switched capacitor with series inductance as a function of β/ω_0: L and C constant, load resistance (R) varied; $0 \leq \omega_0 t \leq 10$. (*a*) Underdamped circuit ($\beta/\omega_0 = 0.1$). (*b*) Critically damped circuit ($\beta/\omega_0 = 1$). (*c*) Overdamped circuit ($\beta/\omega_0 = 5$).

derives from the fact that immediately after switching the total capacitor voltage appears across the inductor rather than the resistor (see Section 9.1).

The solution of Eq. (9.63) is usually written in three different forms, depending on the values of the following parameters:

$$\omega_0 = 1/\sqrt{LC}\,, \qquad\qquad \boxed{9.64}$$

$$\beta = R/2L, \qquad\qquad \boxed{9.65}$$

$$\omega_1 = \sqrt{\omega_0^2 - \beta^2}\,,$$

$$\omega_2 = \sqrt{\beta^2 - \omega_0^2}\,,$$

and

$$\delta = \tan^{-1}(\beta/\omega_1).$$

The solution with $\beta < \omega_0$ is illustrated in Figure 9.24*a*. The circuit behavior is oscillatory. Energy is transferred back and forth between the inductor and capacitor at approximately the characteristic frequency ω_0. There is a small energy loss each oscillation, determined by the damping parameter β. The circuit is *underdamped*. We shall study the underdamped *LRC* circuit in more detail when we consider rf accelerators. The time dependent current is

$$i(t) = (CV_0/\cos\delta)\exp(-\beta t)[\omega_1\sin(\omega_1 t - \delta) + \beta\cos(\omega_1 t - \delta)],$$

$$\cong CV_0\omega_0\exp(-\beta t)\sin(\omega_0 t) \quad (\beta \ll \omega_0). \qquad \boxed{9.66}$$

The voltage on the load resistor is $i(t)R$.

The opposite extreme, *overdamping*, occurs when $\omega_0 < \beta$. As indicated in Figure 9.24b, the circuit is dominated by the resistance and does not oscillate. The monopolar voltage pulse on the load rises in a time of approximately L/R and decays exponentially over a time RC. The current following switching in an overdamped circuit is

$$i(t) = [CV_0(\beta^2 - \omega_2^2)/2\omega_2][\exp(\omega_2 t) - \exp(-\omega_2 t)]\exp(-\beta t).$$

$$\boxed{9.67}$$

An *LRC* circuit is *critically damped* when $\omega_0 = \beta$, or

$$R_c = 2\sqrt{L/C}. \qquad \boxed{9.68}$$

The current for a critically damped circuit is

$$i(t) = \beta CV_0(\beta t)\exp(-\beta t). \qquad \boxed{9.69}$$

The time-dependent load voltage is plotted in Figure 9.24c.

The waveforms in Figures 9.24a, b, and c have the same values of L and C with different choices of R. Note that the transfer of energy from the capacitor to the load resistor is accomplished most rapidly for the critically damped circuit. Thus, the power extracted from a pulsed voltage modulator is maximum when $R = R_c$. The quantity $2\sqrt{L/C}$, which has units of ohms, is called the *characteristic impedance* of the voltage modulator. Energy transfer is optimized when the load resistance is *matched* to the modulator impedance, as specified by Eq. (9.68). The peak power flow in a critically damped circuit occurs at a time $t = 1/\beta$. The maximum voltage at this time is $V_{max} = 0.736V_0$, the maximum current is $0.368V_0/\sqrt{L/C}$, and the maximum power in the load is $0.271V_0^2/\sqrt{L/C}$. The maximum power from this simple pulsed power generator is limited by the parasitic inductance of the circuit.

It is possible, in principle, to use an inductor for energy storage in a pulsed power circuit. A magnetic energy storage circuit is illustrated in Figure 9.25. In this case, a normally closed switch must be opened to transfer the energy to a load. We could also include the effects of parasitic capacitance in the circuit.

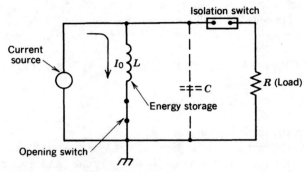

Figure 9.25 Pulse modulator with inductive energy storage and opening switch.

Charging of this capacitance limits the current risetime in the load, analogous to the inductance in the capacitive storage circuit. The main advantage of magnetic energy storage for pulsed power applications is that a much higher energy density can be stored. Consider, for instance, a Mylar insulated capacitor with $\varepsilon/\varepsilon_0 \cong 3$ and a field stress of 200 kV/cm. Applying Eq. (5.19), the electrical energy density is only 10^4 J/m^3 (0.01 J/cm^3). In contrast, the energy density in a vacuum inductor with a field of 2 T (20 kG) is 3×10^6 J/m^3 (3 J/cm^3). The reason magnetic storage is not regularly used is the lack of a suitable fast opening switch. It is relatively easy to make a fast closing switch that conducts no current until a self-sustained discharge is initiated, but it is difficult to interrupt a large current and hold off the subsequent inductive voltage spike.

9.7 IMPULSE GENERATORS

The single-capacitor modulator is suitable for voltages less than 100 kV, but is seldom used at higher voltages. Transformer-based power supplies for direct charging are very large above 100 kV. Furthermore, it is difficult to obtain commercial high-energy density capacitors at high voltage. *Impulse generators* are usually used for pulsed high voltage in the 0.1–10 MV range. These generators consist of a number of capacitors charged in parallel to moderate voltage levels (~ 50 kV). The capacitors are switched to a series configuration by simultaneously triggered shorting switches. The voltages of the capacitors add in the series configuration.

We will consider two widely used circuits, the *Marx generator*[†] and the *LC generator*. Impulse generators that produce submicrosecond pulses require less

[†] E. Marx, *Electrotech. Z.*, **45**, 652 (1925).

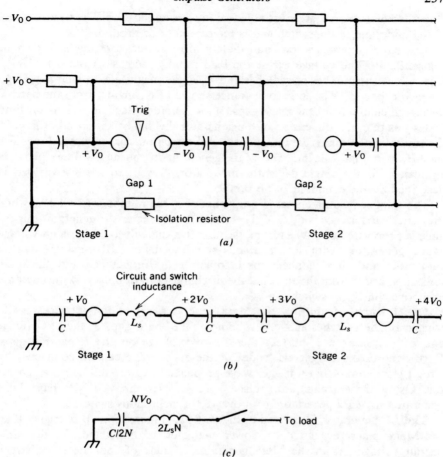

Figure 9.26 Marx generator. (*a*) Complete circuit diagram of a typical Marx generator showing charging lines. (*b*) Circuit model applicable for fast energy transfer following switching. (*c*) Simple lumped element circuit model for fast energy transfer neglecting electromagnetic wave phenomena.

insulation than a single capacitor with a steady-state charge at the full output voltage. The peak voltage is applied for only a short time, and breakdown paths do not have time to form in the liquid or gas insulation. High-field stress means that compact systems can be designed. Small systems have lower parasitic inductance and can therefore achieve higher output power.

A typical configuration for a Marx generator is illustrated in Figure 9.26. Positive and negative power supplies ($\pm V_0$) are used, charging the capacitors as shown in Figure 9.26a. Charging current is carried to the capacitors through isolation resistors (or inductors) that act as open circuits during the fast output pulse. The capacitor stack is interrupted by high-voltage shorting switches.

Gas-filled distortion spark gaps are ideal for this application. The trigger enters at the midplane of the switch and is referenced to dc ground.

The circuit configuration immediately after switch shorting is shown in Figure 9.26b. The voltage across the load rapidly increases from 0 to $2NV_0$, where N is the number of switches and V_0 is the magnitude of charge voltage on each capacitor. The series inductance shown arises mainly from the narrow discharge channels in the spark gaps. If we ignore voltage variations on time scales less than the dimension of the generator divided by the speed of light in the insulating medium, the inductances and capacitors can be lumped together as shown in Figure 9.26c. The Marx generator in the high-voltage phase is equivalent to the single capacitor modulator. Output to a resistive load is described by the equations of Section 9.6.

The total series inductance is proportional to the number of switches, while the series capacitance is $C_0/2N$. Therefore, the characteristic generator impedance is proportional to N. This implies that it is difficult to design high-voltage Marx generators with low characteristic impedance. High-energy density capacitors and short connections help lower the inductance, but the main limitation arises from the fact that the discharge current must flow through the inductive spark gap switches.

One favorable feature of the Marx generator is that it is unnecessary to trigger all the switches actively. If some of the spark gaps at the low-voltage end of the stack are shorted, there is overvoltage on the remaining gaps. Furthermore, the trigger electrodes of the spark gaps can be connected by circuit elements so that a trigger wave propagates rapidly through the generator. Using these techniques, pulsed voltages exceeding 10 MV have been generated in Marx generators with over 100 synchronous switches.

The LC generator, illustrated in Figure 9.27, is more difficult to trigger than the Marx generator, but it has lower characteristic impedance for the same output voltage. As in the Marx generator, a stack of capacitors is charged slowly in a parallel configuration by a positive–negative voltage supply (Fig. 9.27a). The main difference is that the switches are external to the main power flow circuit. Transition from a parallel to a series configuration is accomplished in the following way. Half of the capacitors are connected to external switched circuits with a series inductance. When the switches are triggered, each LC circuit begins a harmonic oscillation. After one half-cycle, the polarity on the switched capacitors is reversed. At this time, the voltages of all capacitors add in series, as shown in Figure 9.27b. The voltage at the output varies as

$$V_{\text{out}}(t) = (2NV_0)\left[1 - \cos\left(\sqrt{2/LC}\,t\right)\right]. \qquad (9.70)$$

where C is the capacitance of a single capacitor, N is the number of switches, and V_0 is the magnitude of the charge voltage.

The load must be isolated from the generator by a high-voltage switch during the mode change from parallel to series. The isolation switch is usually a low-inductance spark gap. It can be actively triggered or it can be adjusted

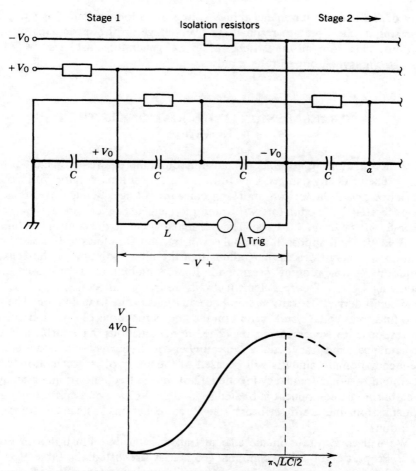

Figure 9.27 Circuit diagram of a typical *LC* generator, showing voltage at point *a* following switching through the external inductor.

for self-breakdown near the peak output voltage. Energy transfer to the load should take place at $t = \sqrt{LC/2}$ so that no energy remains in the external inductors. The transfer must be rapid compared to the mode change time or energy will return to the inductors, reducing the generator efficiency. Typically, an *LC* generator may have a mode change time of 6 μs and a transfer time to the load of 0.3 μs.

The equivalent circuit for the *LC* generator in the series state is the same as that for the Marx generator (Fig. 9.26c). The main difference is that the series inductances are reduced by elimination of the switches. The main disadvantages of the *LC* generator compared to the Marx generator are that the switching sequence is more complex, a low-inductance output switch is required, and the circuit remains at high voltage for a longer time. Triggering one reversing circuit does not result in an overvoltage on the spark gaps of other sections. Therefore, all the switches in an *LC* generator must be actively fired with strong, synchronized trigger pulses.

9.8   TRANSMISSION LINE EQUATIONS IN THE TIME DOMAIN

Most accelerator applications for pulse modulators require a constant-voltage pulse. The critically damped waveform is the closest a modulator with a single capacitor and inductor can approach constant voltage. Better waveforms can be generated by modulators with multiple elements. Such circuits are called *pulse-forming networks* (PFNs). The *transmission line* is the continuous limit of a PFN. We shall approach the analysis of transmission lines in this section by a lumped element description rather than the direct solution of the Maxwell equations. Application of transmission lines as modulators is discussed in the following section. Discrete element PFNs are treated in Section 9.11.

We will derive the transmission line equations in the time domain. The goal is to find total voltage and current on the line as functions of time and position in response to specified inputs. The input functions have arbitrary time dependence, and may contain a number of frequency components. The frequency-domain analysis will be used in the study of rf accelerators which operate at a single frequency (Section 12.6). In the frequency-domain analysis, each harmonic component is treated separately. Voltage and current along the transmission line are described by algebraic equations instead of differential equations.

We will concentrate on the coaxial transmission line, illustrated in Figure 9.28. Properties of other common geometries are listed in Table 9.3. The coaxial line consists of an inner conducting cylinder (of radius R_i) and a

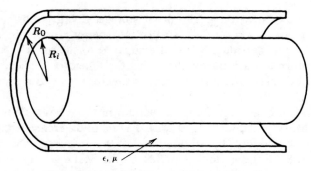

Figure 9.28 Section of a coaxial transmission line.

TABLE 9.3 Properties of Common Transmission Lines, TEM Modes[a]

1 Coaxial transmission line:

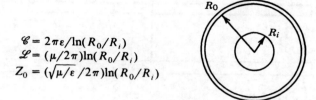

$$\mathscr{C} = 2\pi\varepsilon/\ln(R_0/R_i)$$
$$\mathscr{L} = (\mu/2\pi)\ln(R_0/R_i)$$
$$Z_0 = (\sqrt{\mu/\varepsilon}/2\pi)\ln(R_0/R_i)$$

2 Two-wire transmission line

$$\mathscr{C} = \pi\varepsilon/\cosh^{-1}(D/d)$$
$$\mathscr{L} = \mu/\pi \cosh^{-1}(D/d)$$
$$Z_0 = (\sqrt{\mu/\varepsilon}/\pi)\cosh^{-1}(D/d)$$

3 Isolated parallel plates $(d \ll D)$

$$\mathscr{C} = \varepsilon D/d$$
$$\mathscr{L} = \mu d/D$$
$$Z_0 = \sqrt{\mu/\varepsilon}\,(d/D)$$

4 Stripline $(d \ll D)$

$$\mathscr{C} = 2\varepsilon D/d$$
$$\mathscr{L} = \mu d/2D$$
$$Z_0 = \sqrt{\mu/\varepsilon}\,(d/2D)$$

[a] $\mathscr{C}$ = capacitance per unit length (farads/meter); $\mathscr{L}$ = inductance per unit length (henries/meter); Z_0 = characteristic impedance (ohms).

grounded outer cylinder (R_0) separated by a medium with dielectric constant ε. We assume a linear magnetic permeability, μ. Figure 9.29a shows a sectional view of a line divided into differential elements of length (Δz). Each element has a capacitance between the center and outer conductors proportional to the length of the element. If $\mathscr{C}$ is the capacitance per length, the capacitance of an element is $\mathscr{C}\Delta z$. Magnetic fields are produced by current flow along the center conductor. Each differential element also has a series inductance, $\mathscr{L}\Delta z$, where $\mathscr{L}$ is the inductance per unit length. The circuit model of Figure 9.29b can be applied as a model of the transmission line. The quantities $\mathscr{C}$ and $\mathscr{L}$ for

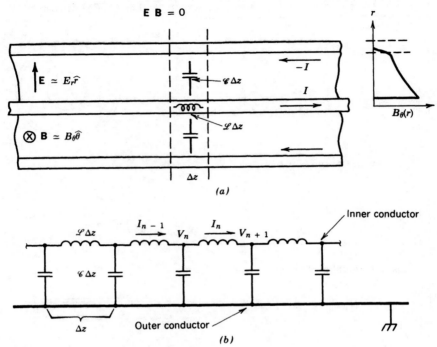

Figure 9.29 Coaxial transmission line. (*a*) Physical basis for lumped circuit element model of TEM wave propagation. (*b*) Lumped circuit element analog of a coaxial transmission line.

cylindrical geometry are given by

$$\mathscr{C} = 2\pi\varepsilon/\ln\frac{R_0}{R_i} \ (\text{F/m}),$$

9.71

$$\mathscr{L} = \frac{\mu}{2\pi}\ln\frac{R_0}{R_i} \ (\text{H/m}).$$

9.72

Before solving the circuit of Figure 9.29*b*, we should consider carefully the physical basis for the correspondence between the circuit model and the coaxial line. The following observations about the nature of electric and magnetic fields in the line are illustrated in Figure 9.29*a*.

1. Fast-pulsed magnetic fields cannot penetrate into a good conducting material. We shall study this effect, the magnetic skin depth, in Section 10.2. Since there is no magnetic field inside the center conductor, the longitudinal current must flow on the outer surface.

2. The outer conductor is thick compared to the magnetic skin depth. Therefore, magnetic fields resulting from current flow on the inner conductor

are not observed outside the line. Real current flowing on the outside of the inner conductor is balanced by negative current flow on the inside of the outer conductor. Magnetic fields are confined between the two cylinders.

3.   To an external observer, the outer conductor is an equipotential surface. There is no electrostatic voltage gradient along the conductor, and there is no inductively generated voltage since there are no magnetic fields outside the conductor. Furthermore, the inside surface of the outer conductor is at the same potential as the outside surface since there are no magnetic fields in the volume between the surfaces.

4.   If the distance over which current on the inner conductor varies is large compared to $(R_0 - R_i)$, then the only component of magnetic field is toroidal, B_θ.

5.   Similarly, if the voltage on the inner conductor varies over a long distance scale, then there is only a radial component of electric field.

Observations 1, 2, and 3 imply that we can treat the outer conductor as an ideal ground; voltage variations occur along the inner conductor. The two quantities of interest that determine the fields in the line are the current flow along the inner conductor and the voltage of the inner conductor with respect to the grounded outer conductor. Observations 4 and 5 define conditions under which electromagnetic effects can be described by the simple lumped capacitor and inductor model. The condition is that $\omega \ll v/(R_0 - R_i)$, where ω is the highest frequency component of the signal and $v = 1/\sqrt{\varepsilon\mu}$. At higher frequency, complex electromagnetic modes can occur that are not well described by a single capacitor and inductor in a length element.

Voltage differences along the center conductor are inductive. They are supported by changes in magnetic flux in the region between the two cylinders. Differences in current along the center conductor result from displacement current between the inner and outer conductors. The interaction of voltage, real current, and displacement current is illustrated in Figure 9.30a. The figure indicates the motion of a step input in voltage and current down a transmission line. The magnetic field behind the front is constant; the flux change to support the voltage differences between the inner and outer conductors comes about from the motion of the front. There is a rapid change of voltage at the pulse front. This supports a radial displacement current equal to the current of the pulse. Current returns along the other conductor to complete the circuit. The balance between these effects determines the relationship between voltage and current and the propagation velocity of the pulse.

Referring to Figure 9.29b, the voltage at point n is equal to the total current that has flowed into the point divided by the capacitance. Using the sign convention shown, this statement is expressed by

$$\int dt(I_{n-1} - I_n)/(\mathscr{C}\Delta z) = V_n. \qquad (9.73)$$

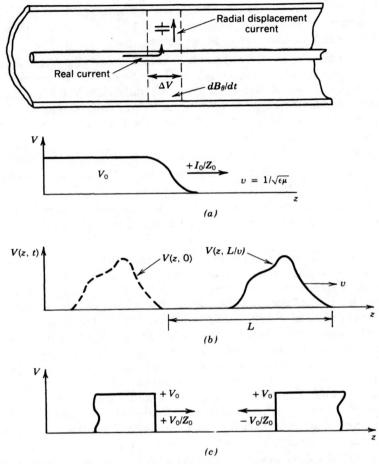

Figure 9.30 Coaxial transmission line. (*a*) Distribution of voltage, real current, and displacement current for a sharp rising step pulse propagating along a transmission line. (*b*) Distributions of voltage on a transmission line at time $t = 0$ (dashed line) and time $t = L/v$ (solid line). (*c*) Polarity conventions for positive voltage impulses traveling in the $+z$ and $-z$ directions.

Taking the derivative,

$$-\mathscr{C}\Delta z(\partial V_n/\partial t) = I_n - I_{n-1}. \qquad (9.74)$$

This is expressed as a partial differential equation since we are viewing the change in voltage with time at a constant position. The difference in voltage between two points is the inductive voltage between them, or

$$\mathscr{L}\Delta z(\partial I_n/\partial t) = V_n - V_{n+1}. \qquad (9.75)$$

If Δz becomes small, the discrete voltages approach a continuous function, $V(z, t)$. The voltage difference between two points can be approximated in terms of this function as

$$V_{n+1} \cong V_n + [\partial V(z_n, t)/\partial z]\, \Delta z. \tag{9.76}$$

Similarly,

$$I_{n-1} \cong I_n - [\partial I(z_n, t)/\partial z]\, \Delta z. \tag{9.77}$$

Substituting the results of Eqs. (9.76) and (9.77) into Eqs. (9.74) and (9.75) gives the continuous partial differential equations

$$\partial V/\partial z = -\mathscr{L}(\partial I/\partial t), \tag{9.78}$$

$$\partial I/\partial z = -\mathscr{C}(\partial V/\partial t). \tag{9.79}$$

Equations (9.78) and (9.79) are called the *telegraphist's equations*. They can be combined to give wave equations for V and I of the form

$$\partial^2 V/\partial z^2 = (\mathscr{L}\mathscr{C})(\partial^2 V/\partial t^2). \tag{9.80}$$

Equation (9.80) is a mathematical expression of the properties of a transmission line. It has the following physical implications:

1. It can easily be verified that any function of the form

$$V(z, t) = F(t \pm z/v) \tag{9.81}$$

is a solution of Eq. (9.80) if

$$v = 1/\sqrt{\mathscr{L}\mathscr{C}}. \qquad \boxed{9.82}$$

The spatial variation of voltage along the line can be measured at a particular time t by an array of probes. A measurement at a time $t + \Delta t$ would show the same voltage variation but translated a distance $v\,\Delta t$ either upstream or downstream. This property is illustrated in Figure 9.30b. Another way to phrase this result is that a voltage pulse propagates in the transmission line at velocity v without a change in shape. Pulses can travel in either the $+z$ or $-z$ directions, depending on the input conditions.

2. The current in the center conductor is also described by Eq. (9.80) so that

$$I(z, t) = G(z \pm vt). \tag{9.83}$$

Measurements of current distribution show the same velocity of propagation.

3.   The velocity of propagation in the coaxial transmission line can be found by substituting Eqs. (9.71) and (9.72) into Eq. (9.82):

$$v = 1/\sqrt{\varepsilon\mu} = c/\sqrt{(\varepsilon/\varepsilon_0)(\mu/\mu_0)}\,. \tag{9.84}$$

This velocity is the speed of light in the medium. The geometric factors in $\mathscr{C}$ and $\mathscr{L}$ cancel for all transmission lines so that the propagation velocity is determined only by the properties of the medium filling the line.

4.   Inspection of Eqs. (9.78) and (9.79) shows that voltage is linearly proportional to the current at all points in the line, independent of the function form of the pulse shape. In other words,

$$V = IZ_0, \qquad \boxed{9.85}$$

where Z_0 is a real number. The quantity Z_0 is called the *characteristic impedance* of the line. Its value depends on the geometry of the line. The characteristic impedance for a coaxial transmission line is

$$Z_0 = \sqrt{\mathscr{L}/\mathscr{C}} = \sqrt{\mu/\varepsilon}\ln(R_0/R_i)/2\pi. \qquad \boxed{9.86}$$

Rewriting Eq. (9.86) in practical units, we find that

$$Z_0 = \frac{60\ln(R_0/R_i)\sqrt{\mu/\mu_0}}{\sqrt{\varepsilon/\varepsilon_0}}\ (\Omega). \tag{9.87}$$

5.   Polarities of voltage and current often cause confusion. Conventions are illustrated in Figure 9.30c. A positive current is directed in the $+z$ direction; a negative current moves in the $-z$ direction. A positive-going current waveform creates a magnetic flux change that results in a positive voltage on the center conductor.

9.9   TRANSMISSION LINES AS PULSED POWER MODULATORS

A transmission line is a distributed capacitor. Energy is stored in the line when the center conductor is charged to high voltage. Transmission lines have the property that they can produce a constant-voltage output pulse when discharged into a resistive load. In order to understand this, we shall first consider the properties of signal propagation on finite length lines with resistive loads at an end. A load at the end of a transmission line is called a *termination*.

Figure 9.31 shows a transmission line extending from $z = 0$ to infinity driven by a voltage generator. The generator determines the entrance boundary

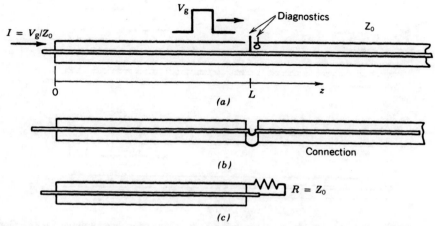

Figure 9.31 Matched terminations of a transmission line. (a) Infinite length line with characteristic impedance Z_0. (b) Finite length line connected to a semiinfinite length line with the same characteristic impedance. (c) Transmission line terminated by a matched resistor ($R = Z_0$).

condition on the line, V_g ($z = 0, t$). The generator supplies a current V_g/Z_0. The assumption of an infinite line means that there are no negatively directed waves. There is a positively directed wave produced by the generator. A voltage probe a distance L from the generator measures a signal $V_s(t') = V_g(t' - L/v)$. Similarly, a current probe gives a signal $I_s(t') = V_g(t' - L/v)/Z_0$.

Assume that the line is split at the point L and the pieces are reconnected together with a good coaxial connector. This change makes no difference in wave propagation on the line or the signals observed at L. Proceeding one step further, the infinite length of line beyond the split could be replaced with a resistor with value $R = Z_0$. The important point is that, in terms of observations at L, wave propagation with the resistive termination is indistinguishable from that with the infinite line. The boundary condition at the connection point is the same, $V(L, t) = I(L, t)Z_0$. In both cases, the energy of the pulse passes through the connector and does not return. In the case of the line, pulses propagate to infinity. With the resistor, the pulse energy is dissipated. The resistor with $R = Z_0$ is a *matched termination*.

We must consider the properties of two other terminations in order to understand the transmission line as an energy storage element. One of them is an open circuit, illustrated in Figure 9.32a. Assume the voltage generator produces a sharp rising step pulse with voltage $+V_0$ and current $+I_0$. When the step function reaches the open circuit, it can propagate no further. The open-circuit condition requires that a probe at position $z = L$ measures zero current. This occurs if the boundary reflects the original wave as shown, giving a pulse propagating in the negative z direction with positive voltage and negative current. When this pulse is added to the positive-going pulse from the generator, the net current is zero and the net voltage is $+2V_0$.

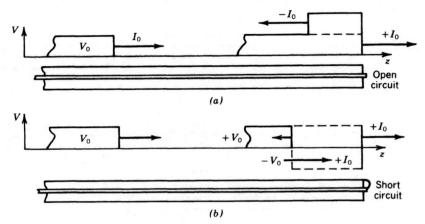

Figure 9.32 Reflection of step-function pulses at a transmission line termination. (*a*) Open circuit. (*b*) Short circuit.

Similar considerations apply to the short-circuit termination (Fig. 9.32*b*). A voltage probe at $z = L$ always measures zero voltage. The interaction of the incoming wave with a short-circuit termination generates a negative-going wave with negative voltage and positive current. In other words, the termination acts as an inverting transformer. These reflection properties can be confirmed by a direct numerical solution of Eqs. (9.78) and (9.79) with appropriate boundary conditions.

The circuit for a transmission line pulsed voltage modulator is shown in Figure 9.33. The center conductor of a line of length L is charged to high voltage by a power supply. The supply is connected through a large resistance or inductance; the connection can be considered an open circuit during the output pulse. The other end of the line is connected through a shorting switch to the load. We assume the load is a matched termination, $R = Z_0$.

Consider, first, the state of the charged line before switching. The center conductor has voltage V_0. There is no net current flow. Reference to Figure 9.33 shows that the standing voltage waveform of this state can be decomposed into two oppositely directed propagating pulses. The pulses are square voltage pulses with length L and magnitude $\frac{1}{2}V_0$. The positively directed pulse reflects from the switch open-circuit termination to generate a negatively directed pulse that fills in behind the original negative-going pulse. The same process occurs at the other open-circuit termination; therefore, the properties of the line are static.

The resolution of a static charge into two traveling pulses helps in determining the time-dependent behavior of the circuit following switching. We are free to choose the boundaries of the positive and negative pulses at any location; let us assume that they are at $z = 0$ and $z = L$ at the time of switching, $t = 0$. Following switching, the positive-going pulse travels into the termination while

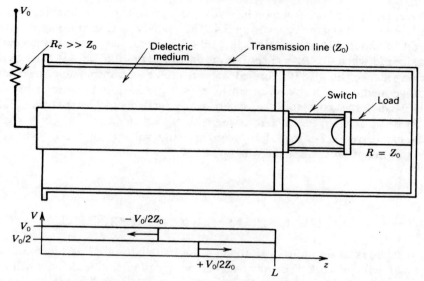

Figure 9.33 Coaxial transmission line used as a square-pulse generator.

the negative pulse moves away from it. The positive pulse produces a flat-top voltage waveform of duration $\Delta t = L/v$ and magnitude $\frac{1}{2}V_0$ in the load. At the same time, the negative-going pulse reflects from the open circuit at the charging end and becomes a positive-going pulse following immediately behind the original. The negative pulse deposits its energy in the load during the time $L/v \le t \le 2L/v$. In summary, the discharge of a transmission line through a shorting switch into a matched resistive load produces a constant-voltage pulse. The magnitude of the pulse is $\frac{1}{2}V_0$ and the duration is

$$\Delta t_{\mathrm{p}} = 2L/v.$$

$\boxed{9.88}$

where v is the pulse propagation velocity.

The total capacitance of a transmission line of length L can be expressed in terms of the single transit time for electromagnetic pulses ($\Delta t = L/v$) and the characteristic impedance as

$$C = \Delta t/Z_0.$$

$\boxed{9.89}$

It is easily demonstrated that energy is conserved when the line is discharged into a matched termination by comparing $\frac{1}{2}CV_0^2$ to the time integral of power into the load. The total series inductance of a transmission line is

$$L = Z_0 \Delta t.$$

$\boxed{9.90}$

Transformer oil is a common insulator for high-voltage transmission lines. It has a relative dielectric constant of 3.4. The velocity of electromagnetic pulses in oil is about 0.16 m/ns. Voltage hold-off in a coaxial transmission line is maximized when $R_i/R_0 = 1/e$. For oil this translates into a 33-Ω characteristic impedance [Eq. (9.87)]. Purified water is used as a transmission line energy storage medium in high-power density pulsed modulators because of its high relative dielectric constant ($\varepsilon/\varepsilon_0 \cong 80$). Water is conductive, so that water lines must be pulse charged. In comparison with oil, a water line with a $1/e$ radius ratio can drive a 6.8-Ω load. For the same charge voltage, the energy density in a water line is 24 times higher than in an oil line.

9.10   SERIES TRANSMISSION LINE CIRCUITS

Two features of the transmission line pulse modulator are often inconvenient for high-voltage work. First, the matched pulse has an amplitude only half that of the charge voltage. Second, the power transfer switch must be located between the high-voltage center conductor and the load. The switch is boosted to high voltage; this makes trigger isolation difficult. The problems are solved by the Blumlein transmission line configuration.[†] The circuit consists of two (or more) coupled transmission lines. Fast-shorting switches cause voltage reversal in half the lines for a time equal to the double transit time of electromagnetic pulses. The result is that output pulses are produced at or above the dc charge voltage, depending on the number of stages. The Blumlein line circuit is the distributed element equivalent of the *LC* generator.

We shall analyze the two-stage transmission line driving a matched resistive load. A circuit with nested coaxial transmission lines is illustrated in Figure 9.34*a*. The three cylinders are labeled OC (outer conductor), IC (intermediate conductor), and CC (center conductor). The diameters of the cylinders are usually chosen so that the characteristic impedance of the inner line is equal to that of the outer line. This holds when $R_{OC}/R_{IC} = R_{IC}/R_{CC}$. We neglect end effects and assume that both lines have the same electromagnetic transit time.

A high-voltage feed penetrates the outer cylinder to charge the intermediate cylinder. The center conductor is connected to ground through an isolation element. The isolator acts as a short circuit over long times but approximates an open circuit during the output pulse. It is usually a simple inductor, although we shall investigate a more complex isolator when we study linear induction accelerators (Section 10.5). A shorting switch between the IC and OC is located at the end opposite the load.

The equivalent circuit of the two-stage Blumlein line is shown in Figure 9.34*b*. In order to analyze the pulse output of the circuit, we make the

---

[†]A. D. Blumlein, U.S. Patent No. 2,465,840 (1948).

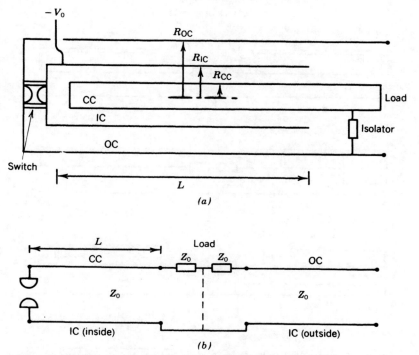

Figure 9.34 Nested coaxial transmission lines in the Blumlein configuration. (*a*) Schematic view and definition of quantities. (*b*) Equivalent circuit.

following assumptions:

1. The middle conductor is taken as a reference to analyze traveling voltage pulses.
2. The isolation element is an ideal open circuit during the pulse output.
3. The load resistance is $Z_1 = 2Z_0$, where Z_0 is the characteristic impedance of the individual nested lines.
4. An imaginary connection is attached from the intermediate conductor to the midpoint of the load during the output pulse. This connection is indicated by dotted lines in Figure 9.34*b*.

In the steady state, the intermediate conductor is charged to $-V_0$. The other two electrodes have positive voltage relative to the intermediate conductor. The static charge can be represented in the inner and outer lines as two oppositely directed positive voltage pulses of length L and magnitude $\frac{1}{2}V_0$. Pulses arriving at the end connected to the load are partially reflected and partially transmitted to the other line. Transmission is the same in both directions so that a steady state is maintained. Assume that the boundaries of positive and negative pulses are aligned as shown in Figure 9.35*a* at switching time $t = 0^-$.

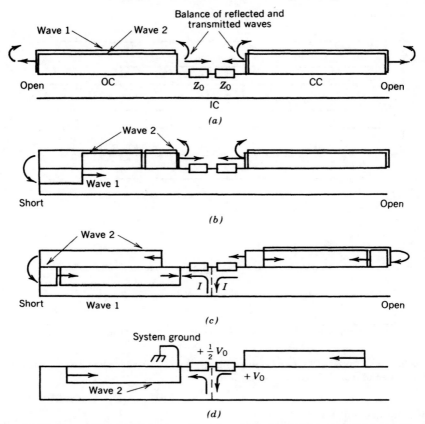

Figure 9.35 Propagation and reflection of step-function pulses following switching of a Blumlein line. (*a*) Resolution of static charge voltage prior to switching into interleaved traveling pulses. (*b*) Pulse polarities and directions during interval $0 < t < L/v$. (*c*) Pulses during interval $L/v < t < 2L/v$. (*d*) Pulses during interval $2L/v < t < 3L/v$.

The following events take place when the shorting switch is activated:

1. There is a time lag before information about the shorting is communicated to the load. During this time, the $+z$-directed pulse in the outer line and the $-z$-directed pulse in the inner line continue to move toward the load. The balance of transmission and reflection causes a positive-polarity reflected pulse to move backward in each line as though the midpoint were an open circuit (Fig. 9.35*b*).

2. The $+z$-directed pulse in the inner line moves to the open-circuit end opposite the load and is reflected with positive polarity. The major activity in the circuit is associated with the $-z$-directed pulse on the outer line. This pulse encounters the short circuit and reflects with negative polarity.

3. The negative-polarity pulse arrives at the load at $\Delta t = L/v$. We make the imaginary connection at this time. Power from the positive pulse in the inner line and the negative pulse in the outer line is deposited in the two matched loads (Fig. 9.35c). The voltages on the loads have magnitude $\frac{1}{2}V_0$ and the same polarity; therefore, there is a net voltage of V_0 between the connection points of the center and outer conductors to the load.

4. The situation of Figure 9.35c holds for $\Delta t < t < 2 \Delta t$. During this time, the other pulse components in the two lines reflect from the boundaries opposite the load. The polarity of the reflected wave is positive in the inner line and negative in the outer line.

5. The pulses that originally moved away from the load deposit their energy during the time $2 \Delta t < t < 3 \Delta t$ (Fig. 9.35d).

6. Inspection of Figures 9.35c and d shows that currents from the two lines are equal and opposite in the imaginary conductor during the output pulse. Since the connection carries no current, we can remove it without changing the circuit behavior.

The above model resolves the Blumlein line into two independent transmission lines driving a series load. The voltage in one line is reversed by reflection at a short circuit. Voltages that cancel in the static state add in the switched mode. In summary, the Blumlein line circuit has the following characteristics:

1. An output voltage pulse of magnitude V_0 is applied to a matched load between the center conductor and the outer conductor for the double transit time of an individual line, $2L/v$.
2. The voltage pulse is delayed from the switch time by an interval L/v.
3. The matched impedance for a two-stage Blumlein line is $2Z_0$.
4. A negatively charged intermediate conductor results in a positive output pulse when the switch is located between the intermediate and outer conductors. The output pulse is negative if a shorting switch is located between the intermediate and center conductors.

The Blumlein line configuration is more difficult to construct than a simple transmission line. Furthermore, the Blumlein line has no advantage with respect to energy storage density. Neglecting the thickness of the middle conductor and its voltage grading structures, it is easy to show that the output impedance, stored energy density, pulselength, and output voltage are the same as that for an equal volume transmission line (with $R_i = R_{CC}$ and $R_0 = R_{OC}$) charged to $2V_0$. The main advantage of the Blumlein line is that requirements on the charging circuit are relaxed. It is much less costly to build a 1-MV Marx generator than 2-MV Marx generator with the same stored energy. Furthermore, in the geometry of Figure 9.34, the switch can be a trigatron with the triggered electrode on the ground side. This removes the problem of trigger line isolation.

9.11   PULSE-FORMING NETWORKS

Transmission lines are well suited for output pulselengths in the range 5 ns $< \Delta t_p <$ 200 ns, but they are impractical for pulselengths above 1 μs. Discrete element circuits are usually used for long pulselengths. They achieve better output waveforms than the critically damped circuit by using more capacitors and inductors. Discrete element circuits that provide a shaped waveform are called *pulse-forming networks*.

The derivations of Section 9.8 suggest that the circuit of Figure 9.36 can provide a pulse with an approximately constant voltage. A transmission line is emulated by a finite number N of inductor–capacitor units. Following the derivation of Section 9.8, the resistance of a matched load is

$$Z_0 = \sqrt{L/C}, \qquad (9.91)$$

The quantities L and C are the inductance and capacitance of discrete elements. We shall call Z_0 the impedance of the PFN. The single transit time of an electromagnetic pulse through the network is approximately $N\sqrt{LC}$. The output voltage pulse has average magnitude $\frac{1}{2}V_0$ and duration $\Delta t_p \cong 2N\sqrt{LC}$.

The output pulse of a five-element network into a matched resistive load is shown in Figure 9.37. Although the general features are as expected, there is substantial overshoot at the beginning of the pulse and an undershoot at the end. In addition, there are voltage oscillations during the pulse. In some applications, these imperfections are not tolerable. For instance, the pulse modulator may be used to drive an ion injector where the beam optics depends critically on the voltage.

A Fourier analysis of the circuit of Figure 9.36 indicates the basis for the poor pulse shaping. The circuit generates N Fourier components with relative amplitudes optimized to replicate a sharp-edged square voltage pulse. In the Fourier series expansion of a square pulse, the magnitude of the terms of order n decreases only as the inverse of the first power of n, $a_n \sim 1/n$. Thus, many

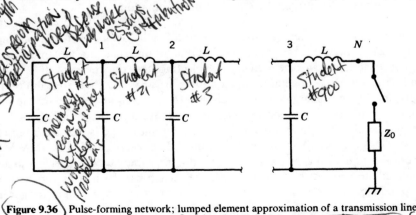

Figure 9.36   Pulse-forming network; lumped element approximation of a transmission line.

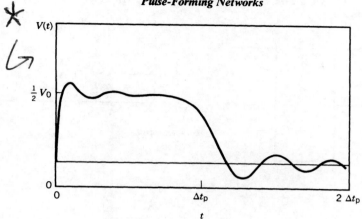

Figure 9.37 Load voltage, five-element PFN discharged into a matched load.

terms are needed for an accurate representation. In other words, a large number of elements is needed in the circuit of Figure 9.36 for a relatively constant output voltage. Such a division increases the size and cost of the modulator.

A different approach to the PFN, developed by Guillemin,[†] provides much better pulse shaping with fewer elements. He recognized that the slow convergence of the network of Figure 9.36 is a consequence of approximating a discontinuous waveform. In most applications, the main concern is a good voltage flat-top, and gradual voltage variation on the rise and fall of the pulse can be tolerated. The key is to work in reverse from a smooth waveform to derive a generating circuit.

We can utilize a Fourier series analysis if we apply the following procedure. Consider, first, an ideal transmission line discharged into a short-circuit load (Fig. 9.38a). The current oscillates between $+2I_0$ and $-2I_0$, where I_0 is the output current when the line is discharged into a matched load, $I_0 = V_0/2Z_0$. The periodic bipolar waveform can be analyzed by a Fourier series. When the circuit is connected to a matched load, there is a single square pulse (Fig. 9.38b). By analogy, if we could determine a circuit that produced a different bipolar current waveform with peak amplitude $2I_0$, such as the trapezoidal pulse of Figure 9.38c, then we expect that it would produce a trapezoidal pulse (Fig. 9.38d) of amplitude I_0 when connected to a resistance $R = V_0/2I_0$. We shall verify this analogy by direct computation.

It remains to determine a circuit that will produce the waveform of Figure 9.38c when discharged into a short. The Fourier series representation of a trapezoidal current pulse with magnitude $2I_0$, total length Δt_p and rise and fall

[†]E. A. Guillemin, *Communications Networks, Vol. II*, Wiley, New York, 1935.

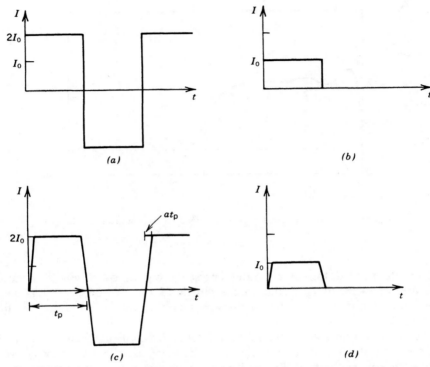

Figure 9.38 Design concept for Guillemin networks. (*a*) Output current of an ideal transmission line discharged into a short circuit. (*b*) Output current of an ideal transmission line discharged into a matched resistor. (*c*) Trapezoidal pulse of a Guillemin network discharged into a short circuit. (*d*) Guillemin network discharged into a matched resistor.

times $a \, \Delta t_p$ is

$$i(t) = 2I_0 \sum_n (4/n\pi)[\sin(n\pi a)/(n\pi a)]\sin(n\pi t/\Delta t_p). \qquad (9.92)$$

Consider the circuit of Figure 9.39. It consists of a number of parallel *LC* sections. If the PFN is discharged into a short circuit, the current flow through each of the sections is independent of the others. This occurs because the voltage across each section is zero. The current in a particular section after switching is

$$i_n(t) = i_{n0}\sin\left(t/\sqrt{L_n C_n}\right). \qquad (9.93)$$

We choose the inductance and capacitance of sections so that their free harmonic oscillations are at the frequency of the Fourier components in Eq.

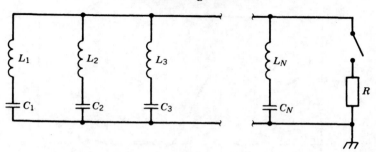

Figure 9.39 Basic Guillemin network.

(9.92). In other words,

$$\sqrt{L_1 C_1} = \Delta t_p/\pi, \qquad \sqrt{L_2 C_2} = \Delta t_p/2\pi, \qquad \sqrt{L_3 C_3} = \Delta t_p/3\pi, \dots . \quad (9.94)$$

We know from our study of the undamped LC circuit (Section 9.6) that the magnitude of the current flowing through any one of the sections after switching is

$$i_{n0} = V_0/\sqrt{L_n/C_n} . \qquad (9.95)$$

If Z_0 is the desired characteristic impedance, the magnitudes are matched to the Fourier series if

$$V_0/\sqrt{L_n/C_n} = (V_0/Z_0)(4/n\pi)[\sin(n\pi a)/n\pi a], \qquad (9.96)$$

where we have substituted $V_0/Z_0 = 2I_0$. Equations (9.94) and (9.96) can be solved to give the appropriate component values of the PFN:

$$L_n = (Z_0 \Delta t_p/4)/[\sin(n\pi a)/\pi a],$$

$$C_n = (4\Delta t_p/n^2\pi^2 Z_0)[\sin(n\pi a)/n\pi a]. \qquad \boxed{9.97}$$

Figure 9.40 shows a voltage pulse on a matched resistor ($R = Z_0$) for three and five-element Guillemin networks using the circuit values determined from Eqs. (9.97). Note the improvement compared to the pulseshape of Figure 9.37. Equation (9.92) shows that the Fourier expansion converges as $1/N^2$. A better flat-top can be obtained by approximating smoother pulses. For instance, the series converges as $1/N^3$ if the pulse is assumed to have parabolic edges with no discontinuity in slope. Alternate circuits to that of Figure 9.39 can be derived. The main disadvantage of the Guillemin network is that the capacitors all have different values. It is usually difficult to procure commercial high-voltage capacitors with the required capacitances.

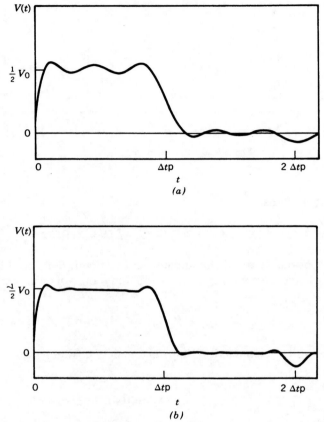

Figure 9.40 Load voltage, Guillemin network discharged into a matched resistor. (*a*) Three-element network. (*b*) Five-element network.

9.12 PULSED POWER COMPRESSION

Electron and ion beams in the megampere range have been generated in pulsed power diodes.[†] Such beams have application to inertial fusion and studies of materials at high temperature, pressure, and radiation levels. High-power pulse modulators are needed to drive the diodes. Pulsed power research is largely centered on extending the limits of the output power of voltage generators. At present, single-unit generators have been built that can apply power in the TW (10^{12} W) range to a load. Typical output parameters of such a modulator are 1 MA of current at 1 MV in a 50-ns pulse. Parallel generators have reached levels of 10 TW.

[†]See T. H. Martin and M. F. Rose, Eds., *Proc. 4th IEEE Pulsed Power Conference*, IEEE, Piscataway, New Jersey (83CH1908-3), 1983.

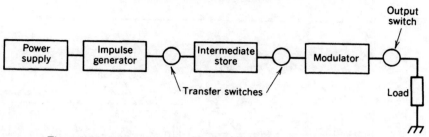

Figure 9.41 Schematic diagram of a pulsed power compression sequence.

Power compression is the technique that has allowed the generation of such high output. A general power compression circuit is illustrated in Figure 9.41. Energy is transferred from one stage of energy storage to the next in an increasingly rapid sequence. Each storage stage has higher energy density and lower inductance. Even though energy is lost in each transfer, the decrease in transfer time is sufficient to raise the peak power.

The first power compression stage in Figure 9.41 is one we are already familiar with. A dc source charges an impulse generator such as a Marx generator. The output power from the Marx generator may be a factor of 10^7 higher than the average power from the source. In the example, the Marx generator transfers its energy to a low-inductance water-filled capacitor. A shorting switch then passes the energy to a low-impedance transmission line. A low-inductance multichannel switch then connects the line to a vacuum load. Power multiplication in stages subsequent to the Marx generator are not so dramatic; gains become increasingly difficult. Figure 9.42 is a scale drawing of a pulsed power compression system to generate 300-kV, 300-kA, 80-ns pulses. The system consists of a low inductance, 1-MV, *LC* generator that pulse-charges a 1.5-Ω, water-filled transmission line. Energy from the line is transferred by a multielement, high-pressure gas switch. The pulse is matched to a 1-Ω electron beam load by a coaxial line transformer.

The highest power levels have been achieved with multiple stages of capacitive energy storage connected by sequenced shorting switches. Triggered gas-filled switches are used in the early stages. Gas switches have too much inductance for later stages, so that self-breakdown between electrodes in a highly stressed liquid medium is used. Discharges in liquids absorb considerable energy. The resulting shock wave rapidly erodes electrodes. Therefore, machines of this type are fired typically 1–10 times per day and may need repair after 10–100 shots. In Section 9.13, we shall study a potential method for low-inductance switching at high repetition rate, saturable core magnetic switching. Characteristics of the EAGLE pulsed power generator are summarized in Table 9.4.

Transfer of energy between capacitors forms the basis of most pulsed power generators. A model for the transfer is illustrated in Figure 9.43. A capacitor is

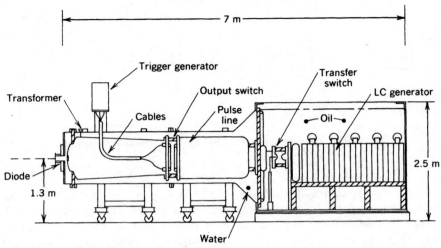

Figure 9.42 Neptune C, low-impedance pulsed power generator using overstressed water dielectric. (Courtesy H. Milde, Ion Physics Corporation.)

charged to voltage V_0, and then energy is switched through an inductance to a second capacitor by a shorting switch. The inductance may be introduced purposely or may represent the inevitable parasitic inductance associated with current flow. Current in the circuit is described by the equation

$$-V_0 + \int i \, dt/C_1 + L(di/dt) + \int i \, dt/C_2 = 0$$

or

$$L(d^2i/dt^2) = -iC_1C_2/(C_1 + C_2). \tag{9.98}$$

If the switch is closed at $t = 0$, then the initial conditions are $i(0) = 0$ and $di(0)/dt = V_0/L$. The solution of Eq. (9.98) is

$$i(t) = \frac{V_0 \sin \omega t}{\left[L(C_1 + C_2)/C_1C_2\right]^{1/2}}, \tag{9.99}$$

where

$$\omega = \left[LC_1C_2/(C_1 + C_2)\right]^{-1/2}.$$

The quantities of interest are the time-dependent charge voltages on the two capacitors.

$$V_1(t) = V_0 - \int i/C_1, \tag{9.100a}$$

TABLE 9.4   Parameters: EAGLE Pulsed Power Generator[a]

EAGLE module

1. Power compression cycle[b]

Element	Energy Transfer Time (μs)	Output Switch
Marx generator	1.3	Gas switches (25), active trigger
Transfer capacitor water-filled	0.3	Triggered gas switch, high-voltage isolation
Charging pulseline	0.1	Multipin, self-breaking water switch
Pulse-forming line	0.1	Multipin, self-breaking water switch

2. Measured parameters

Marx generator, stored energy	500 kJ
Marx generator voltage	3.15 MV
Energy transfer efficiency	50%
Output power	4.5×10^{12} W
Output energy	275 kJ
Output current (1.9-Ω load)	1.6 MA
Pulselength (FWHM power)	75 ns
Output voltage	3 MV

ROULETTE-X (Proposed)

Function	Generation of Bremsstrahlung radiation and soft X rays for nuclear weapons effect simulation
Number of EAGLE modules	20
Design power at load	7×10^{13} W
Pulselength (FWHM power)	75 ns
Output voltage	2–2.5 MV
Output energy	5 MJ
Output current	20–22 MA

[a] Physics International Company
[b] Configuration: Marx generator with three stages of power compression using overstressed water dielectric. First two elements perform power compression in time, last two perform compression in space for multimodule compatibility.

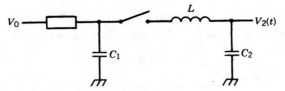

Figure 9.43   Peaking capacitor circuit.

and

$$V_2(t) = \int i/C_2.$$ (9.100b)

Substituting Eq. (9.99) into Eqs. (9.100a) and (9.100b), we find that

$$V_1 = V_0\{1 - [C_2/(C_1 + C_2)](1 - \cos \omega t)\}.$$ $\boxed{9.101}$

$$V_2 = V_0[C_1/C_1 + C_2)](1 - \cos \omega t).$$ $\boxed{9.102}$

Waveforms are plotted in Figure 9.44 for $C_2 = C_1$ and $C_2 \ll C_1$. The first case is the optimum choice for a high-efficiency power compression circuit. A complete transfer of energy from the first to the second capacitor occurs at time $t = \pi/\omega$. In the second case, the energy transfer is inefficient but the second capacitor is driven to twice the charge voltage of the first. For this reason, the circuit of Figure 9.43 is often called the *peaking capacitor circuit*.

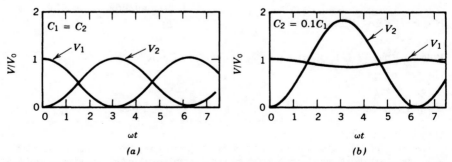

Figure 9.44   Voltage waveforms on storage capacitor and charged capacitor in a peaking capacitor circuit. (a) $C_2 = C_1$. (b) $C_2 = 0.1C_1$.

9.13 PULSED POWER SWITCHING BY SATURABLE CORE INDUCTORS

Although the process of pulse shaping by saturable core ferromagnetic inductors has been used for many years in low-voltage circuits, application to high-power circuits is a recent area of interest. Magnetic switches can be constructed with low inductance and high transfer efficiency. Energy losses are distributed evenly over the core mass, and there is no deionization time as there is in gas or liquid breakdown switches. These factors give magnetic switches potential capability for high repetition rate operation.

A two-stage power compression circuit with a magnetic switch is shown in Figure 9.45a. Capacitors C_0 and C_1 constitute a peaking circuit. Energy is transferred by a normal shorting switch. The energy in C_1 is then routed to a load by a saturable core inductor switch. The circuit achieves power compression if the switch out time from C_1 is short compared to the initial transfer. We choose $C_0 = C_1$ for high efficiency.

In order to understand the operation of the circuit of Figure 9.45a, we must refer to the hysteresis curve of Figure 5.12. We assume that the ferromagnetic core of the inductor is ideal; its properties are described by a static hysteresis

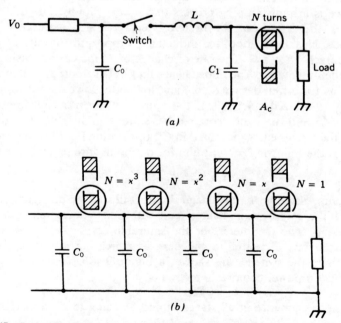

(a)

(b)

Figure 9.45 Saturable core magnetic switching. (*a*) Peaking capacitor circuit with magnetic output switch. (*b*) Multistage power compression circuit with uniform capacitors and ferromagnetic cores.

curve. This means that there are no real currents flowing in the core material; magnetic effects are produced solely by alignment of atomic currents. Methods of core construction to assure this condition are described in Chapter 10. We further assume that at some time before $t = 0$ a separate circuit has pulsed a negative current through the ferromagnetic core sufficient to bias it to $-B_s$. After the current is turned off, the core settles to a state with flux level $-B_r$. This process is called *core reset*.

Consider the sequence of events after the shorting switch is activated. At early times, the right-hand portion of the circuit is approximately an open circuit because of the high inductance of the winding around the high μ core. Energy flows from C_0 to C_1; the voltage on C_1 is given by Eq. (9.102). Further, Faraday's law implies that

$$V_1(t) = NA_c(dB/dt),\tag{9.103}$$

where A_c is the cross-sectional area of the core and N is the number of windings in the inductor. The core reaches saturation when

$$\int V_1(t)\,dt = NA_c(B_s + B_r).\tag{9.104}$$

After saturation, the inductance L_2 decreases by a large factor, approaching the vacuum inductance of the winding. The transition from high to low inductance is a bootstrapping process that occurs rapidly. Originally, translation along the H axis of the hysteresis curve is slow because of the high inductance. Near saturation, the inductance drops and the rate of change of leakage current increases. The rate of change of H increases, causing a further drop in the inductance. The impedance change of the output switch may be as fast as 5 ns for a well-designed core and low-inductance output winding.

Energy is utilized efficiently if the output switch has high impedance for $0 \le t \le \pi/\omega$ and the switch core reaches saturation at the end of the interval when all the circuit energy is stored in C_1. Integrating Eq. (9.103) over this time scale gives the following prescription for optimum core parameters:

$$NA_c(B_s + B_r) = V_0\pi/2\omega.\tag{9.105}$$

Power compression in the two-stage circuit is illustrated in Figure 9.46*a*.

A multiple compression stage circuit with magnetic switching is shown in Figure 9.45*b*. The parameters of the saturable cores are chosen to transfer energy from one capacitor to the next at increasing power levels. Current waveforms in the switches are shown in Figure 9.46*b*. There are a number of constraints on power compression circuits:

1. The capacitance in all stages should be equal to C_0 for high efficiency.
2. Switching of each stage should occur at peak energy transfer. This means that

$$\pi\sqrt{L_{n-1}^s C_0/2} = 2N_n A_c(B_s + B_r).\qquad \boxed{9.106}$$

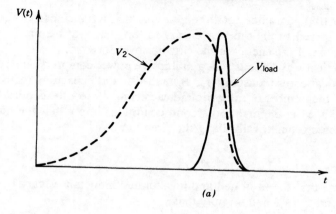

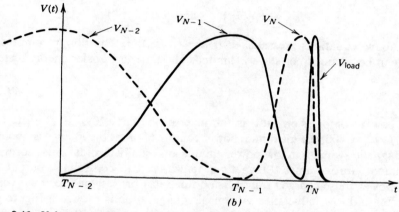

Figure 9.46 Voltage waveforms—saturable core magnetic switching. (*a*) Voltage on charged capacitor and load in single-stage circuit. (*b*) Voltages on last three capacitors and load in multistage network.

In Eq. (9.106), L_{n-1}^s is the saturated (or vacuum) inductance of the winding around core $n - 1$ and N_n is the number of turns around core n.

3. The unsaturated inductance of winding n should be much larger than the saturated inductance of winding $n - 1$, or

$$L_n^u \gg L_{n-1}^s. \qquad (9.107)$$

Equation (9.107) implies that *prepulse* is small; power does not move forward in the circuit until switching time.

4. The saturated inductance of winding n should be small compared to the saturated inductance of winding $n - 1$,

$$L_n^u \ll L_{n-1}^u. \qquad (9.108)$$

Equation (9.108) guarantees that the time for energy transfer from C_n to C_{n+1} is short compared to the time for energy to flow backward from C_n to C_{n-1}; therefore, most of the energy in the circuit moves forward.

One practical way to construct a multistage power compression circuit is to utilize identical cores (constant A_c, B_s, and B_r) and capacitors (constant C_0) and to vary the number of turns around each core to meet the conditions listed above. In this case, the first and second conditions imply that the number of turns decreases geometrically along the circuit or

$$N_n = N_{n-1}/\kappa.$$

If we assume that the saturated and unsaturated inductances vary by a factor of μ/μ_0, conditions 3 and 4 imply that

$$\kappa^2 \ll (\mu/\mu_0) \quad \text{and} \quad \kappa^2 \gg 1.$$

The above equations are satisfied if $\kappa = (\mu/\mu_0)^{1/2}$. The power compression that can be attained per stage is limited. The time to transfer energy scales as

$$\tau_n/\tau_{n-1} \cong (L_n/L_{n-1})^{1/2} \cong 1/\kappa. \tag{9.109}$$

The power compression ratio is the inverse of Eq. (9.109). For ferrite cores with $\mu/\mu_0 \cong 400$, the maximum power compression ratio per stage is about 20.

Magnetic switches in low-power circuits are usually toroidal pulse cores with insulated wire windings. Magnetic switches used to control the output of a high-power pulse-charged transmission line must be designed for high-voltage standoff and low inductance. A typical configuration is shown in Figure 9.47. Saturable core inductors can also be used for pulselength shortening and

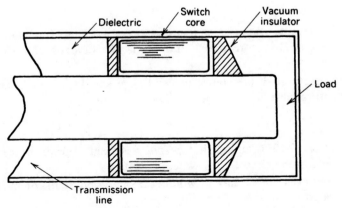

Figure 9.47 Pulse-charged coaxial transmission line with a low-inductance, saturable core magnetic output switch.

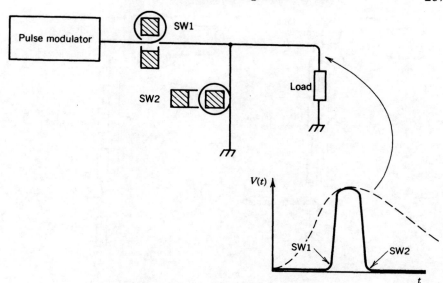

Figure 9.48 Pulse shaping with saturable core magnetic switches.

risetime sharpening if efficiency is not a prime concern. The circuit illustrated in Figure 9.48 produces a short, fast-rising voltage pulse from a slow pulse generator.

9.14 DIAGNOSTICS FOR PULSED VOLTAGES AND CURRENTS

Accurate measurements of acceleration voltage and beam current are essential in applications to charged particle acceleration. The problem is difficult for pulsed beams because the frequency response of the diagnostic must be taken into account. In this section, we shall consider some basic diagnostic methods.

The diagnostic devices we shall discuss respond to electric or magnetic fields over a limited region of space. For instance, measurement of the current through a resistor gives the instantaneous spatially averaged electric field over the dimension of the resistor. Confusion often arises when electric field measurements are used to infer a voltage. In particle acceleration applications, voltage usually means the energy a charged particle gains or loses crossing the region of measurement. Signals from voltage diagnostics must be carefully interpreted at high frequency because of two effects.

1. Particle energy gain is not equal to $\int \mathbf{E} \cdot \mathbf{dx}$ (where the integral is taken at a fixed time) when the time interval for the particle to cross the region is

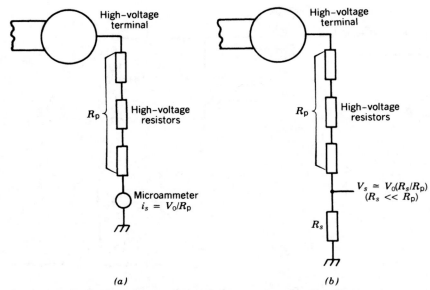

Figure 9.49   High-voltage measurements with resistor strings. (*a*) Resistive shunt. (*b*) Resistive divider.

comparable to or less than the period of voltage oscillations. This is called the *transit time effect* and is treated in Section 14.4.

2.   The voltage difference between two points is not a useful concept when the wavelength of electromagnetic oscillations is less than or comparable to the dimension of the region. In this case, electric field may vary over the length of the diagnostic. For example, in the presence of a bipolar electric field pattern, a voltage diagnostic may generate no signal.

Similarly, magnetic field measurements can be used to infer current flow at low frequency, but caution should be exercised in interpretation at high frequency.

A.   Steady-State Voltage Measurement

The only time-independent device that can measure a dc voltage is the resistive shunt (Fig. 9.49*a*). The resistive shunt consists of a resistor string of value R_p attached between the measurement point and ground. The quantity R_p must be high enough to prevent overloading the voltage source or overheating the resistors. The net voltage can be inferred by measuring current flow through the chain with a microammeter,

$$i_s = V_0/R_p. \tag{9.110}$$

Another approach is to measure the voltage across a resistor R_s at the low-voltage end of the chain. This configuration is called a *resistive divider* (Fig. 9.49b). The *division ratio*, or the signal voltage divided by the measured voltage, is

$$V_s/V_0 = R_s/R_p. \qquad \boxed{9.111}$$

Capacitive voltage measurements of dc voltages are possible if there is a time variation of capacitance between the source and the diagnostic. The charge stored in a capacitance between a high-voltage electrode and a diagnostic plate (C_p) is

$$Q = C_p V_0.$$

The total time derivative of the charge is

$$i_s = dQ/dt = C_p(dV_0/dt) + V_0(dC_p/dt). \qquad (9.112)$$

In pulsed voltage measurements, C_p is maintained constant and a diagnostic current is generated by dV_0/dt. Measurements of dc voltages can be performed by varying the capacitance as indicated by the second term on the right-hand side of Eq. (9.112).

The voltage probe of Figure 9.50 has an electrode exposed to electric fields from a high-voltage terminal. A rotating grounded disc with a window changes the mutual capacitance between the probe and high-voltage electrodes, inducing a current. The following example of a voltage measurement on a Van de Graaff accelerator illustrates the magnitude of the signal current. Assume the high-voltage electrode is a sphere of radius R_0 in a grounded spherical chamber of radius R_1. With gas insulation, the total capacitance between the outer and inner electrodes is

$$C_t \cong 4\pi\varepsilon_0 R_0 R_1/(R_1 - R_0). \qquad (9.113)$$

We assume further that the probe electrode and rotating window are near ground potential and that their surfaces are almost flush with the surface of the outer electrode. Field lines between the high-voltage electrode and the probe are radial; therefore, the mutual capacitance is equal to C_t times the fraction of outer electrode area occupied by the probe. If A_p is the area of the hole in the rotating plate, then the capacitance between the high-voltage electrode and probe varies between

$$0 \le C_p \le C_t A_p/4\pi R_1^2.$$

As an example, take $V_0 = 1$ MV, $R_0 = 0.3$ m, $R_1 = 1$ m, and $A_p = 8 \times 10^{-3}$ m^2. A plate, rotating at 1000 rmp ($\omega = 100$ s^{-1}), obscures the probe half

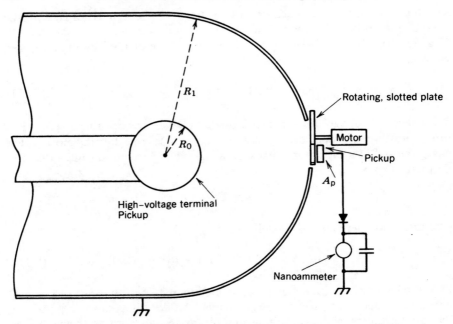

Figure 9.50   Generating voltmeter; high-voltage probe with time-varying capacitance.

the time. The total spherical capacitance is $C_t = 100$ pF; the mutual capacitance between terminal and probe is 0.07 pF. The signal current is approximately $i_s \cong V_0(\omega C/2)\cos \omega t$. Substituting values, the magnitude of the ac current is 3.5 μA, an easily measured quantity.

B.   Resistive Dividers for Pulsed Voltages

The electrostatic approximation is usually applicable to the measurements of output voltage for pulsed power modulators. For example, electromagnetic waves travel 15 m in vacuum during a 50-ns pulse. The interpretation of voltage monitor outputs is straightforward as long as the dimensions of the load and leads are small compared to this distance.

Resistive dividers are well suited to pulsed voltage measurements, but some care must be exercised to compensate for frequency-dependent effects. Consider the divider illustrated in Figure 9.49. Only the resistance values are important for dc measurements, but we must include effects of capacitance and inductance in the structure for pulsed measurements. A more detailed circuit model for the resistive divider is shown in Figure 9.51. There is inductance associated with current flow through the resistors and a shunt capacitance between points on the divider.

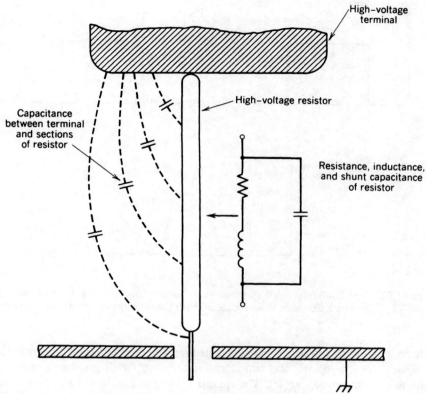

Figure 9.51 Effect of stray capacitance, shunt capacitance, and series inductance on resistive voltage divider.

A particularly unfavorable situation occurs when the primary resistor chain is a water solution resistor and the signal resistor is an ordinary carbon-composition resistor. Water resistors have high dielectric strength and good energy-absorbing ability, but they also have a large shunt capacitance, C_p. Therefore, on time scales less than $R_0 C_p$, the circuit acts as a differentiator. A square input pulse gives the pulse shape shown in Figure 9.52b.

The above example illustrates a general problem of pulsed voltage attenuators; under some circumstances, they may not produce a true replication of the input waveform. We can understand the problem by considering the diagnostic as a device that transforms an input waveform to an output waveform. We can represent the process mathematically by expressing the input signal as a Fourier integral,

$$V_0(t) = \int a(\omega)\exp(j\omega t), \tag{9.114}$$

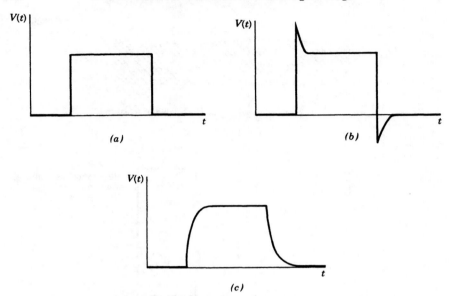

Figure 9.52   Waveforms from resistive voltage divider. (*a*) Ideal output voltage from a square-pulse input. (*b*) Output with significant shunt capacitance. (*c*) Output with significant series inductance.

where $a(\omega)$ is a complex number. Neglecting the electromagnetic transit time through the diagnostic, the transform function of a diagnostic system is a function of frequency, $g(\omega)$. The quantity $g(\omega)$ is a complex number, representing changes in amplitude and phase. The output signal is

$$V_s(t) = \int a(\omega)g(\omega)\exp(j\omega t). \qquad (9.115)$$

In an ideal diagnostic, the transformation function is a real number independent of frequency over the range of interest. When this condition does not hold, the shape of the output signal differs from the input. In the case of the water solution resistive divider, $g(\omega)$ is constant at low frequency but increases in magnitude at high frequency. The result is the spiked appearance of Figure 9.52*b*. Conversely, if the inductance of the divider stack is a dominant factor, high frequencies are inhibited. The pulse risetime is limited to about L/R_0, producing the pulse shown in Figure 9.52*c*.

Pulse shapes can often be reconstructed by computer if the transfer function for a diagnostic and the associated cabling is known. Nonetheless, the best practice is to design the diagnostic for flat frequency response. Devices with frequency variations must be compensated. For example, consider the water solution resistive divider. A method of compensation is evident if we recognize the fact that pulsed voltage dividers can be constructed with capacitors or

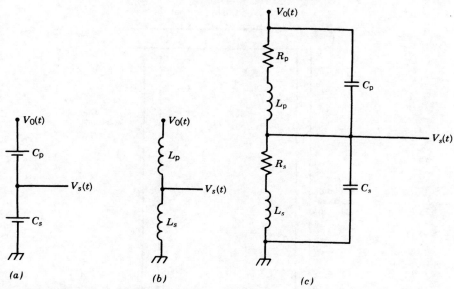

Figure 9.53 Voltage dividers. (*a*) Capacitive. (*b*) Inductive. (*c*) Balanced voltage divider.

inductors. In the capacitive divider (Fig. 9.53*a*), the division ratio is

$$V_s/V_0 = C_p/C_s, \qquad \boxed{9.116}$$

and for an inductive divider (Fig. 9.53*b*),

$$V_s/V_0 = L_s/L_p. \qquad \boxed{9.117}$$

A *balanced divider* is illustrated in Figure 9.53*c*. Extra circuit components (C_s and L_s) have been added to the probe of Figure 9.51 to compensate for the capacitance and inductance of the probe. The divider is balanced if

$$R_s/R_p = C_p/C_s = L_s/L_p. \qquad \boxed{9.118}$$

The component C_s pulls down the high-frequency components passed by C_p, eliminating the overshoot of Figure 9.52*a*. Similarly, the inductance L_s boosts low-frequency components which were overattenuated by L_p to improve the risetime of Figure 9.52*b*. Sometimes, C_p and L_p are not known exactly, so that variable C_s are L_s are incorporated. These components are adjusted to give the best output for a fast-rising square input pulse.

Balanced dividers can also be achieved utilizing geometric symmetry. A balanced water solution resistive divider is shown in Figure 9.54. The signal

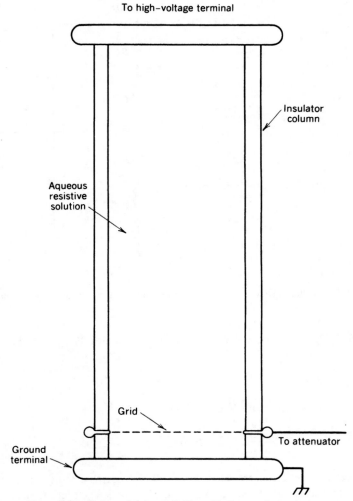

Figure 9.54 Balanced voltage divider with water solution resistor.

pick-off point is a screen near the ground plane. The probe resistor and signal resistor, R_p and R_s, share the same solution. The division ratio is the ratio of the length R_s to the length of R_p. The high dielectric constant of the water solution assures that electric field lines in the water are parallel to the column [see Eq. (5.6)]. Inspection of Figure 9.54 shows that the resistances, capacitances, and inductances are all in the proper ratio for a balanced divider when the solution resistance, resistor diameter, and electric field are uniform along the length of the resistor.

C.   Capacitive Dividers for Pulsed Voltages

High-voltage dividers can be designed with predominantly capacitive coupling to the high-voltage electrode. Capacitive probes have the advantage that there is no direct connection and there are negligible electric field perturbations. Two probes for voltage measurements in high-voltage transmission lines are shown in Figures 9.55a and b.

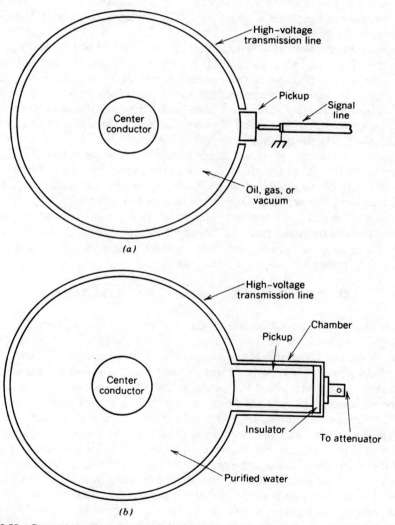

Figure 9.55   Capacitive voltage monitors in high-voltage transmission line. (a) Capacitive dV/dt pickup. (b) Balanced capacitive divider.

The first probe is suitable for use in an oil or gas-filled line where the resistance of the medium is effectively infinite and the mutual capacitance between terminal and probe is low. The capacitance between the probe and the wall, C_s is purposely made small. The probe is connected directly to a signal transmission line of impedance Z_0. The signal voltage on the line is

$$V_s \cong (C_P Z_0) \, dV_0/dt. \tag{9.119}$$

This signal is processed by an integration circuit (Section 9.1). The minimum time response of the probe is about $Z_0 C_s$. The maximum time is determined by the integration circuit.

The second probe is a self-integrating probe for use in water-filled transmission lines. The coupling capacitance is high in water lines; in addition, there is real current flow because of the nonzero conductance. In this situation, the capacitance between the probe and the wall is made high by locating the probe in a reentrant chamber. The probe acts as a capacitive divider rather than a coupling capacitor. If the space between the probe electrode and the wall of the chamber is filled with a nonconductive dielectric, the divider is unbalanced because of the water resistance. Balance can be achieved if the water of the transmission line is used as the dielectric in the probe chamber. The minimum time response of the divider is set by electromagnetic pulse transit time effects in the probe chamber. Low-frequency response is determined by the input resistance of the circuit attached to the probe. If R_d is the resistance, then the signal of a square input pulse will droop as $\exp(-t/R_d C_s)$. Usually, a high series R_d is inserted between the transmission line and the probe to minimize droop and further attenuate the signal voltage.

D. Pulsed Current Measurements with Series Resistors

Measurements of fast-pulsed beam flux or currents in modulators are usually performed with either series resistors, magnetic pickup loops, or Rogowski loops. A typical series resistor configuration for measuring the current from a pulsed electron gun is shown in Figure 9.56a. Power is supplied from a voltage generator through a coaxial transmission line. The anode is connected to the return conductor through a series resistor called a *current-viewing resistor* (CVR). Current is inferred from the voltage across the resistor. The CVR has low resistance, R_s, and very low series inductance. In many cases, diagnostic cables and power leads are connected to the anode. In this situation, some circuit current flows through the ground connectors of the cables, ultimately to return to the generator. This current does not contribute to a signal on the CVR. The current loss is minimized if the inductance associated with the loss paths is large compared to the series inductance of the CVR. When this condition holds, most of the primary circuit current flows through the CVR for times short compared to L/R_s, where L is the inductance along the connecting cables.

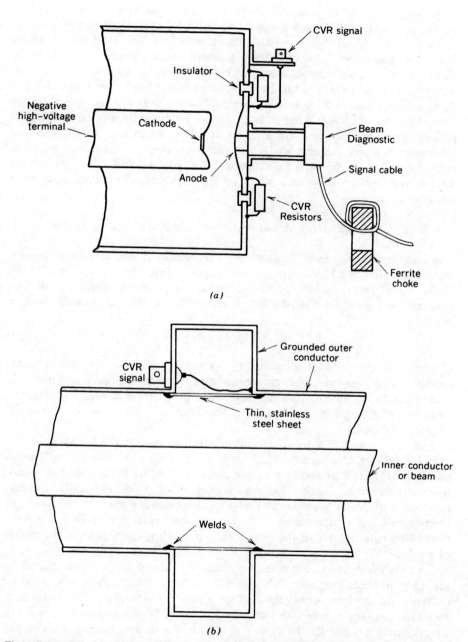

Figure 9.56 Current-viewing resistors. (*a*) Typical arrangement for current measurements from a pulsed electron injector with inductive isolation for beam diagnostics. (*b*) Low-inductance CVR for high-power transmission line or beamline.

A design used for commercial CVRs is shown in Figure 9.56*b*. The diagnostic is mounted on the return conductor of a coaxial transmission line. It consists of a toroidal chamber with a thin sheet of stainless steel welded across the opening to preserve the cylindrical geometry. In the absence of the resistive sheet, current flow generates a magnetic field inside the chamber; there is an inductance L_s associated with the convoluted path of return current. When a sheet with resistance R_s is added, current flows mainly through the sheet for times less than L_s/R_s. When this condition holds, voltage measured at the pick-off point is approximately equal to iR_s. The response of the CVR can be extended to lower frequencies by the inclusion of a ferromagnetic material such as ferrite in the toroidal chamber. The principle of operation of the ferrite-filled probe will be clear when we study inductive linear accelerators in the next chapter.

E.   Pulsed Current Measurements with Magnetic Pickup Loops

The magnetic pickup loop measures the magnetic field associated with current flow. It consists of a loop normal to the magnetic field of area A with N turns. If the signal from the loop is passed through a passive integrator with time constant RC, then the magnetic field is related to the integrator output voltage by

$$V_s = NAB/RC. \tag{9.120}$$

Current can be inferred if the geometry is known. For instance, if the current is carried by the center conductor of a transmission line or a cylindrically symmetric beam on axis, the magnetic field at a probe located a distance r from the axis is given by Eq. (4.40). The loop is oriented to sense toroidal magnetic fields.

The net current of a beam can be determined even if the beam moves off axis by adding the signals of a number of loops at different azimuthal positions. Figure 9.57 shows a simple circuit to add and integrate the signals from a number of loops. Magnetic pickup loops at diametrically opposite positions can detect beam motion off-axis along the diameter. In this case, the loop signals are subtracted to determine the difference signal. This is accomplished by rotating one of the loops 180° and adding the signals in the circuit of Figure 9.57.

The high-frequency response of a magnetic pickup loop is determined by the time it takes magnetic fields to fill the loop. The loop is diamagnetic; currents induced by changing flux produce a counterflux. The minimum response time for a loop with inductance L and a series resistance R is L/R. The resistance is usually the input impedance of a transmission line connected to a loop. Sensitive magnetic pickup loops with many turns generally have slow time response.

The magnetic pickup loop has a useful application in pulsed voltage systems when the voltage diagnostic cannot be attached at the load. For instance, in

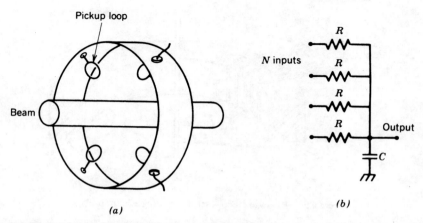

$$RC/N \gg \Delta t$$

Figure 9.57 (*a*) Multiple magnetic pickup loops for measurements of net beam current or beam position. (*b*) Passive integrator with provision for adding (or subtracting) probe signals.

driving a high current electron extractor with a transmission line modulator, it may be necessary to measure voltage in the line rather than at the vacuum load. If there is inductance L between the measurement point and the acceleration gap, the actual voltage on the gap is

$$V_g(t) = V_0(t) - L_p(di/dt). \tag{9.121}$$

The quantity V_0 is the voltage at the measurement point and V_g is the desired voltage. A useful measurement can often be achieved by adding an inductive correction to the signal voltage. An unintegrated magnetic pickup loop signal (proportional to $-di/dt$) is added to the uncorrected voltage through a variable attenuator. The attenuator is adjusted for zero signal when the modulator drives a shorted load. The technique is not reliable for time scales comparable to or less than the electromagnetic transit time between the measurement point and the load. A transmission line analysis must be used to infer high-frequency correction for the voltage signal.

F. Rogowski Loop

The final current diagnostic we will study is the Rogowski loop (Fig. 9.58). It is a multiturn toroidal inductor that encircles the current to be measured. When the dimension of the loops is small compared to the major radius, Eq. (4.39) implies that the net flux linking the series of loops does not depend on the position of the measured current inside the loop. Adopting this approximation, we assume the magnetic field is constant over the cross-sectional area of the windings and approximate the windings as a straight solenoid.

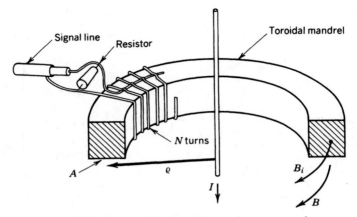

Figure 9.58   Rogowski loop—self-integrating current probe.

The quantity $B(t)$ is the magnetic field produced by the measured current and $B_i(t)$ is the total magnetic field inside the loop. The loop has N windings and a series resistance R at the output. The current flowing in the loop circuit is the total loop voltage divided by R or

$$i(t) = \frac{NA(dB_i/dt)}{R}.$$ (9.122)

The field inside the loop is the applied field minus the diagmagnetic field associated with the loop current or

$$B_i(t) = B(t) - \mu_0 i(t)(N/2\pi\rho).$$ (9.123)

Substituting Eq. (9.122) into (9.123),

$$(\mu_0 N^2 A/2\pi\rho R)(dB_i/dt) + B_i = (L/R)(dB_i/dt) + B_i = B(t).$$ (9.124)

The second form is derived by substituting the expression for the inductance of a toroidal winding, Eq. (9.18). In the limit that the time scale for variations of the measured current I is short compared to L/R, the first term on the left-hand side is large compared to the second term. In this limit very little of the magnetic field generated by the current to be measured penetrates into the winding. Dropping the second term and substituting Eqs. (9.122) and (4.40), we find that

$$i = I/N.$$ $\boxed{9.125}$

where i is the loop output current and I is the measure current.

In summary, the Rogowski loop has the following properties:

1. The output signal is unaffected by the distribution of current encircled by the loop.

2. The Rogowski loop is a self-integrating current monitor for time scales $\Delta t \ll L/R$.

3. The probe can respond to high-frequency variations of current.

[handwritten: Σ count × copper wire cross sectional area;]

The low-frequency response of the Rogowski loop can be extended by increasing the winding inductance at the expense of reduced signal. The inductance is increased a factor μ/μ_0 by winding the loop on a ferrite torus. In this case, Eq. (9.125) becomes

$$i = \frac{I}{N(\mu/\mu_0)}.$$

$$\boxed{9.126}$$

G. Electro-optical Diagnostics

The diagnostics we have considered are basic devices that are incorporated on almost all pulsed voltage systems. There has been considerable recent interest in the use of electro-optical techniques to measure rapidly varying electric and magnetic fields. The main reason is the increasing use of digital data acquisition systems. Optical connections isolate sensitive computers from electromagnetic noise, a particular problem for pulsed voltage systems.

A diagnostic for measuring magnetic fields is illustrated in Figure 9.59. Linearly polarized light from a laser is directed through a single-mode, fiber-optic cable. The linearly polarized light wave can be resolved into two circularly polarized waves. A magnetic field parallel to the cable affects the rotational motion of electrons in the cable medium. The consequence is that the index of refraction for right and left circularly polarized waves is different. There is a phase difference between the waves at the end of the cable. The phase difference is proportional to the strength of the magnetic field integrated along the length of the cable, $\int \mathbf{B} \cdot \mathbf{dl}$. When the waves are recombined, the direction of the resulting linearly polarized wave is rotated with respect to the initial wave. This effect is called *Faraday rotation*. The technique can be used for current diagnostics by winding one or more turns of the fiber-optic cable around the measured current path. Equation (4.39) implies that the rotation angle is proportional to the current. Electric fields can be sensed through the *Kerr effect* or the *Pockels effect* in an optical medium. Interferometric techniques are used to measure changes in the index of refraction of the medium induced by the fields.

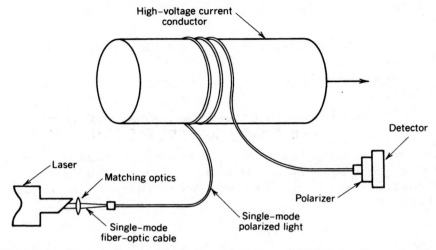

Figure 9.59 Current probe using Faraday rotation of laser-generated light in a fiber-optic cable.

The main drawbacks of electro-optical diagnostics techniques is the cost of the equipment and the complex analyses necessary to unfold the data. In the case of the Faraday rotation measurement, bifringence in ordinary fiber-optic materials complicates the interpretation of the rotation, and extensive computations must be performed to determine the measured current. Electro-optical devices have some unique applications that justify the additional effort, such as the direct measurement of intense propagating microwave fields.

10

Linear Induction Accelerators

The maximum beam energy achievable with electrostatic accelerators is in the range of 10 to 30 MeV. In order to produce higher-energy beams, the electric fields associated with changing magnetic flux must be used. In many high-energy accelerators, the field geometry is such that inductive fields cancel electrostatic fields over most the accelerator except the beamline. The beam senses a large net accelerating field, while electrostatic potential differences in the accelerating structure are kept to manageable levels. This process is called *inductive isolation*. The concept is the basis of linear induction accelerators.[†] The main application of linear induction accelerators has been the generation of pulsed high-current electron beams.

In this chapter and the next we shall study the two major types of nonresonant, high-energy accelerators, the linear induction accelerator and the betatron. The principle of energy transfer from pulse modulator to beam is identical for the two accelerators; they differ mainly in geometry and methods of particle transport. The linear induction accelerator and betatron have the following features in common:

1. They use ferromagnetic inductors for broadband isolation.
2. They are driven by high-power pulse modulators.
3. They can, in principle, produce high-power beams.
4. They are both equivalent to a step-up transformer with the beam acting as the secondary.

[†] N. C. Christofilos et al., *Rev. Sci. Instrum.*, **35**, 886 (1964).

In the linear induction accelerator, the beam is a single turn secondary with multiple parallel primary inputs from high-voltage modulators. In the betatron, there is usually one pulsed power primary input. The beam acts as a multiturn secondary because it is wrapped in a circle.

The linear induction accelerator is treated first since operation of the induction cavity is relatively easy to understand. Section 10.1 describes the simplest form of inductive cavity with an ideal ferromagnetic isolator. Section 10.2 deals with the problems involved in designing isolation cores for short voltage pulses. The limitations of available ferromagnetic materials must be understood in order to build efficient accelerators with good voltage waveform. Section 10.3 describes more complex cavity geometries. The main purpose is to achieve voltage step-up in a single cavity. Deviations from ideal behavior in induction cavities are described in Sections 10.4 and 10.5. Subjects include flux forcing to minimize unequal saturation in cores, core reset for maximum flux swing, and compensation circuits to achieve uniform accelerating voltage. Section 10.6 derives the electric field in a complex induction cavity. The goal is to arrive at a physical understanding of the distribution of electrostatic and inductive fields to determine insulation requirements. Limitations on the average longitudinal gradient of an induction accelerator are also reviewed. The chapter concludes with a discussion of induction accelerations without ferromagnetic cores. Although these accelerators are of limited practical use, they make an interesting study in the application of transmission line principles.

10.1 SIMPLE INDUCTION CAVITY

We can understand the principle of an induction cavity by proceeding stepwise from the electrostatic accelerator. A schematic of a pulsed electrostatic acceleration gap is shown in Figure 10.1a. A modulator supplies a voltage pulse of magnitude V_0. The pulse is conveyed to the acceleration gap through one or more high-voltage transmission lines. If the beam particles have positive charge $(+q)$, the transmission lines carry voltage to elevate the particle source to positive potential. The particles are extracted at ground potential with kinetic energy qV_0. The energy transfer efficiency is optimized when the characteristic impedance of the generator and the parallel impedance of the transmission lines equals V_0/I. The quantity I is the constant-beam current. Note the current path in Figure 10.1a. Current flows from the modulator, along the transmission line center-conductors, through the beam load on axis, and returns through the transmission line ground conductor.

A major problem in electrostatic accelerators is controlling and supplying power to the particle source. The source and its associated power supplies are at high potential with respect to the laboratory. It is more convenient to keep both the source and the extracted beam at ground potential. To accomplish this, consider adding a conducting cylinder between the high-voltage and

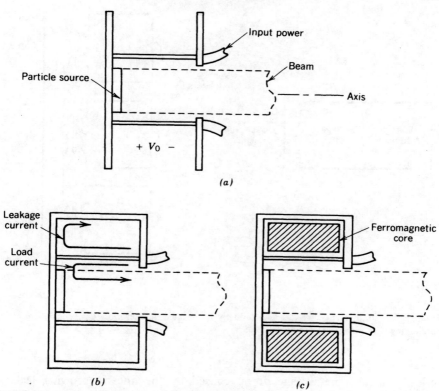

Figure 10.1 Inductively isolated injector cavity. (*a*) Electrostatic injector. (*b*) Shorted electrostatic injector. (*c*) Injector with high-inductance leakage path.

ground plates to define a toroidal cavity (Fig. 10.1*b*). The source and extraction point are at the same potential, but the system is difficult to operate since the transmission line output is almost short circuited. Most of the current flows around the outer ground shield; this is *leakage current*. There is a small voltage across the acceleration gap because the toroidal cavity has an inductance L_1. The leakage inductance is given by Eq. (9.15) if we take R_i as the radius of the power feeds and R_0 as the radius of the cylinder. Thus, a small fraction of the total current flows in the load. The goal is clearly to reduce the leakage current compared to the load current; this is accomplished by increasing L_1.

In the final configuration (Fig. 10.1*c*), the toroidal volume occupied by magnetic field from leakage current is filled with ferromagnetic material. If we approximate the ferromagnetic torus as an ideal inductor, the leakage inductance is increased by a factor μ/μ_0. This factor may exceed 1000. The leakage current is greatly reduced, so that most of the circuit current flows in the load. At constant voltage, the cavity appears almost as a resistive load to the pulse modulator. The voltage waveform is approximately a square pulse of magni-

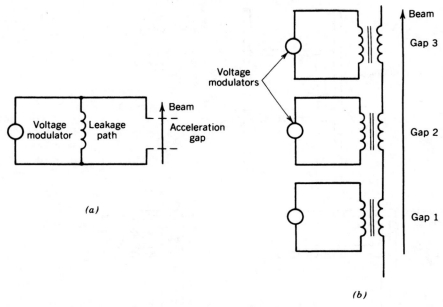

Figure 10.2 Equivalent circuits. (*a*) Inductively isolated acceleration cavity. (*b*) Induction accelerator with a series of cavities.

tude V_0 with some voltage droop caused by the linearly growing leakage current. The equivalent circuit model of the induction cavity is shown in Figure 10.2*a*; it is identical to the equivalent circuit of a 1 : 1 transformer (Fig. 9.7).

The geometry of Figure 10.1*c* is the simplest possible inductive linear accelerator cavity. A complete understanding of the geometry will clarify the operation of more complex cavities.

1. The load current does not encircle the ferromagnetic core. This means that the integral $\int \mathbf{H} \cdot \mathbf{dl}$ from load current is zero through the core. In other words, there is little interaction between the load current and the core. The properties of the core set no limitation on the amount of beam current that can be accelerated.

2. To an external observer, both the particle source and the extraction point appear to be at ground potential during the voltage pulse. Nonetheless, particles emerge from the cavity with kinetic energy gain qV_0.

3. The sole purpose of the ferromagnetic core is to reduce leakage current.

4. There is an electrostaticlike voltage across the acceleration gap. Electric fields in the gap are identical to those we have derived for the electrostatic accelerator of Figure 10.1*a*. The inductive core introduces no novel accelerating field components.

5. Changing magnetic flux generates inductive electric fields in the core. The inductive field at the outer radius of the core is equal and opposite to the electrostatic field; therefore, there is no net electric field between the plates at the outer radius, consistent with the fact that they are connected by a conducting cylinder. The ferromagnetic core provides inductive isolation for the cavity.

When voltage is applied to the cavity, the leakage current is small until the ferromagnetic core becomes saturated. After saturation, the differential magnetic permeability approaches μ_0 and the cavity becomes a low-inductance load. The product of voltage and time is limited. We have seen a similar constraint in the transformer [Eq. (9.29)]. If the voltage pulse has constant magnitude V_0 and duration t_p, then

$$V_0 t_p = \Delta B A_c,$$

10.1

where A_c is the cross-sectional area of the core. The quantity ΔB is the change of magnetic field in the core; it must be less than $2B_s$. Typical operating parameters for an induction cavity with a ferrite core are $V_0 = 250$ kV and $t_p = 50$ ns. Ferrites typically have a saturation field of 0.2–0.3 T. Therefore, the core must have a cross-sectional area greater than 0.025 m². The most common configuration for an inductive linear accelerator is shown in Figure 10.3. The beam passes through a series of individual cavities. There is

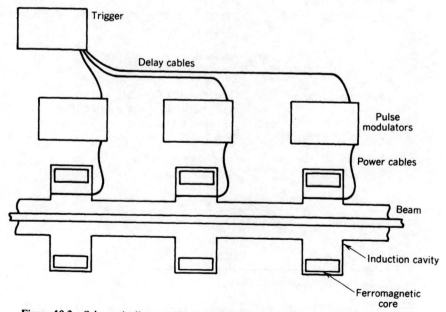

Figure 10.3 Schematic diagram of a linear induction acceleration with cavities in series.

TABLE 10.1   Parameters of the ATA[a] Accelerator

Function

Research on the propagation of self-focused electron beams in gas, free electron laser driver.

Configuration

Length	85 m
Average gradient	0.59 MV/m

Injector

Voltage	2.5 MV
Voltage source	10 stacked 250-kV induction cavities
Electron source diameter	0.25 m
Configuration	Triode, extraction by high-voltage control grid
Cathode	Cold cathode, whisker emission

Main accelerator

Voltage gain	47.5 MeV
Voltage per stage	0.25 MV
Number of stages	190
Repetition rate	5 Hz
	10 kHz (10 pulse burst)

Acceleration cavity

Voltage	250 kV
Power modulator	Coaxial, water-filled Blumlein line pulse charged through a 10 : 1 step-up transformer
Modulator impedance	12 Ω
Modulator switch	Gas-blown spark gap
Inductive isolator	Ferrite toroids

Output beam

Pulselength	70 ns
Energy	50 MeV
Peak current	10 kA

[a] Advanced Test Accelerator, Lawrence Livermore Laboratory.

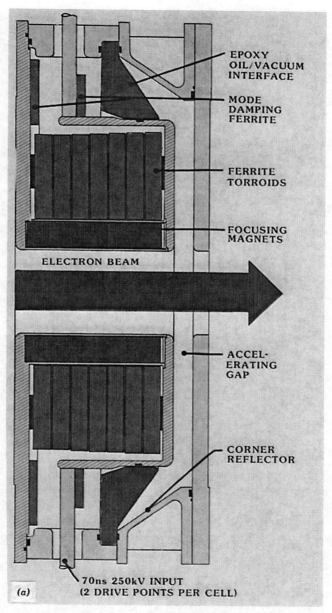

EPOXY
OIL/VACUUM
INTERFACE

MODE
DAMPING
FERRITE

FERRITE
TORROIDS

FOCUSING
MAGNETS

ELECTRON BEAM

ACCEL-
ERATING
GAP

CORNER
REFLECTOR

70ns 250kV INPUT
(2 DRIVE POINTS PER CELL)

(a)

Figure 10.4   Scale drawings, Advanced Test Accelerator. (a) 250 kV, ferrite isolated induction cell
with mode damping ferrites to minimize beam breakup instabilities. (b) 2.5 MV, 10-cell block.
(Courtesy D. Prono, Lawrence Livermore Laboratory.)

289

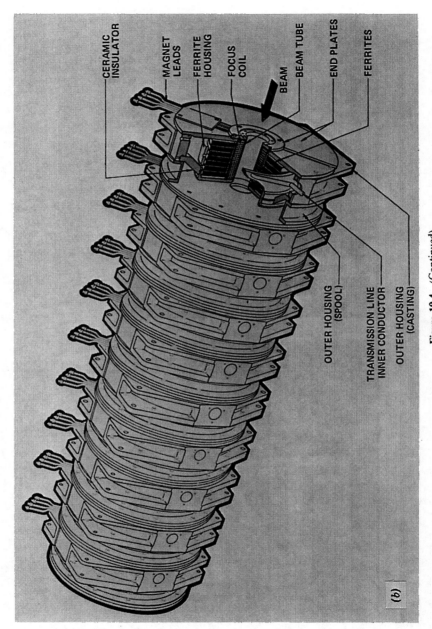

CERAMIC INSULATOR

MAGNET LEADS

FERRITE HOUSING

FOCUS COIL

BEAM

BEAM TUBE

END PLATES

FERRITES

OUTER HOUSING (SPOOL)

TRANSMISSION LINE INNER CONDUCTOR

OUTER HOUSING (CASTING)

(b)

Figure 10.4 (Continued).

290

no electrostatic voltage difference in the system higher than V_0. Any final beam energy consistent with cost and successful beam transport can be attained by adding more cavities. The equivalent circuit of an induction accelerator is shown in Figure 10.2*b*. Characteristics of the ATA machine, the highest energy induction accelerator constructed to data, are summarized in Table 10.1. A single-acceleration cavity and a 10-cavity block of the ATA accelerator are illustrated in Figures 10.4*a* and 10.4*b*, respectively.

10.2 TIME-DEPENDENT RESPONSE OF FERROMAGNETIC MATERIALS

We have seen in Section 5.3 that ferromagnetic materials have atomic currents that align themselves with applied fields. The magnetic field is amplified inside the material. The alignment of atomic currents is equivalent to a macroscopic current that flows on the surface of the material. When changes in applied field are slow, atomic currents are the dominant currents in the material. In this case, the magnetic response of the material follows the static hysteresis curve (Fig. 5.12).

Voltage pulselengths in linear induction accelerators are short. We must include effects arising from the fact that most ferromagnetic materials are conductors. Inductive electric fields can generate real currents; real currents differ from atomic currents in that electrons move through the material. Real current driven by changes of magnetic flux is called *eddy current*. The contribution of eddy currents must be taken into account to determine the total magnetic fields in materials. In ferromagnetic materials, eddy current may prevent penetration of applied magnetic field so that magnetic moments in inner regions are not aligned. In this case, the response of the material deviates from the static hysteresis curve. Another problem is that resistive losses are associated with eddy currents. Depending on the the type of material and geometry of construction, magnetic cores have a maximum usable frequency. At higher frequencies, resistive losses increase and the effective core inductance drops.

Eddy currents in inductive isolators and transformer cores are minimized by laminated core construction. Thin sheets of steel are separated by insulators. Most common ac cores are designed to operate at 60 Hz. In contrast, the maximum-frequency components in inductive accelerator pulses range from 1 to 100 MHz. Therefore, core design is critical for fast pulses. The frequency response is extended either by using very thin laminations or using alternatives to steel, such as ferrites.

The *skin depth* is a measure of the distance magnetic fields penetrate into materials as a function of frequency. We can estimate the skin depth in ferromagnetic materials in the geometry of Figure 10.5. A lamination of high μ material with infinite axial extent is surrounded by a pulse coil excited by a step-function current waveform. The coil carries an applied current per length

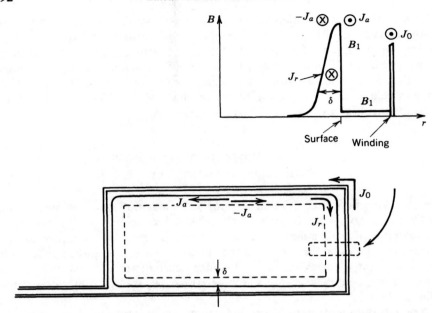

Figure 10.5 Eddy currents and atomic currents in a ferromagnetic lamination subject to a pulsed external magnetic field.

J_a (A/m) for $t > 0$. The applied magnetic field outside the high μ material is

$$B_1 = \mu_0 J_0. \qquad (10.2)$$

A real return current J_r flows in the conducting sample in the opposite direction from the applied current. We assume this current flows in an active layer on the surface of the material of thickness δ. The magnetic field decreases across this layer and approaches zero inside the material. The total magnetic field as a function of depth in the material follows the variation of Figure 10.5. The return current is distributed through the active layer, while the atomic currents are concentrated at the layer surfaces. The atomic current J_a is the result of aligned dipoles in the region of applied magnetic field penetration; it amplifies the field in the active layer by a factor of μ/μ_0. The field just inside the material surface is $(\mu/\mu_0)B_1$. The magnetic field inward from the active layer is zero since the return current cancels the field produced by the applied currents. This implies that

$$J_r \approx -J_0. \qquad (10.3)$$

We can estimate the skin depth by making a global balance between resistive effects (which impede the return current) and inductive effects (which drive the return current). The active layer is assumed to penetrate a small distance into the lamination. The lamination has circumference C. If the material is an imperfect conductor with volume resistivity ρ (Ω-m), the resistive voltage around the circumference from the flow of real current is

$$V_r \cong J_r \rho C / \delta. \tag{10.4}$$

The return current is driven by an electromotive force (emf) equal and opposite to V_r. The emf is equal to the rate of change of flux enclosed within a loop at the location under consideration. Since the peak magnetic field $(\mu/\mu_0) B_1$ is limited, the change of enclosed flux must come about from the motion of the active layer into the material. If the layer moves inward a distance δ, then the change of flux inside a circumferential loop is $\Delta\Phi \cong (\mu/\mu_0) B_1 C \delta$. Taking the time derivative and using Eq. (4.42), we find the inductive voltage

$$V_i \cong \mu C J_r \, d\delta/dt.$$

Setting V_i equal to V_r gives

$$C\mu J_r (d\delta/dt) \cong J_r \rho C / \delta. \tag{10.5}$$

The circumference cancels out. The solution of Equation (10.5) gives the skin depth

$$\delta = \sqrt{2\rho t/\mu}. \qquad \boxed{10.6}$$

The magnetic field moves into the material a distance proportional to the square root of time if the applied field is a step function. A more familiar expression for the skin depth holds when the applied field is harmonic, $B_1 \sim \cos(\omega t)$:

$$\delta = \sqrt{2\rho/\mu\omega}. \qquad \boxed{10.7}$$

In this case, the depth of the layer is constant; the driving emf is generated by the time variation of magnetic field.

The two materials commonly used in pulse cores are ferrites and steel. The materials differ mainly in their volume resistivity. Ferrites are ceramiclike compounds of iron-bearing materials with volume resistivity on the order of 10^4 Ω-m. Silicon steel is the most common transformer material. It is magnetically soft; the area of its hysteresis loop is small, minimizing hysteresis losses. Silicon steel has a relatively high resistivity compared to other steels, 45×10^{-8} Ω-m. Nickel steel has a somewhat higher resistivity, but it is expensive. Recently, noncrystalline iron compounds have been developed. They are known by the tradename Metglas.[†] Metglas is produced in thin ribbons by

[†] Trademark of Allied Corporation.

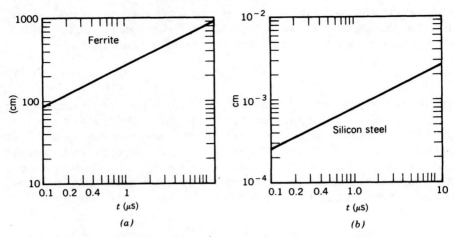

Figure 10.6   Typical values of small-signal skin depth. (*a*) Ferrite. (*b*) Silicon steel.

injecting molten iron compounds onto a cooled, rapidly rotating drum. The rapid cooling prevents the formation of crystal structures. Metglas alloy 2605SC has a volume resistivity of 125×10^{-8} Ω-m. Typical small-signal skin depths for silicon steel and ferrites as a function of applied field duration are plotted in Figure 10.6. There is a large difference between the materials; this difference is reflected in the construction of cores and the analysis of time-dependent effects.

An understanding of the time-dependent response of ferromagnetic materials is necessary to determine leakage currents in induction linear accelerators. The leakage current affects the efficiency of the accelerator and determines the compensation circuits necessary for waveform shaping. We begin with ferrites. In a typical ferrite isolated accelerator, the pulselength is 30–80 ns and the core dimension is ≤ 0.5 m. Reference to Figure 10.6 shows that the skin depth is larger than the core; therefore, to first approximation, we can neglect eddy currents and consider only the time variation of atomic currents. A typical geometry for a ferrite core accelerator cavity is shown in Figure 10.7. The toroidal ferrite cores are contained between two cylinders of radii R_i and R_0. The leakage current circuit approximates a coaxial transmission line filled with high μ material. The transmission line has length d; it is shorted at the end opposite the pulsed power input. We shall analyze the transmission line behavior of the leakage current circuit with the assumption that $(R_0 - R_i)/R_0$ $\ll 1$. Radial variations of toroidal magnetic field in the cores are neglected.

The transmission line of length d has impedance Z_c given by Eq. (9.86) and a relatively long transit time $\Delta t = d\sqrt{\varepsilon\mu}$. Consider, first, application of a low-voltage step-function pulse. A voltage wave from a low-impedance generator of magnitude V_0 travels through the core at velocity $c/\sqrt{(\varepsilon/\varepsilon_0)(\mu/\mu_0)}$ carrying current V_0/Z_c. The wave is reflected at time Δt with inverse polarity

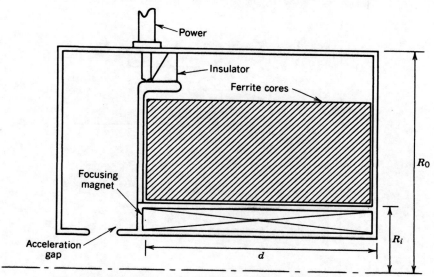

Figure 10.7 Geometry of an idealized ferrite core acceleration cavity.

from the short-circuit termination. The inverted wave arrives back at the input at time $2\,\Delta t$. In order to match the input voltage, two voltage waves, each carrying current V_0/Z_c, are launched on the line. The net leakage current during the interval $2\,\Delta t \le t \le 4\,\Delta t$ is $3V_0/Z_c$. Subsequent wave reflections result in the leakage current variation illustrated in Figure 10.8a. The dashed line in the figure shows the current corresponding to an ideal lumped inductor with $L = Z_c\,\Delta t$. The core approximates a lumped inductor in the limit of low voltage and long pulselength. The leakage current diverges when it reaches the value $i_s = 2\pi R_i H_s$. The quantity H_s is the saturation magnetizing force.

Next, suppose that the voltage is raised to V_s so that the current during the initial wave transit is

$$i = V_s/Z_c = i_s = 2\pi R_i H_s. \tag{10.8}$$

The wave travels through the core at the same velocity as before. The main difference is that the magnetic material behind the wavefront is saturated. When the wave reaches the termination at time Δt, the entire core is saturated. Subsequently, the leakage circuit acts as a vacuum transmission line. The quantities Z_c and Δt decrease by a factor of $\sqrt{\mu/\mu_0}$, and the current increases rapidly as inverted waves reflect from the short circuit. The leakage current for this case is plotted in Figure 10.8b. The volt-second product before saturation again satisfies Eq. (10.1).

At higher applied voltage, electromagnetic disturbances propagate into the core as a *saturation wave*. The wave velocity is controlled by the saturation of

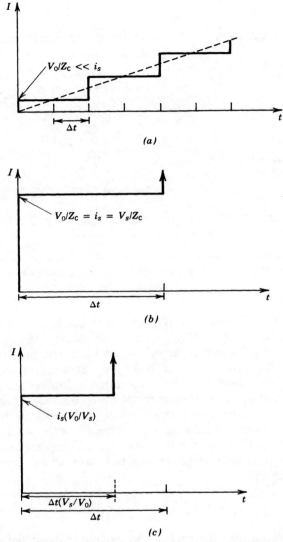

Figure 10.8 Time-dependent leakage current in a ferrite core acceleration cavity. (*a*) Low applied voltage. (*b*) Saturation voltage. (*c*) High voltage.

magnetic material in the region of rising current; the saturation wave moves more rapidly than the speed of electromagnetic pulses in the high μ medium. When $V_0 > V_s$, conservation of the volt-second product implies that the time T for the saturation wave to propagate through the core is related to the small-signal propagation time by

$$T/\Delta t = V_s/V_0. \tag{10.9}$$

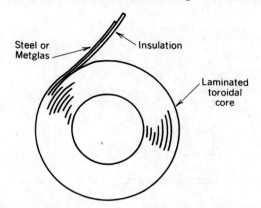

Figure 10.9 Construction of a laminated ferromagnetic isolation core.

The transmission line is charged to voltage V_0 at time T; therefore, a charge CV_0 flowed into the leakage circuit where C is the total capacitance of the leakage circuit. The magnitude of the leakage current accompanying the saturation wave is thus $i = CV_0/T$, or

$$i = i_s(V_0/V_s). \tag{10.10}$$

The leakage current exceeds the value given in Eq. (10.8). Leakage current variation in the high-voltage limit is illustrated in Figure 10.8c.

In contrast to ferrites, the skin depth in steel or Metglas is much smaller than the dimension of the core. The core must therefore be divided into small sections so that the magnetic field penetrates the material. This is accomplished by laminated construction. Thin metal ribbon is wound on a cylindrical mandrel with an intermediate layer of insulation. The result is a toroidal core (Fig. 10.9). In subsequent analyses, we assume that the core is composed of nested cylinders and that there is no radial conduction of real current. In actuality, some current flows from the inside to the outside along the single ribbon. This current is very small since the path has a huge inductance. A laminated pulse core may contain thousands of turns.

Laminated steel cores are effective for pulses in the microsecond range. In the limit that $(R_0 - R_i)/R_0 \ll 1$, the applied magnetic field is the same at each lamination. The loop voltage around a lamination is thus V_0/N, where N is the number of layers. In an actual toroidal core, the applied field is proportional to $1/r$ so that lamination voltage is distributed unevenly; we will consider the consequences of flux variation in Section 10.4.

If the thickness of the lamination is less than the skin depth associated with the pulselength, then magnetic flux is distributed uniformly through a lamination. Even in this limit, it is difficult to calculate the inductance exactly since the magnetic permeability varies as the core field changes from $-B_r$ to B_s. For

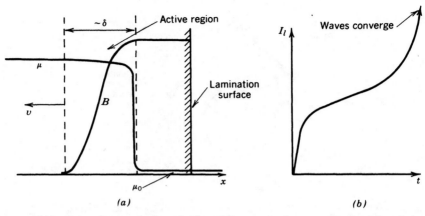

Figure 10.10 Saturation waves in a laminated ferromagnetic core. (*a*) Spatial variation of magnetic field and μ in a lamination. (*b*) Time variation of leakage current.

a first-order estimate of the leakage current, we assume an average permeability $\langle \mu \rangle = B_s/H_s$. The inductance of the core is

$$L = \langle \mu \rangle d \ln(R_0/R_i)/2\pi. \qquad (10.11)$$

The time-dependent leakage current is

$$i_1 \cong (V_0/L)t = \left[2\pi V_0/\langle \mu \rangle d \ln(R_0/R_i) \right] t. \qquad (10.12)$$

The behavior of laminated cores is complex for high applied voltage and short pulselength. When the skin depth is less than half the lamination thickness, magnetic field penetrates in a saturation wave (Fig. 10.10*a*). There is an active layer with large flux change from atomic current alignment. There is a region behind the active layer of saturated magnetic material; the skin depth for field penetration is large in this region because $\mu \cong \mu_0$. Changes in the applied field are communicated rapidly through the saturated region. The active layer moves inward, and the saturation wave converges at the center at a time equal to the volt–second product of the lamination divided by V_0/N.

Although the volt–second product is conserved in the saturation wave regime, the inductance of the core is reduced below the value given by Eq. (10.11). This comes about because only a fraction of the lamination cross-sectional area contributes to flux change at a particular time. The core inductance is reduced by a factor on the order of the width of the active area divided by the half-width of the lamination. Accelerator efficiency is reduced because of increased leakage current and eddy current core heating. Leakage current in the saturation wave regime is illustrated in Figure 10.10*b*.

Core material and lamination thickness should be chosen so that skin depth is greater than half the thickness of the lamination. If this is impossible, the

TABLE 10.2   Properties: Materials for Induction Linear Accelerators

Quantity	Ferrite (TDK PE-14)	Silicon Steel	Metglas Allied (2605SC)
ρ (Ω-m)	10^4	45×10^{-8}	125×10^{-8}
μ/μ_0 (small signal)	300–600	600	1200
$\langle(\mu/\mu_0)\rangle$ $B_s/\mu_0 H_s$ (static)		10^4	10^5
B_s (T)	0.4	1.4	1.6
B_r (T)	0.3	1.2	1.4
ΔB_{max} (T)	0.7	2.6	3.0

severity of leakage current effects can best be estimated by experimental modeling. Saturation wave analyses seldom give an accurate figure for leakage current. Measurements for a single lamination are simple; a loop around the lamination is driven with a pulse of voltage V_0/N and pulselength equal to that of the accelerator. The most reliable method to include the effects of radial field variations is to perform measurements on a full-radius core segment.

Properties of common magnetic materials are listed in Table 10.2. Ferrite cores have the capability for fast response; they are the only materials suitable for the 10–50-ns regime. The main disadvantages of ferrites are that they are expensive and that they have a relatively small available flux swing. This implies large core volumes for a given volt–second product.

Silicon steel is inexpensive and has a large magnetic field change, ~ 3 T. On the other hand, it is a brittle material and cannot be wound in thicknesses < 2 mil (5×10^{-3} cm). Reference to Figure 10.6 shows that silicon steel cores are useful only for pulselengths greater than 1 μs. There has been considerable recent interest in Metglas for induction accelerator cores. Metglas has a larger volume resistivity than silicon steel. Because of the method of its production, it is available in uniform thin ribbons. It has a field change about equal to silicon steel, and it is expected to be fairly inexpensive. Most important, because it is noncrystalline, it is not brittle and can be wound into cores in ribbons as thin as 0.7 mil (1.8×10^{-3} cm). It is possible to construct Metglas cores for short voltage pulses. If there is high load current and leakage current is not a primary concern, Metglas cores can be used for pulses in the 50-ns range.

The distribution of electric field in isolation cores must be known to determine insulation requirements. Electric fields have a simple form in laminated steel cores. Consider the core in the cavity geometry of Figure 10.11 with radial variations of applied field neglected. We know that there is a voltage V_0 between the inner and outer cylinders at the input end (marked α) and there is zero voltage difference on the right-hand side (marked β) because of the connecting radial conductor. Furthermore, the laminated core has zero

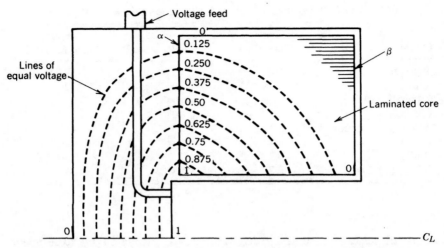

Figure 10.11   Distribution of voltage in a laminated isolation core (no local saturation).

conductivity in the radial direction but has high conductivity in the axial direction. Image charge is distributed on the laminations to assure that E_z equals zero along the surfaces. Therefore, the electric field is almost purely radial in the core. This implies that:

1.   Except for small fringing fields, the electric field is radial along the input edge (α). The voltage drop across each insulating layer on the edge is V_0/N.

2.   Moving into the core, inductive electric fields cancel the electrostatic fields. The net voltage drop across the insulating layers decreases linearly to zero moving from α to β.

Figure 10.11 shows voltage levels in the core relative to the outer conductor. Note that this is not an electrostatic plot; therefore, the equal voltage lines in the core are not normal to the electric fields.

10.3   VOLTAGE MULTIPLICATION GEOMETRIES

Inductive linear accelerator cavities can be configured as step-up transformers. High-current, moderate-voltage modulators can be used to drive a lower-current beam load at high voltage. Step-up cavities are commonly used for high-voltage electron beam injectors. Problems of beam transport and stability in subsequent acceleration sections are reduced if the injector voltage is high. Multi-MV electrostatic pulse generators are bulky and difficult to operate, but

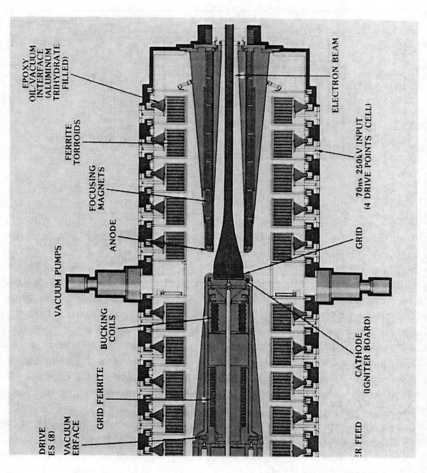

EPOXY
OIL-VACUUM
INTERFACE
(ALUMINUM
TRIHYDRATE
FILLED)

FERRITE
TORROIDS

FOCUSING
MAGNETS

ANODE

VACUUM PUMPS

BUCKING
COILS

GRID FERRITE

DRIVE
ES (8)

VACUUM
ERFACE

ELECTRON BEAM

70ns 250kV INPUT
(4 DRIVE POINTS /CELL)

GRID

CATHODE
(IGNITER BOARD)

ER FEED

ongitudinal stacking of induction cells for electron injection high energy. (*a*) Scale 2.5-MeV injector, Advanced Test Accelerator. (Courtesy D. Prono, Lawrence Livermore Current flow in a four-stage cavity.

301

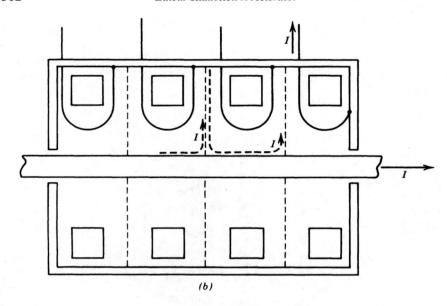

(b)

Figure 10.12   (*Continued*).

0.25-MV modulators are easy to design. The inductive cavity of Figure 10.12*a* uses 10 parallel 0.25-MV pulses to generate a 2.5-MV pulse across an injection gap. Figure 10.12 illustrates *longitudinal core stacking*.

A schematic view of a 4:1 step-up circuit with longitudinal stacking is shown in Figure 10.12*b*. Note that if electrodes are inserted at the positions of the dotted lines, the single gap of Figure 10.12*b* is identical to a four-gap linear induction accelerator. The electrodes carry no current, so that the circuit is unchanged by their presence. We assume that the four input transmission lines of characteristic impedance Z_0 carry pulses with voltage V_0 and current $I_0 = V_0/Z_0$. A single modulator to drive the four lines must have an imped-ance $\frac{1}{4}Z_0$. The high core inductance constrains the net current through the axis of each core to be approximately zero. Therefore, the beam current for a matched circuit is I_0. The voltage across the acceleration gap is $4V_0$. The matched load impedance is therefore $4Z_0$. This is a factor of 16 higher than the primary impedance, as we expect for a 4:1 step-up transformer.

It is also possible to construct voltage step-up cavities with radial core stacking, as shown in Figure 10.13. It is more difficult to understand the power flow in this geometry. For clarity, we will proceed one step at a time, evolving from the basic configuration of Figure 10.1*c* to a dual-core cavity. We assume the beam load is driven by matched transmission lines. There are two main constraints if the leakage circuits have infinite impedance: (1) all the current in the system must be accounted for and (2) there is no net axial current through the centers of either of the cores.

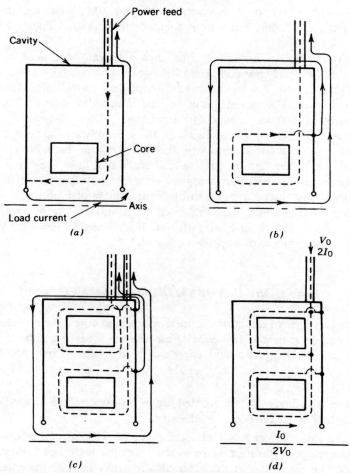

Figure 10.13 Acceleration cavity with radially stacked cores. Dashed arrows, current on power feeds; solid arrows, current in cavity. (*a*) Single core, simple power feed. (*b*) Single core with encircling power feed. (*c*) Two cores with dual power feeds. (*d*) Two cores with single power feed and flux-forcing connections.

Figure 10.13*a* shows a cavity with one core and one input transmission line. The difference from Figure 10.1*c* is that the line enters radially; this feature does not affect the behavior of the cavity. Note the current path; the two constraints are satisfied. In Figure 10.13*b*, the power lead is wrapped around the core and connected back to the input side of the inductive cavity. The current cannot flow outward along the wall of the cavity and return immediately along the transmission line outer conductor; this path has high inductance. Rather, the current follows the convoluted path shown, flowing through the on-axis load before returning along the ground lead of the transmission line. Although the circuit of Figure 10.13*b* has a more complex current path and higher parasitic inductance than that of Figure 10.12*a*, the net behavior is the same.

The third step is to add an additional core and an additional transmission line with power feed wrapped around the core. The voltage on the gap is $2V_0$ and the load current is equal to that from one line. Current flow from the two lines is indicated. The current paths are rather complex; current from the first transmission line flows around the inner core, along the cavity wall, and returns through the ground conductor of the second line. The current from the second line flows around the cavity, through the load, back along the cavity wall, and returns through the ground lead of the first line. The cavity of Figure 10.13*c* conserves current and energy. Furthermore, inspection of the current paths shows that the net current through the centers of both cores is zero. An alternative configuration that has been used in accelerators with radially stacked cores is shown in Figure 10.13*d*. Both cores are driven by a single-input transmission line of impedance $V_0/2I_0$.

10.4 CORE SATURATION AND FLUX FORCING

In our discussions of laminated inductive isolation cores, we assumed that the applied magnetic field is the same at all laminae. This is not true in toroidal cores where the magnetic field decreases with radius. Three problems arise from this effect:

1. Electric fields are distributed unevenly among the insulation layers. They are highest at the center of the core.

2. The inner layers reach saturation before the end of the voltage pulse. There is a global saturation wave in the core; the region of saturation grows outward. The result is that the magnetically active area of the core decreases following saturation of the inner lamination. The inductance of the isolation circuit drops at the end of the pulse.

3. During the saturation wave, the circuit voltage is supported by the remaining unsaturated laminations. The field stress is highly nonuniform so that insulation breakdowns may occur.

The first problem can be addressed by using thicker insulation near the core center. The second and third problems are more troublesome, especially in accelerators with radially stacked cores. The effects of unequal saturation on the leakage current and cavity voltage are illustrated in Figure 10.14. The leakage current grows nonlinearly during the latter portion of the pulse making compensation (Section 10.5) difficult. The tail end of the voltage pulse droops. Although the quantity $\int V\, dt$ is conserved, the waveform of Figure 10.14 may be useless for acceleration if a small beam energy spread is required.

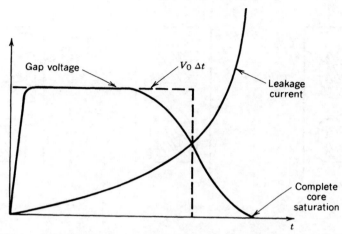

Figure 10.14 Acceleration gap voltage and leakage current waveforms with local saturation of isolation cores.

The problem of unequal saturation could be solved by using a large core to avoid saturation of the inner laminations. This approach is undesirable since core utilization is inefficient. The core volume and cost are increased and the average accelerating gradient is reduced (see Section 10.7). Ideally, the core and cavity should be designed so that the entire volume of the core reaches saturation simultaneously at the end of the voltage pulse. This condition can be approached through *flux forcing*.

Flux forcing was originally developed to equalize saturation in large betatron cores. We will illustrate the process in a two-core radially stacked induction accelerator cavity. In Figure 10.15a, two open conducting loops encircle the cores. There is an applied voltage pulse of magnitude V_0. The voltage between the terminals of the outer loop is smaller than that of the inner loop since the enclosed magnetic flux change is less. The sum of the voltages equals V_0 with polarities as shown in the figure. In Figure 10.14b, the ends of the loops are connected together to form a single figure-8 winding. If the net enclosed magnetic flux in the outer loop were smaller than that of the inner loop, a high current would flow in the winding. Therefore, we conclude that a moderate current is induced in the winding that equalizes the magnetic flux enclosed by the two loops.

The figure-8 winding is called a *flux-forcing strap*. The distribution of current is illustrated in Figure 10.15c. The inner loop of the flux-forcing strap reduces the magnetic flux in the inner core, while the outer loop current adds flux at the outer core. If both cores in Figure 10.15 have the same cross-sectional area, they reach saturation at the same time since $d\Phi/dt$ is the same inside both loops. Of course, local saturation can still occur in a single core.

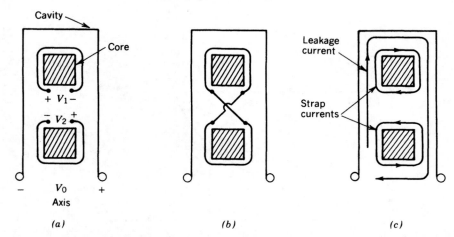

Figure 10.15 Flux-forcing straps. (*a*) Voltage on unconnected straps. (*b*) Figure-8 winding for flux forcing. (*c*) Current flow in flow-forcing straps relative to leakage current.

Nonetheless, the severity of saturation is reduced for two reasons:

1.   The ratio of the inner to outer radius is closer to unity for a single core section than for the entire stack.

2.   Even if there is early saturation at the inside of the inner core, the drop in leakage circuit inductance is averaged over the stack.

Figure 10.16 illustrates an induction cavity with four core sections. The cavity was designed to achieve a high average longitudinal gradient in a long pulse linear induction accelerator. A large-diameter core stack was used for high cross-sectional area. There are two interesting aspects of the cavity:

1.   The cores are driven in parallel from a single-pulse modulator. There is a voltage step-up by a factor of 4.

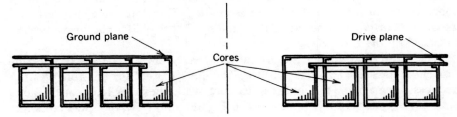

Figure 10.16 Core and power feed geometry for a 4 : 1 step-up cavity with flux forcing. (Courtesy M. Wilson, National Bureau of Standards.)

2. The parallel drive configuration assures that the loop voltage around each core is the same; therefore, the current distribution in the driving loops provides automatic flux forcing.

10.5   CORE RESET AND COMPENSATION CIRCUITS

Following a voltage pulse, the ferromagnetic core of an inductive accelerator has magnetic flux equal to $+B_r$. The core must be *reset* to $-B_r$ before the next pulse; otherwise, the cavity will be short-circuited soon after the voltage is applied.

Reverse biasing of the core is accomplished with a *reset circuit*. The reset circuit must have the following characteristics:

1. It is able to achieve an inverse voltage–time product greater than $(B_s + B_r)A_c$.

2. It can supply a unidirectional reverse current through the core axis of magnitude

$$I_s > 2\pi R_0 H_s. \qquad \boxed{10.13}$$

The quantity H_s is the magnetizing force of the core material and R_0 is the outer core radius. Higher currents are required for fast-pulsed reset.

3. The reset circuit has high voltage isolation so that it does not absorb power during the primary pulse.

A long pulse induction cavity with reset circuit is shown in Figure 10.17. Note that a *damping resistor* in parallel with the beam load is included in the cavity. Damping resistors are incorporated in most induction accelerators. Their purpose is to prevent overvoltage if there is an error in beam arrival time. Some of the available modulator energy is lost in the damping resistor. Induction cavities have typical energy utilization efficiencies of 20–50%.

Reset is performed by an RC circuit connected to the cavity through a mechanical high-voltage relay. The relay acts as an isolator and a switch. The reset capacitor C_r is charged to voltage $-V_r$; the reset resistance R_r is small compared to the damping resistor R_d. There is an inductance L_r associated with the flow of reset current. Neglecting L_r and the current through the leakage circuit, the first condition above is satisfied if

$$\int V_r \exp(-t/R_d C_r)\, dt > (B_s + B_r)A_c. \qquad (10.14)$$

If the inequality of Eq. (10.14) is well satisfied, the second condition is fulfilled if

$$V_r/R_d > I_s. \qquad (10.15)$$

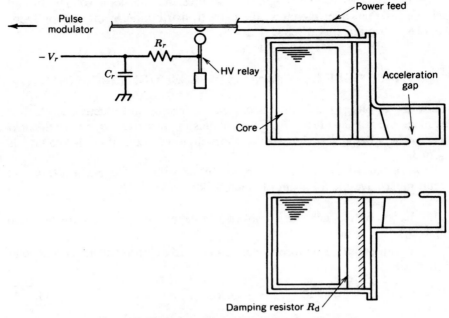

Figure 10.17   Reset circuit for a long-pulse induction cavity.

Equation (10.15) implies that the reset resistance should be as low as possible. There is a minimum value of R_r that comes about when the effect of the inductance L_r is included. Without the resistance, the circuit is underdamped; the reset voltage oscillates. The behavior of the LC circuit with saturable core inductor is complex; depending on the magnetic flux in the core following the voltage pulse and the charge voltage on the reset circuit, the core may remain with flux anywhere between $+B_r$ and $-B_r$ after the reset. Thus, it is important to assure that reset current flows only in the proper direction. The optimum choice of R_r leads to a critically damped reset circuit:

$$R_s \geq 2\sqrt{L_r/C_r}. \tag{10.16}$$

when Eq. (10.16) is satisfied, the circuit generates a unidirectional pulse with maximum output current.

The following example illustrates typical parameters for an induction cavity and reset circuit. An induction linear accelerator cavity supports a 100-kV, 1-μs pulse. The beam current is 2 kA. The laminated core is constructed of 2-mil silicon steel ribbon. The skin depth for the pulselength in silicon steel is about 0.33 mil. The core is therefore in the saturation wave regime. The available flux change in the laminations is 2.6 T. The radially sectioned core has a length of 8 in. (0.205 m). We assume about 30% of the volume of the isolation cavity is occupied by insulation and power straps. Taking into account the required volt–second product, the area of the core assembly is

$$A_c = (10^5 \text{ V})(10^{-6} \text{ s})/(0.7)(2.6 \text{ T}) = 0.055 \text{ m}^2.$$

If the inner radius of the core assembly is 6 in. (0.154 m), then the outer radius must be 16.4 in. (0.42 m). These parameters illustrate two features of long-pulse induction accelerators: (1) there is a large difference between R_0 and R_i, so that flux forcing must be used for a good impedance match, and (2) the cores are large. The mass of the core assembly in this example (excluding insulation) is 544 kG. The beam impedance is 50 Ω. Assume that the damping resistor is 25 Ω and the charge voltage on the reset circuit is 1000 V. Equation (10.14) implies that

$$C_r > (0.1 \text{ V-s})/(30 \text{ }\Omega)(1000 \text{ V}) = 3.3 \text{ }\mu\text{F}.$$

We choose $C_r = 10 \text{ }\mu\text{F}$; the core is reverse saturated 120 μs after closing the isolation relay. The capacitor voltage at this time is 670 V. Referring to the hysteresis curve of Figure 5.13, the required reset current is 670 A. This implies that $R_d \leq 1 \text{ }\Omega$. Assuming that the parasitic inductance of the reset circuit is about 1 μH, the resistance for a critically damped circuit is 0.632 Ω; therefore, the reset circuit is overdamped, as required.

A convenient method of core reset is possible for short-pulse induction cavities driven by pulse-charged Blumlein line modulators. In the system of Figure 10.18a, a Marx generator is used to charge an oil- or water-filled Blumlein line. The Blumlein line provides a flat-top voltage pulse to a ferrite or Metglas core cavity. A gas-filled spark gap shorts the intermediate conductor to the outer conductor to initiate the pulse. In this configuration, the intermediate conductor is charged negative for a positive output pulse. The crux of the auto-reset process is to use the negative current flowing from the center conductor of the Blumlein line during the pulse–charge cycle to reset the core. Reset occurs just before the main voltage pulse. Auto-reset eliminates the need for a separate reset circuit and high-voltage isolator. A further advantage is that the core can be driven to $-B_s$ just before the pulse.

Figure 10.18b shows the main circuit components. The impulse generator has capacitance C_g and inductance L_g. It has an open circuit voltage of $-V_0$. If the charge cycle is long compared to an electromagnetic transit time, the inner and outer parts of the Blumlein line can be treated as lumped capacitors of value C_1. The outer conductor is grounded; the inner conductor is connected to ground through the induction cavity. We assume there is a damping resistor R_d that shunts some of the reset current. We will estimate $V_c(t)$ (the negative voltage on the core during the charge cycle) and determine if the quantity $\int V_c \, dt$ exceeds the volt-second product of the core.

Assume that $V_c(t)$ is small compared to the voltage on the intermediate conductor. In this case, the voltage on the intermediate conductor is approxi-

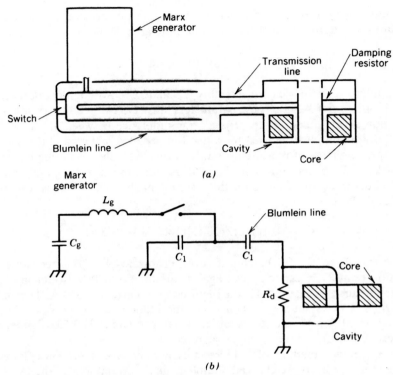

Figure 10.18 Auto-reset in an induction cavity driven by a Blumlein line. (*a*) System geometry. (*b*) Equivalent circuit.

mated by Eq. (9.102) with $C_2 = 2C_1$. Assuming that $C_g = 2C_1$, the voltage on the intermediate conductor is

$$V_2 = \tfrac{1}{2}[V_0(1 - \cos \omega t)], \qquad (10.17)$$

where $\omega = 1/\sqrt{L_g C_1}$. The charging current to the cavity is $i_c \cong C_1(dV_2/dt)$. Assuming most of this current flows through the damping resistor, the reset voltage on the core during the Blumlein line charge is

$$V_c \cong -(V_0/2)\big(R_d/\sqrt{L/C_1}\big)\sin \omega t. \qquad (10.18)$$

If the Blumlein is triggered at the time of completed energy transfer, $t = \pi/\omega$, then the volt–second integral during the charge phase is

$$\int V_c \, dt = V_0(C_1 R_d). \qquad (10.19)$$

The integral of Eq. (10.19) must exceed the volt–second product of the core for successful reset. This criterion can be written in a convenient form in terms of Δt_p, the length of the main voltage pulse in the cavity. Assuming a square pulse of magnitude V_0, the volt–second product of the core should be $V_0 \Delta t_p$. Substituting this expression in Eq. (10.19) gives the following condition:

$$\Delta t_p \le C_1 R_d. \tag{10.20}$$

Furthermore, we can use Eq. (9.89) to find the pulselength in terms of the line capacitance. Equation (10.20) reduces to the following simple criterion for auto-reset:

$$Z_1 \le R_d. \tag{10.21}$$

The quantity Z_1 is the net output impedance of the Blumlein line, equal to twice the impedance of the component transmission lines. Clearly, Eq. (10.21) is always satisfied.

In summary, auto-reset always occurs during the charge cycle if (1) the Marx generator is matched to the Blumlein line, (2) the core volt–second product is matched to the output voltage pulse, and (3) the Blumlein line impedance is matched to the combination of beam and damping resistor. Reset occurs earlier in the charge cycle as R_d is increased. The condition of Eq. (10.21) holds only for the simple circuit of Figure 10.18. More complex cases may occur; for instance, in some accelerators the cavity and Blumlein line are separated by a long transmission line which acts as a capacitance during the charge cycle. Premature core saturation shorts the reset circuit and can lead to voltage reversal on the connecting line. The resulting negative voltage applied to the cavity subtracts from the available volt–second product.

Flat voltage waveforms are usually desirable. In electron accelerators, voltage control assures an output beam with small energy spread. Voltage waveform shaping is essential for induction accelerators used for nonrelativistic particles. In this case, a rising voltage pulse is required for longitudinal beam confinement (see Section 13.5). Power is usually supplied from a pulse modulator which generates a square pulse in a matched load. There are two primary causes of waveform distortion: (1) beam loading and (2) transformer droop. We will concentrate on transformer droop in the remainder of this section.

The equivalent circuit of an induction linear accelerator cavity is shown in Figure 10.19a. The driving modulator maintains constant voltage if the current to the cavity is constant. Current is divided between the beam load, the damping resistor, and the leakage inductance. The leakage current increases with time; therefore, the cavity does not present a matched load at all times and the voltage droops. The goal is to compensate leakage current by inserting an element with a rising impedance.

A simple compensation circuit is shown in Figure 10.19b. A series capacitor is added to the damping resistor so that the impedance of the damping circuit

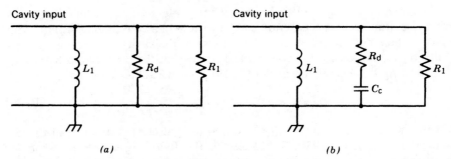

Figure 10.19 Compensation of voltage droop in an induction cavity. (*a*) Equivalent circuit of cavity with damping resistor R_d, no compensation. (*b*) Simple compensation circuit.

rises with time. We can estimate the value of capacitance that must be added to keep V_0 constant by making the following simplifying assumptions: (1) the leakage path inductance is assumed constant over the pulselength, Δt_p; (2) the voltage on the compensation capacitor C_c is small compared to V_0; (3) the leakage current is small compared to the total current supplied by the modulator; and (4) the beam current is constant. The problem resolves into balancing the increase in leakage current by a decrease in damping current.

With the above assumptions, the time-dependent leakage current is

$$i_1 \cong V_0 t / L_1 \tag{10.22}$$

for a voltage pulse initiated at $t = 0$. The current through the damping circuit is approximately

$$i_d \cong (V_0 / R_d)(1 - t / R_d C_c). \tag{10.23}$$

Balancing the time-dependent parts gives the following condition for constant circuit current:

$$R_d C_c = L_1 / R_d. \tag{10.24}$$

As an example, consider the long-pulse cavity that we have already discussed in this section. Taking the average value of μ / μ_0 as 10,000 and applying Eq. (9.15), the inductance of the leakage path for ideal ferromagnetic material is

$$L_1 = (\mu / \mu_0)(\mu_0 / 2\pi) d \ln(R_0 / R_i) \cong 400 \ \mu\text{H}.$$

The core actually operates in the saturation regime with a skin depth about one-third the half-thickness of the lamination; we can make a rough estimate of the leakage current by dividing the inductance by three, $L_1 = 133 \ \mu\text{H}$. With a damping resistance of $R_d = 30 \ \Omega$ and a cavity voltage of 100 kV, the modulator supplies a current greater than 5.3 kA. The leakage current is maximum at the end of the pulse. Equation (10.22) implies that $i_1 = 0.25$ kA at $t = \Delta t_p$, so that the third assumption above is valid. The compensating capacitance is predicted from Eq. (10.24) to be $C_c = 0.15$ μF. The maximum voltage on the capacitor is 22 kV: thus, assumption 2 is also satisfied. Capacitors are generally available in the voltage and capacitance range required, so that the compensation method of Figure 10.19 is feasible.

10.6 INDUCTION CAVITY DESIGN: FIELD STRESS AND AVERAGE GRADIENT

At first glance, it is difficult to visualize the distribution of voltage in induction cavities since electrostatic and inductive electric fields act in concord. In order to clarify field distributions, we shall consider the specific example of the electron injector illustrated in Figure 10.20. The configuration is the most

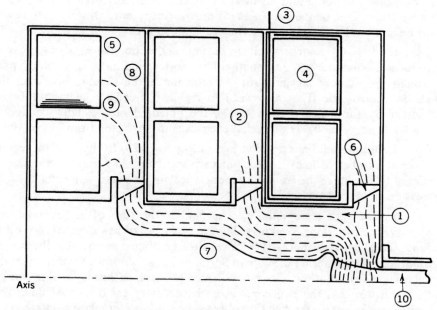

Figure 10.20 Equipotential lines in six-core injector cavity with radial and longitudinal stacking. (1) Vacuum region, acceleration gap. (2) Transformer oil insulation. (3) Insulated power feed. (4) Laminated isolation core. (5) Exposed face of laminated core. (6) Vacuum insulator with optimized shaping. (7) Shaped negative high-voltage electrode with electron source. (8) Grounded face of laminated core. (9) Location of equipotential line at voltage V_0 with respect to outer wall when flux-forcing straps used. (10) Extracted electron beam.

complex one that would normally be encountered in practice. It has laminated cores, longitudinal stacking, radial stacking, and flux-forcing straps. We shall develop an electric field map for times during which all core laminations are unsaturated. In the following discussion, bracketed numbers are keyed to points in the figure.

1. Three cavities are combined to provide 3 : 1 longitudinal voltage step-up. The load circuit region (1) is maintained at high vacuum for electron transport. The leakage circuit region (2) is filled with transformer oil for good insulation of the core and high-voltage leads. The vacuum insulator (6) is shaped for optimum resistance to surface breakdown.

2. Power is supplied through transmission lines entering the cavity radially (3). At least two diametrically opposed lines should be used in each cavity. Current distribution from a single power feed has a strong azimuthal asymmetry. If a single line entered from the top, the magnetic field associated with load current flow would be concentrated at the top, causing a downward deflection of the electron beam.

3. The cores (4) are constructed by interleaving continuous ribbons of ferromagnetic material and insulator; laminations are orientated as shown in Figure 10.20. One of the end faces (5) must be exposed; otherwise, the cavity will be shorted by conduction across the laminations.

4. If V_0 is the matched voltage output of the modulator, a single cavity produces a voltage $2V_0$. Equipotential lines corresponding to this voltage pass through the vacuum insulator (6). At radii inside the vacuum insulator, the field is electrostatic. The inner vacuum region has coaxial geometry. To an observer on the center conductor (7), the potential of the outer conductor appears to increase by $2V_0$ crossing each vacuum insulator from left to right.

5. Equipotential lines are sketched in Figure 10.20. In the vacuum region, an equal number of lines is added at each insulator. The center conductor is tapered to minimize the secondary inductance and to preserve a constant field stress on the metal surfaces.

6. In the core region, the electric field is the sum of electrostatic and inductive contributions. We know that the two types of fields cancel each other along the shorted wall (8). We have already discussed the distribution of equipotential lines inside the core in Section 10.2, so we will concentrate on the potential distribution on the exposed face (5). The inclusion of flux-forcing straps assures that the two cores in each subcavity enclose equal flux. This implies that the point marked (9) between the cores is at a relative potential V_0.

7. Each lamination in the cores isolates a voltage proportional to the applied magnetic field $B_1(r)$. Furthermore, electrostatic fields in the core are radial. Therefore, the electric field along the exposed face of an individual core has a $1/r$ variation, and potential varies as $\Phi(r) \sim \ln(r)$. The electrostatic potential distribution in the oil insulation has been sketched by connecting equipotential lines to the specified potential on the exposed core surface (5).

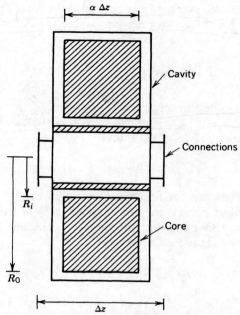

Figure 10.21 Geometric parameters for calculation of average gradient of linear induction accelerator.

In summary, although the electric field distribution in compound inductive cavities is complex, we can estimate them by analyzing the problem in parts. The distribution could be determined exactly with a computer code to solve the Laplace equation in the presence of dielectrics. There is a specified boundary condition on ϕ along the core face. It should be noted that the above derivation gives, at best, a first-order estimate of field distributions. Effects of core nonlinearities and unequal saturation complicate the situation considerably.

Average longitudinal gradient is one of the main figures of merit of an accelerator. Much of the equipment associated with a linear accelerator, such as the accelerator tunnel, vacuum systems, and focusing system power supplies, have a cost that scales linearly with the accelerator length. Thus, if the output energy is specified, there is an advantage to achieving a high average gradient.

In the induction accelerator, the average gradient is constrained by the magnetic properties and geometry of the ferromagnetic core. Referring to Figure 10.21, assume the core has inner and outer radii R_i and R_0, and define κ as the ratio of the radii, $\kappa = R_0/R_i$. The cavity has length Δz of which the core fills a fraction α. If particles gain energy ΔE in eV in a cavity with pulselength t_p, the volt–second constraint implies the following difference equation:

$$\Delta E t_p = \Delta B (R_0 - R_i) \alpha \Delta z. \tag{10.25}$$

where ΔB is the volume-averaged flux change in the core. The pulselength may vary along the length of the machine; Eq. (10.25) can be written as an integral equation:

$$\int_{E_i}^{E_f} dE\, t_p(E) = \alpha\, \Delta B\, R_0(\kappa - 1)L, \qquad (10.26)$$

where E_f is the final beam energy in electron volts, E_i is the injection energy, and L is the total length. Accelerators for relativistic electrons have constant pulselength. Equation (10.26) implies that

$$(E_f - E_i)/L = \alpha\, \Delta B\, R_0(\kappa - 1)/\Delta t_p \ (\text{V/m}). \qquad (10.27)$$

In proposed accelerators for nonrelativistic ion beams,[†] the pulselength is shortened as the beam energy is raised to maintain a constant beam length and space charge density. One possible variation is to take the pulselength inversely proportional to the longitudinal velocity:

$$t_p(E) = t_{pf}\left[E_f/E(z)\right]^{1/2}, \qquad (10.28)$$

The quantity t_{pf} is the pulselength of the output beam. Inserting Eq. (10.28) into Eq. (10.26) gives

$$\left(E_f - \sqrt{E_f E_i}\right)/L = \alpha\, \Delta B\, R_0(\kappa - 1)/2t_{pf}. \qquad (10.29)$$

In the limit that $E_i \ll E_f$, the expressions of Eqs. (10.27) and (10.29) are approximately equal to the average longitudinal gradient. In terms of the quantities defined, the total volume of ferromagnetic cores in the accelerator is

$$V = \pi(\kappa^2 - 1)\alpha L R_0^2. \qquad (10.30)$$

Equations (10.27), (10.29), and (10.30) have the following implications:

1.   High gradients are achieved with a large magnetic field swing in the core material (ΔB) and the tightest possible core packing α.

2.   The shortest pulselength gives the highest gradient. Properties of the core material and the inductance of the pulse modulators determine the minimum t_p. High-current ferrite accelerators have a minimum practical pulselength of about 50 ns. The figure is about 1 μs for silicon steel cores and 100 ns for Metglas cores with thin laminations. Because of their high flux swing and fast pulse response, Metglas cores open new possibilities for high-gradient-induction linear accelerators.

3.   The average gradient is maximized when κ approaches infinity. On the other hand, the net core volume is minimized when L approaches infinity.

[†]See A. Faltens, E. Hoyer, and D. Keefe, *Proc. 4th Intl. Conf. High Power Electron and Ion Beam Research and Technology*, Ecole Polytechnique, 1981, p. 751.

There is a crossover point of minimum accelerator cost at certain values of κ and L, depending on the relative cost of cores versus other components.

Substitution of some typical parameters in Eq. (10.27) will indicate the maximum gradient that can be achieved with a linear induction accelerator. Take $t_p = 100$ ns pulse and a Metglas core with $\Delta B = 2.5$ T, $R_i = 0.1$ m and $R_0 = 0.5$ m. Vacuum ports, power feeds, and insulators must be accommodated in the cavity in addition to the cores; we will take $\alpha = 0.5$. These numbers imply a gradient of 5 MV/m. This gradient is within a factor of 2–4 of those achieved in rf linear electron accelerators. Higher gradients are unlikely because induction cavities have vacuum insulators exposed to the full accelerating electric fields.

10.7 CORELESS INDUCTION ACCELERATORS

The ferromagnetic cores of induction accelerator cavities are massive, and the volt–second product limitation restricts average longitudinal gradient. There has been considerable effort devoted to the development of linear induction accelerators without ferromagnetic cores.[†] These devices incorporate transmission lines within the cavity. They achieve inductive isolation through the flux change accompanying propagation of voltage pulses through the lines.

In order to understand the coreless induction cavity, we must be familiar with the radial transmission line (Fig. 10.22). This geometry has much in common with the transmission lines we have already studied except that voltage pulses propagate radially. Consider the conical section electrode as the ground conductor; the radial plate is the "center" conductor. The structure has minimum radius R_i; we take the inner radius as the input point of the line. If voltage is applied to the center conductor, a voltage pulse travels outward at a velocity determined by the medium in the line. The voltage pulse maintains constant shape. If the line extends radially to infinity, the pulse never returns. If the line has a finite radius (R_0), the pulse is reflected and travels back to the center.

It is not difficult to show that the structure of Figure 10.22 has constant characteristic impedance as a function of radius. (In other words, the ratio of the voltage and current associated with a radially traveling pulse is constant.) In order to carry out the analysis, we assume that α, the angle of the conical electrode, is small. In this case, the electric field lines are primarily in the axial direction. There is a capacitance per unit of length in the radial direction given by

$$\mathscr{C}\,\Delta r \cong (2\pi r\,\Delta r)\varepsilon/r\tan\alpha = (2\pi\varepsilon/\tan\alpha)\,\Delta r. \qquad (10.31)$$

In order to determine the inductance per unit of radial length, we must consider current paths for the voltage pulse. Inspection of Figure 10.22 shows

[†]A. I. Pavlovskii, A. I. Gerasimov, D. I. Zenkov, V. S. Bosamykin, A. P. Klementev, and V. A. Tananakin, *Sov. At. En.*, **28**, 549 (1970).

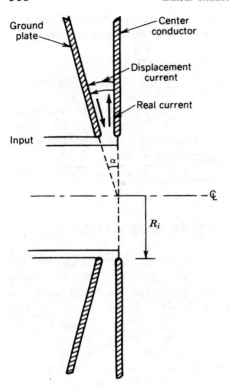

Figure 10.22 Current flow in radial transmission line.

that current flows axially through the power feed, outward along the radial plate, and axially back across the gap as displacement current. The current then returns along the ground conductor to the input point. If the pulse has azimuthal symmetry, the only component of magnetic field is B_θ. The toroidal magnetic field is determined by the combination of axial feed current, axial displacement current, and the axial component of ground return current. Radial current does not produce toroidal magnetic fields. The combination of axial current components results in a field of the form

$$B_\theta = \mu I / 2\pi r \qquad (10.32)$$

confined inside the transmission line behind the pulse front. The region of magnetic field expands as the pulse moves outward, so there is an inductance associated with the pulse. The magnetic field energy per element of radial section length is

$$U_m \, dr = \left(B_\theta^2 / 2\mu \right)(2\pi r \, dr)(r \tan \alpha) = \mu I^2 \tan \alpha \, dr / 4\pi. \qquad (10.33)$$

This energy is equal to $\frac{1}{2}\mathscr{L} I^2 dr$, so that

$$\mathscr{L} = \mu \tan \alpha / 2\pi. \qquad (10.34)$$

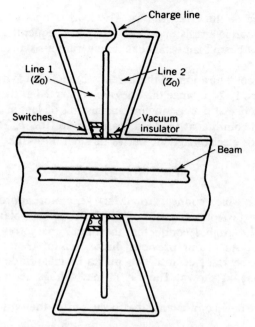

Figure 10.23 Coreless induction cavity.

To summarize, both the capacitance and inductance per radial length element are constant. The velocity of wave propagation is

$$v = 1/\sqrt{\mathscr{L}\mathscr{C}} = 1/\sqrt{\varepsilon\mu},$$

(10.35)

as expected. The characteristic impedance is

$$Z_0 = \sqrt{\mathscr{L}/\mathscr{C}} = (\tan\alpha/2\pi)\sqrt{\mu/\varepsilon}.$$

$\boxed{10.36}$

For purified water dielectric, Eq. (10.34) becomes

$$Z_0 = 3.4\tan\alpha \ (\Omega).$$

(10.37)

The basic coreless induction cavity of Figure 10.23 is composed of two radial transmission lines. There are open-circuit terminations at both the inner and outer radius; the high-voltage radial electrode is supported by insulators and direct current charged to high voltage. The region at the outer radius is shaped to provide a matched transition; a wave traveling outward in one line propagates without reflection through the transition and travels radially inward down the other line. There is a low-inductance, azimuthally symmetric shorting switch in one line at the inner radius. The load is on the axis. We assume initially that the load is a resistor with $R = Z_0$; subsequently, we will consider the possibility of a beam load which has time variations synchronized to pulses in the cavity.

The electrical configuration of the coreless induction cavity is shown schematically in Figure 10.24a. Since the lines are connected at the outer radius, we can redraw the schematic as a single transmission of length $2(R_0 - R_i)$ that doubles back and connects at the load, as in Figure 10.24b. The quantity Δt is the transit time for electromagnetic pulses through both lines, or

$$\Delta t = 2(R_0 - R_i)/v. \qquad (10.38)$$

The charging feed, which connects to a Marx generator, approximates an open circuit during the fast output pulse of the lines. Wave polarities are defined with respect to the ground conductor; the initial charge on the high-voltage electrode is $+V_0$. As in our previous discussions of transmission lines, the static charge can be resolved into two pulses of magnitude $\frac{1}{2}V_0$ traveling in opposite directions, as shown. There is no net voltage across the load in the charged state.

Consider the sequence of events that occurs after the switch is activated at $t = 0$.

1.   The point marked A is shorted to the radial electrode. During the time $0 < t < \Delta t$, the radial wave traveling counterclockwise encounters the short circuit and is reflected with reversed polarity. At the same time, the clockwise pulse travels across the short circuit and backward through the matched resistance, resulting in a voltage $-\frac{1}{2}V_0$ across the load. Charged particles gain an energy $\frac{1}{2}qV_0$ traveling from point A to point B. The difference in potential between A and B arises from the flux change associated with the traveling pulses.

2.   At time $t = \Delta t$, the clockwise-going positive wave is completely dissipated in the load resistor. The head of the reflected negative wave arrives at the load. During the time $\Delta t < t < 2\,\Delta t$, this wave produces a positive voltage of magnitude $\frac{1}{2}V_0$ from point B to point A. The total waveform at the load is shown in Figure 10.25a.

The bipolar waveform of Figure 10.25a is clearly not very useful for particle acceleration. Only half of the stored energy can be used. Better coupling is achieved by using the beam load as a switched resistor. The beam load is connected only when the beam is in the gap. Consider the following situation. The beam, with a current $V_0/2Z_0$, does not arrive at the acceleration gap until time Δt. In the interval $0 < t < \Delta t$, the gap is an open circuit. The status of the reflecting waves is illustrated in Figure 10.24c. The counterclockwise wave reflects with inverted polarity; the clockwise wave reflects from the open circuit gap with the same polarity. The voltage from B to A is the open-circuit voltage

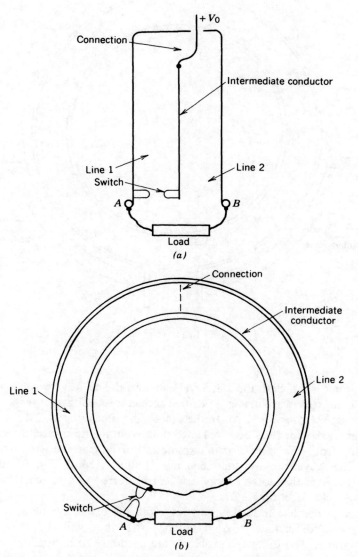

Figure 10.24 Coreless induction cavity. (*a*) Simplified geometry. (*b*) Equivalent geometry with ideal outer connection between lines. (*c*) Resolution of static charge into traveling square pulses; polarity and propagation direction of pulses immediately following switch shorting.

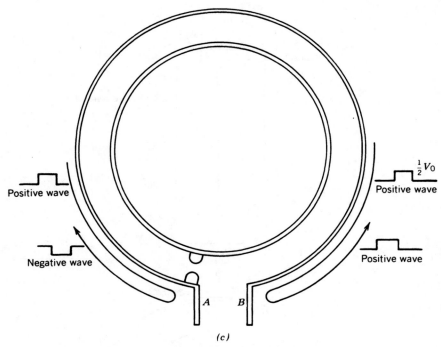

Positive wave

$\frac{1}{2}V_0$
Positive wave

Negative wave

Positive wave

A B

(c)

Figure 10.24   (*Continued*).

V_0. At time $t = \Delta t$, both the matched beam and the negative wave arrive at the gap. The negative wave makes a positive accelerating voltage of magnitude $\frac{1}{2}V_0$ during the time $\Delta t < t < 2\,\Delta t$. In the succeeding time interval, $2\,\Delta t < t < 3\,\Delta t$, the other wave which has been reflected once from the open-circuit termination and once from the short-circuit termination arrives at the gap to drive the beam. The waveform for this sequence is illustrated in Figure 10.25*b*. In theory, 100% of the stored energy can be transferred to a matched beam load at voltage $\frac{1}{2}V_0$ for a time $2\,\Delta t$.

Although coreless induction cavities avoid the use of ferromagnetic cores, technological difficulties make it unlikely that they will supplant standard configurations. The following problems are encountered in applications:

1. The pulselength is limited by the electromagnetic transit time in the structure. Even with a high dielectric constant material such as water, the radial transmission lines must have an outer diameter greater than 9 ft for an 80-ns pulse.

2. Energy storage is inefficient for large-diameter lines. The maximum electric field must be chosen to avoid breakdown at the smallest radius. The stored energy density of electric fields [Eq. (5.19)] decreases as $1/r^2$ moving out in radius.

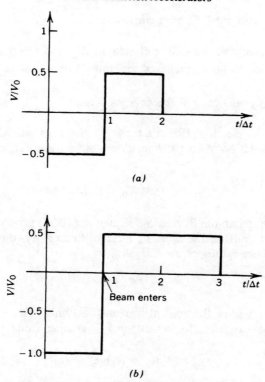

Figure 10.25 Voltage waveform in coreless induction cavity. (*a*) Simple matched load resistor. (*b*) Synchronized beam load with matched impedance.

3.   Synchronous low-inductance switching at high voltage with good azimuthal symmetry is difficult.

4.   Parasitic inductance in the load circuit tends to be larger than the primary inductance of the leakage path, since the load is a beam with radius that is small compared to the radius of the shorting switches. The parasitic inductance degrades the gap pulseshape and the efficiency of the accelerator.

5.   The switch sequence for high-efficiency energy transfer means that damping resistors cannot be used in parallel with the gap to protect the cavity.

6.   The vacuum insulators must be designed to withstand an overvoltage by a factor of 2 during the open-circuit phase of wave reflection.

One of the main reasons for interest in coreless induction accelerators was the hope that they could achieve higher average accelerating gradients than ferromagnetic cavity accelerators. In fact, a careful analysis shows that coreless induction accelerators have a significant disadvantage in terms of average gradient compared to accelerators with ferromagnetic isolation. We will make the comparison with the following constraints:

1. The cavities have the same pulselength Δt_p and beam current I_0.
2. Regions of focusing magnets, pumping ports, and insulators are not included.
3. The energy efficiency of the cavities is high.

We have seen in Section 10.6 that the gradient of an ideal cavity with ferromagnetic isolation of length l with core outer radius R_0 and inner radius R_i is

$$V_0/l = \Delta B (R_0 - R_i)/\Delta t_p. \qquad (10.39)$$

where ΔB is the maximum flux swing. Equation (10.39) proceeds directly from the volt–second limit on the core. In a radial line cavity, the cavity length is related to the outer radius of the line, R_0, by

$$l = 2R_0 \tan \alpha, \qquad (10.40)$$

where α is the angle of the conical transmission lines. The voltage pulse in a high-efficiency cavity with charge voltage V_0 has magnitude $\frac{1}{2} V_0$ and duration

$$\tau_p = 4(R_0 - R_i)\sqrt{\varepsilon/\varepsilon_0}/c. \qquad (10.41)$$

where R_i is the inner radius of the transmission lines. We further require that the beam load is matched to the cavity:

$$Z_0 = V_0/2I_0.$$

The characteristic impedance is given by Eq. (10.36). Combining Eq. (10.36) with Eqs. (10.40) and (10.41) gives the following expression for voltage gradient:

$$V_0/l = (\mu_0/\pi)(1 - R_i/R_0)I_0/\Delta t_p. \qquad (10.42)$$

Equation (10.42) has some interesting implications for coreless induction cavities. First, gradient in an efficient accelerator is proportional to beam current. Second, the gradient for a given pulselength is relatively insensitive to the outer radius of the line. This reflects the fact that the energy storage density is low at large radii. Third, the gradient for ideal cavities does not depend on the filling medium of the transmission lines. Within limits of practical construction, an oil-filled line has the same figure of merit as a water-filled line.

In order to compare the ferromagnetic core and coreless cavities, assume the following conditions. The pulselength is 100 ns and the beam current is 50 kA (the highest current that has presently been transported a significant distance in a multistage accelerator). The ferromagnetic core has $R_0 = 0.5$ m, $R_i = 0.1$ m, and $\Delta B = 2.5$ T. The coreless cavity has $R_i/R_0 \ll 1$. Equation (10.39) implies that the maximum theoretical gradient of the Metglas cavity is 10 Mv/m, while Eq. (10.42) gives an upper limit for the coreless cavity of only 0.16 Mv/m, a factor of 63 lower. Similar results can be obtained for any coreless configuration. Claims for high gradient in coreless accelerators usually are the result of implicit assumptions of extremely short pulselengths (10–20 ns) and low system efficiency.

11

Betatrons

The betatron[†] is a circular induction accelerator used for electron acceleration. The word betatron derives from the fact that high-energy electrons are often called β particles. Like the linear induction accelerator, the betatron is the circuit equivalent of a step-up transformer. The main difference from the linear induction accelerator is that magnetic bending and focusing fields are added to confine electrons to circular orbits around the isolation core. The beam acts as a multiturn secondary. A single-pulsed power modulator operating at a few kilovolts drives the input; the output beam energy may exceed 100 MeV. The maximum electron kinetic energy achieved by betatrons is about 300 MeV. The energy limit is determined in part by the practical size of pulsed magnets and in part by synchrotron radiation.

General principles of the betatron are introduced in Section 11.1. The similarities between the power circuits of the linear induction accelerator, the recirculating induction linear accelerator, and the betatron are emphasized. An expression is derived for the maximum energy from a betatron; neglecting radiation, the limit depends only on the properties of the ferromagnetic core.

Two areas of accelerator physics must be studied in detail in order to understand the betatron; the theory of particle orbits in a gradient-type magnetic field and properties of magnetic circuits. Regarding orbits, the simple theory of betatron oscillations introduced in Section 7.3 must be extended. The amplitude of transverse-orbit oscillations and conditions for constant main-orbit radius must be determined for highly relativistic particles in a slowly

[†] D. W. Kerst, *Phys. Rev.*, **58**, 841 (1940).

changing magnetic field. Section 11.2 treats main orbit equilibria. The main orbit in the betatron has a constant radius during the acceleration cycle. The orbit exists when the well-known *betatron condition* is satisfied. The confinement properties of the system for nonideal orbits are subsequently discussed. The derivations demonstrate two properties of orbits: (1) particles injected on a circular orbit inside or outside the main orbit approach the main orbit during acceleration and (2) the amplitude of transverse oscillations decreases during the acceleration cycle. Section 11.3 addresses the first effect, motion of the instantaneous circle. Section 11.5 discusses damping of relativistic betatron oscillations during acceleration. As an introduction, Section 11.4 reviews the properties of periodic particle motions under the influence of slowly changing forces. The laws governing reversible compressions, both for nonrelativistic and relativistic particles, are discussed. The results are applicable to a wide variety of accelerators and particle confinement devices. Section 11.6 covers injection and extraction of electrons from the machine.

Section 11.7 surveys betatron magnet circuits, proceeding from simple low-energy devices to high-energy accelerators with optimal use of the core. The betatron magnet provides fields for particle acceleration, beam bending, and particle confinement. The magnet must be carefully designed in order to fulfill these functions simultaneously. Ferromagnetic materials are an integral part of all betatrons except the smallest laboratory devices. Thus, the available flux change is limited by the saturation properties of iron. Within these limits, the magnet circuit is designed to achieve the highest beam kinetic energy for a given stored modulator energy.

Even with good magnet design, existing betatrons are inefficient. Conventional betatrons rely on gradients of the bending field for focusing and utilize low-energy electron injection. The self-electric field of the beam limits the amount of charge that can be contained during the low-energy phase of the acceleration cycle. Usually, the beam current is much smaller than the driving circuit leakage current. Consequently, energy losses from hysteresis and eddy currents in the core are much larger than the net beam energy. Efficiency is increased by high beam current. Some strategies for high-current transport are discussed in Section 11.6. The two most promising options are (1) addition of supplemental focusing that is effective at low energy and (2) high-energy electron injection using a linear induction accelerator as a preaccelerator. In principle, betatrons can produce beam powers comparable to linear induction accelerators with a considerable reduction in isolation core mass.

11.1 PRINCIPLES OF THE BETATRON

Figure 11.1 illustrates the basic betatron geometry. A toroidal vacuum chamber encircles the core of a large magnet. The magnetic field is produced by pulsed coils; the magnetic flux inside the radius of the vacuum chamber

Return flux yoke

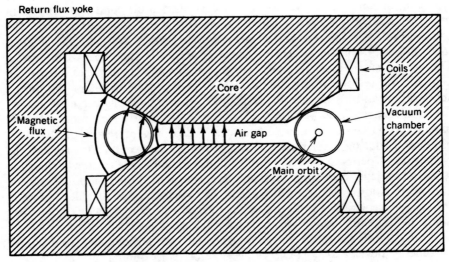

Figure 11.1 Schematic diagram of betatron with air gap.

changes with time. Increasing flux generates an azimuthal electric field which accelerates electrons in the chamber.

In the absence of an air gap, there is little magnetic flux outside the core. An air gap is included to divert some of the magnetic flux into the vacuum chamber. By the proper choice of gap width, the vertical magnetic field can be adjusted to confine electrons to a circular orbit in the vacuum chamber. As shown in Figure 11.1, the confining field lines are curved. The resultant field has a positive field index. As we found in Section 7.3, the field can focus in both the horizontal and vertical directions.

In summary, the simple betatron of Figure 11.1 has the following elements:

1. A pulsed magnet circuit to accelerate electrons by inductive fields.
2. An air gap to force magnetic field into the beam transport region; electrons follow circular orbits in the bending field.
3. Shaped magnetic fields for beam focusing.

At first glance, the betatron appears quite different from the linear induction accelerator. Nonetheless, we can show that the power circuits of the two devices are similar. To begin, consider the induction accelerator illustrated in Figure 11.2*a*. The geometry is often called a *recirculating induction linac*. The transport tube is bent so that the beam passes through the same cavity a number of times. This allows higher beam kinetic energy for a given volt–second product of the isolation cores. The transport tubes are made of metal; each cavity has separate vacuum insulators and high-voltage feeds. There are

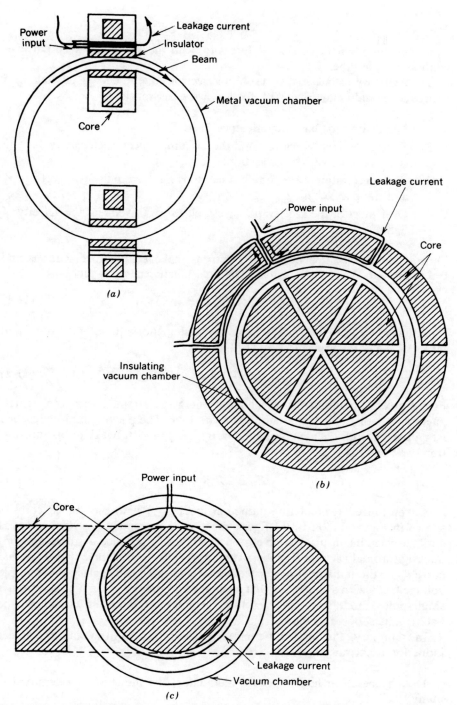

Figure 11.2 Equivalence between betatrons and linear induction accelerators. (*a*) Recirculating induction linac with two acceleration cavities. (*b*) Recirculating induction linac with isolation cores that fill the available area inside beam orbit. (*c*) Betatron with single core and single power feed point.

supplemental magnetic or electric forces to bend the orbits and keep particles confined in the tube.

To begin, we calculate the maximum electron kinetic energy possible in a recirculating induction linac with the following assumptions:

1. The beam tube has circumference $2\pi R$.
2. There are N cavities around the circumference; each cavity has an isolation core with cross-sectional area A_c.
3. The accelerating waveform in a cavity is a square pulse with voltage V_0 and the pulselength t_p.
4. Over most of the acceleration cycle, electrons travel near the velocity of light.

During the acceleration cycle, the electrons make $ct_p/2\pi R$ revolutions and travel through $Nct_p/2\pi R$ cavities. The final kinetic energy is therefore

$$E_b = V_0 t_p Nc/2\pi R \text{ (eV)}. \tag{11.1}$$

Equation (11.1) can be rewritten by expressing the volt–second product in terms of the core properties [Eq. (10.1)]:

$$E_b \leq 2B_s NA_c c/2\pi R. \tag{11.2}$$

For a given circumference, the highest energy is attained with the tightest packing of isolation cores around the beam tube. The packing limit is reached when the cores fill the area inside the beam, $NA_c = \pi R^2$. Making this substitution, we find that

$$E_b \leq 2B_s Rc/2. \qquad \boxed{11.3}$$

An optimized recirculating induction accelerator with pie-shaped cores is shown in Figure 11.2b. In the figure, much of the structure has been removed and the vacuum insulators have been extended to produce a single nonconducting toroidal vacuum chamber. The final step is to recognize that the radial currents of the individual power feeds cancel out; we can replace the multiple voltage feeds with a single line that encircles the core. Power is supplied from a single-pulse modulator. The resulting geometry, the power circuit of the betatron, is shown in Figure 11.2c.

In summary, the main differences between the betatron and the linear induction accelerator are as follows:

1.   The betatron has one pulse modulator; the induction accelerator has many.

2.   The beam in an induction accelerator makes a single pass through the machine. The equivalent circuit is a transformer with a single-turn secondary

and multiple parallel primary windings. In the betatron, the beam makes many revolutions around the core. The circuit representing this machine is a single primary with a multiturn secondary.

3.   Because of recirculation, average gradient is not a concern in the betatron. Therefore, low accelerating voltages and relatively long pulselengths (matched to the available volt–second product of the core) are used. The circuit of Figure 11.2c requires a slow voltage pulse because it has significantly higher inductance than the driving circuits of Figure 11.2b.

4.   Shaping of the voltage pulse shape is not important in the betatron. The beam is distributed uniformly around the transport tube; there is no need for longitudinal confinement. The betatron magnet is usually driven by a bipolar, harmonic voltage waveform that cycles the core between $-B_s$ and B_s.

The slow acceleration cycle and small circuital voltage allow a number of options for construction of the transport tube. The tube may be composed of metal interrupted azimuthally by one or more insulating rings. It is also possible to use a metal chamber constructed of thin stainless steel; the wall resistance must be high enough to keep inductively driven return currents small.

Equation 11.3 is also applicable to the betatron. As an example of kinetic energy limits, take $R = 1$ m and $B_s = 1.5$ T. The maximum kinetic energy is less than 450 MeV. Equation (11.3) has an important implication for the scaling of betatron output energy. The beam energy increases linearly with the radius of the central core, while the volume of core and flux return yoke increase as R^3. Cost escalates rapidly with energy; this is one of the main reasons why betatrons are limited to moderate beam energy.

As a final topic, we shall consider why betatrons have little potential for ion acceleration. In the discussion, ion dynamics is treated nonrelativistically. Assume an ion of mass m_i is contained in a betatron with radius R; the emf around the core is V_0. The energy ions gain in a time interval Δt is eV_0 multiplied by the number of revolutions, or

$$dE_b' = (eV_0/2\pi R)(2E_b'/m_i)^{1/2}\,\Delta t. \qquad (11.4)$$

Equation (11.4) can be rearranged to give

$$dE_b'/E_b'^{1/2} = (V_0/2\pi R)(2e^2/m_i)^{1/2}\,\Delta t. \qquad (11.5)$$

Integrating Eq. (11.5) (with the assumption that the final ion energy E_b is much larger than the injection energy), we find that

$$E_b^{1/2} \leq (e/2m_i)^{1/2}(V_0 t_p/2\pi R)\ \text{(eV)}. \qquad (11.6)$$

Substituting for the volt–second product and assuming a core area πR^2, Eq.

(11.6) can be rewritten

$$E_b \leq (2e/m_i)(B_s R/2)^2. \tag{11.7}$$

With the same magnet parameters as above ($R = 1$ m, $B_s = 1.5$ T), Eq. (11.7) implies that the maximum energy for deuterons is only 54 MeV.

Comparing Eq. (11.7) to Eq. (11.3), we find that the ratio of maximum obtainable energies for ions compared to electrons is

$$E_b \text{ (ions)}/E_b \text{ (electrons)} \cong v_{if}/c, \tag{11.8}$$

where v_{if} is the final ion velocity. Equation (11.8) has a simple interpretation. During the same acceleration cycle, the nonrelativistic ions make fewer revolutions around the core than electrons and gain a correspondingly smaller energy.

11.2 EQUILIBRIUM OF THE MAIN BETATRON ORBIT

The magnitude of the magnetic field at the orbit radius of electrons in a betatron is determined by the shape of the magnet poles. The equilibrium orbit has the following properties: (1) the orbit is circular with a radius equal to that of the major radius R of the vacuum chamber and (2) the orbit is centered in the symmetry plane of the field with no vertical oscillations. This trajectory is called the *main orbit*. We will consider other possible orbits in terms of perturbations about the main orbit.

The vertical field at R is designated $B_z(R)$. Equation (3.38) implies that $B_z(R)$ and R are related by

$$R = \gamma m_e v_\theta / e B_z(R) = p_\theta / e B_z(R). \tag{11.9}$$

The quantity p_θ is the total momentum of particles on the main orbit. The magnetic field varies with time. The azimuthal electric field acting on electrons is

$$\int \mathbf{E} \cdot \mathbf{dl} = d\Phi/dt = 2\pi R E_\theta, \tag{11.10}$$

where Φ is the magnetic flux enclosed within the particle orbit. Particle motion on the main orbit is described by the following equations:

$$v_r = 0, \qquad dp_r/dt = 0,$$

$$v_z = 0, \qquad dp_z/dt = 0,$$

$$dp_\theta/dt = eE_\theta = (e/2\pi R)\, d\Phi/dt. \tag{11.11}$$

Equation (11.11) is obtained from Eq. (3.34) by setting $v_r = 0$. We assume that

R does not vary in time; consequently, Eq. (11.11) can be integrated directly to give

$$p_\theta = e[\Phi(t) - \Phi(0)]/2\pi R = (e/2\pi R)\,\Delta\Phi. \qquad (11.12)$$

Combining Eqs. (11.9) and (11.12),

$$B_z(R) = \Delta\Phi/2\pi R^2. \qquad \boxed{11.13}$$

Equation (11.13) is the well-known *betatron condition*. The betatron pole piece is designed so that vertical field at the average beam radius is equal to one-half the flux change in the core divided by the area inside the particle orbit. The betatron condition has a simple interpretation for the machine illustrated in Figure 11.1. Electrons are injected at low energy when the orbital field and the flux in the core are near zero. The bending field and accelerating field are produced by the same coils, so that they are always proportional if there is no local saturation of the core iron. The main orbit has radius R throughout the acceleration cycle if the vertical field at R is equal to one-half the average field enclosed by the orbit. This condition holds both in the nonrelativistic and relativistic regimes. The acceleration cycle is illustrated in Figure 11.3.

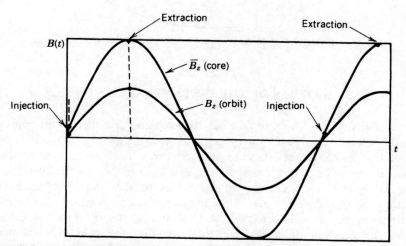

Figure 11.3 Acceleration cycle of air gap betatron; average magnetic field inside electron orbit and vertical magnetic field at main orbit as a function of time.

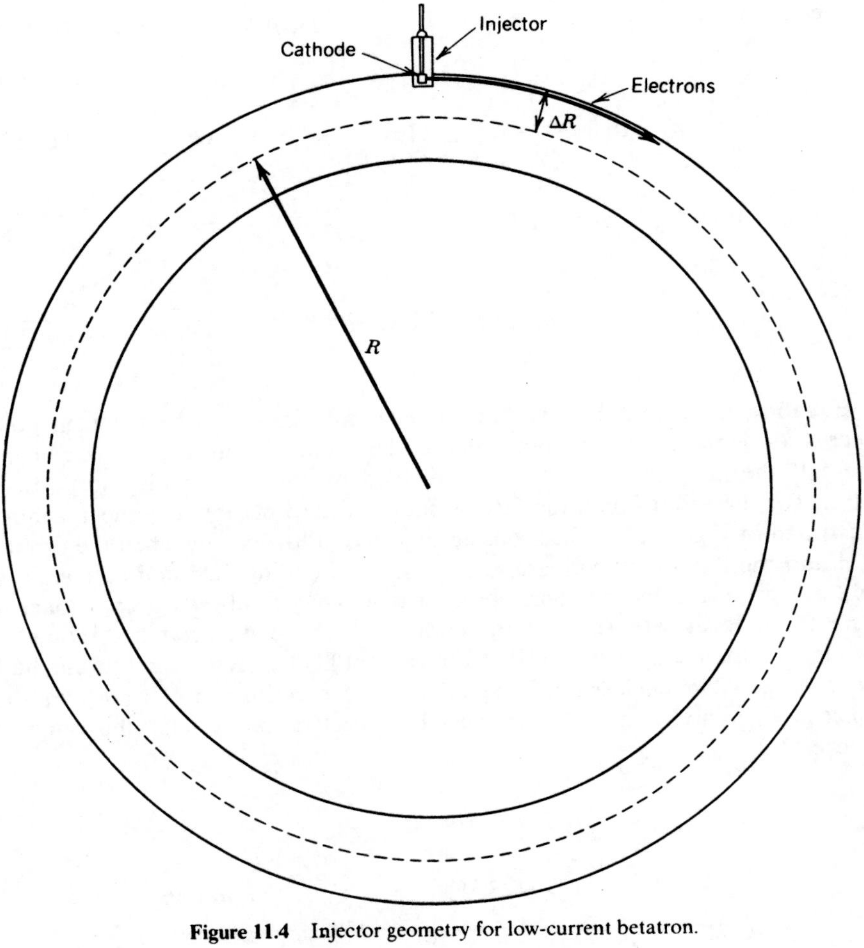

Figure 11.4 Injector geometry for low-current betatron.

11.3 MOTION OF THE INSTANTANEOUS CIRCLE

The standard electron injector of a betatron consists of a thermionic source at high dc voltage (20–120 kV) with extractor electrodes (Fig. 11.4). It is clear that such a device cannot extend to the main orbit. The injector is located at a radius inside or outside the main orbit and is displaced vertically from the symmetry plane. The extractor voltage is set so that the electrons have a circular orbit of radius $R + \Delta R$ in the magnetic field at injection. The betatron condition is not satisfied on this orbit; therefore, the orbit radius changes during the acceleration cycle. We shall see that the orbit asymptotically approaches the main orbit as the electron energy increases. The circular orbit with slowly varying radius is referred to as the *instantaneous circle*.

Let p_0 be the momentum of a particle on the main orbit and p_1 be the momentum of a particle injected a distance ΔR from the main orbit on the instantaneous circle. At injection, the momenta and magnetic fields are related by

$$p_0(0) = eB_z(R)R, \tag{11.14}$$

$$p_1(0) = p_0(0) + \delta p(0) = eB_z(R + \Delta R)(R + \Delta R). \tag{11.15}$$

The time variation of flux enclosed within the instantaneous circle is

$$d\Phi_1/dt = 2\pi R^2 [dB_z(R)/dt] = \int_R^{R+\Delta R} 2\pi r\, dr\, [dB_z(r)/dt]. \tag{11.16}$$

Equation (11.13) has been used in the first term to express the magnetic flux in the region $0 < r < R$. Assume that field variations are small over the region near the main orbit so that $B_z(r) \cong B_z(R)$. To first order in Δr, Eq. (11.16) can be rewritten

$$d\Phi_1/dt \cong 2\pi R^2 [dB_z(R)/dt] + 2\pi R\, \Delta R(dB_z/dt)$$

$$= 2\pi Rr [dB_z(R)/dt]. \tag{11.17}$$

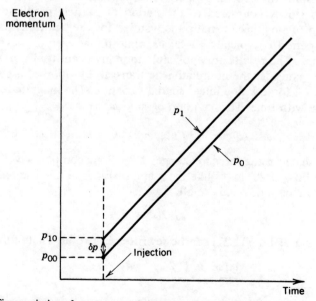

Figure 11.5 Time variation of momentum of electron on main orbit (p_0) and electron injected on instantaneous circle outside main orbit (p_1).

The equation of motion for an electron on the instantaneous circle is

$$dp_1/dt = (e/2\pi r)(d\Phi_1/dt) \cong eR\left[dB_z(R)/dt\right]. \qquad (11.18)$$

We recognize that the expression on the right-hand side is equal to dp_0/dt [Eq. (11.14)].

The main conclusion is that particles on the instantaneous circle gain momentum at the same rate as particles on the main orbit, as illustrated in Figure 11.5. The ratio of the radius of the instantaneous circle to that of the main orbit is equal to the relative momentum difference, or

$$\delta p/p_0 \cong \Delta R/R. \qquad (11.19)$$

The radius difference is proportional to $1/p_0$ since δp is constant by Eq. (11.18). Therefore, the instantaneous circle approaches the main orbit as the electron energy increases.

11.4 REVERSIBLE COMPRESSION OF TRANSVERSE PARTICLE ORBITS

As we saw in Section 7.3, the focusing strength of magnetic field gradients is proportional to the magnitude of the bending field. In order to describe the betatron, the derivations of particle transport in continuous focusing systems must be extended to include time-varying focusing forces. As an introduction, we will consider the general properties of periodic orbits when the confining force varies slowly compared to the period of particle oscillations. The approximation of slow field variation is justified for the betatron; the transverse oscillation period is typically 10–20 ns while the acceleration cycle is on the order of 1 ms. The results are applicable to many beam transport systems.

To begin, consider the nonrelativistic transverse motion of a particle under the action of a force with a linear spatial variation. The magnitude of the force may change with time. The equation of motion is

$$d^2x/dt^2 = -\left[F(t)/mx_0\right]x = -\omega(t)^2x. \qquad (11.20)$$

If the time scale for the force to change, ΔT, is long compared to $1/\omega$, then the solution of Eq. (11.20) looks like the graph of Figure 11.6. The relative change in ω over one period, $\Delta\omega$, is small:

$$\Delta\omega/\omega \ll 1. \qquad (11.21)$$

The condition of Eq. (11.21) can be rewritten in two alternate forms:

$$\frac{(d\omega/dt)(1/\omega)}{\omega} = \frac{d\omega/dt}{\omega^2} \ll 1, \qquad (11.22)$$

$$\frac{1}{\omega\Delta T} \ll 1. \qquad (11.23)$$

$x(t)$

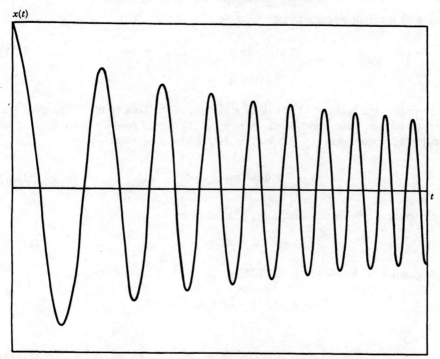

Figure 11.6 Time-dependent displacement of particle acted on by a focusing force that increases with time. Force varies linearly with x and increases a factor of 5 over time interval shown.

Equations (11.21)–(11.23) give the condition for a *reversible compression* (or *reversible expansion*). The meaning of reversible will be evident when we consider properties of the particle orbits.

Following Figure 11.6, an approximate solution to Eq. (11.20) should be oscillatory with a slow variation of amplitude. We assume a form

$$x(t) = A(t)\sin[\Phi(t)]. \tag{11.24}$$

The quantities $A(t)$ and $\Phi(t)$ are determined by substituting Eq. (11.24) into Eq. (11.20) and dropping terms of order $(1/\omega\,\Delta T)^2$ or higher.

Calculating the derivatives and substituting,

$$x = \ddot{A}\sin\Phi + 2\dot{A}\dot{\Phi}\cos\Phi - A\sin\Phi(\dot{\Phi})^2 + A\ddot{\Phi}\cos\Phi = -\omega^2 A\sin\Phi.$$

The solution must hold at all values of Φ. Therefore, the $\sin\Phi$ and $\cos\Phi$ terms must be individually equal, or

$$\ddot{A} - A(\dot{\Phi})^2 = -\omega^2 A, \qquad (11.25)$$

$$2\dot{A}\dot{\Phi} + A\ddot{\Phi} = 0. \qquad (11.26)$$

The first term in Eq. (11.25) is of order $A/\Delta T^2$. This term is less than the expression on the right-hand side by a factor $(1/\omega \Delta T)^2$, so it can be neglected. Equation (11.25) becomes $d\Phi/dt = \omega$; therefore,

$$\Phi = \int \omega \, dt + \Phi_0. \qquad (11.27)$$

Substituting this expression in Eq. (11.26) gives

$$2\dot{A}/A = -\ddot{\Phi}/\Phi \cong -\dot{\omega}/\omega. \qquad (11.28)$$

Integrating both sides of Eq. (11.28),

$$\ln(\omega) = -2\ln(A) + \text{const.}$$

or

$$\omega A^2 = \text{const.} \qquad \boxed{11.29}$$

The approximate solution of Eq. (11.20) is

$$x(t) \cong A_0\sqrt{\omega_0/\omega} \sin\left(\int \omega \, dt + \Phi_0\right). \qquad (11.30)$$

Taking the derivative of Eq. (11.30), the particle velocity is

$$v_x \cong A_0\sqrt{\omega \omega_0}\left[\cos\left(\int \omega \, dt + \Phi_0\right) - (\dot{\omega}/2\omega^2)\sin\left(\int \omega \, dt + \Phi_0\right)\right]$$

$$\cong A_0\sqrt{\omega \omega_0} \cos\left(\int \omega \, dt + \Phi_0\right). \qquad (11.31)$$

Having solved the problem mathematically, let us consider the physical implications of the results.

1. At a particular time, the particle orbits approximate harmonic orbits with an angular frequency ω determined by the magnitude of the force. The amplitude and angular frequency of the oscillations changes slowly with time.

2. As the force increases, the amplitude of particle oscillations decreases, $x_{max} \sim 1/\sqrt{\omega}$. This process is called compression of the orbit.

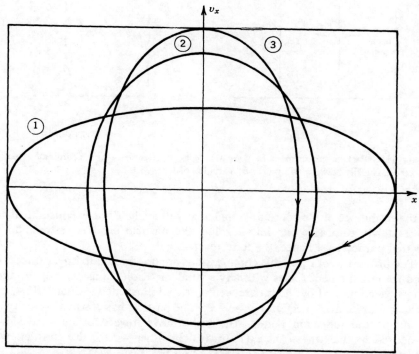

Figure 11.7 Variation of particle phase space orbits during reversible compression. (1) Initial orbit influenced by force that varies linearly with x. (2) Magnitude of force increased by a factor of 8. (3) Magnitude of force increased by a factor of 16.

3. The particle velocity is approximately 90° out of phase with the displacement.

4. The magnitudes of the velocity and displacement are related by

$$v_{x,max} \cong \omega x_{max}.$$

$$\boxed{11.32}$$

5. The product of the displacement and velocity is conserved in a reversible process, or

$$x_{max} v_{x,max} \cong \text{const.}$$

$$\boxed{11.33}$$

Figure 11.7 gives a graphical interpretation of the above conclusions. Particle orbits are plotted in phase space with x and v_x as axes. Inspection of Eqs. (11.30) and (11.31) shows that particle orbits acted on by a linear force are ellipses in phase space. Orbits are plotted in Figure 11.7 for a slow increase in focusing force (reversible compression). Although the oscillation amplitude

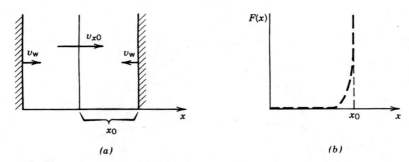

Figure 11.8 Particle confinement in a square well. (*a*) Geometry and coordinates, particle reflected elastically between two walls. (*b*) Variation of force with x.

changes, the net phase space area included within the orbit is constant. If the force slowly returns to its initial value, the particle orbit is restored to its original parameters; hence, the term reversible.

The properties of reversible compressions are not limited to linear forces but hold for confinement forces with any spatial variation. Consider, for instance, a particle contained by the square-well potential illustrated in Figure 11.8. The force is infinite at $x = x_0$ and $x = -x_0$. The particle has constant velocity v_{x0} except at the reflection points. The walls move inward or outward slowly compared to the time scale $x_0(t)/v_{x0}(t)$. In other words, the constant wall velocity v_w is small compared to $v_{x0}(t)$ at all times.

Particles reflect from the wall elastically. Conservation of momentum implies that the magnitude of v_{x0} is constant if the wall is stationary. If the wall moves inward at velocity v_w, the particle velocity after a collision is increased by an amount

$$\Delta v_{x0} = 2 v_w. \tag{11.34}$$

In a time interval Δt, a particle collides with the walls $v_{x0}(t)\,\Delta t/2x_0(t)$ times. Averaging over many collisions, we can write the following differential equation:

$$dv_{x0}/dt \cong 2v_w v_{x0}(t)/2x_0(t). \tag{11.35}$$

The equation of the wall position is

$$x_0(t) = x_0(0) - v_w(t)t,$$

or

$$dx_0 = -v_w\,dt. \tag{11.36}$$

Substituting into Eq. (11.35), we find

$$dv_{x0}/v_{x0} = -dx_0/x_0,$$

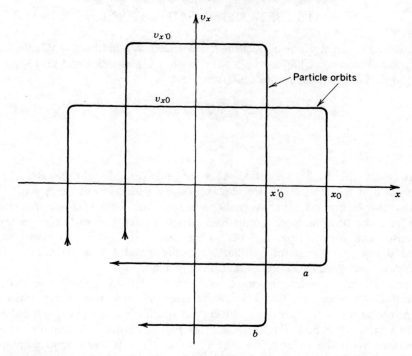

Figure 11.9 Phase space orbits of particles confined between two reflecting walls that move inward slowly. (*a*) Initial orbit. (*b*) Orbit with distance between walls decreased by a factor of 2.

or

$$x_0(t)v_{x0}(t) \cong \text{const.} \qquad (11.37)$$

This is the same result that we found for the harmonic potential. Similarly, defining the periodic frequency $\omega = v_{x0}(t)/x_0(t)$, Eq. (11.37) implies that

$$x_0 \sim 1/\sqrt{\omega}. \qquad (11.38)$$

or

$$x_0^2\omega = \text{const.,}$$

as before. A phase space plot of particle orbits in a highly nonlinear focusing system during a reversible compression is given in Figure 11.9.

11.5 BETATRON OSCILLATIONS

Reviewing the conclusions of Section 7.3, particles in a gradient magnetic field perform harmonic oscillations about the main orbit in the radial and vertical directions. The frequencies of oscillation are

$$\omega_r = \omega_g\sqrt{1 - n}, \tag{11.39}$$

$$\omega_z = \omega_g\sqrt{n}, \tag{11.40}$$

where n is the field index and $\omega_g = eB_z(R)/\gamma m_e$. In the betatron, the magnitude of the magnetic field increases (ω_g is a function of time) while the relative shape remains constant (n is constant). The focusing force increases; therefore, the amplitude of oscillations in the radial and vertical directions decreases and particles move closer to the main orbit. This process is often called damping of betatron oscillations, although this is a misnomer. The process is reversible and no dissipation is involved.

The mathematical description of betatron oscillations is similar to that of Section 11.4 except that the variation of electron mass with energy must be taken into account for relativistically correct results. We shall consider motion in the vertical direction; the derivation for radial motion is a straightforward extension. With the assumption that $v_z \ll v_\theta$, the transverse approximation (Section 2.10) can be applied. This means that vertical motions do not influence the value of γ.

The vertical equation of motion for a linear force can be written

$$dp_z(t)/dt = d\left[m(t)v_z(t)\right]/dt = -m(t)\omega_z(t)^2 z. \tag{11.41}$$

Expanding the time derivative, Eq. (11.41) becomes

$$\ddot{z} + \dot{m}\dot{z}/m + \omega_z^2 z = 0. \tag{11.42}$$

Again, we seek a solution of the form

$$z = A(t)\sin\Phi_z(t). \tag{11.43}$$

Substituting in Eq. 11.42,

$$\left(\omega_z^2 - \dot{\Phi}_z^2\right)A\sin\Phi_z + \left(A\ddot{\Phi}_z + 2\dot{A}\dot{\Phi}_z + A\dot{m}\dot{\Phi}_z/m\right)\cos\Phi_z$$
$$+ \left(\ddot{A} + \dot{A}\dot{m}/m\right)\sin\Phi_z = 0. \tag{11.44}$$

We can show by dimensional arguments that the third term of Eq. (11.44) is smaller than the first term by a factor of $(1/\omega\Delta T)^2$, where ΔT is the time scale of the acceleration cycle. Therefore, to first order, the first term is approximately equal to zero:

$$\omega_z^2 = \dot{\Phi}_z^2. \tag{11.45}$$

Equation (11.45) gives the same result as the nonrelativistic derivation [Eq. (11.27)]:

$$\Phi_z = \int \omega_z \, dt = \Phi_0. \tag{11.46}$$

Setting the second term equal to zero gives

$$A\ddot{\Phi}_z + 2\dot{A}\dot{\Phi}_z + A(\dot{m}/m)\Phi_z = 0. \tag{11.47}$$

We can show that Eq. (11.47) is equivalent to

$$d\left(A^2 m \omega_z\right) = 0,$$

or

$$A^2 m \omega_z = \text{const.} \tag{11.48}$$

There are some interesting implications associated with the above derivation. As before, the vertical displacements and velocity are 90° out of phase with magnitudes related by

$$v_{z,\text{max}} \cong z_{\text{max}} \omega_z = A\omega_z. \tag{11.49}$$

The conservation law for a relativistic reversible compression is

$$m v_{z,\text{max}} z_{\text{max}} = z_{\text{max}} p_{z,\text{max}} = \text{const.} \quad \boxed{11.50}$$

For relativistic particles, the area circumscribed by an orbit is constant if it is plotted in phase space axes of displacement and momentum rather than displacement and velocity.

11.6 ELECTRON INJECTION AND EXTRACTION

Particle injection into linear accelerators is not difficult. In contrast, injection is a significant problem for circular accelerators, particularly those with constant beam radius such as the betatron. This is one of the reasons why high current electron beams have not yet been accelerated in betatrons. The conventional betatron electron source consists of a thermionic cathode located in the vacuum chamber (Fig. 11.4) capable of emitting 1–2 A current. The cathode is biased to high negative potential and electrons are extracted and focused by

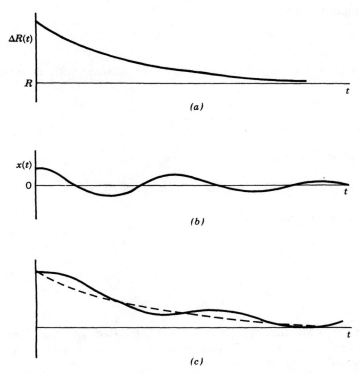

Figure 11.10 Schematic view of orbit of electron injected into low-current betatron. (*a*) Motion of instantaneous circle toward main orbit. (*b*) Reversibly compressed betatron oscillations. (*c*) Composite orbit.

shaped electrodes. The emerging beam has a large spread in particle direction. The source is pulsed on for a few microseconds at the time when electrons will travel on an instantaneous circle orbit in the rising bending magnetic field.

Following injection, the combined effects of inward motion of the instantaneous circle and damping of betatron oscillations carries electrons away from the injector so that some are trapped. The process is illustrated in Figure 11.10. Without such effects, the electrons would eventually strike the back of the injector. The fraction of electrons trapped is increased if the injector is displaced vertically from the main orbit. Because of the vertical oscillations, particles may travel many revolutions before striking the injector, even in the absence of radial motion.

As an example, consider a 300-MeV betatron with main orbit radius of 1 m operating at 180 Hz. The rate of energy gain is about 7 keV/turn. If the injection energy is 100 keV and the initial instantaneous circle has radius 1.05 m, then Eq. (11.19) implies that the orbit moves radially inward a distance 0.24 cm in a single turn. If vertical oscillations allow the particles 5–10 turns, this radial motion is sufficient to trap a substantial number of electrons.

The main limit on trapping in a high-energy betatron appears to result from beam space charge effects. Focusing is weak at injection because of the low applied magnetic field. In the example above, the injection field is only 10^{-3} T. Estimating the space charge force and specifying a balance with the vertical focusing force leads to a predicted equilibrium current of less than 1 A for a beam with 4 cm vertical extent. This figure is consistent with the maximum current observed in betatrons. The dominant role of space charge in limiting injection current is consistent with the fact the trapped current increases significantly with increased injector voltage. The injection efficiency for high-energy betatrons with an internal, electrostatic injector is typically only a few percent.

Trapping mechanisms are not as easily explained in small, low-energy betatrons. In a machine with output energy of 20 MeV, motion of the instantaneous circle is predicted to be on the order of only 2×10^{-3} cm. Nonetheless, the trapped current is observed to be much higher than that predicted from single-particle orbit dynamics combined with the probability of missing the injector. The most widely accepted explanation is that collective particle effects are responsible for the enhanced trapping. There is a substantial inductance associated with the changes of beam current around the ferromagnetic core. The increasing beam current during the injection pulses induces a back emf that is larger than the accelerating emf of the core. The inductive electric field decelerates electrons. The effect is almost independent of radius, so that particle orbits shrink toward the main orbit much more rapidly than predicted by the arguments of Section 11.3. This explanation is supported by the fact that trapping in low-energy betatrons is improved considerably when *orbit contraction coils* are incorporated in the machine. These rapidly pulsed coils enhance the self-field effects by inducing a back emf.

Extraction of electron beams from betatrons is accomplished with a magnetic peeler, illustrated in Figure 11.11. This device is a magnetic field shunt located on an azimuth outside the radius where $n = 1$. It cannot be located too close to the main orbit since the associated magnetic field perturbation would cause particle loss during the low-energy phase of the acceleration cycle. If particles are forced past the $n = 1$ radius, radial focusing is lost and they spiral

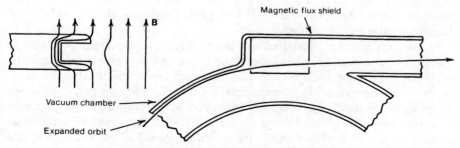

Figure 11.11 Extraction of low-current electron beam from betatron.

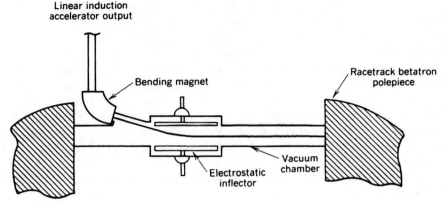

Figure 11.12 Injection of high-current electron beam from linear induction accelerator into racetrack betatron.

outward into the peeler. There are a number of options for inducing radial motion of the betatron beam. One possibility is an orbit expander coil. The expander coil is activated at the peak of the electron energy. It subtracts from the bending field in the beam chamber, causing the beam radius to expand. Another method of moving electrons out in radius is to induce betatron oscillations by resonant fields. Electric or magnetic fields oscillating at $\sqrt{1 - n}\,\omega_g$ are generated by coils or plates at particular azimuthal positions. If the growth of betatron oscillations is rapid, the beam spills out at a specific azimuth.

The maximum current that can be contained in a betatron is determined by a balance between the mutual repulsion between electrons and the focusing forces. In terms of space–charge equilibrium, the gradient focusing strength in a betatron at peak field (~ 1 T) is sufficient to contain a high-energy (~ 300 MeV) electron beam with current in excess of 10 kA. A high-energy electron beam is stiff and largely confined by its own magnetic fields; therefore, an extension of conventional betatron extraction techniques would be sufficient to extract the beam from the machine. Containing the beam during the low-energy portion of the acceleration phase is the major impediment to a high-current, high-efficiency betatrons. Two methods appear feasible to improve the operation of betatrons: (1) high-energy injection and (2) addition of supplemental focusing devices.

In the first method, illustrated in Figure 11.12, a high-current, high-energy beam from a linear induction accelerator is injected in a single turn into the betatron. To facilitate injection, the betatron could be constructed in a *racetrack* configuration. The circular machine is split into two parts connected by straight sections. Injection and extraction are performed in the straight sections, which are free of bending fields. The betatron performs the final portion of the acceleration cycle (for example, from 100 to 300 MeV). The current limit in the betatron is high for two reasons: (1) the bending field and

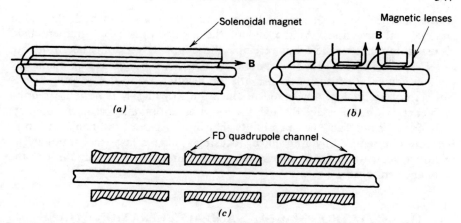

Figure 11.13 Methods to supplement focusing of low-energy electrons in betatron. (*a*) Magnet winding around vacuum chamber to produce uniform toroidal magnetic field. (*b*) Periodic array of magnetic lenses (with alternating field polarity) displaced around vacuum torus. (*c*) FD (or FODO) quadrupole lens array.

its gradients are large and (2) the self-magnetic field force of the relativistic beam almost balances the self-electric field repulsion so that space charge effects are of reduced importance. The beam is directed along the main orbit by a pulsed electrostatic inflector. The radial inflector field is activated only during a single transit of the beam around the accelerator; otherwise, it would deflect the trapped beam onto an exit orbit similar to the extrance orbit. The combination of induction linear accelerator and betatron is a good symbiosis for high-flux electron beams. The induction accelerator, with its strong solenoidal focusing magnets, solves the problem of injection and low-energy transport. The betatron provides the bulk of the particle acceleration. The combined accelerator would have a size and core volume much smaller than that of a 300-MeV linear induction accelerator.

A second approach to high-flux betatrons is to supplement gradient focusing with axi-centered focusing lenses arrayed around the toroidal vacuum chamber. Some options, illustrated in Figure 11.13, include (1) a bent solenoidal field (toroidal field), (2) discrete solenoidal magnetic lenses with reversing applied field direction, and (3) an array of magnetic quadrupole lenses in an FD configuration. The study of alternate focusing methods in betatrons is an active area of research. There are some difficult technological problems to be solved. For instance, injection into a betatron with a strong toroidal field is considerably more difficult than injection into a standard geometry, even at low current. The main problem in any strong focusing betatron is the fact that the beam must pass through the $\nu = 1$ condition (see Section 7.2). When the low-energy beam is injected, the strong space charge forces require strong supplementary focusing. Strong focusing implies that the betatron wavelength is less than the circumference of the machine; thus, $\nu > 1$ in both the radial and vertical directions. At the end of the acceleration cycle, gradient field focusing dominates. The orbits resemble those in a conventional betatron with $\nu < 1$. Passage through the resonance condition could be avoided by increasing the supplementary focusing fields with the bending fields and keeping $\nu > 1$. This is not technologically practical since the focusing system would require high energy input. Passage through the $\nu = 1$ condition may result in complete loss of the beam. There is a possibility that the severity of resonance instabilities could be reduced by a nonlinear focusing system, a fast acceleration cycle, or tuned electrostatic lenses that sweep the focusing system rapidly through the resonance condition.

11.7 BETATRON MAGNETS AND ACCELERATION CYCLES

The kinetic energy limit of betatrons is tied closely to the saturation properties of iron. Although air core betatrons have been operated successfully, they are impractical except for small research devices because of the large circulating energy and power losses involved. The volume of magnetic field outside the iron core should be minimized for the highest accelerator efficiency and lowest cost. With these factors in mind, we will review some of the types of betatron magnets that have been developed. The order will be roughly historical, proceeding from the simplest circuits at low energy to the highest energy attained.

An early betatron for electrons at 20 MeV is illustrated in Figure 11.1. The acceleration cycle is illustrated in Figure 11.3. The core flux and bending field are part of the same magnetic circuit; therefore, they are proportional to one another. A betatron driving circuit is illustrated in Figure 11.14. The inductance represents the betatron core and windings; a resistor has been included to represent energy loss through winding resistivity, hysteresis, and eddy currents. The beam load is also indicated; at current typical of conventional betatrons, the impedance of the beam load is high. The beam current is much smaller than the leakage current. In order to keep the power consumed by the betatron at a reasonable level, the core inductor is often combined with a capacitor bank to form a resonant circuit. The leakage current is supported as reactive current in the resonant circuit; a fraction of the energy of the underdamped LC circuit is lost on each cycle to resistive losses and beam acceleration. The stored energy of the capacitor bank is topped up on each cycle by a driving circuit with high-power vacuum tubes.

The components of the resonant circuit fulfill the following conditions:

1. The circuit has the desired resonant frequency, or

$$f = 2\pi/\sqrt{LC}\,.$$

Typically, betatrons operate at 180 Hz.

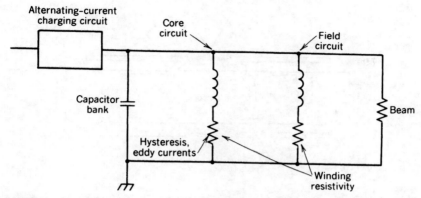

Figure 11.14 Alternating-current power circuit for a low-current betatron operated without core saturation.

2. The stored energy in the capacitor bank, $\frac{1}{2}CV_0^2$, equals the total magnetic field energy at the peak of the acceleration cycle,

$$U_m = \int dx^3 (B^2/2\mu).$$

3. The ampere turns in the coil box are sufficient to produce the field in the air gap.

The above conditions can be combined to determine the capacitor bank voltage and number of turns in the coil box given the operating parameters of the betatron.

The betatron of Figure 11.1 has a major drawback for application to high-energy beams. Most of the energy in the drive circuit is utilized to produce magnetic flux in the central air gap. This translates into a large capacitor bank to store energy and increased resistive losses because of the high NI of the coil. In order to extend the betatron to higher energy and keep power consumption low enough to run on a continuous basis, it is clearly advantageous to eliminate the air gap. One solution is illustrated in Figure 11.15*a*. The magnetic flux at the electron orbit is produced by a separate magnet circuit. The beam transport circuit has its own flux-guiding core and magnet windings. The size of the capacitor bank is reduced considerably, and power losses are typically only one-third those that would occur with a single-magnet circuit. The disadvantage of the design is the increased complexity of assembly and increased volume of the main circuit core in order to accommodate the bending field circuit.

An interesting problem associated with the betatron of Figure 11.15*a* is how to drive the two magnetic circuits with close tracking between the transport

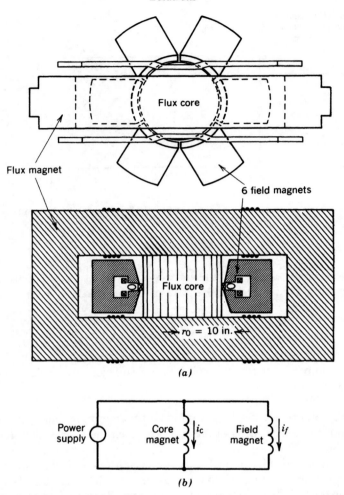

Figure 11.15 High-energy (80-MeV) betatron with no air gap. (*a*) Geometry of device, showing separate magnet circuits for accelerating flux and bending field. (*b*) Parallel drive of magnetic circuits guarantees proportionality of magnetic flux. (M. S. Livingston and J. P. Blewett, *Particle Accelerators*, used by permission, McGraw-Hill Book Co.)

fields and acceleration flux. An effective solution is to connect both magnets to the same power supply in parallel, as shown in Figure 11.15b. Since the voltage across the windings must be the same, the flux change through both windings is the same. Thus, if the number of turns and geometry of the windings are chosen properly, the ratio of bending field and core flux will be correct throughout the acceleration cycle, independent of the effective μ values in the two cores. This is another application of *flux forcing* (see Section 10.4).

The magnet design of Figure 11.16a represents another stage of improvement. It is much simpler than the magnet of Figure 11.15a and still produces a

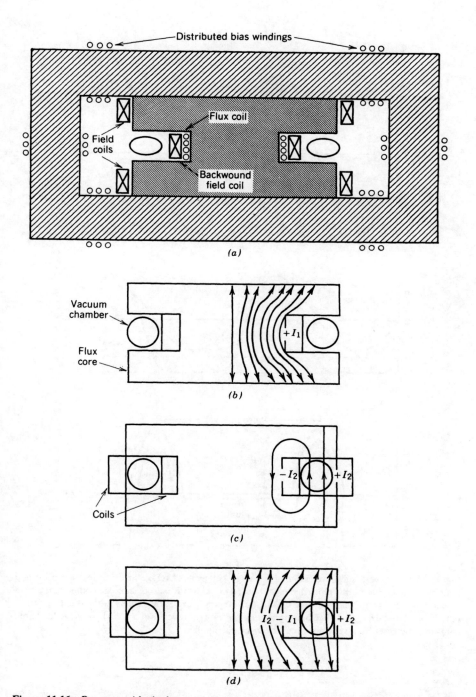

Figure 11.16 Betatron with single magnetic circuit, bias windings, and no air gap. (*b*) Magnetic field lines in flux core produced by flux winding. (*c*) Magnetic field lines in flux core and vacuum torus produced by combination of external field coil and backwound field coil. (*d*) Composite field lines showing accelerating flux and bending field.

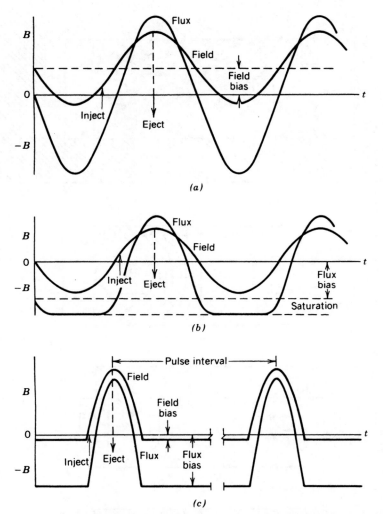

Figure 11.17 Acceleration cycles for high-energy betatrons. Average field inside main orbit and bending field at main orbit plotted versus time. (*a*) Cycle with field biasing, dc component added to bending field. (*b*) Cycle with flux biasing and ac power drive; flux core saturated half the time. (*c*) Cycle of betatron with flux biasing and pulsed power drive from a switched capacitor bank; core must be reset by a separate circuit between each pulse.

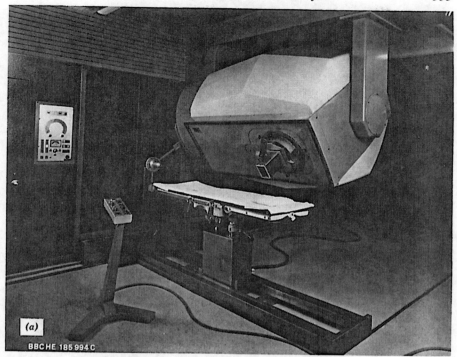

(a)

BBC HE 185 994 C

Figure 11.18 Betatron for radiation therapy (45 MeV). (*a*) Photograph of system; accelerator and treatment table. (*b*) Cross section of accelerator: gamma ray mode (top) and electron beam mode (bottom). (1) Lead shielding. (2) Transformer yokes. (3) Suspension and rotation mechanisms. (4) X-ray target. (5) Central magnet core. (6) Electron guns. (7) X-ray equalizers. (8) Lead shutter. (9) X-ray monitoring system. (10) Collimator. (11) Movable yoke for servicing. (12) Ten scatterer system for electron beam. (13) Electron radiation monitoring system. (14) Variable localizer. (Courtesy BBC Brown, Boveri and Company.)

bending field without an air gap. In order to understand how this configuration works, we shall approach the circuit in parts and then determine the total magnetic field by superposition. First, consider a single coil inside the radius of the vacuum chamber, as shown in Figure 11.16*b*. All the magnetic flux flows through the central core as shown. In the second stage (Fig. 11.16*c*), we consider the field produced by a winding inside the vacuum chamber carrying current $-NI$ and a windings outside the chamber carrying current $+NI$. This produces a bending field at the main orbit, and flux returns through the core as shown. In the final configuration, Figure 11.16*d*, the external windings are present and the windings on the flux coil are reduced by $-NI$ ampere turns to generate the net field. Proper choice of the number of turns on the flux coil versus the field coils plus shimming of the bending field gap assures that the betatron condition is satisfied.

A further improvement to the magnet of Figure 11.16 to reach higher beam energy is to utilize the full available flux swing of the central core during

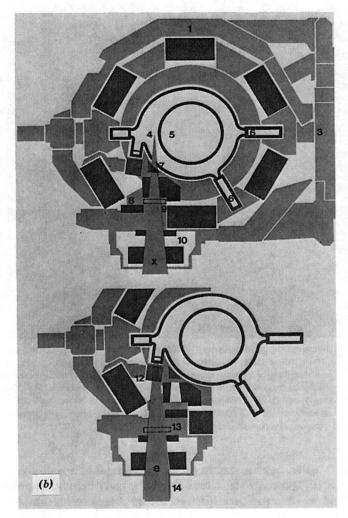

Figure 11.18 *(Continued)*.

acceleration. In the previous acceleration cycles we have discussed, the core field changes from 0 to $+B_s$ while the bending field changes from 0 to $\frac{1}{2}B_s$. Inspection of Eq. (11.13) shows that the betatron condition is expressed in terms of the change of included flux, not the absolute value. An acceleration cycle in which the core magnetic field changes from $-B_s$ to $+B_s$ while the bending field changes from 0 to B_s satisfies the betatron condition and doubles the final electron energy for a given core size.

There are two methods to achieve an acceleration cycle with full flux swing, *field biasing* and *flux biasing*. Field biasing is illustrated in Figure 11.17a. A dc component of magnitude $+\frac{1}{2}B_s$ is added to the bending field. Acceleration takes place over a half-cycle of the ac waveform. For flux biasing, dc bias windings are added to the core circuit to maintain the core at $-B_s$. Bias windings are illustrated in Figure 11.16. The field and flux coils are energized in parallel to produce the accelerating waveform illustrated in Figure 11.17b. Acceleration takes place over one quarter-cycle. The main technological difficulty associated with flux biasing is that the core is driven to saturation, resulting in increased hystersis and eddy current losses. Also, during the negative half-cycle, the core has $\mu = \mu_0$ so that the circuit inductance varies considerably. Betatrons with flux biasing are usually driven by pulse power modulators rather than resonant circuits. A pulsed acceleration cycle is shown in Figure 11.17c.

A modern commercial betatron for radiation therapy is illustrated in Figure 11.18a. The machine accelerates electrons to a maximum kinetic energy of 45 MeV to generate deeply penetrating radiation. Electrons can be extracted directly or used to generate forward-directed gamma rays on an internal target. The 12,000-kG machine and the treatment table can be moved to a variety of positions to achieve precise dose profiles. A cross section of the betatron (Fig. 11.18b) illustrates operation in the gamma ray and electron modes.

12

Resonant Cavities and Waveguides

This chapter initiates our study of resonant accelerators. The category includes rf (radio-frequency) linear accelerators, cyclotrons, microtrons, and synchrotrons. Resonant accelerators have the following features in common:

1. Applied electric fields are harmonic. The continuous wave (CW) approximation is valid; a frequency-domain analysis is the most convenient to use. In some accelerators, the frequency of the accelerating field changes over the acceleration cycle; these changes are always slow compared to the oscillation period.

2. The longitudinal motion of accelerated particles is closely coupled to accelerating field variations.

3. The frequency of electromagnetic oscillations is often in the microwave regime. This implies that the wavelength of field variations is comparable to the scale length of accelerator structures. The full set of the Maxwell equations must be used.

Microwave theory relevant to accelerators is reviewed in this chapter. Chapter 13 describes the coupling of longitudinal particle dynamics to electromagnetic waves and introduces the concept of phase stability. The theoretical tools of this chapter and Chapter 13 will facilitate the study of specific resonant accelerators in Chapters 14 and 15.

As an introduction to frequency-domain analysis, Section 12.1 reviews complex exponential representation of harmonic functions. The concept of complex impedance for the analysis of passive element circuits is emphasized.

Section 12.2 concentrates on a lumped element model for the fundamental mode of a resonant cavity. The Maxwell equations are solved directly in Section 12.3 to determine the characteristics of electromagnetic oscillations in resonant cavities. Attention is centered on the TM_{010} mode since it is the most useful mode for particle acceleration. Physical properties of resonators are discussed in Section 12.4. Subjects include the Q value of a cavity and effects of competing modes. Methods of extracting energy from and coupling energy to resonant cavities are discussed in Section 12.5.

Section 12.6 develops the frequency-domain analysis of transmission lines. There are three reasons to extend the analysis of transmission lines. First, an understanding of transmission lines helps to illuminate properties of resonant cavities and waveguides. Second, transmission lines are often used to transmit power to accelerator cavities. Finally, the transmission line equations illustrate methods to match power sources to loads with reactive components, such as resonant cavities. In this application, a transmission line acts to transform the impedance of a single-frequency input. Section 12.7 treats the cylindrical resonant cavity as a radial transmission line with an open-circuit termination at the inner radius and a short-circuit termination at the outer radius.

Section 12.8 reviews the theory of the cylindrical waveguide. Waveguides are extended hollow metal structures of uniform cross section. Traveling waves are contained and transported in a waveguide; the frequency and field distribution is determined by the shape and dimensions of the guide. A lumped circuit element model is used to demonstrate approximate characteristics of guided wave propagation, such as dispersion and cutoff. The waveguide equations are then solved exactly.

The final two sections treat the topic of slow-wave structures, waveguides with boundaries that vary periodically in the longitudinal direction. They transport waves with phase velocity equal to or less than the speed of light. The waves are therefore useful for continuous acceleration of synchronized charged particles. A variety of models are used to illustrate the physics of the iris-loaded waveguide, a structure incorporated in many traveling wave accelerators. The interpretation of dispersion relationships is discussed in Section 12.10. Plots of frequency versus wavenumber yield the phase velocity and group velocity of traveling waves. It is essential to determine these quantities in order to design high-energy resonant accelerators. As an example, the dispersion relationship of the iris-loaded waveguide is derived.

12.1 COMPLEX EXPONENTIAL NOTATION AND IMPEDANCE

Circuits consisting of a harmonic voltage source driving resistors, capacitors, and inductors, are described by an equation of the form

$$\alpha(di/dt) + \beta i + \gamma \int i\, dt = V_0 \cos \omega t. \qquad (12.1)$$

The solution of Eq. (12.1) has homogeneous and particular parts. Transitory behavior must include the homogenous part. Only the particular part need be included if we restrict our attention to CW (continuous wave) excitation. The particular solution has the form

$$i(t) = I_0\cos(\omega t + \phi).\tag{12.2}$$

I_0 and ϕ depend on the magnitude of the driving voltage, the elements of the circuit, and ω. Since Eq. (12.1) describes a physical system, the solution must reflect a physical answer. Therefore, I_0 and ϕ are real numbers. They can be determined by direct substitution of Eq. (12.2) into Eq. (12.1). In most cases, this entails considerable manipulation of trigonometric identities.

The mathematics involved in determining the particular solution of Eq. (12.1) and other circuit equations with a single driving frequency can be simplified considerably through the use of the complex exponential notation for trigonometric functions. In using complex exponential notation, we must remember the following facts:

1.   All physical problems must have an answer that is a real number. Complex numbers have no physical meaning.

2.   Complex numbers are a convenient mathematical method for handling trigonometric functions. In the solution of a physical problem, complex numbers can always be grouped to form real numbers.

3.   The answers to physical problems are often written in terms of complex numbers. This convention is used because the results can be written more compactly and because there are well-defined rules for extracting the real-number solution.

The following equations relate complex exponential functions to trigonometric functions:

$$\cos \omega t = [\exp(j\omega t) + \exp(-j\omega t)]/2, \qquad \boxed{12.3}$$

$$\sin \omega t = [\exp(j\omega t) - \exp(-j\omega t)]/2j, \qquad \boxed{12.4}$$

where $j = \sqrt{-1}$. The symbol j is used to avoid confusion with the current, i. The inverse relationship is

$$\exp(j\omega t) = \cos \omega t + j \sin \omega t. \qquad \boxed{12.5}$$

In Eq. (12.1), the expression $V_0[\exp(j\omega t) + \exp(-j\omega t)]/2$ is substituted for the voltage, and the current is assumed to have the form

$$i(t) = A \exp(j\omega t) + B \exp(-j\omega t).\tag{12.6}$$

The coefficients A and B may be complex numbers if there is a phase difference between the voltage and the current. They are determined by substituting Eq. (12.6) into Eq. (12.1) and recognizing that the terms involving $\exp(j\omega t)$ and $\exp(-j\omega t)$ must be separately equal if the solution is to hold at all times. This procedure yields

$$A = \left(\tfrac{1}{2}j\omega V_0\right)/\left(-\alpha\omega^2 + j\omega\beta + \gamma\right), \tag{12.7}$$

$$B = \left(-\tfrac{1}{2}j\omega V_0\right)/\left(-\alpha\omega^2 - j\omega\beta + \gamma\right). \tag{12.8}$$

The *complex conjugate* of a complex number is the number with $-j$ substituted for j. Note that B is the complex conjugate of A. The relationship is denoted $B = A^*$.

Equations (12.7) and (12.8) represent a formal mathematical solution of the problem; we must rewrite the solution in terms of real numbers to understand the physical behavior of the system described by Eq. (12.1). Expressing Eq. (12.2) in complex notation and setting the result equal to Eq. (12.6), we find that

$$A \exp(j\omega t) + A^*\exp(-j\omega t) = \tfrac{1}{2}I_0[\exp(j\omega t)\exp(j\phi)$$

$$+ \exp(-j\omega t)\exp(-j\phi)]. \tag{12.9}$$

Terms involving $\exp(j\omega t)$ and $\exp(-j\omega t)$ must be separately equal. This implies that

$$A = \tfrac{1}{2}I_0\exp(j\phi) = \tfrac{1}{2}I_0(\cos\phi + j\sin\phi) \tag{12.10}$$

by Eq. (12.5). The magnitude of the real solution is determined by multiplying Eq. (12.10) by its complex conjugate:

$$A \cdot A^* = \tfrac{1}{4}I_0^2[\exp(j\phi)\exp(-j\phi)],$$

or

$$I_0 = 2\sqrt{A \cdot A^*}. \tag{12.11}$$

Inspection of Eq. (12.10) shows that the phase shift is given by

$$\phi = \tan^{-1}\mathrm{Im}(A)/\mathrm{Re}(A). \tag{12.12}$$

Returning to Eq. (12.1), the solution is

$$I_0 = V_0\omega/\left[(\gamma - \alpha\omega^2)^2 + \omega^2\beta^2\right]^{1/2},$$

$$\phi = \tan^{-1}(\gamma - \alpha\omega^2)/\omega\beta.$$

This is the familiar resonance solution for a driven, damped harmonic oscillator.

Part of the effort in solving the above problem was redundant. Since the coefficient of the second part of the solution must equal the complex conjugate of the first, we could have used a trial solution of the form

$$i(t) = A \exp(j\omega t). \tag{12.13}$$

We arrive at the correct answer if we remember that Eq. (12.13) represents only half of a valid solution. Once A is determined, the real solution can be extracted by applying the rules of Eq. (12.11) and (12.12). Similarly, in describing an electromagnetic wave traveling in the $+z$ direction, we will use the form $E \sim E_0 \exp[j(\omega t - kz)]$. The form is a shortened notation for the function $E \sim E_0 \exp[j(\omega t - kz)] + E_0^* \exp[-j(\omega t - kz)] \sim E_0 \cos(\omega t - kz + \phi)$, where E_0 is a real number. The function for a wave traveling in the negative z direction is abbreviated $E \sim E_0 \exp[j(kz + \omega t)]$.

Complex exponential notation is useful for solving lumped element circuits with CW excitation. In this circumstance, voltages and currents in the circuit vary harmonically at the driving frequency and differ only in amplitude and phase. In complex exponential notation, the voltage and current in a section of a circuit are related by

$$V/I = Z. \tag{12.14}$$

The quantity Z, the *impedance*, is a complex number that contains information on amplitude and phase. Impedance is a function of frequency.

The impedance of a resistor R is simply

$$Z_R = R. \qquad \boxed{12.15}$$

A real impedance implies that the voltage and current are in phase as shown in Figure 12.1a. The time-averaged value of VI through a resistor is nonzero; a resistor absorbs energy.

The impedance of a capacitor can be calculated from Eq. (9.5). If the voltage across the capacitor is

$$V(t) = V_0 \cos \omega t, \tag{12.16}$$

then the current is

$$i(t) = C\, dV/dt = -\omega C V_0 \sin \omega t = \omega C V_0 \cos(\omega t + \pi/2). \tag{12.17}$$

Equation (12.17) specifies the magnitude and amplitude of voltage across versus current through a capacitor. There is a 90° phase shift between the voltage and current; the current *leads* the voltage, as shown in Figure 12.1b. The capacitor is a reactive element; the time average of $V(t)i(t)$ is zero. In

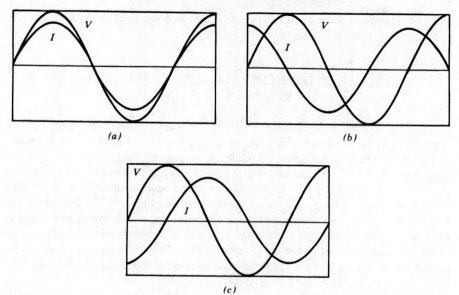

Figure 12.1 Variations of voltage and current in simple circuit elements driven at single frequency. (*a*) Resistor. (*b*) Capacitor. (*c*) Inductor.

complex exponential notation, the impedance can be expressed as a single complex number

$$Z_C = (1/\omega C)\exp(j\pi/2) = -j/\omega C \qquad \boxed{12.18}$$

if the convention of Eq. (12.13) is adopted. The impedance of a capacitor has negative imaginary part. This implies that the current leads the voltage. The impedance is inversely proportional to frequency; a capacitor acts like a short circuit at high frequency.

The impedance of an inductor can be extracted from the equation

$$V(t) = L[di(t)/dt].$$

Again, taking voltage in the form of Eq. (12.16), the current is

$$i(t) = \frac{V_0 \sin \omega t}{\omega L} = -\frac{V_0 \cos(\omega t - \pi/2)}{\omega L}.$$

The current lags the voltage, as shown in Figure 12.1c. The complex impedance of an inductor is

$$Z_L = j\omega L. \qquad \boxed{12.19}$$

The impedance of an inductor is proportional to frequency; inductors act like open circuits at high frequency.

12.2 LUMPED CIRCUIT ELEMENT ANALOGY FOR A RESONANT CAVITY

A *resonant cavity* is a volume enclosed by metal walls that supports an electromagnetic oscillation. In accelerator applications, the oscillating electric fields accelerate charged particles while the oscillating magnetic fields provide inductive isolation. To initiate the study of electromagnetic oscillations, we shall use the concepts developed in the previous section to solve a number of lumped element circuits. The first, diagrammed in Figure 12.2, illustrates the process of inductive isolation in a resonant circuit. A harmonic voltage generator with output $V(t) = V_0\exp(j\omega t)$ drives a parallel combination of a resistor, capacitor, and inductor. Combinations of impedances are governed by the same rules that apply to parallel and series combinations of resistors. The total circuit impedance at the voltage generator is

$$Z(\omega) = V_0\exp(j\omega t)/I_0\exp(j\omega t)$$

$$= (1/Z_R + 1/Z_C + 1/Z_L)^{-1}. \qquad (12.20)$$

The quantity I_0 is generally a complex number.

Consider the part of the circuit of Figure 12.2 enclosed in dashed lines: a capacitor in parallel with an inductor. The impedance is

$$Z(\omega) = (j\omega C + 1/j\omega L)^{-1} = .j\omega L/(1 - \omega^2 LC). \qquad (12.21)$$

The impedance is purely imaginary; therefore, the load is reactive. At low frequency ($\omega < 1/\sqrt{LC}$) the impedance is positive, implying that the circuit is inductive. In other words, current flow through the inductor dominates the behavior of the circuit. At high frequency, the impedance is negative and the circuit acts as a capacitive load. When $\omega = \omega_0 = 1/\sqrt{LC}$, the impedance of the combined capacitor and inductor becomes infinite. This condition is called

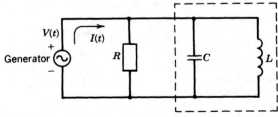

Figure 12.2 Driven *RLC* circuit with shunt resistance. Reactive section of circuit indicated by dashed line.

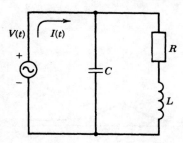

Figure 12.3 Driven *RLC* circuit; inductor with series resistance.

resonance; the quantity ω_0 is the resonant frequency. In this circumstance, the reactive part of the total circuit of Figure 12.2 draws no current when a voltage is applied across the resistor. All current from the generator flows into the resistive load. The reactive part of the circuit draws no current at $\omega = \omega_0$ because current through the inductor is supplied completely by displacement current through the capacitor. At resonance, the net current from the generator is minimized for a given voltage. This is the optimum condition for energy transfer if the generator has nonzero output impedance.

The circuit of Figure 12.3 illustrates power losses in resonant circuits. Again, an inductor and capacitor are combined in parallel. The difference is that the inductor is imperfect. There are resistive losses associated with current flow. The losses are represented by a series resistor. The impedance of the circuit is

$$Z(\omega) = \left[\, j\omega C + 1/(\, j\omega L + R)\right]^{-1} = \left[\, j\omega L + R\right]/\left[(1 - \omega^2 LC) + j\omega RC\right].$$

Converting the denominator in the above equation to a real number, we find that the magnitude of the impedance is proportional to

$$Z(\omega) \sim 1/\left[(1 - \omega^2/\omega_0^2)^2 + (\omega RC)^2\right]. \tag{12.22}$$

Figure 12.4 shows a plot of total current flowing in the reactive part of the circuit versus current input from the generator. Two cases are plotted: resonant circuits with low damping and high damping. Note that the impedance is no longer infinite at $\omega = 1/\sqrt{LC}$. For a cavity with resistive losses, power must be supplied continuously to support oscillations. A circuit is in resonance when large reactive currents flow in response to input from a harmonic power generator. In other words, the amplitude of electromagnetic oscillations is high. Inspection of Figure 12.4 and Eq. (12.22) shows that there is a finite response width for a driven damped resonant circuit. The frequency width, $\Delta\omega = \omega - \omega_0$, to reduce the peak impedance by a factor of 5 is

$$\frac{\Delta\omega}{\omega_0} \cong \frac{R}{\sqrt{L/C}}. \tag{12.23}$$

Resonant circuits are highly underdamped; therefore, $\Delta\omega/\omega \ll 1$.

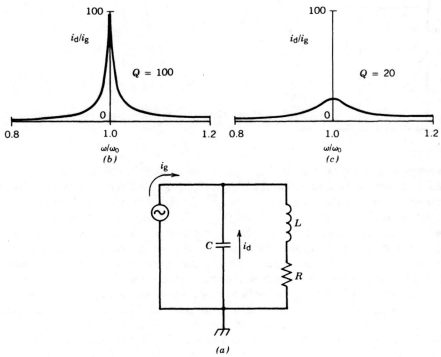

Figure 12.4 Resonant drive of *RLC* circuit. (*a*) Circuit diagram; circle represents ac generator. (*b*) Ratio of reactive current to current input from generator (i_d/i_g); $Q = 100$. (*c*) i_d/i_g; $Q = 20$.

Resonant circuit damping is parametrized by the quantity Q. The circuit Q is defined as

$$Q = \frac{\omega_0 \text{ (energy stored in the resonant circuit)}}{\text{(time averaged power loss)}}$$

$$= \frac{\pi \text{ (energy stored in the resonant circuit)}}{\text{(energy lost per half cycle)}}. \qquad \boxed{12.24}$$

In the limit of low damping near resonance, the reactive current exchanged between the inductor and capacitor of the circuit of Figure 12.3 is much larger than the current input from the generator. The reactive current is $i(t) = I_0 \exp(j\omega t)$, where I_0 is a slowly decreasing function of time. The circuit energy, U, is equal to the energy stored in the inductor at peak current:

$$U = \tfrac{1}{2} L I_0 I_0^*. \qquad (12.25)$$

Energy is lost to the resistor. The power lost to the resistor (averaged over a cycle) is

$$\langle P \rangle = \langle i(t)^2 \rangle R = \tfrac{1}{2} I_0 I_0^* R. \tag{12.26}$$

Substituting Eqs. (12.25) and (12.26) into Eq. (12.24), the Q value for the LRC circuit of Figure 12.3 is

$$Q \cong \frac{\omega_0 L}{R} = \frac{\sqrt{L/C}}{R}. \tag{12.27}$$

In an underdamped circuit, the characteristic impedance of the LC circuit is large compared to the resistance, so that $Q \gg 1$.

Energy balance can be used to determine the impedance that the circuit of Figure 12.3 presents to the generator at resonance. The input voltage V_0 is equal to the voltage across the capacitor. The input voltage is related to the stored energy in the circuit by

$$U = \frac{C V_0^2}{2} = \frac{V_0^2}{2\omega_0 \sqrt{L/C}}. \tag{12.28}$$

By the definition of Q, the input voltage is related to the average power loss by

$$\langle P \rangle = \frac{V_0^2}{2Q\sqrt{L/C}}. \tag{12.29}$$

Defining the resistive input impedance so that

$$\langle P \rangle = \langle V^2(t) \rangle / R_{\text{in}} = V_0^2 / 2R_{\text{in}}, \tag{12.30}$$

we find at resonance ($\omega \cong \omega_0$) that

$$R_{\text{in}} \cong Q\sqrt{L/C} = \frac{\left(\sqrt{L/C}\right)^2}{R}. \tag{12.31}$$

The same result can be obtained directly from the general impedance expression in the limit $\sqrt{L/C} \gg R$. The impedance is much larger than R. This reflects the fact that the reactive current is much larger than the current from the generator. In terms of Q, the resonance width of an imperfect oscillating circuit [Eq. (12.23)] can be written

$$\Delta \omega / \omega_0 \cong 1/Q. \qquad \boxed{12.32}$$

Resonant cavities used for particle acceleration have many features in common with the circuits we have studied in this section. Figure 12.5 illustrates a particularly easy case to analyze, the reentrant cavity. This cavity is used in systems with space constraints, such as klystrons. It oscillates at relatively low frequency for its size. The reentrant cavity can be divided into predominantly

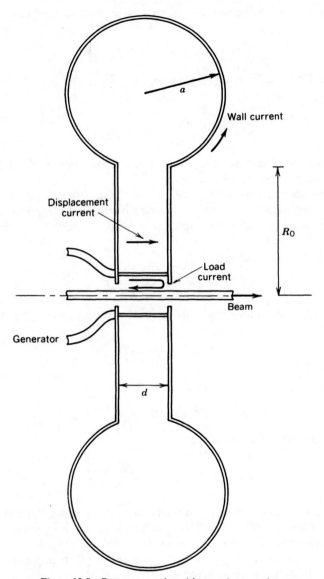

Figure 12.5 Reentrant cavity with on-axis power input.

capacitive and predominantly inductive regions. In the central region, there is a narrow gap. The capacitance is large and the inductance is small. A harmonic voltage generator connected at the center of the cavity induces displacement current. The enlarged outer region acts as a single-turn inductor. Real current flows around the wall to complete the circuit. If the walls are not superconducting, the inductor has a series resistance.

Assume that there is a load, such as a beam, on the axis of the cavity. Neglecting cavity resistance, the circuit is the same as that of Figure 12.2. If the generator frequency is low, most of the input current flows around the metal wall (leakage current). The cavity is almost a short circuit. At high frequency, most of the current flows across the capacitor as displacement current. At the resonance frequency of the cavity, the cavity impedance is infinite and all the generator energy is directed into the load. In this case, the cavity can be useful for particle acceleration. When the cavity walls have resistivity, the cavity acts as a high impedance in parallel to the beam load. The generator must supply energy for cavity losses as well as energy to accelerate the beam.

The resonant cavity accelerator has much in common with the cavity of an induction linear accelerator. The goal is to accelerate particles to high energy without generating large electrostatic voltages. The outside of the accelerator is a conductor; voltage appears only on the beamline. Electrostatic voltage is canceled on the outside of the accelerator by inductively generated fields. The major difference is that leakage current is inhibited in the induction linear accelerator by ferromagnetic inductors. In the resonant accelerator, a large leakage current is maintained by reactive elements. The linear induction accelerator has effective inductive isolation over a wide frequency range; the resonant accelerator operates at a single frequency. The voltage on the axis of a resonant cavity is bipolar. Therefore, particles are accelerated only during the proper half-cycle. If an accelerator is constructed by stacking a series of resonant cavities, the crossing times for particles must be synchronized to the cavity oscillations.

The resonant frequency of the reentrant cavity can be estimated easily. Dimensions are illustrated in Figure 12.5. The capacitance of the central region is

$$C \cong \varepsilon_0 \pi R_0^2 / d,$$

and the inductance is

$$L \cong \mu_0 \pi a^2 / 2\pi (R_0 + a).$$

The resonant angular frequency is

$$\omega_0 = 1/\sqrt{LC} \cong \left[2\pi (R_0 + a) d / \varepsilon_0 \mu_0 R_0^2 a^2 \pi^2 \right]^{1/2}$$

$$= c \left[2(R_0 + a) d / R_0^2 a^2 \pi \right]^{1/2}. \tag{12.33}$$

12.3 RESONANT MODES OF A CYLINDRICAL CAVITY

The resonant modes of a cavity are the natural modes for electromagnetic oscillations. Once excited, a resonant mode will continue indefinitely in the absence of resistivity with no further input of energy. In this section, we shall

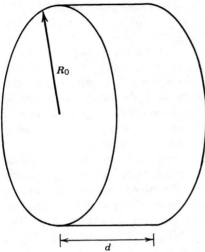

Figure 12.6 Geometry of cylindrical resonant cavity.

calculate modes of the most common resonant structure encountered in particle accelerator applications, the cylindrical cavity (Fig. 12.6). The cavity length is denoted d and radius R_0. In the initial treatment of resonant modes, we shall neglect the perturbing effects of power feeds, holes for beam transport, and wall resistivity. The cylindrical cavity has some features in common with the reentrant cavity of Section 2.2. A capacitance between the upstream and downstream walls carries displacement current. The circuit is completed by return current along the walls. Inductance is associated with the flow of current. The main difference from the reentrant cavity is that regions of electric field and magnetic field are intermixed. In this case, a direct solution of the Maxwell equations is more effective than an extension of the lumped element analogy. This approach demonstrates that resonant cavities can support a variety of oscillation modes besides the low-frequency mode that we identified for the reentrant cavity.

We seek solutions for electric and magnetic fields that vary in time according to $\exp(j\omega t)$. We must use the full set of coupled Maxwell equations [Eqs. (3.11)–(3.14)]. We allow the possibility of a uniform fill of dielectric or ferromagnetic material; these materials are assumed to be linear, characterized by parameters ε and μ. The field equations are

$$\nabla \times \mathbf{E} + \partial \mathbf{B}/\partial t = 0, \tag{12.34}$$

$$\nabla \cdot \mathbf{E} = 0, \tag{12.35}$$

$$\nabla \times \mathbf{B} - (\varepsilon\mu)\, \partial \mathbf{E}/\partial t = 0, \tag{12.36}$$

$$\nabla \cdot \mathbf{B} = 0. \tag{12.37}$$

Applying the vector identity $\nabla \times (\nabla \times \mathbf{V}) = \nabla(\nabla \cdot \mathbf{V}) - \nabla^2\mathbf{V}$, Eqs. (12.34)–(12.37) can be rewritten as

$$\nabla^2\mathbf{E} - \left(1/v^2\right)\partial^2\mathbf{E}/\partial t^2 = 0, \qquad \boxed{12.38}$$

$$\nabla^2\mathbf{B} - \left(1/v^2\right)\partial^2\mathbf{B}/\partial t^2 = 0, \qquad \boxed{12.39}$$

where v is the velocity of light in the cavity medium,

$$v = c/\left(\varepsilon\mu/\varepsilon_0\mu_0\right)^{1/2}. \qquad (12.40)$$

The features of electromagnetic oscillations can be found by solving either Eq. (12.38) or (12.39) for $\mathbf{E}$ or $\mathbf{B}$. The associated magnetic or electric fields can then be determined by substitution into Eq. (12.34) or (12.36). Metal boundaries constrain the spatial variations of fields. The wave equations have solutions only for certain discrete values of frequency. The values of resonant frequencies depend on how capacitance and inductance are partitioned in the mode.

The general solutions of Eqs. (12.38) and (12.39) in various cavity geometries are discussed in texts on electrodynamics. We shall concentrate only on resonant modes of a cylindrical cavity that are useful for particle acceleration. We shall solve Eq. (12.38) for the electric field since there are easily identified boundary conditions. The following assumptions are adopted:

1. Modes of interest have azimuthal symmetry ($\partial/\partial\theta = 0$).
2. The electric field has no longitudinal variation, or $\partial\mathbf{E}/\partial z = 0$.
3. The only component of electric field is longitudinal, E_z.
4. Fields vary in time as $\exp(j\omega t)$.

The last two assumptions imply that the electric field has the form

$$\mathbf{E} = E_z(r)\exp(j\omega t)\hat{z}. \qquad (12.41)$$

Using the cylindrical coordinate form of the Laplacian operator, dropping terms involving azimuthal and longitudinal derivatives, and substituting Eq. (12.41), we find that the class of resonant modes under consideration satisfies the equation

$$\frac{d^2E_z(r)}{dr^2} + \frac{1}{r}\frac{dE_z(r)}{dr} + \frac{\omega^2}{v^2}E_z(r) = 0. \qquad (12.42)$$

Equation (12.42) is expressed in terms of total derivatives since there are only radial variations. Equation (12.42) is a special form of the Bessel equation. The solution can be expressed in terms of the zero-order Bessel functions, $J_0(k_n r)$ and $Y_0(k_n r)$. The Y_0 function is eliminated by the requirement that E_z has a

TABLE 12.1   Parameters of TM_{0n0} Modes

Mode	k_n	ω_n
TM_{010}	$2.405/R_0$	$2.405/\sqrt{\varepsilon\mu}\,R_0$
TM_{020}	$5.520/R_0$	$5.520/\sqrt{\varepsilon\mu}\,R_0$
TM_{030}	$8.654/R_0$	$8.654/\sqrt{\varepsilon\mu}\,R_0$
TM_{040}	$11.792/R_0$	$11.792/\sqrt{\varepsilon\mu}\,R_0$

finite value on the axis. The solution is

$$E_{zn}(r, t) = E_{0n} J_0(k_n r)\exp(j\omega_n t), \tag{12.43}$$

where E_{0n} is the magnitude of the field on the axis.

The second boundary condition is that the electric field parallel to the metal wall at $r = R_0$ must be zero, or $E_{zn}(R_0, t) = 0$. This implies that only certain values of k_n give valid solutions. Allowed values of k_n are determined by the zeros of J_0 (Table 12.1). A plot of $E_z(r)$ for $n = 1$ is given in Figure 12.7. Substituting Eq. (12.43) into Eq. (12.42), the angular frequency is related to the

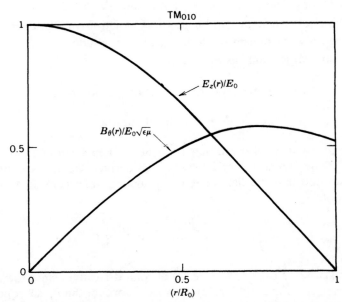

Figure 12.7 Normalized electric axial electric field and azimuthal magnetic field as a function of radius; TM_{010} mode in a cylindrical cavity.

wavenumber k_n by

$$\omega_n = vk_n. \tag{12.44}$$

Angular frequency values are tabulated in Table 12.1.

The magnetic field of the modes can be calculated from Eq. (12.34),

$$\partial \mathbf{B}/\partial t = -(\nabla \times \mathbf{E}). \tag{12.45}$$

Magnetic field is directed along the θ direction. Assuming time variation $\exp(j\omega t)$ and substituting from Eq. (12.43),

$$j\omega_n B_{\theta n} = dE_{zn}/dr = E_{0n}\, dJ_0(k_n r)/dr. \tag{12.46}$$

Rewriting Eq. (12.46),

$$B_{\theta n}(r) = -j\sqrt{\varepsilon\mu}\, E_{0n} J_1(k_n r). \tag{12.47}$$

Magnetic field variation for the TM_{010} mode is plotted in Figure 12.7. The magnetic field is zero on the axis. Moving outward in radius, B_θ increases linearly. It is proportional to the integral of axial displacement current from 0 to r. Toward the outer radius, there is little additional contribution of the displacement current. The $1/r$ factor [see Eq. (4.40)] dominates, and the magnitude of B_θ decreases toward the wall.

12.4 PROPERTIES OF THE CYLINDRICAL RESONANT CAVITY

In this section, we consider some of the physical implications of the solutions for resonant oscillations in a cylindrical cavity. The oscillations treated in the previous section are called TM_{0n0} modes. The term TM (transverse magnetic) indicates that magnetic fields are normal to the longitudinal direction. The other class of oscillations, TE modes, have longitudinal components of $\mathbf{B}$, and $E_z = 0$. The first number in the subscript is the azimuthal mode number; it is zero for azimuthally symmetric modes. The second number is the radial mode number. The radial mode number minus one is the number of nodes in the radial variation of E_z. The third number is the longitudinal mode number. It is zero in the example of Section 12.3 since E_z is constant along the z direction. The wavenumber and frequency of TM_{0n0} modes depends only on R_0, not d. This is not generally true for other types of modes.

TM_{0n0} modes are optimal for particle acceleration. The longitudinal electric field is uniform along the propagation direction of the beam and its magnitude is maximum on axis. The transverse magnetic field is zero on axis; this is important for electron acceleration where transverse magnetic fields could deflect the beam. TM modes with nonzero longitudinal wavenumber ($p \neq 0$)

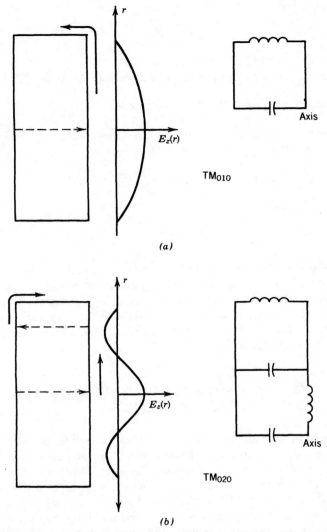

Figure 12.8 Spatial variations of axial electric field and equivalent circuits for electromagnetic oscillations in a cylindrical resonant cavity. Dashed line indicates displacement current, solid line indicates real current flow in the cavity walls. (*a*) TM_{010} mode. (*b*) TM_{020} mode.

have axial electric field of the form $E_z(0, z) \sim \sin(p\pi x/d)$; it is clear that the acceleration of particles crossing the cavity is reduced for these modes.

Figure 12.8 clarifies the nature of TM_{0n0} modes in terms of lumped circuit element approximations. Displacement currents and real currents are indicated along with equivalent circuit models. At values of n greater than 1, the cavity is divided into n interacting resonant LC circuits. The capacitance and inductance of each circuit is reduced by a factor of about $1/n$; therefore, the resonant frequency of the combination of elements is increased by a factor close to n.

Resonant cavities are usually constructed from copper or copper-plated steel for the highest conductivity. Nonetheless, effects of resistivity are significant because of the large reactive current. Resistive energy loss from the flow of real current in the walls is concentrated in the inductive regions of the cavity; hence, the circuit of Figure 12.3 is a good first-order model of an imperfect cavity. Current penetrates into the wall a distance equal to the skin depth [Eq. (10.7)]. Power loss is calculated with the assumption that the modes approximate those of an ideal cavity. The surface current per length on the walls is $J_s = B_\theta(r, z, t)/\mu_0$. Assuming that the current is distributed over a skin depth, power loss can be summed over the surface of the cavity. Power loss clearly depends on mode structure through the distribution of magnetic fields. The Q value for the TM_{010} mode of a cylindrical resonant cavity is

$$Q = \frac{(d/\delta)}{1 + d/R_0} \qquad \boxed{12.48}$$

where the skin depth δ is a function of the frequency and wall material. In a copper cavity oscillating at $f = 1$ GHz, the skin depth is only 2 μm. This means that the inner wall of the cavity must be carefully plated or polished; otherwise, current flow will be severely perturbed by surface irregularities raising the cavity Q. With a skin depth of 2 μm, Eq. (12.48) implies a Q value of 3×10^4 in a cylindrical resonant cavity of radius 12 cm and length 4 cm. This is a very high value compared to resonant circuits composed of lumped elements. Equation (12.32) implies that the bandwidth for exciting a resonance is only

$$\Delta f/f_0 \cong 1/Q = 3 \times 10^{-5}.$$

A rf power source that drives a resonant cavity must operate with very stable output frequency. For $f_0 = 1$ GHz, the allowed frequency drift is less than ± 33 kHz.

The total power lost to the cavity walls can be determined from Eq. (12.24) if the stored energy in the cavity, U, is known. The quantity U can be calculated from Eq. (12.43) for the TM_{010} mode; we assume the calculation is performed at the time when magnetic fields are zero.

$$U = \int_0^d dz \int_0^{R_0} 2\pi r \, dr \left(\varepsilon E_0^2/2\right) J_0^2(2.405r/R_0)$$

$$= \left(\pi R_0^2 d\right)\left(\varepsilon E_0^2/2\right) J_1^2(2.405). \qquad (12.49)$$

A cylindrical cavity can support a variety of resonant modes, generally at higher frequency than the fundamental accelerating mode. Higher-order modes

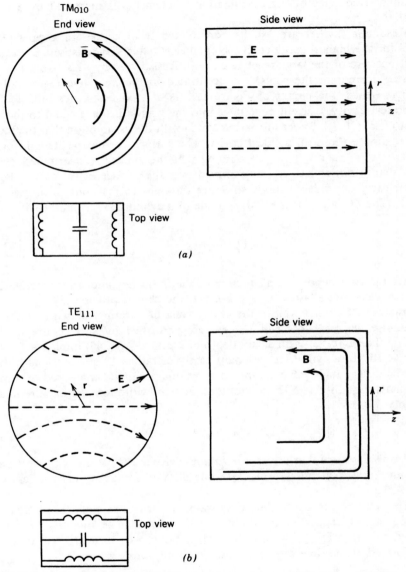

Figure 12.9 Field variations and distribution of capacitance and inductance in cylindrical resonant cavity. (a) TM_{010} mode. (b) TE_{111} mode.

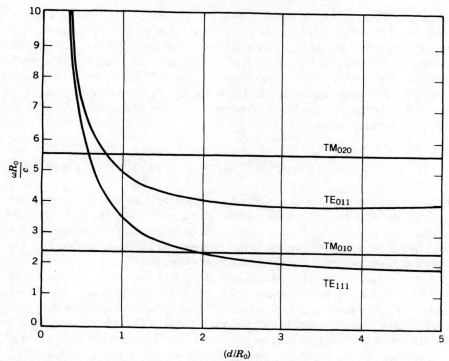

Figure 12.10 Variation of resonant angular frequency with geometry of cylindrical resonant cavity for various low-frequency modes.

are generally undesirable. They do not contribute to particle acceleration; the energy shunted into higher-order modes is wasted. Sometimes, they interfere with particle acceleration; modes with transverse field components may induce beam deflections and particle losses.

As an example of an alternate mode, consider the lowest frequency TE mode, the TE_{111} mode. Figure 12.9 shows a sketch of the electric and magnetic fields. The displacement current oscillates from side to side across the diameter of the cavity. Magnetic fields are wrapped around the displacement current and have components in the axial direction. The distribution of capacitance and inductance for TE_{111} oscillations is also shown in Figure 12.9. The mode frequency depends on the cavity length (Fig. 12.10). As d increases over the range $d \ll R_0$ to $d \geq R_0$, there is a large increase in the capacitance of the cavity for displacement current flow across a diameter. Thus, the resonant frequency drops. For $d \gg R_0$, return current flows mainly back along the circular wall of the cavity. Therefore, the ratio of electric to magnetic field energy in the cavity approaches a constant value, independent of d. Inspection of Figure 2.10 shows that in long cavities, the TE_{111} mode has a lower frequency than the TM_{010}. Care must be taken not to excite the TE_{111} mode in parameter regions where there is *mode degeneracy*. The term *degeneracy* indicates that two modes have the same resonant frequency. Mode selection is a major problem in the complex structures used in linear ion accelerators. Generally, the cavities are long cylinders with internal structures; the mode plot is considerably more complex than Figure 12.10. There is a greater possibility for mode degeneracy and power coupling between modes. In some cases, it is necessary to add metal structures to the cavity, which selectively damp competing modes.

12.5   POWER EXCHANGE WITH RESONANT CAVITIES

Power must be coupled into resonant cavities to maintain electromagnetic oscillations when there is resistive damping or a beam load. The topic of power coupling to resonant cavities involves detailed application of microwave theory. In this section, the approach is to achieve an understanding of basic power coupling processes by studying three simple examples.

We have already been introduced to a cavity with the power feed located on axis. The feed drives a beam and supplies energy lost to the cavity walls. In this case, power is electrically coupled to the cavity, since the current in the power feeds interacts predominantly with electric fields. Although this geometry is never used for driving accelerator cavities, there is a practical application of the inverse process of driving cavity oscillations by a beam. Figure 12.11a shows a klystron, a microwave generator. An on-axis electron beam is injected across the cavity. The electron beam has time-varying current with a strong Fourier component at the resonant frequency of the cavity, ω_0. We will consider only this component of the current and represent it as a harmonic current source. The cavity has a finite Q, resulting from wall resistance and extraction of microwave energy.

The complete circuit model for the TM_{010} mode is shown in Figure 12.11b. The impedance presented to the component of the driving beam current with frequency ω is

$$Z = (j\omega L + R)/[(1 - \omega^2 LC) + j\omega RC]. \qquad (12.50)$$

Assuming that $\omega = \omega_0 = 1/\sqrt{LC}$ and that the cavity has high Q, Eq. (12.50) reduces to

$$Z \cong L/RC = Z_0^2/R = QZ_0, \qquad (12.51)$$

with Q given by Eq. (12.27). The impedance is resistive; the voltage oscillation induced is in phase with the driving current so that energy extraction is maximized. Equation (12.51) shows that the cavity acts as a step-down

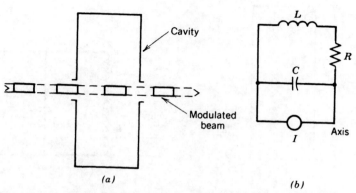

(a) *(b)*

Figure 12.11 Electromagnetic oscillations in resonant cavity driven by modulated beam on axis (klystron). (*a*) Geometry. (*b*) Equivalent circuit.

transformer when the power feed is on axis. Power at low current and high voltage (impedance Z_0^2/R) drives a high current through resistance R.

In applications to high-energy accelerators, the aim is to use resonant cavities as step-up transformers. Ideally, power should be inserted at low impedance and coupled to a low-current beam at high voltage. This process is accomplished when energy is *magnetically coupled* into a cavity. With magnetic coupling, the power input is close to the outer radius of the cavity; therefore, interaction is predominantly through magnetic fields. A method for coupling energy to the TM_{010} mode is illustrated in Figure 12.12*a*. A loop is formed on the end of a transmission line. The loop is orientated to encircle the azimuthal magnetic flux of the TM_{010} mode. (A loop optimized to drive the TE_{111} mode would be rotated 90° to couple to radial magnetic fields.)

We shall first consider the inverse problem of extracting the energy of a TM_{010} oscillation through the loop. Assume that the loop couples only a small fraction of the cavity energy per oscillation period. In this case, the magnetic fields of the cavity are close to the unperturbed distribution. The magnetic field at the loop position, ρ, is

$$B(t) = B_0(\rho, t)\cos \omega_0 t, \tag{12.52}$$

where $B_0(\rho, t)$ is a slowly varying function of time. The spatial variation is given by Eq. (12.47). The loop is attached to a transmission line that is terminated by a matched resistor R.

The voltage induced at the loop output depends on whether the loop current significantly affects the magnetic flux inside the loop. As we saw in the

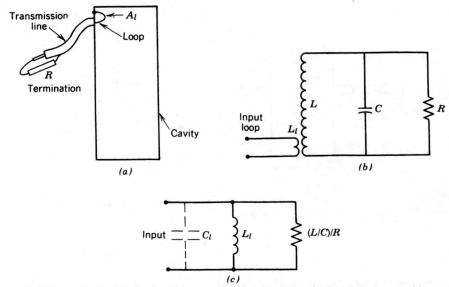

Figure 12.12 Resonant cavity interactions with magnetic coupling loop. (*a*) Geometry of loop set up to extract energy from cavity. (*b*) Equivalent circuit model to give input impedance of loop coupled to cavity. (*c*) Circuit of part *b* at resonance (transformer replaced by its equivalent circuit).

discussion of the Rogowski loop (Section 9.14), the magnetic field inside the loop is close to the applied field when $L/R \ll 1/\omega_0$, where L is the loop inductance and $1/\omega_0$ is the time scale for magnetic field variations. In this limit, the magnitude of the induced voltage around a loop of area A_1 is $V \cong A_1\omega_0 B_0$. The extracted power is

$$P \cong (A_1 B_0 \omega)^2 / 2R. \tag{12.53}$$

Power coupled out of the cavity increases as A_1^2 in this regime. At the opposite extreme ($L/R \gg 1/\omega_0$), the loop voltage is shifted 90° in phase with respect to the magnetic field. Application of Eq. (9.124) shows that the extracted power is approximately

$$P \cong [A_1 B_0/(L/R)]^2 / 2R. \tag{12.54}$$

Since the loop inductance is proportional to A_1, the power is independent of the loop area in this limit. Increasing A_1 increases perturbations of the cavity modes without increasing power output. The optimum size for the coupling loop corresponds to maximum power transfer with minimum perturbation, or $L/R \sim 1/\omega_0$.

Note that in Eqs. (12.53) and (12.54), power loss from the cavity is proportional to B_0^2 and is therefore proportional to the stored energy in the cavity, U. The quantity U is governed by the equation

$$dU/dt \sim -U. \tag{12.55}$$

The stored energy decays exponentially; therefore, losses to the loop can be characterized by a Q factor Q_1. If there are also resistive losses in the cavity characterized by Q_c, then the total cavity Q is

$$Q = (1/Q_1 + 1/Q_c)^{-1}. \qquad \boxed{12.56}$$

We can now proceed to develop a simple circuit model to describe power transfer through a magnetically coupled loop into a cavity with a resistive load on axis. The treatment is based on our study of the transformer (Section 9.2). The equivalent circuit model is illustrated in Figure 12.12b. The quantity R represents the on-axis load. We consider the loop as the primary and the flow of current around the outside of the cavity as the secondary. The primary and secondary are linked together through shared magnetic flux. The loop area is much smaller than the cross-sectional area occupied by cavity magnetic fields. An alternate view of this situation is that there is a large secondary inductance, only part of which is linked to the primary.

Following the derivation of Section 9.2, we can construct the equivalent circuit seen from the primary input (Fig. 12.12c). The part of the cavity magnetic field enclosed in the loop is represented by L_1; the secondary series inductance is $L - L_1$. We assume that energy transfer per oscillation period is small and that $L_1 \ll L$. Therefore, the magnetic fields are close to those of an unperturbed cavity. This assumption allows a simple estimate of L_1.

To begin, we neglect the effect of the shunt inductance L_1 in the circuit of Figure 12.12c and calculate the impedance the cavity presents at the loop input. The result is

$$Z = [j\omega L + R(1 - \omega^2 LC)]/(1 + j\omega RC). \tag{12.57}$$

Damping must be small for an oscillatory solution. This is true if the load resistance is high, or $R \gg \sqrt{L/C}$. Assuming this limit and taking $\omega = \omega_0$, Eq. (12.57) becomes

$$Z \cong R[(L/C)/R^2]. \tag{12.58}$$

Equation (12.58) shows that the cavity presents a purely resistive load with impedance much smaller than R. The combination of coupling loop and cavity act as a step-up transformer.

We must still consider the effect of the primary inductance in the circuit of Figure 12.12c. The best match to typical power sources occurs when the total input impedance is resistive. A simple method of matching is to add a shunt capacitor C_1, with a value chosen so that $C_1 L_1 = LC$. In this case, the parallel combination of L_1 and C_1 has infinite impedance at resonance, and the total load is $(L/C)/R$. Matching can also be performed by adjustment of the transmission line leading to the cavity. We shall see in Section 12.6 that transmission lines can act as impedance transformers. The total impedance will appear to be a pure resistance at the generator for input *at a specific frequency* if the generator is connected to the cavity through a transmission line of the proper length and characteristic impedance.

12.6 TRANSMISSION LINES IN THE FREQUENCY DOMAIN

In the treatment of the transmission line in Section 9.8, we considered propagating voltage pulses with arbitrary waveform. The pulses can be resolved into frequency components by Fourier analysis. If the waveform is limited to a single frequency, the description of electromagnetic signal propagation on a transmission line is considerably simplified. In complex exponential notation, current is proportional to voltage. The proportionality constant is a complex number, containing information on wave amplitude and phase. The advantage is that wave propagation problems can be solved algebraically, rather than through differential equations.

Voltage waveforms in a transmission line move at a velocity $v = 1/\sqrt{\varepsilon\mu}$ along the line. A harmonic disturbance in a transmission line may have components that travel in the positive or negative directions. A single-frequency voltage oscillation measured by a stationary observer has the form

$$V(z, t) = V_+ \exp[j\omega(t - z/v)] + V_- \exp[j\omega(t + z/v)]. \quad (12.59)$$

Equation (12.59) states that points of constant V move along the line at speed v in either the positive or negative z directions. As we found in Section 9.9, the current associated with a wave traveling in either the positive or negative direction is proportional to the voltage. The constant of proportionality is a real number, Z_0. The total current associated with the voltage disturbance of Eq. (12.59) is

$$I(z, t) = (V_+/Z_0)\exp[j\omega(t - z/v)] - (V_-/Z_0)\exp[j\omega(t + z/v)]. \quad (12.60)$$

Note the minus sign in the second term of Eq. (12.60). This is included to preserve the convention that current is positive when positive waves move in the $+z$ direction. A voltage wave with positive voltage moving in the $-z$

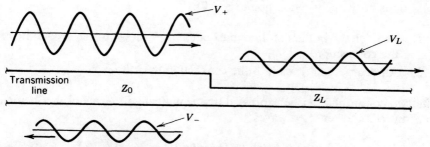

Figure 12.13 Incident, transmitted, and reflected waves at junction of two transmission lines.

direction has negative current. The total impedance at a point is, by definition

$$Z = V(z, t)/I(z, t). \qquad (12.61)$$

If there are components of $V(z, t)$ moving in both the positive and negative directions, Z may not be a real number. Phase differences arise because the sum of V_+ and V_- may not be in phase with the sum of I_+ and I_-.

We will illustrate transmission line properties in the frequency-domain by the calculation of wave reflections at a discontinuity. The geometry is illustrated in Figure 12.13. Two infinite length transmission lines are connected together at $z = 0$. Voltage waves at angular ω frequency travel down the line with characteristic impedance Z_0 toward the line with impedance Z_L. If $Z_L = Z_0$, the waves travel onward with no change and disappear down the second line. If $Z_L \neq Z_0$, we must consider the possibility that wave reflections take place at the discontinuity. In this case, three wave components must be included:

1. The incident voltage wave, of form $V_+\exp[\,j\omega(t - z/v)]$, is specified. The current of the wave is $(V_+/Z_0)\exp[\,j\omega(t - z/v)]$.

2. Some of the incident wave energy may continue through the connection into the second line. The wave moves in the $+z$ direction and is represented by $V_L\exp[\,j\omega(z - t/v')]$. The current of the *transmitted wave* is $(V_L/Z_L)\exp[\,j\omega(t - z/v')]$. There is no negatively directed wave in the second line since the line has infinite length.

3. Some wave energy may be reflected at the connection, leading to a backward-directed wave in the first line. The voltage and current of the *reflected wave* are $V_-\exp[\,j\omega(t + z/v)]$ and $-(V_-/Z_0)\exp[\,j\omega(t + z/v)]$.

The magnitudes of the transmitted and reflected waves are related to the incident wave and the properties of the lines by applying the following conditions at the connection point ($z = 0$):

1. The voltage in the first line must equal the voltage in the second line at the connection.
2. All charge that flows into the connection must flow out.

The two conditions can be expressed mathematically in terms of the incident, transmitted, and reflected waves.

$$V_+\exp(j\omega t) + V_-\exp(j\omega t) = V_L\exp(j\omega t), \qquad (12.62)$$

$$(V_+/Z_0)\exp(j\omega t) - (V_-/Z_0)\exp(j\omega t) = (V_L/Z_L)\exp(j\omega t).$$

$$(12.63)$$

Canceling the time dependence, Eqs. (12.62) and (12.63) can be solved to relate the reflected and transmitted voltages to the incident voltage:

$$\rho = (V_-/V_+) = (Z_L - Z_0)/(Z_L + Z_0), \qquad \boxed{12.64}$$

$$T = (V_L/V_0) = 2Z_L/(Z_L + Z_0). \qquad \boxed{12.65}$$

Equations (12.64) and (12.65) define the *reflection coefficient* ρ and the *transmission coefficient* T. The results are independent of frequency; therefore, they apply to transmission and reflection of voltage pulses with many frequency components. Finally, Eqs. (12.64) and (12.65) also hold for reflection and absorption of waves at a resistive termination, since an infinite length transmission line is indistinguishable from a resistor with $R = Z_L$.

A short-circuit termination has $Z_L = 0$. In this case, $R = -1$ and $T = 0$. The wave is reflected with inverted polarity, in agreement with Section 9.10. There is no transmitted wave. When $Z_L \to \infty$, there is again no transmitted wave and the reflected wave has the same voltage as the incident wave. Finally, if $Z_L = Z_0$, there is no reflected wave and $T = 1$; the lines are matched.

As a final topic, we consider transformations of impedance along a transmission line. As shown in Figure 12.14, assume there is a load Z_L at $z = 0$ at the end of a transmission line of length l and characteristic impedance Z_0. The load may consist of any combination of resistors, inductors, and capacitors; therefore, Z_L may be a complex number. A power source, located at the point $z = -l$, produces a harmonic input voltage, $V_0\exp(j\omega t)$. The goal is to determine how much current the source must supply in order to support the input voltage. This is equivalent to calculating the impedance $Z(-l)$.

The impedance at the generator is generally different from Z_L. In this sense, the transmission line is an *impedance transformer*. This property is useful for matching power generators to loads that contain reactive elements. In this

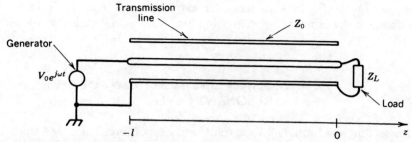

Figure 12.14 Connection of ac voltage generator to a load through transmission line.

section, we shall find a mathematical expression for the transformed impedance. In the next section, we shall investigate some of the implications of the
result.

Voltage waves are represented as in Eq. (12.59). Both a positive wave
traveling from the generator to the load and a reflected wave must be included.
All time variations have the form $\exp(j\omega t)$. Factoring out the time dependence, the voltage and current at $z = 0$ are

$$V(0) = V_+ + V_-, \tag{12.66}$$

$$I(0) = V_+/Z_0 - V_-/Z_0. \tag{12.67}$$

The voltage and current at $z = -l$ are

$$V(-l) = V_+\exp(+l\omega/v) + V_-\exp(-l\omega/v), \tag{12.68}$$

$$I(-l) = (V_+/Z_0)\exp(+l\omega/v) - (V_-/Z_0)\exp(-l\omega/v). \tag{12.69}$$

Furthermore, the treatment of reflections at a line termination [Eq. (12.64)]
implies that

$$V_-/V_+ = (Z_L - Z_0)/(Z_L + Z_0). \tag{12.70}$$

Taking $Z(-l) = V(-l)/I(-l)$, and substituting from Eq. (12.70), we find
that

$$
\begin{aligned}
Z(-l) &= Z_0 \frac{\exp(l\omega/v) + (Z_L - Z_0)\exp(-l\omega/v)/(Z_L + Z_0)}{\exp(l\omega/v) - (Z_L - Z_0)\exp(-l\omega/v)/(Z_L + Z_0)} \\[2mm]
&= Z_0 \frac{Z_L[\exp(l\omega/v) + \exp(-l\omega/v)] + Z_0[\exp(l\omega/v) - \exp(-l\omega/v)]}{Z_L[\exp(l\omega/v) - \exp(-l\omega/v)] + Z_0[\exp(l\omega/v) + \exp(-l\omega/v)]} \\[2mm]
&= Z_0 \frac{Z_L\cos(2\pi l/\lambda) + jZ_0\sin(2\pi l/\lambda)}{Z_0\cos(2\pi l/\lambda) + jZ_L\sin(2\pi l/\lambda)},
\end{aligned}
$$

$$\boxed{12.71}$$

where $\omega/v = 2\pi/\lambda$. In summary, the expressions of Eq. (12.71) give the impedance at the input of a transmission line of length l terminated by a load Z_L.

12.7 TRANSMISSION LINE TREATMENT OF THE RESONANT CAVITY

In this section, the formula for the transformation of impedance by a transmission line [Eq. (12.71)] is applied to problems related to resonant cavities. To begin, consider terminations at the end of a transmission line with characteristic impedance Z_0 and length l. The termination, Z_L, is located at $z = 0$ and the voltage generator at $z = -l$. If Z_L is a resistor with $R = Z_0$, Eq. (12.71) reduces to $Z(-l) = Z_0$, independent of the length of the line. In this case, there is no reflected wave. The important property of the matched transmission line is that the voltage wave at the termination is identical to the input voltage wave delayed by time interval l/v. Matched lines are used to conduct diagnostic signals without distortion.

Another interesting case is the short-circuit termination, $Z_L = 0$. The impedance at the line input is

$$Z(-l) = jZ_0\tan(2\pi l/\lambda). \tag{12.72}$$

The input impedance is zero when $l = 0, \lambda/2, \lambda, 3\lambda/2, \ldots$. An interesting result is that the shorted line has infinite input impedance (open circuit) when

$$l = \lambda/4, 3\lambda/4, 5\lambda/4, \ldots. \tag{12.73}$$

A line with length given by Eq. (12.73) is called a *quarter wave line*.

Figure 12.15 illustrates the analogy between a cylindrical resonant cavity and a quarter wave line. A shorted radial transmission line of length l has power input at frequency ω at the inner diameter. Power flow is similar to that of Figure 12.2. If the frequency of the input power matches one of the resonant frequencies of the line, then the line has an infinite impedance and power is transferred completely to the load on axis. The resonant frequencies of the radial transmission line are

$$\omega_1 = \pi v/2l, \qquad \omega_2 = 3\pi v/2l, \qquad \omega_3 = 5\pi v/2l, \ldots. \tag{12.74}$$

These frequencies differ somewhat from those of Table 12.1 because of geometric differences between the cavities.

The quarter wave line has positive and negative-going waves. The positive wave reflects at the short-circuit termination giving a negative-going wave with 180° phase shift. The voltages of the waves subtract at the termination and add at the input ($z = -l$). The summation of the voltage waves is a standing-wave

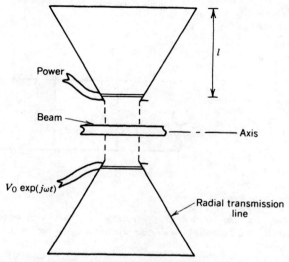

Figure 12.15 Resonant acceleration cavity formed from shorted radial transmission line.

pattern:

$$V(z, t) = V_0 \sin(-\pi z / 2l) \exp(j\omega t). \tag{12.75}$$

At resonance, the current of the two waves at $z = l$ is equal and opposite. The line draws no current and has infinite impedance. At angular frequencies below ω_1, inspection of Eq. (12.72) shows that $Z \sim +j$; thus, the shorted transmission line acts like an inductor. For frequencies above ω_1, the line has $Z \sim -j$; it appears to be a capacitive load. This behavior repeats cyclically about higher resonant frequencies.

A common application of transmission lines is *power matching* from a harmonic voltage generator to a load containing reactive elements. We have already studied one example of power matching, coupling of energy into a resonant cavity by a magnetic loop (Section 12.5). Another example is illustrated in Figure 12.16. An ac generator drives an acceleration gap. Assume, for simplicity, that the beam load is modeled as a resistor R. The generator efficiency is optimized when the total load is resistive. If the load has reactive components, the generator must supply displacement currents that lead to internal power dissipation. Reactances have significant effects at high frequency. For instance, displacement current is transported through the capacitance between the electrodes of the accelerating gaps, C_p. The displacement current is comparable to the load current when $\omega \sim 1/RC_p$. In principle, it is unnecessary for the power supply to support displacement currents since energy is not absorbed by reactances. The strategy is to add circuit elements than can support the reactive current, leaving the generator to supply power only to the

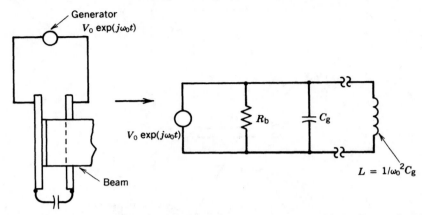

Figure 12.16 Power matching of harmonic generators to loads with reactive and resistive components. Geometry and equivalent circuit of acceleration gap at high frequency.

resistive load. This is accomplished in the acceleration gap by adding a shunt inductance with value $L = 1/\omega_0^2 C_g$, where ω_0 is the generator frequency. The improvement of the Wideroe linac by the addition of resonant cavities (Section 14.2) is an example of this type of matching.

Section 12.5 shows that a coupling loop in a resonant cavity is a resistive load at the driving frequency if the proper shunt capacitance is added. Matching can also be accomplished by adjusting the length of the transmission line connecting the generator to the loop. At certain values of line length, the reactances of the transmission line act in concert with the reactances of the loop to support displacement current internally. The procedure for finding the correct length consists of adjusting parameters in Eq. (12.71) with Z_1 equal to the loop impedance until the imaginary part of the right-hand side is equal to zero. In this circumstance, the generator sees a purely resistive load. The search for a match is aided by use of the Smith chart; the procedure is reviewed in most texts on microwaves.

12.8 WAVEGUIDES

Resonant cavities have finite extent in the axial direction. Electromagnetic waves are reflected at the axial boundaries, giving rise to the standing-wave patterns that constitute resonant modes. We shall remove the boundaries in this section and study electromagnetic oscillations that travel in the axial direction. A structure that contains a propagating electromagnetic wave is called a *waveguide*. Consideration is limited to metal structures with uniform cross section and infinite extent in the z direction. In particular, we will concentrate on the cylindrical waveguide, which is simply a hollow tube.

Waveguides transport electromagnetic energy. Waveguides are often used in accelerators to couple power from a microwave source to resonant cavities. Furthermore, it is possible to transport particle beams in a waveguide in synchronism with the wave phase velocity so that they continually gain energy. Waveguides used for direct particle acceleration must support slow waves with phase velocity equal to or less than the speed of light. Slow-wave structures have complex boundaries that vary periodically in the axial direction; the treatment of slow waves is deferred to Section 12.9.

Single-frequency waves in a guide have fields of the form $\exp[\,j(\omega t - kz)]$ or $\exp[\,j(\omega t + kz)]$. Electromagnetic oscillations move along the waveguide at velocity ω/k. In contrast to transmission lines, waveguides do not have a center conductor. This difference influences the nature of propagating waves in the following ways:

1. The phase velocity in a waveguide varies with frequency. A structure with frequency-dependent phase velocity exhibits *dispersion*. Propagation in transmission lines is dispersionless.

2. Waves of any frequency can propagate in a transmission line. In contrast, low-frequency waves cannot propagate in a waveguide. The limiting frequency is called the *cutoff frequency*.

3. The phase velocity of waves in a waveguide is greater than the speed of light. This does not violate the principles of relativity since information can be carried only by modulation of wave amplitude or frequency. The propagation velocity of frequency modulations is the *group velocity*, which is always less than the speed of light in a waveguide.

The properties of waveguides are easily demonstrated by a lumped circuit element analogy. We can evolve a circuit model for a waveguide by starting from the transmission line model introduced in Section 9.9. A coaxial transmission line is illustrated in Figure 12.17a. At frequencies low compared to $1/(R_0 - R_i)\sqrt{\epsilon\mu}$, the field pattern is the familiar one with radial electric fields and azimuthal magnetic fields. This field is a TEM (transverse electric and magnetic) mode; both the electric and magnetic fields are transverse to the direction of propagation. Longitudinal current is purely real, carried by the center conductor. Displacement current flows radially; longitudinal voltage differences result from inductive fields. The equivalent circuit model for a section of line is shown in Figure 12.17a.

The field pattern may be modified when the radius of the center conductor is reduced and the frequency is increased. Consider the limit where the wavelength of the electromagnetic disturbance, $\lambda = 2\pi/k$, is comparable to or less than the outer radius of the line. In this case, voltage varies along the high-inductance center conductor on a length scale $\leq R_0$. Electric field lines may directly connect regions along the outer conductor (Fig. 12.17b). The field pattern is no longer a TEM mode since there are longitudinal components of

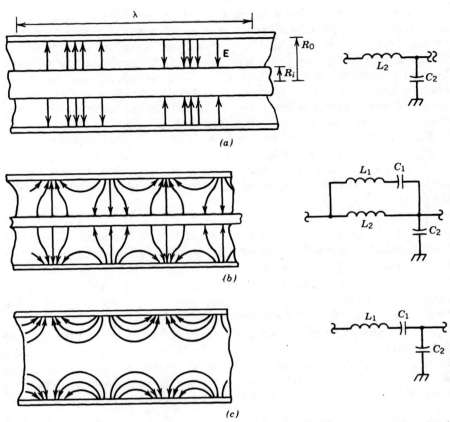

Figure 12.17 Propagating waves in coaxial transmission lines and circular waveguides, electric field patterns, and equivalent circuits. (a) TEM mode in transmission line, low frequency. (b) TM mode in transmission line, high frequency. (c) TM_{10} mode in waveguide.

electric field. Furthermore, a portion of the longitudinal current flow in the transmission line is carried by displacement current. An equivalent circuit model for the coaxial transmission line at high frequency is shown in Figure 12.17b. The capacitance between the inner and outer conductors, C_2, is reduced. The flow of real current through inductor L_2 is supplemented by axial displacement current through the series combination of C_1 and L_1. The inductance L_1 is included because displacement currents generate magnetic fields.

As the diameter of the center conductor is reduced, increasing L_2, a greater fraction of the axial current is carried by displacement current. The limit where $R_i \rightarrow 0$ is illustrated in Figure 12.17c. All axial current flow is via displacement current; L_2 is removed from the mode. The field pattern and equivalent circuit model are shown. We can use the impedance formalism to find the appropriate wave equations for the circuit of Figure 12.17c. Assume that there is a wave moving in the $+z$ direction and take variations of voltage and current as

$$V = V_+ \exp[j(\omega t - kz)] \quad \text{and} \quad I = I_+ \exp[j(\omega t - kz)].$$

The waveguide is separated into sections of length Δz. The inductance of a section is $\mathscr{L}_1 \Delta z$, where $\mathscr{L}_1$ is the inductance per unit length. The quantity C_2 equals $\mathscr{C}_2 \Delta z$, where $\mathscr{C}_2$ is the shunt capacitance per unit length in farads per meters. The series capacitance is inversely proportional to length, so that $C_1 = \mathscr{C}_1/\Delta z$, where $\mathscr{C}_1$ is the series capacitance of a unit length. The quantity $\mathscr{C}_1$ has units of farad-meters. The voltage drop across an element is the impedance of the element multiplied by the current or

$$\Delta V = -I(-j\Delta z/\omega\mathscr{C}_1 + j\omega\mathscr{L}_1 \Delta z),$$

or

$$\partial V/\partial z = -(-j/\omega\mathscr{C}_1 + j\omega\mathscr{L}_1)I. \tag{12.76}$$

The change in longitudinal current occurring over an element is equal to the current that is lost through C_2 to ground or

$$\Delta I = -j\mathscr{C}_2 \Delta z \, \omega V,$$

or

$$\partial I/\partial z = -j\omega\mathscr{C}_2 V. \tag{12.77}$$

Equations (12.76) and (12.77) can be combined to the single-wave equation

$$\partial^2 V/\partial z^2 = -k^2 V = j\omega\mathscr{C}_2(-j/\omega\mathscr{C}_1 + j\omega\mathscr{L}_1)V$$

$$= (\mathscr{C}_2/\mathscr{C}_1 - \omega^2\mathscr{L}_1\mathscr{C}_2)V. \tag{12.78}$$

Solving for k and letting $\omega_c = 1/\sqrt{\mathscr{L}_1\mathscr{C}_1}$, we find that

$$k = (\mathscr{C}_2/\mathscr{C}_1)^{1/2}(\omega^2/\omega_c^2 - 1)^{1/2}. \tag{12.79}$$

Equation (12.79) relates the wavelength of the electromagnetic disturbance in the cylindrical waveguide to the frequency of the waves. Equation (12.79) is a *dispersion relationship*. It determines the phase velocity of waves in the guide as a function of frequency:

$$\omega/k = (\mathscr{C}_1/\mathscr{C}_2)^{1/2}\omega_c/(1 - \omega_c^2/\omega^2)^{1/2}.$$

Note that the phase velocity is dispersive. It is minimum at high frequency and approaches infinity as $\omega \rightarrow \omega_c$. Furthermore, there is a *cutoff frequency*, ω_c, below which waves cannot propagate. The wavenumber is imaginary below ω_c. This implies that the amplitude of low-frequency waves decreases along the guide. Low-frequency waves are reflected near the input of the waveguide; the waveguide appears to be a short circuit.

The above circuit model applies to a propagating wave in the TM_{01} mode. The term TM refers to the fact that magnetic fields are transverse; only electric fields have a longitudinal component. The leading zero indicates that there is azimuthal symmetry; the 1 indicates that the mode has the simplest possible radial variation of fields. There are an infinite number of higher-order modes that can occur in a cylindrical transmission line. We will concentrate on the TM_{01} mode since it has the optimum field variations for particle acceleration. The mathematical methods can easily be extended to other modes.

We will now calculate properties of azimuthally symmetric modes in a cylindrical waveguide by direct solution of the field equations. Again, we seek propagating disturbances of the form

$$\mathbf{E}(r, \theta, z, t) = \mathbf{E}(r, \theta)\exp[j(\omega t - kz)], \qquad (12.80)$$

$$\mathbf{B}(r, \theta, z, t) = \mathbf{B}(r, \theta)\exp[j(\omega t - kz)]. \qquad (12.81)$$

With the above variation and the condition that there are no free charges or current in the waveguide, the Maxwell equations [Eqs. (3.11) and (3.12)] are

$$\nabla \times \mathbf{E} = -j\omega\mathbf{B}, \qquad (12.82)$$

$$\nabla \times \mathbf{B} = j\omega\varepsilon\mu\mathbf{E}. \qquad (12.83)$$

Equations (12.82) and (12.83) can be combined to give the two wave equations

$$\nabla^2\mathbf{E} = -k_0^2\mathbf{E}, \qquad (12.84)$$

$$\nabla^2\mathbf{B} = -k_0^2\mathbf{B}, \qquad (12.85)$$

where $k_0 = \sqrt{\varepsilon\mu}\,\omega = \omega/c$.

The quantity k_0 is the *free-space wavenumber*; it is equal to $2\pi/\lambda_0$, where λ_0 is the wavelength of electromagnetic waves in the filling medium of the waveguide in the absence of the boundaries.

In principle, either Eq. (12.84) or (12.85) could be solved for the three components of $\mathbf{E}$ or $\mathbf{B}$, and then the corresponding components of $\mathbf{B}$ or $\mathbf{E}$ found through Eq. (12.82) or (12.83). The process is complicated by the boundary conditions that must be satisfied at the wall radius, R:

$$\mathbf{E}_{\parallel}(R_0) = 0, \qquad (12.86)$$

$$\mathbf{B}_{\perp}(R_0) = 0. \qquad (12.87)$$

Equations (12.86) and (12.87) refer to the vector sum of components; the boundary conditions couple the equations for different components. An organized approach is necessary to make the calculation tractable.

We will treat only solutions with azimuthal symmetry. Setting $\partial/\partial\theta = 0$, the component forms of Eqs. (12.82) and (12.83) are

$$jkE_\theta = -j\omega B_r, \tag{12.88}$$

$$(1/r)\,\partial(rE_\theta)/\partial r = -j\omega B_z, \tag{12.89}$$

$$-jkE_r - \partial E_z/\partial r = -j\omega B_\theta; \tag{12.90}$$

$$jkB_\theta = j(k_0^2/\omega)E_r, \tag{12.91}$$

$$(1/r)\,\partial(rB_\theta)/\partial r = j(k_0^2/\omega)E_z, \tag{12.92}$$

$$-jkB_r - \partial B_z/\partial r = j(k_0^2/\omega)E_\theta. \tag{12.93}$$

These equations can be manipulated algebraically so that the transverse fields are proportional to derivatives of the longitudinal components:

$$B_r = -jk(\partial B_z/\partial r)/(k_0^2 - k^2), \tag{12.94}$$

$$E_r = -jk(\partial E_z/\partial r)/(k_0^2 - k^2), \tag{12.95}$$

$$B_\theta = -j(k_0^2/\omega)(\partial E_z/\partial r)/(k_0^2 - k^2), \tag{12.96}$$

$$E_\theta = j\omega(\partial B_z/\partial r)/(k_0^2 - k^2). \tag{12.97}$$

Notice that there is no solution if both B_z and E_z equal zero; a waveguide cannot support a TEM mode. Equations (12.94)–(12.97) suggest a method to simplify the boundary conditions on the wave equations. Solutions are divided into two categories: waves that have $E_z = 0$ and waves that have $B_z = 0$. The first type is called a TE wave, and the second type is called a TM wave. The first type has transverse field components B_r and E_θ. The only component of magnetic field perpendicular to the metal wall is B_r. Setting $B_r = 0$ at the wall implies the simple, decoupled boundary condition

$$\partial B_z(R_0)/\partial r = 0. \tag{12.98}$$

Equation (12.98) implies that $E_\theta(R_0) = 0$ and $B_r(R_0) = 0$. The wave equation for the axial component of **B** [Eq. (12.85)] can be solved easily with the above boundary condition. Given B_z, the other field components can be calculated from Eqs. (12.94) and (12.97).

For TM modes, the transverse field components are E_r and B_θ. The only component of electric field parallel to the wall is E_z, so that the boundary condition is

$$E_z(R_0) = 0. \tag{12.99}$$

Equation (12.84) can be used to find E_z; then the transverse field components are determined from Eqs. (12.95) and (12.96). The solutions for TE and TM waves are independent. Therefore, any solution with longitudinal components of both E_z and B_z can be generated as a linear combination of TE and TM waves. The wave equation for E_z of a TM mode is

$$\nabla^2 E_z = (1/r)(\partial/\partial r)(\partial E_z/\partial r) - k^2 E_z$$

$$= -k_0^2 E_z, \tag{12.100}$$

with $E_z(R_0) = 0$. The longitudinal contribution to the Laplacian follows from the assumed form of the propagating wave solution. Equation (12.100) is a special form of the Bessel equation. The solution is

$$E_z(r, z, t) = E_0 J_0\left[\sqrt{(k_0^2 - k^2)}\, r\right]\exp[j(\omega t - kz)]. \tag{12.101}$$

The boundary condition of Eq. (12.99) constraints the wavenumber in terms of the free-space wavenumber:

$$k_0^2 - k^2 = x_n^2/R_0^2, \qquad \boxed{12.102}$$

where $x_n = 2.405, 5.520, \ldots$. Equation (12.102) yields the following dispersion relationship for TM_{0n} modes in a cylindrical waveguide:

$$k = \left(\epsilon\mu\omega^2 - x_n^2/R_0^2\right)^{1/2}. \qquad \boxed{12.103}$$

The mathematical solution has a number of physical implications. First, the wavenumber of low-frequency waves is imaginary so there is no propagation. The cutoff frequency of the TM_{01} mode is

$$\omega_c = 2.405/\sqrt{\epsilon\mu}\, R_0. \tag{12.104}$$

Near cutoff, the wavelength in the guide approaches infinity. The free-space wavelength of a TM_{01} electromagnetic wave at frequency ω_c is

$$\lambda_0 \cong 2.61 R_0.$$

The free-space wavelength is about equal to the waveguide diameter; waves with longer wavelengths are shorted out by the metal waveguide walls.

The wavelength in the guide is

$$\lambda = \lambda_0/\sqrt{(1 - \omega_c^2/\omega^2)}. \tag{12.105}$$

The phase velocity is

$$\omega/k = 1/\sqrt{\varepsilon\mu}\,\sqrt{(1 - \omega_c^2/\omega^2)}\,.$$

$$\boxed{12.106}$$

Note that the phase velocity in a vacuum waveguide is always greater than the speed of light.

The solution of the field equations indicates that there are higher-order TM_{0n} waves. The cutoff frequency for these modes is higher. In the frequency range $2.405/\sqrt{\varepsilon\mu}\,R_0$ to $5.520/\sqrt{\varepsilon\mu}\,R_0$, the only TM mode that can propagate is the TM_{01} mode. On the other hand, a complete solution for all modes shows that the TE_{11} has the cutoff frequency, $\omega_c = 1.841/\sqrt{\varepsilon\mu}\,R_0$, which is lower than that of the TM_{01} mode. Precautions must be taken not to excite the TE_{11} mode since:

1. the waves consume rf power without contributing to particle acceleration and
2. the on-axis radial electric and magnetic field components can cause deflections of the charged particle beam.

12.9 SLOW-WAVE STRUCTURES

The guided waves discussed in Section 12.7 cannot be used for particle acceleration since they have phase velocity greater than c. It is necessary to generate *slow waves*, with phase velocity less than c. It is easy to show that slow waves cannot be propagated in waveguides with simple boundaries. Consider, for instance, waves with electric field of the form $\exp[\,j(\omega t - kz)]$ with $\omega/k < c$ in a uniform cylindrical pipe of radius R_0. Since the wave velocity is assumed less than the speed of light, we can make a transformation to a frame moving at speed $u_z = \omega/k$. In this frame, the wall is unchanged and the wave appears to stand still. In the wave rest frame, the electric field is static. Since there are no displacement currents, there is no magnetic field. The electrostatic field must be derivable from a potential. This is not consistent with the fact that the wave is surrounded by a metal pipe at constant potential. The only possible static field solution inside the pipe is $\mathbf{E} = 0$.

Slow waves can propagate when the waveguide has periodic boundaries. The properties of slow waves can be derived by a formal mathematical treatment of wave solutions in a periodic structure. In this section, we shall take a more physical approach, examining some special cases to understand how periodic structures support the boundary conditions consistent with slow waves. To begin, we consider the effects of the addition of periodic structures to the transmission line of Figure 12.18a. If the region between electrodes is a vacuum, TEM waves propagate with $\omega/k = c$. The line has a capacitance ($\mathscr{C}$) and inductance ($\mathscr{L}$) per unit length given by Eqs. (9.71) and (9.72). We found

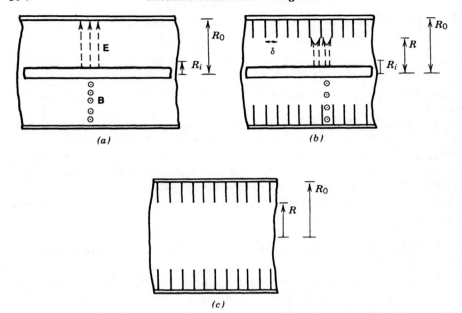

Figure 12.18 Slow-wave structures. (*a*) Electric and magnetic fields of TEM mode in coaxial transmission line. (*b*) Modified TEM mode in coaxial transmission line with capacitive loading. (*c*) Iris-loaded waveguide.

in Section 9.8 that the phase velocity of waves in a transmission line is related to these quantities by

$$\omega/k = 1/\sqrt{\mathscr{L}\mathscr{C}}. \tag{12.107}$$

Consider reconstructing the line as shown in Figure 12.18*b*. Annular metal pieces called *irises* are attached to the outer conductor. The irises have inner radius R and spacing δ.

The field electric field patterns for a TEM wave are sketched in Figure 12.18*b* in the limit that the wavelength is long compared to δ. The magnetic fields are almost identical to those of the standard transmission line except for field exclusion from the irises; this effect is small if the irises are thin. In contrast, radial electric fields cannot penetrate into the region between irises. The electric fields are restricted to the region between the inner conductor and inner radius of the irises. The result is that the inductance per unit length is almost unchanged, but $\mathscr{C}$ is significantly increased. The capacitance per unit length is approximately

$$\mathscr{C} \cong 2\pi\varepsilon/\ln(R/R_i). \tag{12.108}$$

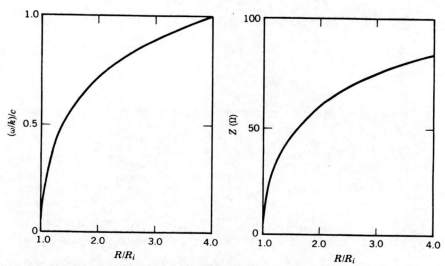

Figure 12.19 Phase velocity and characteristic impedance of capacitively loaded transmission line as function of R/R_i. ($\varepsilon = \varepsilon_0$, $R_i/R_0 = 0.25$).

The phase velocity as a function of R/R_0 is

$$\omega/k \cong c\left[\ln(R/R_i)/\ln(R_0/R_i)\right]^{1/2} \tag{12.109}$$

The characteristic impedance for TEM waves becomes

$$Z = \sqrt{\mathscr{L}/\mathscr{C}} = Z_0\left[\ln(R_0/R_i)/\ln(R/R_i)\right]^{1/2}. \tag{12.110}$$

The phase velocity and characteristic impedance are plotted in Figure 12.19 as a function of R/R_i. Note the following features:

1. The phase velocity decreases with increasing volume enclosed between the irises.
2. The phase velocity is less than the speed of light.
3. The characteristic impedance decreases with smaller iris inner radius.
4. In the long wavelength limit ($\lambda \gg \delta$), the phase velocity is independent of frequency. This is not true when $\lambda \leq \delta$. A general treatment of the capacitively loaded transmission line is given in Section 12.10.

A similar approach can be used to describe propagation of TM_{01} modes in an *iris-loaded waveguide* (Fig. 12.18c). At long wavelength the inductance L_1

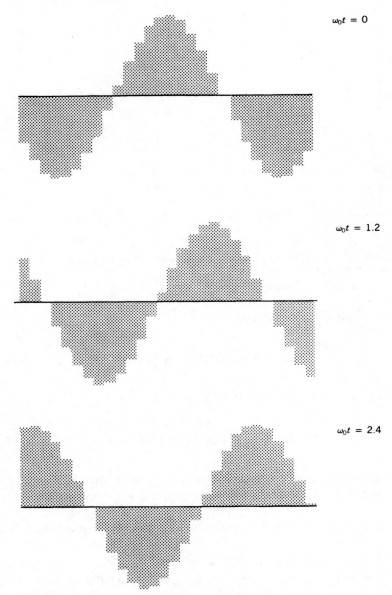

$\omega_0 t = 0$

$\omega_0 t = 1.2$

$\omega_0 t = 2.4$

Figure 12.20 Electric field amplitudes in array of individually phased resonant cavities (27 cavities, with oscillations separated by constant phase difference $\Delta\phi = -0.3$ rad). Plots at times given by $\omega_0 t = 0, 1.2, 2.4, 3.6, 4.8, 6$.

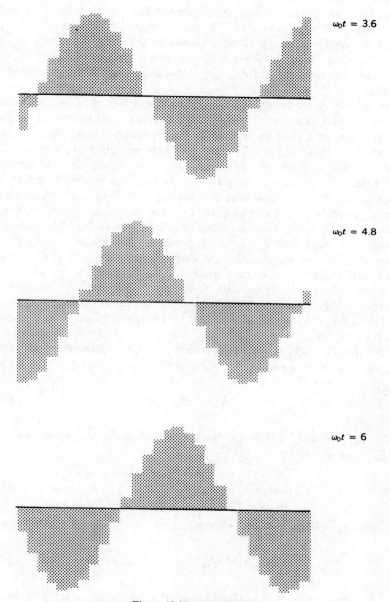

$\omega_0 t = 3.6$

$\omega_0 t = 4.8$

$\omega_0 t = 6$

Figure 12.20 (*Continued*).

397

is almost unchanged by the presence of irises, but the capacitances C_1 and C_2 of the lumped element model is increased. The phase velocity is reduced. Depending on the geometry of the irises, the phase velocity may be pulled down below c. Capacitive loading also reduces the cutoff frequency. ω_c. In the limit of strong loading ($R \ll R_0$), the cutoff frequency for TM_{01} waves approaches the frequency of the TM_{010} mode in a cylindrical resonant cavity of radius R_0.

The following model demonstrates how the irises of a loaded waveguide produce the proper boundary fields to support an electrostatic field pattern in the rest frame of a slow wave. Consider an iris-loaded waveguide in the limit that $R \ll R_0$ (Fig. 12.18c). The sections between irises are similar to cylindrical resonant cavities. A traveling wave moves along the axis through the small holes; this wave carries little energy and has negligible effect on the individual cavities. Assume that cavities are driven in the TM_{010} mode by external power feeds; the phase of the electromagnetic oscillation can be adjusted in each cavity. Such a geometry is called an *individually phased cavity array*. In the limit $\lambda \gg \delta$, the cavity fields at R are almost pure E_z fields. These fields can be matched to the longitudinal electric field of a traveling wave to determine the wave properties.

Assume that δ is longitudinally uniform and that there is a constant phase difference $-\Delta\phi$ between adjacent cavities. The input voltage has frequency $\omega \cong 2.405c/R_0$. Figure 12.20 is a plot of electric field at R in a number of adjacent cavities separated by a constant phase interval at different times. Observe that the field at a particular time is a finite difference approximation to a sine wave with wavelength

$$\lambda = 2\pi\delta/\Delta\phi. \tag{12.111}$$

Comparison of plots at different times shows that the waveform moves in the $+z$ direction at velocity

$$v \text{ (phase)} = \omega_0/k - (2.405/\Delta\phi)(\delta/R_0)c. \tag{12.112}$$

The phase velocity is high at long wavelength. A slow wave results when

$$\Delta\phi > 2.405\delta/R_0 \quad \text{or} \quad \lambda < R_0/2.405. \tag{12.113}$$

In the rest frame of a slow wave, the boundary electric fields at R approximate a static sinusoidal field pattern. Although the fields oscillate inside the individual cavities between irises, the electric field at R appears to be static to an observer moving at velocity ω/k. Magnetic fields are confined within the cavities. The reactive boundaries, therefore, are consistent with an axial variation of electrostatic potential in the wave rest frame.

12.10   DISPERSION RELATIONSHIP FOR THE IRIS-LOADED WAVEGUIDE

The dispersion relationship $\omega = \omega(k)$ is an equation relating frequency and wavenumber for a propagating wave. In this section, we shall consider the implications of dispersion relationships for electromagnetic waves propagating in metal structures. We are already familiar with one quantity derived from the dispersion relationship, the phase velocity ω/k. The group velocity v_g is another important parameter. It is the propagation velocity for modulations of frequency or amplitude. Waves with constant amplitude and frequency cannot carry information; information is conveyed by changes in the wave properties. Therefore, the group velocity is the velocity for information transmission. The group velocity is given by

$$v_g = d\omega/dk. \qquad \boxed{12.114}$$

Equation (12.114) can be derived through the calculation of the motion of a pulsed disturbance consisting of a spectrum of wave components. The pulse is Fourier analyzed into frequency components; a Fourier synthesis after a time interval shows that the centroid of the pulse moves if the wavenumber varies with frequency.

As an example of group velocity, consider TEM electromagnetic waves in a transmission line. Frequency and wavenumber are related simply by $\omega = k/\sqrt{\epsilon\mu} = kv$. Both the phase and group velocity are equal to the speed of light in the medium. There is no dispersion; all frequency components of a pulse move at the same rate through the line; therefore, the pulse translates with no distortion. Waves in waveguides have dispersion. In this case, the components of a pulse move at different velocities and a pulse widens as it propagates.

The group velocity has a second important physical interpretation. In most circumstances, the group velocity is equal to the flux of energy in a wave along the direction of propagation divided by the electromagnetic energy density. Therefore, group velocity usually characterizes energy transport in a wave.

Dispersion relationships are often represented as graphs of ω versus k. In this section, we shall construct $\omega-k$ plots for a number of wave transport structures, including the iris-loaded waveguide. The straight-line plot of Figure 12.21*a* corresponds to TEM waves in a vacuum transmission line. The phase velocity is the slope of a line connecting a point on the dispersion curve to the origin. The group velocity is the slope of the dispersion curve. In this case, both velocities are equal to c at all frequencies.

Figure 12.21*b* shows an $\omega-k$ plot for waves passing along the axis of an array of individually phased circular cavities with small coupling holes. The curve is plotted for an outer radius of $R_0 = 0.3$ m and a distance of 0.05 m between irises. The frequency depends only on the cavity properties not the wavelength of the weak coupling wave. Only discrete frequencies correspond-

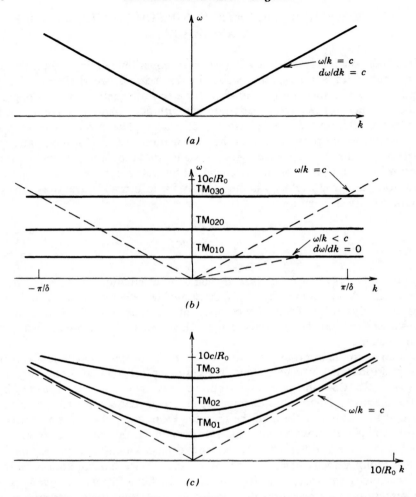

Figure 12.21 Dispersion relationships. (*a*) TEM waves in a transmission line. (*b*) Weakly coupled waves between individually phased cavities with constant phase difference, TM_{0n0} cavity modes. Point indicates possible conditions for a slow wave. (*c*) TM_{0n} modes in a circular waveguide.

ing to cavity resonances are allowed. The reactive boundary conditions for azimuthally symmetric slow waves can be generated by any TM_{0n0} mode. Choice of the relative phase, $\Delta\phi$, determines k for the propagating wave. Phase velocity and group velocity are indicated in Figure 12.21*b*. The line corresponding to $\omega/k = c$ has also been plotted. At short wavelengths (large k), the phase velocity can be less than c. Note that since ω is not a function of k, the group velocity is zero. Therefore, the traveling wave does not transport energy between the cavities. This is consistent with the assumption of small coupling

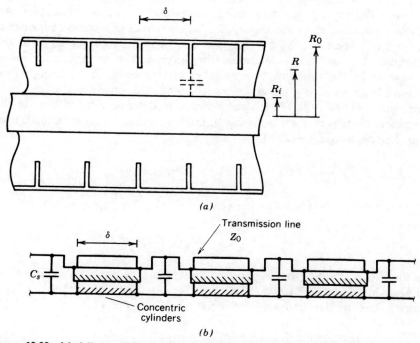

Figure 12.22 Modeling capacitively loaded transmission line as uniform transmission line with periodic shunt capacitance. (*a*) Geometry. (*b*) Equivalent circuit.

holes. The physical model of Section 12.8 is not applicable for wavelengths less than 2δ; this limit has also been indicated on the $\omega-k$ graph.

The third example is the uniform circular waveguide. Figure 12.21*c* shows a plot of Eq. (12.103) for a choice of $R_0 = 0.3$ m. Curves are included for the TM_{01}, TM_{02}, and TM_{03} modes. Observe that wavenumbers are undefined for frequency less than ω_c. The group velocity approaches zero in the limit that $\omega \rightarrow \omega_c$. When $\omega = \omega_c$, energy cannot be transported into the waveguide since $k = 0$. The group velocity is nonzero at short wavelengths (large k). The boundaries have little effect when $\lambda \ll R$; in this limit, the $\omega-k$ plot approaches that of free-space waves, $\omega/k = c$. At long wavelength (small k), the oscillation frequencies approach those of TM_{0n0} modes in an axially bounded cavity with radius R. The phase velocity in a waveguide is minimum at long wavelength; it can never be less than c.

As a fourth example, consider the dispersion relationship for waves propagating in the capacitively loaded transmission line of Figure 12.22*a*. This example illustrates some general properties of waves in periodic structures and gives an opportunity to examine methods for analyzing periodic structures mathematically. The capacitively loaded transmission line can be considered as a transmission line with periodic impedance discontinuities. The discontinuities arise from the capacitance between the irises and the center conductor. An equivalent circuit is shown in Figure 12.22b; it consists of a series of transmission lines of impedance Z_0 and length δ with a shunt capacitance C_s at the junctions. The goal is to determine the wavenumber of harmonic waves propagating in the structure as a function of frequency. Propagating waves may have both positive-going and negative-going components.

Equations (12.68) and (12.69) can be used to determine the change in the voltage and current of a wave passing through a section of transmission line of length δ. Rewriting Eq. (12.68),

$$V(z + \delta) = V_+\exp(-\omega\delta/v) + V_-\exp(\omega\delta/v)$$

$$= (V_+ + V_-)\cos(\omega\delta/v) - j(V_+ - V_-)\sin(\omega\delta/v)$$

$$= V(z)\cos(\omega\delta/v) - jZ_0I(z)\sin(\omega\delta/v). \qquad (12.115)$$

The final form results from expanding the complex exponentials [Eq. (12.5)] and applying Eqs. (12.66) and (12.67). A time variation $\exp(j\omega t)$ is implicitly assumed. In a similar manner, Eq. (12.69) can be modified to

$$I(z + \delta) = I(z)\cos(\omega\delta/v) - jV(z)\sin(\omega\delta/v)/Z_0. \qquad (12.116)$$

Equations (12.115) and (12.116) can be united in a single matrix equation,

$$\begin{pmatrix} V(z + \delta) \\ I(z + \delta) \end{pmatrix} = \begin{bmatrix} \cos(\omega\delta/v) & -jZ_0\sin(\omega\delta/v) \\ -j\sin(\omega\delta/v)/Z_0 & \cos(\omega\delta/v) \end{bmatrix} \begin{pmatrix} V(z) \\ I(z) \end{pmatrix}. \qquad (12.117)$$

The shunt capacitance causes the following changes in voltage and current propagating across the junction:

$$V' = V(z + \delta), \qquad (12.118)$$

$$I' = I(z + \delta) - j\omega C_s V(z + \delta). \qquad (12.119)$$

In matrix notation, Eqs. (12.118) and (12.119) can be written,

$$\begin{pmatrix} V' \\ I' \end{pmatrix} = \begin{bmatrix} 1 & 0 \\ -j\omega C_s & 1 \end{bmatrix} \begin{pmatrix} V(z + \delta) \\ I(z + \delta) \end{pmatrix}. \qquad (12.120)$$

The total change in voltage and current passing through one cell of the capacitively loaded transmission line is determined by multiplication of the matrices in Eqs. (12.117) and (12.120):

$$\begin{pmatrix} V' \\ I' \end{pmatrix} =$$

$$\begin{bmatrix} \cos(\omega\delta/v) & -jZ_0\sin(\omega\delta/v) \\ -j[\omega C_s\cos(\omega\delta/v) + \sin(\omega\delta/v)/Z_0] & \cos(\omega\delta/v) - \omega CZ_0\sin(\omega\delta/v) \end{bmatrix}\begin{pmatrix} V \\ I \end{pmatrix}.$$

$$(12.121)$$

Applying the results of Section 8.6, the voltage and current at the cell boundaries vary harmonically along the length of the loaded transmission line with phase advance given by $\cos\mu = \text{Tr}\,\mathbf{M}/2$, where $\mathbf{M}$ is the transfer matrix for a cell [Eq. (12.121)]. If k is the wavenumber of the propagating wave, the phase advance over a cell of length δ is $\mu = k\delta$. Taking the trace of the matrix of Eq. (12.121) gives the following dispersion relationship for TEM waves in a capacitively loaded transmission line:

$$\cos(k\delta) = \cos(\omega\delta/v) - (C_sZ_0v/2\delta)(\omega\delta/v)\sin(\omega\delta/v). \quad (12.122)$$

Equation (12.122) is plotted in Figure 12.23 for three choices of $(C_sZ_0v/2\delta)$. In the limit of no loading ($C_s = 0$), the dispersion relationship reduces to that of an unloaded line; both the group and phase velocities equal v (the velocity of light in the medium filling the line). With loading, the phase velocity is reduced below v (slow waves). The long wavelength (small k) results agree with the analysis of Section 12.9; the phase velocity and group velocity are independent of frequency. The wave characteristics deviate considerably from those of a TEM wave in an unloaded line when k approaches π/δ. The group velocity approaches zero when $\lambda/2 \rightarrow \delta$. In this case, the wave is a standing-wave pattern with equal components of positive-going and negative-going waves. The feature is explained below in terms of constructive interference of wave reflections from the line discontinuities. The form of Eq. (12.122) implies that the dispersion plot repeats periodically for higher values of wavenumber.

The final example of dispersion curves is the iris-loaded waveguide. The ω–k diagram is important in designing traveling wave particle accelerators; the phase velocity must match the particle velocity at all points in the accelerator, and the group velocity must be high enough to transport power through the structure effectively. In this calculation, we will determine how the size of the aperture (R) affects a TM_{01} wave moving through the coupling holes. We will limit attention to the long wavelength limit ($\lambda > 2\delta$). The iris spacing and outer radius are assumed constant. We have already treated two special cases, $R/R_0 = 1$ (uniform circular waveguide) and $R/R_0 \cong 0$ (independently phased array). Curves for these limits are plotted on Figure 12.24. Consider an intermediate case such as $R/R_0 = 0.5$. At long wavelength, inspection of the curves for the limiting cases infers that the frequency approaches $\omega =$

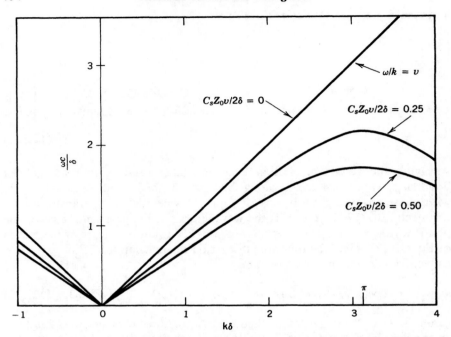

Figure 12.23 Dispersion relationship for capacitively loaded transmission line as function of $C_s Z_0 v / 2\delta$. C_s, shunt capacitance between iris and inner conductor; Z_0, characteristic impedance of uniform line; v, speed of light in line medium; δ, distance between irises.

$2.405c/R_0$. This behavior can be understood if we consider a long wavelength TM_{01} mode in an ordinary waveguide of radius R_0. The magnetic field is azimuthal, while the electric field is predominantly axial. The addition of thin irises has negligible effect on the electric and magnetic field lines since the oscillating fields induce little net current flow on the irises. In the long wavelength limit, electric fields of the TM_{01} mode are relatively unaffected by the radial metal plates. Current flow induced on the irises by oscillating magnetic field is almost equal and opposite on the upstream and downstream sides. The only effect is exclusion of magnetic field from the interior of the thin irises.

We can understand the ω–k diagram at short wavelengths by approximating the wave as a free-space plane wave. The irises represent discontinuities in the waveguide along the direction of propagation; some of the wave energy may be reflected at the discontinuity. Depending on the geometry, there is the possibility of constructive interference of the reflected waves. To understand this, assume that transmitted and reflected waves are observed at the point $z = 0$. Irises are located at distances $\delta, 2\delta, 3\delta, \ldots, n\delta$ downstream. A waveform reaches a particular iris at a time $n\delta/(\omega/k)$ after it passes the point $z = 0$. A reflected wave from the iris takes a time $n\delta/(\omega/k)$ to return to the

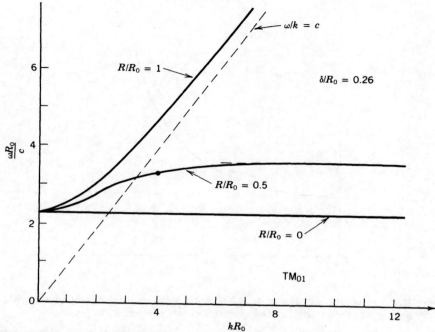

Figure 12.24 Dispersion relationships in the range $0 \leq k \leq \pi/\delta$ for TM_{01} mode propagation in iris-loaded waveguide of radius R_0 as a function of inner radius of iris, R. $R/R_0 = 0$ corresponds to individually phase cavities, $R/R_0 = 1$ corresponds to circular waveguide. Point indicates slow wave with nonzero group velocity.

origin. The sum of reflected waves at $z = 0$ is therefore

$$E_z \text{ (reflected)} \sim \sum \exp\{ j[\omega t - k(2n\delta)] \} = \sum \exp(j\omega t)\cos(2nk\delta).$$

$$(12.123)$$

The summation of Eq. (12.123) diverges when $k = \pi/\delta$. In this case, there is a strong reflected wave. The final state has equipartition of energy between waves traveling in the $+z$ and $-z$ directions; therefore, a standing-wave pattern with zero group velocity is set up.

We can estimate the frequency of the standing wave at $k = \pi/\delta$ by calculating the resonant frequency of a hollow annular cavity with specified inner radius. In the limit $\delta \ll (R_0 - R)$, resonant frequencies of TM_{0n0} modes are determined by solving Eq. (12.42) with boundary conditions $E_z(R_0) = 0$ and $B_\theta(R) = 0$. The latter condition comes about because the axial displacement current between $r = 0$ and $r = R$ is small. The boundary condition can be rewritten as $dE_z(R)/dr = 0$. The resonant frequencies are

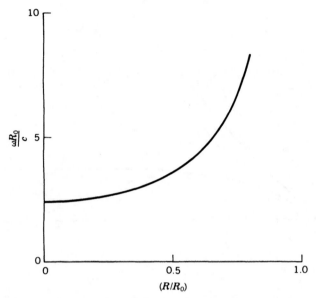

Figure 12.25 Frequency of TM_{010} mode as function of R/R_0 in annular cavity with length d, outer radius R_0, and inner radius R. Open circuit boundary at inner radius; $\delta \ll R$.

determined by the solutions of the transcendental equation:

$$\frac{J_1(\omega R/c)Y_0(\omega R_0/c)}{J_0(\omega R_0/c)Y_1(\omega R/c)} = 1. \qquad (12.124)$$

Resonant frequencies as a function of R/R_0 are plotted in Figure 12.25 for the TM_{01} and TM_{02} modes. For our example [$(R/R_0) = 0.5$], the frequency of the hollow cavity is about 50% higher than that of the complete cavity. This value was incorporated in the plot of Figure 12.24.

Consider some of the implications of Figure 12.24. In the limit of small coupling holes, the cavities are independent. We saw in discussing individually phased cavities that phase velocities much less than the speed of light can be generated by the proper choice of the phasing and δ/R. Although there is latitude to achieve a wide range of phase velocity in the low coupling limit, the low group velocity is a disadvantage. Low group velocity means that energy cannot be coupled between cavities by a traveling wave.

The interdependence of phase and group velocity in a periodic structure enters into the design of rf linear accelerators (Chapter 14). In an accelerator for moderate- to high-energy electrons, the phase velocity is close to c. Inspection of Figure 12.24 shows that this value of phase velocity can be achieved in a structure with substantial coupling holes and a high value of

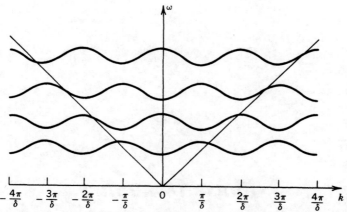

Figure 12.26 Brillouin diagram of wave properties in a capacitively loaded cylindrical waveguide with iris spacing δ.

group velocity. This means that a useful traveling wave can be excited in an extended structure with nonzero wall resistivity by a single power input. The boundary cavities between irises are excited by energy carried by the traveling wave. This approach is not suitable for linear ion accelerators, where the phase velocity must be well below the speed of light. This is the reason linear ion accelerators generally use external rf coupling of individual cavities to synthesize a slow traveling wave on axis.

In the above derivation, we concentrated on TM_{01} waves over the wavenumber range $0 < k < \pi/\delta$. This is the range generally encountered in accelerator applications. We should recognize, nonetheless, that higher-order modes and traveling waves with $\lambda < 2\delta$ can be propagated. The complete ω–k plot for a periodic waveguide structure is called a *Brillouin diagram*.[†] An example is illustrated in Figure 12.26. The periodic repetition of the curve along the k axis is a consequence of the axial periodicity of the waveguide structure. Note the similarities between Figure 12.26 and the dispersion relationship for TEM waves in the capacitively loaded transmission line (Fig. 12.22).

[†] L. Brillouin, *Wave Propagation in Periodic Structures*, Dover, New York, 1953.

13

Phase Dynamics

The axial electric field at a particular location in an rf accelerator has negative polarity half the time. Particles must move in synchronism with variations of electromagnetic fields in order to be accelerated continuously. Synchronization must be effective over long distances to produce high-energy beams. In this chapter, we shall study the longitudinal dynamics of particles moving in traveling electromagnetic waves. Particle motion is summarized in the phase equations, which describe axial displacements of particles relative to the traveling wave. The phase equations lead to the concept of phase stability.[†] Groups of particles can be confined to the accelerating phase of a wave if they have a small enough spread in kinetic energy. Individual particles oscillate about a constant point in the wave called the synchronous phase.

There are a number of important applications of the phase equations:

1. Injected particles are captured efficiently in an rf accelerator only if particles are introduced at the proper phase of the electromagnetic field. The longitudinal acceptance of an accelerator can by calculated from the theory of longitudinal phase dynamics. A knowledge of the allowed kinetic energy spread of injected particles is essential for designing beam injectors and bunchers.

2. There is a trade-off between accelerating gradient and longitudinal acceptance in an rf accelerator. The theory of longitudinal phase dynamics

[†] V. Veksler, *Doklady U.S.S.R.*, **44**, 444 (1944); E. M. McMillan, *Phys. Rev.*, **69** 145 (1945).

predicts beam flux limits as a function of the phase and the properties of the accelerating wave. Effects of space charge can be added to find longitudinal current limits in accelerators.

3. The output from resonant accelerators consists of beam bunches emerging at the frequency of the accelerating wave. Information on the output beam structure is necessary to design debunchers, matching sections to other accelerators, and high-energy physics experiments.

Section 13.1 introduces phase stability. Longitudinal motion is referenced to the hypothetical synchronous particle. Acceleration and inertial forces are balanced for the synchronous particle; it remains at a point of constant phase in a traveling wave. The synchronous particle is the longitudinal analogy of an on-axis particle with no transverse velocity. Particles that deviate in phase or energy from the synchronous particle may either oscillate about the synchronous phase or may fall out of synchronism with the wave. In the former case, the particles are said to be phase stable. Conditions for phase stability are discussed qualitatively in Section 13.1. Equations derived in Section 13.2 give a quantitative description of phase oscillations. The derivation is facilitated by a proof that the fields of all resonant accelerators can be expressed as a sum of traveling waves. Only the wave with phase velocity near the average particle velocity interacts strongly with particles.

The phase equations are solved for nonrelativistic particles in Section 13.3 in the limit that changes in the average particle velocity are slow compared to the period of a phase oscillation. The derivation introduces a number of important concepts such as rf buckets, kinetic energy error, and longitudinal acceptance. A second approximate analytic solution, discussed in Section 13.4, holds when the amplitude of phase oscillations is small. The requirement of negligible velocity change is relaxed. The model predicts reversible compression of longitudinal beam bunches during acceleration. The process is similar to the compression of transverse oscillations in the betatron.

Longitudinal motion of nonrelativistic particles in an induction linac is discussed in Section 13.5. Synchronization is important in the linear induction accelerator even though it is not a resonant device. Pulses of ions must pass through the acceleration gaps during the time that voltage is applied. Section 13.6 discusses longitudinal motion of highly relativistic particles. The material applies to rf electron linacs and linear induction electron accelerators. Solutions of the phase equation are quite different from those for nonrelativistic particles. Time dilation is the major determinant of particle behavior. Time varies slowly in the rest frame of the beam relative to the stationary frame. If the electrons in an rf accelerator are accelerated rapidly, they do not have time to perform a phase oscillation before exiting the machine. In some circumstances, electrons can be captured in the positive half-cycle of a traveling wave and synchronously accelerated to arbitrarily high energy. This process is called electron capture. Relativistic effects are also important in linear induc-

tion electron accelerators; electron beams remain synchronized even in the
presence of large imperfections of voltage waveform.

13.1 SYNCHRONOUS PARTICLES AND PHASE STABILITY

The concept of the synchronous particle and phase stability can be illustrated
easily by considering motion in an accelerator driven by an array of independently phased cavities (Fig. 13.1a). Particles receive longitudinal impulses in
narrow acceleration gaps. The gaps are spaced equal distances apart; the phase
in each cavity is adjusted for the best particle acceleration. Figure 13.1b shows
the time variation of gap voltage in a cavity. The time axis is referenced to the
beginning of the accelerating half-cycle in gap n.

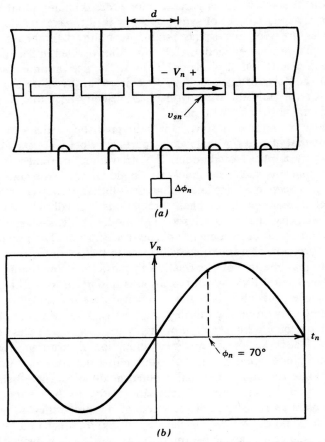

Figure 13.1 Phase of a particle in resonant accelerator. (*a*) Array of uniform cavities with
different rf phase offsets. (*b*) Definition of phase of particle with respect to time-dependent
accelerating voltage in cavity n. Time of crossing of particle with $\phi_n = 70°$ illustrated.

The time at which a test particle crosses gap n is indicated in Figure 13.1b. The *phase* of the particle in defined in terms of the crossing time relative to the cavity waveform. Phase is measured from the beginning of the acceleration half-cycle. Figure 13.1b illustrates a particle with $+70°$ phase. A particle with a phase of $+90°$ crosses at the time of peak cavity voltage; it gains the maximum possible energy.[†]

A *synchronous particle* is defined as a particle that has the same phase in all cavities. Crossing times of a synchronous particle are indicated as solid squares in Figure 13.2a. The *synchronous phase* ϕ_s is the phase of the synchronous particle. The synchronous particle is in longitudinal equilibrium. Acceleration of the particle in the cavities matches the phase difference of electromagnetic oscillations between cavities so that the particle always crosses gaps at the same relative position in the waveform. A synchronous particle exists only if the frequencies of oscillations in all cavities are equal. If frequency varies, the accelerating oscillations will continually shift relative to each other and the diagram of Figure 13.2a will not hold at all times. Particles are accelerated if the synchronous phase is between $0°$ and $180°$.

An accelerator must be properly designed to fullfill conditions for a synchronous particle. In some accelerators, the phase difference is constant while the distance between gaps is chosen to match particle acceleration. In the present example, the phase of oscillations in individual cavities is adjusted to match the particle mass and average accelerating gradient with the distance between gaps fixed. It is not difficult to determine the proper phase differences for nonrelativistic particles. The phase difference between oscillations in cavity $n + 1$ and cavity n is denoted $\Delta\phi_{n+1}$. The accelerating voltages in cavities n through $n + 2$ are defined as

$$V_n = V_0 \sin \omega t,$$

$$V_{n+1} = V_0 \sin(\omega t + \Delta\phi_{n+1}), \tag{13.1}$$

$$V_{n+2} = V_0 \sin(\omega t + \Delta\phi_{n+1} + \Delta\phi_{n+2}),$$

where t is the time and ω is the angular rf frequency. By the definition of ϕ_s, the synchronous particle crosses cavity n at time $\omega t = \phi_s$. Assuming nonrelativistic ions, the change in synchronous particle velocity imparted by cavity n is given by

$$\tfrac{1}{2}mv_{sn}^2 = \tfrac{1}{2}mv_{sn-1}^2 + qV_0 \sin \phi_s. \tag{13.2}$$

where v_{sn} is the particle velocity emerging from gap n. The particle arrives at

[†] In many discussions of linear accelerators, the gap fields are assumed to vary as $E_z(t) = E_{z0}\cos(\omega t)$. Therefore, a synchronous phase value quoted as $-32°$ corresponds to $\phi_s = 58°$ in the convention used in this book.

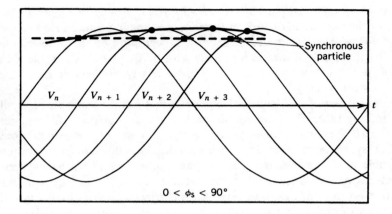

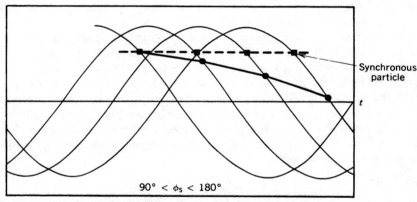

Figure 13.2 Synchronous phase and phase stability. Voltage waveforms illustrated for a number of adjacent acceleration cavities. Squares designate times of crossing for synchronous particle, circles correspond to nonsynchronous particle: (*a*) $0° < \phi_s < 90°$, (*b*) $90° < \phi_s < 180°$.

gap $n + 1$ at time

$$t_{n+1} = d/v_{sn} + \phi_s/\omega, \tag{13.3}$$

where d is the distance between gaps. Since the particle is a synchronous particle, the voltage in cavity $n + 1$ equals $V_0 \sin \phi_s$ at time t_{n+1}; therefore, Eq. (13.1) implies that

$$\omega d/v_{sn} + \phi_s + \Delta\phi_{n+1} = \phi_s,$$

or

$$\Delta\phi_{n+1} = -\omega d/v_{sn}, \tag{13.4}$$

where we have used Eqs. (13.1) and (13.3). Equation (13.4) specifies phase differences between cavity oscillations. Once V_0 and ϕ_s are chosen, the quantities v_{sn} can be calculated.

Particles can never be injected perfectly. Beams always have a spread in longitudinal position and velocity with respect to the synchronous particle. Figure 13.2*a* illustrates crossing times (circles) for a particle with $v \lesssim v_s$ with the synchronous phase chosen so that $0° < \phi_s < 90°$. In the example, the particle crosses cavity n at the same time as the synchronous particle. It crosses cavity $n + 1$ at a later time; therefore, it sees a higher accelerating voltage than the synchronous particle and it receives a higher velocity increment. The process continues until the particle gains enough velocity to overtake and pass the synchronous particle. In subsequent cavities, it sees reduced voltage and slows with respect to the synchronous particle. The result is that particles with parameters near those of the synchronous particle oscillate stably about ϕ_s. These particles constitute a bunch that remains synchronized with accelerating waves; the bunch is *phase stable*.

It is also possible to define a synchronous particle when ϕ_s is in the range $90° < \phi_s < 180°$. Such a case is illustrated in Figure 13.2*b*. The relative phase settings of the cavity are the same as those of Figure 13.2*a* since the voltages are the same at the crossing time of the synchronous particle. The crossing time history of a particle with $v < v_s$ is also plotted in Figure 13.2*b*. This particle arrives at cavity $n + 1$ later than the synchronous particle and sees a reduced voltage. Its arrival time at the subsequent cavity is delayed further because of its reduced velocity. After a few cavities, the particle moves into the decelerating phase of gap voltage. In its subsequent motion, the particle is completely desynchronized from the cavity voltage oscillations; its axial velocity remains approximately constant.

The conclusion is that particle distributions are not phase stable when the synchronous phase is in the range $90° < \phi_s < 18°$. The stable range of ϕ_s for particle acceleration is $0° < \phi_s < 90°$. Similar considerations apply to charged particle deceleration, an important process for microwave generation. The relative phases of the cavity oscillations can be adjusted to define a decelerating synchronous particle. It can be shown that decelerating bunches have phase stability when $0 > \phi_s > -90°$. Particle bunches are dispersed when the synchronous phase is in the range $-90° > \phi_s > -180°$.

Particles accelerated in a traveling electromagnetic wave also can have phase stability. Figure 13.3 shows the electric field as a function of position viewed in the rest frame of a slow wave. The figure illustrates the definition of phase with respect to the wave; the particle shown has $\phi = +70°$. Figure 13.3 (which shows an electric field variation in space at a constant time) should not be confused with Figure 13.1 (which shows an electric field variation in time at a constant position). Note that the phase definition of Figure 13.3 is consistent with Figure 13.1.

The synchronous particle in a traveling wave is defined by

$$dp_z/dt = eE_0\sin\phi_s. \qquad \boxed{13.5}$$

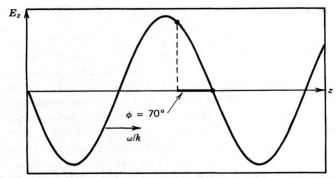

Figure 13.3 Definition of particle phase moving with a traveling wave. Longitudinal electric field plotted as function of position at specific time, position of particle with $\phi = 70°$ indicated.

In order for a synchronous particle to exist, the slow-wave structure must be designed so that the phase velocity of the wave changes to equal the particle velocity at all points in the accelerator. A slow-wave structure must be designed to accelerate particles with a specific charge-to-mass ratio (Z^*e/m_0) if the accelerating electric field E_0 is specified. The above derivations can be modified to show that a particle distribution has phase stability in a traveling wave if the synchronous phase is in the range $0° < \phi_s < 90°$.[†]

13.2 THE PHASE EQUATIONS

The phase equations describe the relative longitudinal motion of particles about the synchronous particle. The general phase equations are applicable to all resonant accelerators; we shall apply them in subsequent chapters to linear accelerators, cyclotrons, and synchrotrons.

It is most convenient to derive continuous differential equations for phase dynamics. We begin by showing that the synchronized accelerating fields in any accelerator can be written as a sum of traveling waves. Only one component (with phase velocity equal to the average particle velocity) interacts strongly with particles. Longitudinal motion is well described by including only the effects of this component. The derivation leads to a unified treatment of both discrete cavity and traveling wave accelerators.

The accelerator of Figure 13.4a has discrete resonant cavities oscillating at ω_0. The cavities drive narrow acceleration gaps. We assume that rf oscillations in the cavities have the same phase. Particles are synchronized to the oscillations by varying the distance between the gaps. The distance between gaps n

---

[†] In the convention common to discussions of linear accelerators, the stable phase range is given as $-90° \le \phi_s \le 0°$.

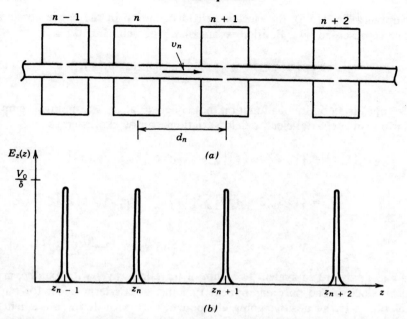

Figure 13.4 Resolution of synchronized gap voltages of standing wave accelerator into traveling wave components. (*a*) Array of uniform, phased cavities with varying intergap distances to preserve synchronization. (*b*) Variation of axial electric field as function of position at time of peak acceleration.

and $n + 1$ is

$$d_n = v_n(2\pi/\omega_0). \tag{13.6}$$

The quantity v_n is the average velocity of particles emerging from gap n. Equation (13.6) implies that the transit time of a synchronous particle between cavities is equal to one oscillation period. The distribution of longitudinal electric fields along the axis is plotted in Figure 13.4*b*. Assuming a peak voltage V_0, E_z can be approximated as a sum of δ functions,

$$E_z(z, t) \cong V_0 \sin \omega t \left[\delta(z - z_1) + \delta(z - z_2) + \cdots + \delta(z - z_n) + \cdots \right]. \tag{13.7}$$

The electric field in a region of width d_n at gap n is represented by the Fourier expansion

$$E_z(z) = (2V_0/d_n) \sum_m \cos\left[m\pi(z - z_n)/d_n \right]. \tag{13.8}$$

Applying Eq. (13.6), the electric field distribution in the entire accelerator can be represented by a Fourier expansion of the delta functions,

$$E_z(z) \cong [V_0\omega_0/\pi v_z(z)] \sum_m \cos[mz\omega_0/v_z(z)], \qquad (13.9)$$

where $v(z)$ is a continuous function that equals v_n at d_n. Assuming a temporal variation $\sin \omega_0 t$, axial electric field variations can be expressed as

$$E_z(z, t) = [V_0\omega_0/\pi v_z(z)] \sum_m \sin \omega_0 t \cos[mz\omega_0/v_z(z)]$$

$$= [V_0\omega_0/2\pi v_z(z)] \sum_m \{\sin[\omega_0 t + mz\omega_0/v_z(z)]$$

$$+ \sin[\omega_0 t - mz\omega_0/v_z(z)]\}. \quad (13.10)$$

The gap fields are equivalent to a sum of traveling waves with axially varying phase velocity. The only component that has a long-term effect for particle acceleration is the positive-going component with $m = 1$. In subsequent discussions, the other wave components are neglected.

The accelerating field of any resonant accelerator can be represented as

$$E(z, t) = E_0(z)\sin(\omega t - z\omega/v_s + \phi_s). \qquad (13.11)$$

The factor $E_0(z)$ represents a long-scale variation of electric field magnitude. In linear accelerators, ω is constant throughout the machine. In cycled circular accelerators such as the synchrocyclotron and synchrotron, ω varies slowly in time. The velocity v_s is the synchronous particle velocity as a function of position. The requirement for the existence of a synchronous particle is that v_s equals ω/k, the phase velocity of the slow wave. The motion of the synchronous particle is determined by Eq. (13.5).

Other particles shift position with respect to the synchronous particle; their phase, ϕ, varies in time. Orbits are parametrized by ϕ rather than by axial position since the phase is almost constant during the acceleration process. In this section, we shall concentrate on a nonrelativistic derivation since phase oscillations are most important in linear ion accelerators. Relativistic results are discussed in Section 13.6.

The longitudinal equation of motion for a nonsynchronous particle is

$$dp_z/dt = qE_0(z)\sin(\omega t - \omega z/v_s + \phi_s)$$

$$= qE_0(z)\sin \phi(z). \qquad (13.12)$$

The particle orbit is expanded about the synchronous particle in terms of the

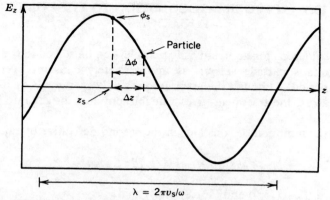

Figure 13.5 Relationship between position difference with respect to synchronous particle, $\Delta z = z - z_s$, and the phase difference, $\Delta\phi = \phi - \phi_s$.

variables

$$z = z_s + \Delta z, \tag{13.13}$$

$$v_z = \dot{z} = v_s + \Delta\dot{z}, \tag{13.14}$$

$$\phi = \phi_s + \Delta\phi, \tag{13.15}$$

where

$$\Delta\phi = \omega t - \omega z / v_s.$$

Inspection of Figure 13.5 shows that the phase difference is related to the position difference by

$$\Delta\phi / 2\pi = -\Delta z / (2\pi v_s / \omega)$$

or

$$\Delta\phi = -\omega \Delta z / v_s. \tag{13.16}$$

Equations (13.12)–(13.16) can be combined to the forms

$$\phi = \phi_s - \omega \Delta z / v_s, \qquad \boxed{13.17}$$

$$d^2 z_s / dt^2 + d^2 \Delta z / dt^2 = (q E_0 / m_0) \sin\phi. \qquad \boxed{13.18}$$

Equations (13.17) and (13.18) relate ϕ to Δz; the relationship is influenced by the parameters of the synchronous orbit. In Sections 13.3, 13.4, and 13.6, analytic approximations will allow the equations to be combined into a single phase equation.

13.3   APPROXIMATE SOLUTION TO THE PHASE EQUATIONS

The phase equations for nonrelativistic particles can be solved in the limit that the synchronous particle velocity is approximately constant over a phase oscillation period. Although the assumption is only marginally valid in linear ion accelerators, the treatment gives valuable physical insight into the phase equations.

With the assumption of constant v_s, the second derivative of Eq. (13.17) is

$$d^2\phi/dt^2 \cong -(\omega/v_s)d^2 \Delta z/dt^2. \tag{13.19}$$

Furthermore, Eqs. (13.5) and (13.18) imply that

$$d^2 \Delta z/dt^2 = (qE_0/m_0)\sin\phi - d^2 z_s/dt^2$$

$$= (qE_0/m_0)(\sin\phi - \sin\phi_s). \tag{13.20}$$

Equations (13.19) and (13.20) are combined to

$$d^2\phi/dt^2 = -(\omega qE_0/m_0 v_s)(\sin\phi - \sin\phi_s). \tag{13.21}$$

Equation (13.21) is a familiar equation in physics; it describes the behavior of a nonlinear oscillator (such as a pendulum with large displacement). Consider, first, the limit of small oscillations ($\Delta\phi/\phi_s \ll 1$). The first sine term becomes

$$\sin\phi = \sin(\phi_s + \Delta\phi) = \sin\phi_s\cos\Delta\phi + \cos\phi_s\sin\Delta\phi$$

$$\cong \sin\phi_s + \Delta\phi\cos\phi_s. \tag{13.22}$$

Equation (13.21) reduces to

$$d^2(\Delta\phi)/dt^2 \cong -(\omega qE_0\cos\phi_s/m_0 v_s)\,\Delta\phi. \tag{13.23}$$

The solution of Eq. (13.23) is

$$\Delta\phi = \Delta\phi_0\cos\omega_z t, \tag{13.24}$$

where ω_z is the phase oscillation frequency

$$\omega_z = \sqrt{qE_0\omega\cos\phi_s/m_0 v_s}. \boxed{13.25}$$

Small-amplitude oscillations are harmonic; this is true for particles confined near a stable equilibrium point of any smoothly varying force.

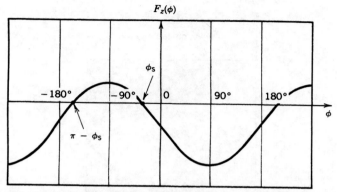

Figure 13.6 Effective force acting on nonsynchronous particles as function of particle phase. Example shows $\phi_s = 30°$.

In order to treat oscillations of arbitrary amplitude, observe that Eq. (13.21) has the form of a force equation. The effective force confines ϕ about ϕ_s. Figure 13.6 shows a plot of the effective restoring force $-(\omega q E_0 m_0 v_s)(\sin \phi - \sin \phi_s)$ as a function of ϕ. This expression is linear near ϕ_s; hence, the harmonic solution of Eq. (13.24). Particles which reach $\phi > \pi - \phi_s$ do not oscillate about ϕ_s. A first integral of Eq. (13.21) can be performed by first multiplying both sides by $2(d\phi/dt)$:

$$(d\phi/dt)^2 = (2\omega q E_0/m_0 v_s)(\cos \phi + \phi \sin \phi_s) + K. \qquad (13.26)$$

We shall determine the integration constant K for the orbit of the oscillating particle with the maximum allowed displacement from ϕ_s. The orbit bounds the distribution of confined particles. Inspection of Figure 13.6 shows that the extreme orbit must have $(d\phi/dt) = 0$ at $\phi = \pi - \phi_s$. Substituting into Eq. (13.26), the phase equation for the boundary orbit is

$$\dot{\phi}^2 = (2\omega q E_0/m_0 v_s)(\cos \phi + \cos \phi_s + (\phi + \phi_s - \pi)\sin \phi_s). \qquad (13.27)$$

The boundary particle oscillates about ϕ_s with maximum phase excursions given by the solution of

$$\cos \phi + \cos \phi_s = \sin \phi_s(\pi - \phi_s - \phi). \qquad (13.28)$$

Figure 13.7 shows the limits of phase oscillations as a function of ϕ_s. The figure illustrates the trade-off between accelerating phase and longitudinal acceptance. A synchronous phase of $0°$ gives stable confinement of particles with a broad range of phase, but there is no acceleration. Although a choice of $\phi_s = 90°$ gives the strongest acceleration, particles with the slightest variation from the synchronous particle are not captured by the accelerating wave.

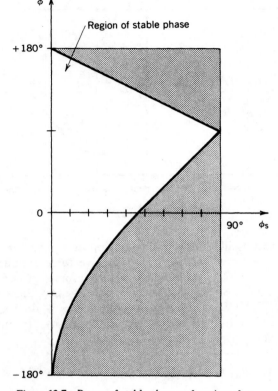

Figure 13.7 Range of stable phase as function of ϕ_s.

Figure 13.8 is a normalized longitudinal acceptance diagram in $\phi\dot{\phi}$ space as a function of ϕ_s derived from the orbit of the boundary particle. The acceptable range of longitudinal orbit parameters for trapped particles contracts as $\phi_s \rightarrow 90°$.

The principles of particle trapping in an accelerating wave and the limits of particle oscillations are well illustrated by a longitudinal potential diagram. Assume a traveling wave with an on-axis electric field given by

$$E_z(z, t) = E_0 \sin(\omega t - \omega z/v_s).$$

The electric field measured by an observer traveling at the non-relativistic velocity v_s is

$$E_z(\Delta z) = E_0 \sin(\omega \Delta z/v_s) \tag{13.29}$$

if the origin of the moving frame is coincident with the point of zero phase and Δz is the distance from the origin. Consider, first, a wave with constant velocity. In this case, $dv_s/dt = 0$ and Eq. (13.5) implies that $\phi_s = 0°$. The

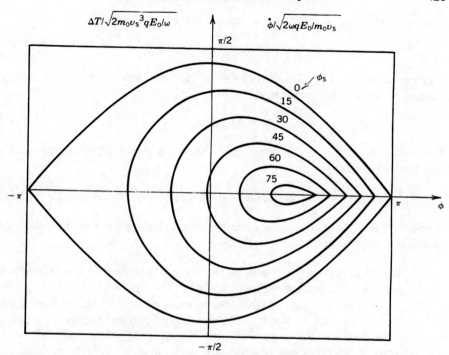

Figure 13.8 Longitudinal acceptance diagram (nonrelativistic motion); particle orbit parameters consistent with synchronized acceleration by a traveling wave as function of ϕ_s. Ordinate is either normalized rate of change of phase or normalized kinetic energy error. $\Delta T = T - T_s$ (kinetic energy), m_0 (particle rest mass), q (particle charge), v_s (synchronous particle velocity), E_0 (peak accelerating field), ω (rf angular frequency).

moving observer determines the following longitudinal variation of potential energy (Fig. 13.9a):

$$U_s(\Delta z) = -\int d\,\Delta z q E_z(\Delta z) = (qE_0 v_s/\omega)[1 - \cos(\omega\,\Delta z/v_s)], \quad (13.30)$$

corresponding to a potential well centered at $\Delta z = 0$. Particles are confined within a single half-cycle of the wave (an *rf bucket*) if they have a rest frame kinetic energy at $\Delta z = 0$ bounded by

$$m_0\,\Delta v_z^2 \leq 2qE_0 v_s/\omega = U_{e,\,max}. \quad (13.31)$$

In order to accelerate particles, the phase velocity of a wave must increase with time. An observer in the frame of an accelerating wave frame sees an addition force acting on particles. It is an inertial force in the negative z

direction. Applying Eq. (13.5), the inertial force is

$$F_i = m_0(dv_s/dt) = qE_0\sin\phi_s. \tag{13.32}$$

Integrating Eq. (13.32) from $\Delta z' = 0$ to Δz, the inertial potential energy relative to the accelerating frame is

$$U_i = qE_0\sin\phi_s\,\Delta z. \tag{13.33}$$

The total potential energy for particles in the wave frame (representing electric and inertial forces) is

$$U_t(\Delta z) = (qE_0v_s/\omega)[1 - \cos(\omega\Delta z/v_s)] + qE_0\sin\phi_s\,\Delta z. \tag{13.34}$$

Figures 13.9b–e show U_t as a function of ϕ_s. The figure has the following physical interpretations:

1. The *rf bucket* is the region of the wave where particle containment is possible. The bucket region is shaded in the figures.
2. A particle with high relative kinetic energy will spill out of the bucket (Fig. 13.9c). In the wave frame, the desynchronized particle appears to move backward with acceleration $-dv_s/dt$ (neglecting the small variations of velocity from interaction with the fields of subsequent buckets). In the stationary frame, the particle drifts forward with approximately the velocity it had at the time of desynchronization.
3. Increased wave acceleration is synonymous with larger ϕ_s. This leads to decreased bucket depth and width.
4. The acceleration limit occurs with $\phi_s = 90°$. At this value, the bucket has zero depth. A wave with higher acceleration will outrun all particles.
5. The conditions for a synchronous particle are satisfied at ϕ_s and $\pi - \phi_s$. The latter value is a point of unstable longitudinal equilibrium.

Equation (13.25) can be applied to derive a validity condition for the assumption of constant v_s over a phase oscillation. Equation (13.5) implies that the change in v_s in time Δt is

$$\Delta v_s \cong (qE_0\sin\phi_s/m_0)\,\Delta t.$$

Taking $\Delta t = 2\pi/\omega_z$, we find that

$$\Delta v_s/v_s \simeq 2\pi\sqrt{(qE_0/m_0v_s\omega)(\sin^2\phi_s/\cos\phi_s)}. \tag{13.35}$$

As an example, assume 20-MeV protons ($v_s = 6.2 \times 10^7$ m/s), $f = 800$ MHz ($\omega = 5 \times 10^9$), $E_0 = 2$ MV/m, and $\phi_s = 70°$. These parameters imply that $\Delta v_s/v_s \cong 0.25$.

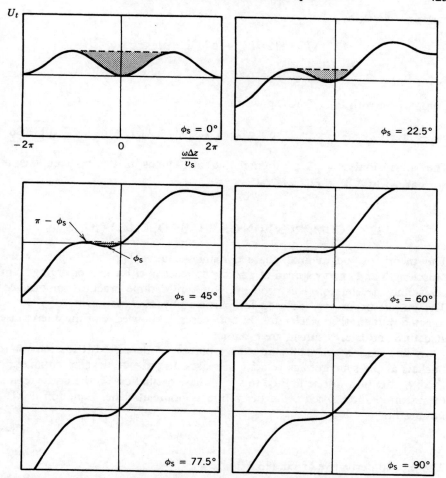

Figure 13.9 Longitudinal potential energy diagrams to illustrate phase stability. Potential energy of particle in accelerating traveling wave as function of ϕ for ϕ_s in the range of 0° to 90°.

The longitudinal acceptance diagram is used to find limits on acceptable beam parameters for injection into an rf accelerator. In a typical injector, a steady-state beam is axially bunched by an acceleration gap oscillating at ω_0. The gap gives the beam a velocity dispersion. Faster particles overtake slower particles. If parameters are chosen correctly, the injected beam is localized to the regions of rf buckets at the accelerator entrance. Bunching involves a trade-off between spatial localization and kinetic energy spread. For injection applications, it is usually more convenient to plot a longitudinal acceptance diagram in terms of phase versus kinetic energy error. The difference in the kinetic energy of a nonrelativistic particle from that of the synchronous particle is given by

$$\Delta T = \tfrac{1}{2}\Big[m_0(\dot{z}_s + \Delta\dot{z})^2 \Big] - \tfrac{1}{2}m_0\dot{z}_s^2$$

$$\cong m_0\dot{z}_s\,\Delta\dot{z}.$$

Comparison with Eq. (13.16) shows that

$$\Delta T \cong \big(mv_s^2/\omega \big)\dot{\phi}. \qquad \boxed{13.36}$$

The dimensionless plot of Figure 13.8 also holds if the vertical axis is normalized to $\Delta T / \sqrt{2m_0 v_s^3 q E_0/\omega}$.

13.4 COMPRESSION OF PHASE OSCILLATIONS

The theory of longitudinal phase dynamics can be applied to predict the pulselength and energy spread of particle bunches in rf buckets emerging from a resonant accelerator. This problem is of considerable practical importance. The output beam may be used for particle physics experiments or may be injected into another accelerator. In both cases, a knowledge of the micropulse structure and energy spread are essential.

In this section, we shall study the evolution of particle distributions in rf buckets as the synchronous velocity increases. In order to develop an analytic theory, attention will be limited to small phase oscillations in the linear region of restoring force. Again, the derivation is nonrelativistic. Equation (13.16) implies that

$$\Delta z = -\big(v_s/\omega_0 \big)\,\Delta\phi. \qquad (13.37)$$

The second derivative of Eq. (13.37) is

$$\Delta\ddot{z} = -(1/\omega_0)\big(\dot{v}_s\,\Delta\dot{\phi} + \ddot{v}_s\,\Delta\phi + v_s\,\Delta\ddot{\phi} + \dot{v}_s\,\Delta\dot{\phi} \big). \qquad (13.38)$$

Equation (13.20) can be rewritten

$$\Delta\ddot{z} \cong \big(qE_0/m_0 \big)\cos\phi_s\,\Delta\phi \qquad (13.39)$$

in the limit that $\Delta\phi \ll \phi_s$. The mathematics is further simplified by taking $\ddot{v}_s = 0$; the wave has constant acceleration. Setting the right-hand sides of Eqs. (13.38) and (13.39) equal gives

$$\Delta\ddot{\phi} + 2\dot{v}_s\,\Delta\dot{\phi}/v_s + \big(\omega_0 q E_0\cos\phi_s/m_0 v_s \big)\,\Delta\phi = 0. \qquad (13.40)$$

By the definition of the synchronous particle,

$$\dot{v}_s = qE_0\sin\phi_s/m_0. \qquad (13.41)$$

This implies that

$$v_s = qE_0 \sin \phi_s t / m_0 \qquad (13.42)$$

if the origin of the time axis corresponds to $v_s = 0$. The quantity t is the duration of time that the particles are in the accelerator. Substituting Eqs. (13.41) and (13.42) into Eq. (13.40) gives

$$\Delta\ddot{\phi} + (2/t)\,\Delta\dot{\phi} + (\dot{\omega}_0/\tan\phi_s t)\,\Delta\phi = 0. \qquad (13.43)$$

Making the substitution $\Psi = \Delta\phi t$, Eq. (13.43) can be written

$$\ddot{\Psi} + \omega_z^2\Psi = 0, \qquad (13.44)$$

where

$$\omega_z = \sqrt{(\omega_0/\tan\phi_s t)}. \qquad (13.45)$$

Equation (13.45) implies that Eq. (13.43) has an oscillatory acceleration solution for $0 < \phi_s < \frac{1}{2}\pi$. The condition for phase stability remains the same.

Equation (13.44) has the same form as Eq. (11-20), which describes the compression of betatron oscillations. Equation (13.44) is a harmonic oscillator equation with a slowly varying frequency. We apply the approximation that changes in the synchronous particle velocity take place slowly compared to the phase oscillation frequency, or $\omega_z T \gg 1$, where T is the time scale for acceleration. The linear phase oscillation frequency decreases as

$$\omega_z \sim 1/\sqrt{t} \sim 1/\sqrt{v_s}, \qquad (13.46)$$

where Eq. (13.42) has been used to relate t and v_s. The quantity Ψ is the analogy of the amplitude of betatron oscillations. Equation (11-29) implies that the product $\Psi^2\omega_z$ is conserved in a reversible compression. Thus, the quantity Ψ increases as

$$\Psi \sim t^{1/4} \sim v_s^{1/4}. \qquad (13.47)$$

Furthermore,

$$\Delta\phi \sim \Psi/t \sim 1/t^{3/4} \sim 1/v_s^{3/4}. \qquad \boxed{13.48}$$

In summary, the following changes take place in the distribution of particles in an rf bucket:

1. Particles injected with velocity v_i and micropulsewidth Δt_i emerge from a constant-frequency accelerator with velocity v_f and micropulsewidth

$$\Delta t_f = \Delta t_i (v_i/v_f)^{3/4}. \qquad \boxed{13.49}$$

Although the spatial extent of the beam bunch is larger, the increase in velocity results in a reduced micropulsewidth.

2. Taking the derivative on the time scale of a phase oscillation, Eq. (13.48) implies that

$$\Delta\dot{\phi} \sim \omega_z/t^{3/4} \sim 1/t^{5/4} \sim 1/v_s^{5/4}.$$

Combining the above equation with Eq. (13.36) implies that the absolute kinetic energy spread of particles in an rf bucket increases as

$$\Delta T_f = \Delta T_i (v_f/v_i)^{3/4}. \qquad \boxed{13.50}$$

Linear ion accelerators produce beams with a fairly large kinetic energy spread. If an application calls for a small energy spread, the particle pulses must be debunched after exiting the accelerator. The relative energy spread scales as

$$\Delta T_f/T_f = (\Delta T_i/T_i)(v_i/v_f)^{5/4}. \qquad (13.51)$$

13.5 LONGITUDINAL DYNAMICS OF IONS IN A LINEAR INDUCTION ACCELERATOR

Although the linear induction accelerator is not a resonant accelerator, longitudinal motions of ions in induction linacs are discussed in this chapter because of the similarities to phase oscillations in rf linear ion accelerators. The treatment is limited to nonrelativistic particles; electron accelerators are discussed in the next section.

The main problem associated with longitudinal dynamics in an induction accelerator is assuring that ions are axially confined so that they cross acceleration gaps during the applied voltage pulses. A secondary concern is maintenance of a good current profile; this is important when beam loading of the pulse modulator is significant. In the following treatment, only single-particle effects are addressed. Cavity voltage waveforms are specified and beam loading is neglected.

To begin, consider a beam pulse of duration Δt_p moving through a gap with constant voltage V_0 (Fig. 13.10). The incoming pulse has axial length l, longitudinal velocity v_s, and velocity spread Δv. The axial length is $l = v_s \Delta t_p$. Assume that the gap is narrow and the beam velocity spread is small ($\Delta v_z/v_s \ll 1$). The beam emerges with an increased velocity v_s'. Every particle entering the gap leaves it immediately, so that the pulselength is not changed. As shown in Figure 13.10, the beam length increases to

$$l' = l(v_s'/v_s). \qquad \boxed{13.52}$$

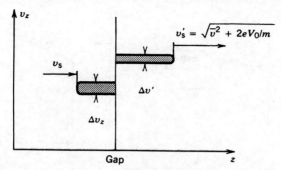

Figure 13.10 Modification of longitudinal beam distribution passing through constant-voltage acceleration gap.

The change in the velocity spread of the beam can be determined from conservation of energy for the highest-energy particles:

$$\tfrac{1}{2}\Big[m_0\big(v'_s + \tfrac{1}{2}\Delta v'\big)^2\Big] = \tfrac{1}{2}\Big[m_0\big(v_s + \tfrac{1}{2}\Delta v\big)^2\Big] + qV_0. \qquad (13.53)$$

Keeping only the first-order terms of Eq. (13.53) and noting that $\tfrac{1}{2}m_0v'^2_s = m_0v^2_s + eV_0$, we find that

$$\Delta v' = \Delta v\big(v_s/v'_s\big). \qquad \boxed{13.54}$$

The longitudinal velocity spread decreases with acceleration. As in any reversible process, the area occupied by the particle distribution in phase space (proportional to $l\,\Delta v$) remains constant.

A flat voltage pulse gives no longitudinal confinement. The longitudinal velocity spread causes the beam to expand, as shown in Figure 13.11*a*. The expansion can be countered by adding a voltage ramp to the accelerating waveform (Fig. 13.11*b*). The accelerator is adjusted so that the synchronous particle in the middle of the beam bunch crosses the gap when the voltage equals V_0. Particles lagging behind the synchronous particle experience a higher gap voltage and gain a larger velocity increment while advanced particles are retarded. This not only confines particles within the bunch but also provides stability for the entire beam pulse. For example, if the voltage in an upstream cavity is low, the centroid of the bunch arrives late in subsequent cavities. With ramped voltage waveforms, the bunch receives extra acceleration and oscillates about the synchronous particle position.

A simple model for beam confinement in an induction linear accelerator can be developed in the limit that (1) the beam crosses many gaps during a phase oscillation and (2) the change in v_s is small during a phase oscillation. Let the quantity Δz be the distance of a particle from the synchronous particle

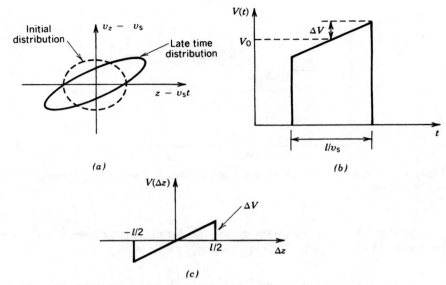

Figure 13.11 Longitudinal focusing of nonrelativistic beams in linear induction accelerators. (*a*) Spatial expansion of beam with spread in longitudinal velocity. (*b*) Accelerating voltage waveform with linear ramp. (*c*) Alternating-current portion of ramped accelerating voltage.

position, or

$$\Delta z = z - v_s t. \tag{13.55}$$

The quantity Δv is the width of the longitudinal velocity distribution at $\Delta z = 0$.

   The dc part of the cavity voltage waveform can be neglected because of the assumption of constant v_s over time scales of interest. The time-varying part of the gap voltage has the waveform of Figure 13.11*c*. The gap voltage that accelerates a particle depends on the position of the particle relative to the synchronous particle:

$$\Delta V = -(2\,\Delta z/l)\,\Delta V_0, \tag{13.56}$$

where ΔV_0 is defined in Figure 13.11*c*. The velocity changes by the amount

$$\Delta v \cong (q/m_0 v_s)(-2\,\Delta z/l)\,\Delta V_0 \tag{13.57}$$

crossing a gap. Equation (13.57) holds in the limit that $\Delta v \ll v_s$. Particles cross $N = v_s\,\Delta t/D$ gaps in time interval Δt if the gaps have uniform spacing D. Therefore, multiplying Eq. (13.57) by N gives the total change of Δv in Δt. The longitudinal equation of motion for ion in an induction linear accelerator is

$$d\,\Delta v/dt = d^2\,\Delta z/dt^2 \cong -(2q\,\Delta V_0/m_0 lD)\,\Delta z. \tag{13.58}$$

Equation (13.58) has solution

$$\Delta z = \Delta z_0 \sin \omega_z t, \tag{13.59}$$

where

$$\omega_z = (2q\,\Delta V_0/m_0 lD)^{1/2}. \tag{13.60}$$

The maximum value of Δv is determined from Eq. (13.59) by substituting $\Delta z_0 = \frac{1}{2}l$:

$$\Delta v \le (q\,\Delta V_0/2m_0)^{1/2}(l/D)^{1/2}. \tag{13.61}$$

Equation (13.61) infers the allowed spread in kinetic energy,

$$\Delta T/T \cong \left[(2q\,\Delta V_0/m_0)^{1/2}/v_s\right](l/D)^{1/2}. \tag{13.62}$$

The longitudinal dynamics of ions in an induction linear accelerator is almost identical to the small $\Delta\phi$ treatment of phase oscillations in an rf accelerator. The time-varying gap electric field in Figure 13.11 can be viewed as an approximation to a sine function expanded about the synchronous particle position. The main difference between the two types of accelerators is in the variation of phase oscillation frequency and velocity spread during beam acceleration. The electric field ramp in an rf linear accelerator is constrained by the condition of constant frequency. The wavelength increases as v_s, and hence the longitudinal confining electric field in the beam rest frame decreases as $1/v_s$. This accounts for the decrease in the phase oscillation frequency of Eq. (13.46). In the induction accelerator, there is the latitude to adjust the longitudinal confinement gradient along the accelerator.

Equation (13.61) implies that the longitudinal acceptance is increased by higher voltage ramp (ΔV_0) and long beam length compared to the distance between gaps (l/D). If v_s varies slowly, Eq. (13.61) implies that the magnitude of ΔV_0 must be reduced along the length of the accelerator to maintain a constant beam pulselength. Constant pulselength implies that $l \sim v_s$ and $\Delta v \sim 1/v_s$; therefore, ΔV_0 must be reduced proportional to $1/v_s^3$ for constant D. On the other hand, if ΔV_0 is constant throughout the accelerator, l and Δv are constant. This means that the pulselength, or the time for the beam bunch to pass through the gap, decreases as $1/v_s$.

The longitudinal shape of the beam bunch is not important when beam loading is negligible. Particles with a randomized velocity distribution acted on by linear forces normally have a bell-shaped density distribution; in this case,

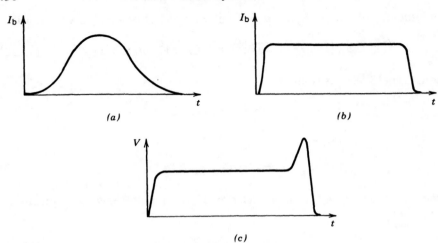

Figure 13.12   Longitudinal dynamics of nonrelativistic beams in linear induction accelerators. (*a*) Typical current profile of beam contained by accelerating voltage with linear ramp. (*b*) Desirable current profile for load matching to pulse modulator. (*c*) Voltage waveform with nonlinear variations to contain beam with flat current profile.

the beam current profile associated with a pulse looks like that of Figure 13.12*a*. When beam loading is significant, it is preferable to have a flat current pulse, like that of Figure 13.12*b*. This can be accomplished by nonlinear longitudinal confinement forces; a confining cavity voltage waveform consistent with the flat current profile is illustrated in Figure 13.12*c*. The design of circuits to generate nonlinear waveforms for beam confinement under varying load conditions is one of the major unanswered questions concerning the feasibility of linear induction ion accelerators.

13.6   PHASE DYNAMICS OF RELATIVISTIC PARTICLES

Straightforward analytic solutions for longitudinal dynamics in rf linear accelerators are possible for highly relativistic particles. The basis of the approach is to take the phase velocity of the accelerating wave exactly equal to c, and to seek solutions in which electrons are captured in the accelerating phase of the wave. In this situation, there is no synchronous phase since the particles (with $v_z < c$) continually move to regions of higher phase in the accelerating wave. We shall first derive the mathematics of electron capture and then consider the physical implications.

Assume electrons are injected into a traveling wave at $z = 0$. The quantity ϕ_0 is the particle phase relative to the wave at the injection point. Equation (13.17) can be rewritten

$$\phi = \phi_0 - \omega \Delta z / c, \tag{13.63}$$

where

$$\Delta z = z - ct. \tag{13.64}$$

The quantity z is the position of the electron after time t, and ct is the distance that the wave travels. Thus, Δz is the distance the electron falls back in the wave during acceleration. Clearly, Δz must be less than $\frac{1}{2}\lambda$ or else the electron will enter the decelerating phase of the wave. Equation (13.63) can also be written

$$\phi = \phi_0 + \omega(t - z/c). \tag{13.65}$$

The derivative of Eq. (13.65) is

$$\dot{\phi} = \omega(1 - \dot{z}/c) = \omega(1 - \beta). \tag{13.66}$$

The relativistic form of Eq. (13.12) for electrons is

$$dp_z/dt = d(\gamma m_e \beta c)/dt = d\left[m_e \beta c/\sqrt{(1 - \beta^2)}\right]/dt = eE_0\sin\phi. \tag{13.67}$$

Equations (13.66) and (13.67) are two equations in the two unknowns ϕ and β. They can be solved by making the substitution

$$\beta = \cos\alpha. \tag{13.68}$$

Equation (13.67) becomes

$$d(\cos\alpha/\sin\alpha)/dt = (eE_0/m_e c)\sin\phi$$

$$= -\dot{\alpha}(1 + \cos^2\alpha/\sin^2\alpha). \tag{13.69}$$

Manipulation of the trigonometric functions yields

$$\dot{\alpha} = -(eE_0/m_e c)\sin\phi\,\sin^2\alpha. \tag{13.70}$$

Equation (13.66) is rewritten

$$\dot{\phi} = (d\phi/d\alpha)\dot{\alpha} = \omega(1 - \cos\alpha). \tag{13.71}$$

Substituting Eq. (13.71) into Eq. (13.70) gives the desired equation,

$$-\sin\phi\,d\phi = (m_e c\omega/eE_0)(1 - \cos\alpha)\,d\alpha/\sin^2\alpha. \tag{13.72}$$

Integrating Eq. (13.72) with the lower limit given by the injection parameters, we find that

$$\cos\phi - \cos\phi_0 = (m_e c\omega/eE_0)\left[\tan(\tfrac{1}{2}\alpha) - \tan(\tfrac{1}{2}\alpha_0)\right]. \tag{13.73}$$

Noting that

$$\tan\left(\tfrac{1}{2}\alpha\right) = \sqrt{(1 - \cos\alpha)/(1 + \cos\alpha)} = \sqrt{(1 - \beta)/(1 + \beta)},$$

the electron phase relative to the accelerating wave is given in terms of β by

$$\cos\phi = \cos\phi_0 + \frac{m_e c\omega}{eE_0}\left\{\left[\frac{1 - \beta}{1 + \beta}\right]^{1/2} - \left[\frac{1 - \beta_0}{1 + \beta_0}\right]^{1/2}\right\}. \quad \boxed{13.74}$$

The solution is not oscillatory; ϕ increases monotonically as β changes from β_0 to 1 and the particle lags behind the wave. Acceleration takes place as long as the particle phase is less than $\phi = 180°$. Note that since there are no phase oscillations, there is no reason to restrict the particle phase to $\phi < 90°$. The important point to realize is that if acceleration takes place fast enough, electrons can be trapped in a single rf bucket and accelerated to arbitrarily high energy. With high enough E_0, the electrons never reach phase $\phi = \pi$. This process is called *electron capture*. In this regime, time dilation dominates so that the particle asymptotically approaches a constant phase, ϕ.

The condition for electron capture can be derived from Eq. (13.74) by assuming that the final particle β is close to unity:

$$\cos\phi_0 - \cos\phi = (m_e c\omega/eE_0)\sqrt{(1 - \beta_0)/(1 + \beta_0)} \leq 1. \quad (13.75)$$

The limit proceeds from the fact that ϕ should approach 90° for optimum acceleration and ϕ_0 must be greater than 0°. Equation (13.75) implies that

$$E_0 > (m_e c\omega/e)\sqrt{(1 - \beta_0)/(1 + \beta_0)}. \quad \boxed{13.76}$$

As an example, consider injection of 1-MeV electrons into a traveling wave accelerator based on a 2-GHz ($\omega = 1.26 \times 10^{10}$ s^{-1}) iris-loaded waveguide. The quantity β_0 equals 0.9411. According to Eq. (13.76), the peak electric field of the wave must exceed 3.7 MV/m. This is a high but feasible value.

The dynamics of relativistic electron capture is illustrated in Figure 13.13. Figure 13.13a shows the relative position of electrons in the accelerating wave as a function of energy in a 1-GeV accelerator. Figure 13.13b graphs energy versus phase. Note that most of the acceleration takes place near the final asymptotic value of phase, ϕ_f. The output beam energy is

$$T \cong E_0 \sin\phi_f L, \quad (13.77)$$

where L is the total length of the accelerator. A choice of $\phi_f = 90°$ gives the highest accelerating gradient.

In the beam rest frame, the accelerator appears to be moving close to the speed of light. The length of the accelerator is shortened by Lorentz contrac-

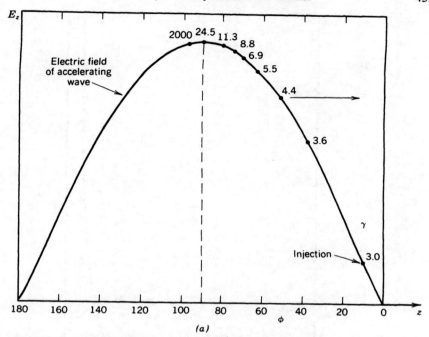

Figure 13.13 Electron acceleration in high-gradient linear accelerator, $f = 3$ GHz, $E_0 = 5$ MV/m. (*a*) Electron phase and relative accelerating field versus γ. (*b*) Kinetic energy versus ϕ.

tion. It is informative to calculate the apparent accelerator length in the beam frame. Let dz be an element of axial length in the accelerator frame and dz' be the length of the element measured in the beam frame. According to Equation (2-24), the length elements are related by

$$dz' = dz/\gamma,$$

or

$$dz' = \left(m_e c^2/E\right) dz, \tag{13.78}$$

where E is the total energy of the electrons. We have seen that the accelerating gradient is almost constant for electrons captured in the rf waveform. The total energy is approximated by

$$E \cong m_e c^2 + T_0 + eE_0 \sin \phi_f z. \tag{13.79}$$

Combining Eqs. (13.78) and (13.79),

$$dz' = dz/\left(1 + T_0/m_e c^2 + eE_0 \sin \phi_f z/m_e c^2\right). \tag{13.80}$$

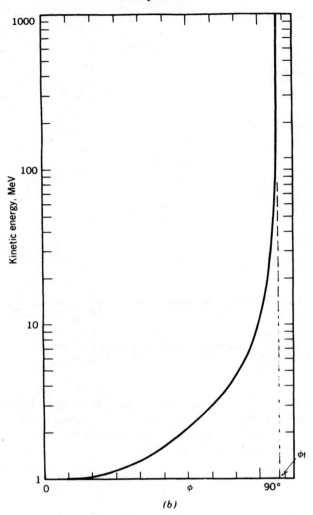

Figure 13.13 (*Continued*).

Integrating Eq. (13.80) from $z = 0$ to $z = L$ gives

$$L' = \left(m_e c^2 / e E_0 \sin \phi_f \right) \ln \left[\frac{1 + T_0/m_e c^2 + e E_0 \sin \phi_f L / m_e c^2}{1 + T_0/m_e c^2} \right]$$

$$= L \left(m_e c^2 / T_f \right) \ln \left(E_f / E_i \right). \qquad \boxed{13.81}$$

As an example, consider electron motion with the acceleration history illustrated in Figure 13.13. The peak accelerating field is 5 MV/m, $f = 3$ GHz, and electrons are injected with kinetic energy 1 MeV. The final phase is near 90° for particles with injection phase angles near 0°. These particles are accelerated mainly at the peak field; the accelerator must have a length $L \simeq 200$ m to generate a 1-GeV beam. Substituting in Eq. (13.81), the apparent accelerator length in the beam frame is only 0.7 m.

Equation (13.81) can also be applied to induction linear electron accelerators if the quantity $E_0 \sin \phi_f$ is replaced by the average accelerating gradient of the machine. Consider, for instance, an induction accelerator with a 50-MeV output beam energy. The injection energy is usually high in such machines; 2.5 MeV is typical. Gradients are lower than rf accelerators because of limits on isolation core packing and breakdown on vacuum insulators in the cavity. An average gradient of 1 MV/m implies a total length $L = 50$ m. Substituting into Eq. (13.81), the apparent length is $L' = 1.5$ m.

The short effective length explains the absence of phase oscillations in relativistic rf linacs. As viewed in the beam frame, the accelerator is passed before there is time for any relative longitudinal motion. The short effective length has an important implication for the design of low-current linear accelerators. The accelerator appears so short that it is unnecessary to add transverse focusing elements for beam confinement; the beam is simply aimed straight through. Radial defocusing of particles (see Chapter 14) is reduced greatly at high γ.

Induction linear electron accelerators are used for high-current pulsed beams. Transverse focusing is required in these machines to prevent space charge expansion of the beam and to reduce the severity of resonant transverse instabilities. Nonetheless, space charge effects and the growth of instabilities are reduced when electrons have high γ. In particular, the radial force from beam space charge decreases as $1/\gamma^2$. This is the main reason that the injector of an electron linear accelerator is designed to operate at high voltage.

We have seen in Section 13.5 that cavity voltage waveforms have a strong effect on the current pulse shape in an induction accelerator for ions. To demonstrate that this is not true for relativistic particles, consider an electron beam in an accelerator with average energy $\gamma m_e c^2$. Assume that there is a spread of energy parametrized by $\pm \Delta \gamma$ resulting from variations in cavity voltage waveforms. We shall demonstrate the effect of energy spread by determining how far a beam pulse travels before there is a significant increase in pulselength. The beam bunch length in the accelerator frame is denoted as L. The total spread in axial velocity is

$$\Delta v_z \cong c\beta^+ - c\beta^-, \tag{13.82}$$

where

$$\beta^+ = \left[1 - 1/(\gamma + \Delta\gamma)^2\right]^{1/2}, \tag{13.83}$$

and

$$\beta^- = \left[1 - 1/(\gamma - \Delta\gamma)^2\right]^{1/2}. \tag{13.84}$$

Applying the binomial theorem, the axial velocity spread is

$$\Delta v_z = 2c(\Delta\gamma/\gamma^3). \qquad \boxed{13.85}$$

Let Δt be the time it takes for the beam to double its length:

$$\Delta t \cong (L/2)/(\Delta v_z/2) = (L/c)(\gamma^3/\Delta\gamma). \qquad (13.86)$$

The beam travels a distance $D = c\,\Delta t$ during this time interval. Substituting from Eq. (13.85), we find that

$$D \cong \gamma^2 L/(\Delta\gamma/\gamma). \qquad (13.87)$$

As an example, consider a 50-ns pulse of 10-MeV electrons with large energy spread, $\Delta\gamma/\gamma = 0.5$. The beam pulse is 15 m long. Equation (13.87) implies that the distance traveled during expansion is 12 km, much longer than any existing or proposed induction accelerator. The implication is that electron beam pulses can be propagated through and synchronized with an induction accelerator even with very poor voltage waveforms. On the other hand, voltage shaping is important if the output beam must have a small energy spread.

14

Radio-Frequency Linear Accelerators

Resonant linear accelerators are usually single-pass machines. Charged particles traverse each section only once; therefore, the kinetic energy of the beam is limited by the length of the accelerator. Strong accelerating electric fields are desirable to achieve the maximum kinetic energy in the shortest length. Although linear accelerators cannot achieve beam output energy as high as circular accelerators, the following advantages dictate their use in a variety of applications: (1) the open geometry makes it easier to inject and extract beams; (2) high-flux beams can be transported because of the increased options for beam handling and high-power rf structures; and (3) the duty cycle is high. The duty cycle is defined as the fraction of time that the operating machine produces beam output.

The operation of resonant linear accelerators is based on electromagnetic oscillations in tuned structures. The structures support a traveling wave component with phase velocity close to the velocity of accelerated particles. The technology for generating the waves and the interactions between waves and particles were described in Chapters 12 and 13. Although the term radio frequency (rf) is usually applied to resonant accelerators, it is somewhat misleading. Although some resonant linear accelerators have been constructed with very large or inductive structures, most present accelerators use resonant cavities or waveguides with dimensions less than 1 m to contain electromagnetic oscillations; they operate in the microwave regime (> 300 MHz).

Linear accelerators are used to generate singly charged light ion beams in the range of 10 to 300 MeV or multiply charged heavy ions up to 4 GeV (17 MeV/nucleon). These accelerators have direct applications such as radiation therapy, nuclear research, production of short-lived isotopes, meson production, materials testing, nuclear fuel breeding, and defense technology. Ion linear accelerators are often used as injectors to form high-energy input beams for large circular accelerators. The recent development of the radio-frequency quadrupole (RFQ), which is effective for low-energy ions, suggests new applications in the 1–10 MeV range, such as high-energy ion implantation in materials. Linear accelerators for electrons are important tools for high-energy physics research since they circumvent the problems of synchrotron radiation that limit beam energy in circular accelerators. Electron linear accelerators are also used as injectors for circular accelerators and storage rings. Applications for high-energy electrons include the generation of synchrotron radiation for materials research and photon beam generation through the free electron laser process.

Linear accelerators for electrons differ greatly in both physical properties and technological realization from ion accelerators. The contrasts arise partly from dissimilar application requirements and partly from the physical properties of the particles. Ions are invariably nonrelativistic; therefore, their velocity changes significantly during acceleration. Resonant linear accelerators for ions are complex machines, often consisting of three or four different types of acceleration units. In contrast, high-gradient electron accelerators for particle physics research have a uniform structure throughout their length. These devices are described in Section 14.1. Electrons are relativistic immediately after injection and have constant velocity through the accelerator. Linear electron accelerators utilize electron capture by strong electric fields of a wave traveling at the velocity of light. Because of the large power dissipation, the machines are operated in a pulsed mode with low-duty cycle. After a description of the general properties of the accelerators, Section 14.1 discusses electron injection, beam breakup instabilities, the design of iris-loaded waveguides with $\omega/k = c$, optimization of power distribution for maximum kinetic energy, and the concept of shunt impedance.

Sections 14.2–14.4 review properties of high-energy linear ion accelerators. The four common configurations of rf ion accelerators are discussed in Sections 14.2 and 14.3: the Wideröe accelerator, the independently phased cavity array, the drift tube linac, and the coupled cavity array. Starting from the basic Wideröe geometry, the rationale for surrounding acceleration gaps with resonant structures is discussed. The configuration of the drift tube linac is derived qualitatively by considering an evolutionary sequence from the Wideröe device. The principles of coupled cavity oscillations are discussed in Section 14.3. Although a coupled cavity array is more difficult to fabricate than a drift tube linac section, the configuration has a number of benefits for high-flux ion beams when operated in a particular mode (the $\frac{1}{2}\pi$ mode).

Coupled cavities have high accelerating gradient, good frequency stability, and strong energy coupling. The latter property is essential for stable electromagnetic oscillations in the presence of significant beam loading. Examples of high-energy ion accelerators are included to illustrate strategies for combining the different types of acceleration units into a high-energy system.

Some factors affecting ion transport in rf linacs are discussed in Section 14.4. Included are the transit-time factor, gap coefficients, and radial defocusing by rf fields. The transit-time factor is important when the time for a particle to cross an acceleration gap is comparable to half the rf period. In this case, the peak energy gain (reflecting the integral of charge times electric field during the transit) is less than the product of charge and peak gap voltage. The transit-time derating factor must be included to determine the synchronous particle orbit. The gap coefficient refers to radial variations of longitudinal electric field. The degree of variation depends on the gap geometry and rf frequency. The spatial dependence of E_z leads to increased energy spread in the output beam or reduced longitudinal acceptance. Section 14.4 concludes with a discussion of the effects of the radial fields of a slow traveling wave on beam containment. The existence and nature of radial fields are derived by a transformation to the rest frame of the wave in which it is described as an electrostatic field pattern. The result is that orbits in cylindrically symmetric rf linacs are radially unstable if the particles are in a phase region of longitudinal stability. Ion linacs must therefore incorporate additional focusing elements (such as an FD quadrupole array) to assure containment of the beam.

Problems of vacuum breakdown in high-gradient rf accelerators are discussed in Section 14.5. The main difference from the discussion of Section 9.5 is the possibility for geometric growth of the number of secondary electrons emitted from metal surfaces when the electron motion is in synchronism with the oscillating electric fields. This process is called multipactoring. Electron multipactoring is sometimes a significant problem in starting up rf cavities; ultimate limits on accelerating gradient in rf accelerators may be set by ion multipactoring.

Section 14.6 describes the RFQ, a recently developed configuration. The RFQ differs almost completely from other rf linac structures. The fields are azimuthally asymmetric and the main mode of excitation of the resonant structure is a TE mode rather than a TM mode. The RFQ has significant advantages for the acceleration of high-flux ion beams in the difficult low-energy regime (0.1–5 MeV). The structure utilizes purely electrostatic focusing from rf fields to achieve simultaneous average transverse and longitudinal containment. The electrode geometries in the device can be fabricated to generate precise field variations over small-scale lengths. This gives the RFQ the capability to perform beam bunching within the accelerator, eliminating the need for a separate buncher and beam transport system. At first glance, the RFQ appears to be difficult to describe theoretically. In reality, the problem is tractable if we divide it into parts and apply material from previous chapters.

The properties of longitudinally uniform RFQs, such as the interdependence of accelerating gradient and transverse acceptance and the design of shaped electrodes, can be derived with little mathematics.

Section 14.7 reviews the racetrack microtron, an accelerator with the ability to produce continuous high-energy electron beams. The racetrack microtron is a hybrid between linear and circular accelerators; it is best classified as a recirculating resonant linear accelerator. The machine consists of a short linac (with a traveling wave component with $\omega/k = c$) and two regions of uniform magnetic field. The magnetic fields direct electrons back to the entrance of the accelerator in synchronism with the rf oscillations. Energy groups of electrons follow separate orbits which require individual focusing and orbit correction elements. Synchrotron radiation limits the beam kinetic energy of microtrons to less than 1 GeV. Beam breakup instabilities are a major problem in microtrons; therefore, the output beam current is low ($< 100 \ \mu\text{A}$). Nonetheless, the high-duty cycle of microtrons means that the time-averaged electron flux is much greater than that from conventional electron linacs.

14.1 ELECTRON LINEAR ACCELERATORS

Radio-frequency linear accelerators are used to generate high-energy electron beams in the range of 2 to 20 GeV. Circular electron accelerators cannot reach high output kinetic energy because of the limits imposed by synchrotron radiation. Linear accelerators for electrons are quite different from ion accelerators. They are high-gradient, traveling wave structures used primarily for particle physics research. Accelerating gradient is the main figure of merit; consequently, the efficiency and duty cycle of electron linacs are low. Other accelerator configurations are used when a high time-averaged flux of electrons at moderate energy is required. One alternative, the racetrack microtron, is described in Section 14.7.

A. General Properties

A block diagram of an electron linac is illustrated in Figure 14.1. The accelerators typically consist of a sequence of identical, iris-loaded slow-wave structures that support traveling waves. The waveguides are driven by high-power klystron microwave amplifiers. The axial electric fields of the waves are high, typically on the order of 8 MV/m. Parameters of the 20-GeV accelerator at the Stanford Linear Accelerator Center are listed in Table 14.1. The accelerator is over 3 km in length; the open aperture for beam transport is only 2 cm in diameter. The successful transport of the beam through such a long, narrow tube is a consequence of the relativistic contraction of the apparent length of the accelerator (Section 13.6). A cross section of the accelerator is illustrated in Figure 14.2. A scale drawing of the rf power distribution system is shown in Figure 14.3.

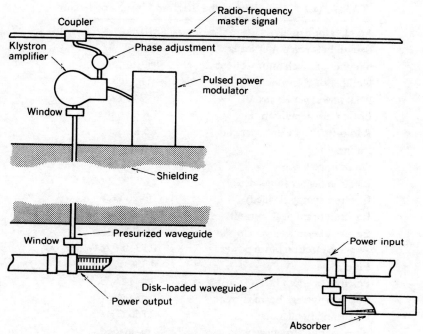

Figure 14.1 High-energy linear electron accelerator.

The features of high-energy electron linear accelerators are determined by the following considerations.

1. Two factors motivate the use of strong accelerating electric fields: (a) high gradient facilitates electron capture (Section 13.6) and (b) the accelerator length for a given final beam energy is minimized.

2. Resistive losses per unit length are large in a high-gradient accelerator since power dissipation in the waveguide walls scales as E_z^2. Dissipation is typically greater than 1 MW/m. Electron linacs must be operated on an intermittent duty cycle with a beam pulselength of a few microseconds.

3. An iris-loaded waveguide with relatively large aperture can support slow waves with $\omega/k = c$. Conduction of rf energy along the waveguide is effective; nonetheless, the waves are attenuated because of the high losses. There is little to be gained by reflecting the traveling waves to produce a standing wave pattern. In practice, the energy of the attenuated wave is extracted from the waveguide at the end of an accelerating section and deposited in an external load. This reduces heating of the waveguides.

4. A pulsed electron beam is injected after the waveguides are filled with rf energy. The beam pulse length is limited by the accelerator duty cycle and by

TABLE 14.1   Parameters of the Stanford Linear Accelerator

Accelerator length	3100 m
Length between power feeds	3.1 m
Number of accelerator sections	960
Number of klystrons	245
Peak power per klystron	6–24 MW
Beam pulse repetition rate	1–360 pulses/s
Radio-frequency pulse length	2.5 μs
Filling time	0.83 μs
Shunt impedance	53 MΩ/m
Electron energy (unloaded)	11.1–22.2 GeV
Electron energy (loaded)	10–20 GeV
Electron beam peak current	25–50 mA
Electron beam average current	15–30 μA
Average electron beam power	0.15–0.6 MW
Efficiency	4.3%
Positron energy	7.4–14.8 GeV
Positron average beam current	0.45 μA
Operating frequency	2.856 GHz
Accelerating structure	Iris-loaded waveguide
Waveguide outer diameter	10.5 cm
Aperture diameter	1.9 cm

the growth of beam breakup instabilities. Relatively high currents (≤ 0.1 A) are injected to maximize the number of electrons available for experiments.

5.   The feasibility of electron linacs is a consequence of technological advances in high-power rf amplifiers. Klystrons can generate short pulses of rf power in the 30-MW range with good frequency stability. High-power klystrons are driven by pulsed power modulators such as the PFN discussed in Section 9.12.

The waveguides of the 2.5-GeV accelerator at the National Laboratory for High Energy Physics (KEK), Tsukuba, Japan, have a diameter of 0.1 m and an operating frequency of 2.856 GHz. The choice of frequency results from the availability of high-power klystrons from the development of the SLAC accelerator. An acceleration unit consists of a high-power coupler, a series of four iris-loaded waveguides, a decoupler, and a load. The individual waveguides are 2 m long. The inner radius of the irises has a linear taper of 75 μm per cell along the length of the guide; this maintains an approximately

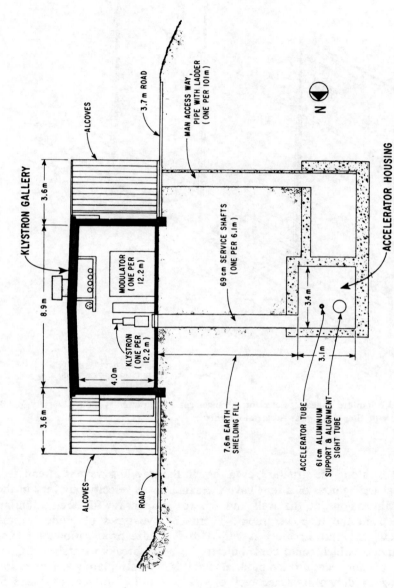

Figure 14.2 Stanford Linear Accelerator; cross section. (Courtesy W. B. Hermmannsfeldt, Stanford Linear Accelerator Center.)

443

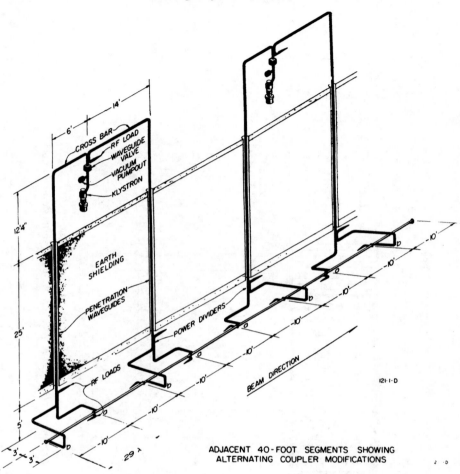

Figure 14.3 Stanford Linear Accelerator; arrangement of rf power system. (Courtesy W. B. Herrmannsfeldt, Stanford Linear Accelerator Center.)

constant E_z along the structure, even though the traveling wave is attenuated. Individual waveguides of a unit have the same phase velocity but vary in the relative dimensions of the wall and iris to compensate for their differing distance from the rf power input. There are five types of guides in the accelerator; the unit structure is varied to minimize propagation of beam-excited modes which could contribute to the beam breakup instability. Construction of the guides utilized modern methods of electroplating and precision machining. A dimensional accuracy of ± 2 μm and a surface roughness of 200 Å was achieved, making postfabrication tuning unnecessary.

B. Injection

The pulsed electron injector of a high-power electron linear accelerator is designed for high voltage (> 200 kV) to facilitate electron capture. The beam pulselength may vary from a few nanoseconds to 1 μs depending on the research application. The high-current beam must be aimed with a precision of a few milliradians to prevent beam excitation of undesired rf modes in the accelerator. Before entering the accelerator, the beam is compressed into micropulses by a buncher. A buncher consists of an rf cavity or a short section of iris-loaded waveguide operating at the same frequency as the main accelerator. Electrons emerge from the buncher cavity with a longitudinal velocity dispersion. Fast particles overtake slow particles, resulting in downstream localization of the beam current to sharp spikes. The electrons must be confined within a small spread in phase angle ($\leq 5°$) to minimize the kinetic energy spread of the output beam.

The micropulses enter the accelerator at a phase between $0°$ and $90°$. As we saw in Section 13.6, the average phase of the pulse increases until the electrons are ultrarelativistic. For the remainder of the acceleration cycle, acceleration takes place near a constant phase called the *asymptotic phase*. The injection phase of the micropulses and the accelerating gradient are adjusted to give an asymptotic phase of $90°$. This choice gives the highest acceleration gradient and the smallest energy spread in the bunch.

Output beam energy uniformity is a concern for high-energy physics experiments. The output energy spread is affected by variations in the traveling wave phase velocity. Dimensional tolerances in the waveguides on the order of 10^{-3} cm must be maintained for a 1% energy spread. The structures must be carefully machined and tuned. The temperature of the waveguides under rf power loading must be precisely controlled to prevent a shift in phase velocity from thermal expansion.

C. Beam Breakup Instability

The theory of Section 13.6 indicated that transverse focusing is unnecessary in an electron linac because of the shortened effective length. This is true only at low beam current; at high current, electrons are subject to the *beam breakup instability*,[†] also known as the *transverse instability* or *pulse shortening*. The instability arises from excitation of TM_{110} cavity modes in the spaces between irises. Features of the TM_{110} mode in a cylindrical cavity are illustrated in Figure 14.4. Note that there are longitudinal electric fields of opposite polarity in the upper and lower portions of the cavity and that there is a transverse

[†] W. K. H. Panofsky and M. Bander, *Rev. Sci. Instrum.*, **39**, 206 (1968); V. K. Neil and R. K. Cooper, *Part. Accel.*, **1**, 111 (1970).

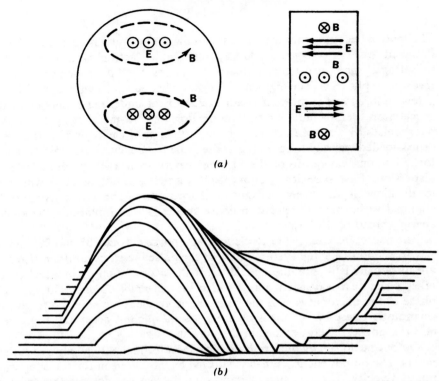

(a)

(b)

Figure 14.4 TM_{110} mode in cylindrical cavity. (*a*) diagram of electric and magnetic field distributions. (*b*) Three-dimensional plot of peak electric field amplitude as function of position.

magnetic field on the axis. An electron micropulse (of sub-nanosecond duration) can be resolved into a broad spectrum of frequencies. If the pulse has relatively high current and is eccentric with respect to the cavity, interaction between the electrons and the longitudinal electric field of the TM_{110} mode takes place. The mode is excited near the entrance of the accelerator by the initial micropulses of the macropulse. The magnetic field of the mode deflects subsequent portions of the macropulse, causing transverse sweeping of the beam at frequency ω_{110}. The sweeping beam can transfer energy continually to TM_{110} excitations in downstream cavities. The result is that beam sweeping grows from the head to the tail of the microsecond duration macropulse and the strength of TM_{110} oscillations grows along the length of the machine. Sweeping motion leads to beam loss. The situation is worsened if the TM_{110} excitation can propagate backward along the iris-loaded waveguide toward the entrance to the accelerator or if the beam makes many passes through the same section of accelerator (as in the microtron). This case is referred to as the *regenerative beam breakup instability*.

The beam breakup instability has the following features.

1.   Growth of the instability is reduced by accurate injection of azimuthally symmetric beams.

2.   The energy available to excite undesired modes is proportional to the beam current. Instabilities are not observed below a certain current; the cutoff depends on the macropulselength and the Q values of the resonant structure.

3.   The amplitude of undesired modes grows with distance along the accelerator and with time. This explains pulse shortening, the loss of late portions of the electron macropulse.

4.   Mode growth is reduced by varying the accelerator structure. The phase velocity for TM_{01} traveling waves is maintained constant, but the resonant frequency for TM_{110} standing waves between irises is changed periodically along the accelerator.

Tranverse focusing elements are necessary in high-energy electron linear accelerators to counteract the transverse energy gained through instabilities. Focusing is performed by solenoid lenses around the waveguides or by magnetic quandrupole lenses between guide sections.

D.   Frequency Equation

The dispersion relationship for traveling waves in an iris-loaded waveguide was introduced in Section 12-10. We shall determine the approximate relationship between the inner and outer radii of the irises for waves with phase velocity $\omega/k = c$ at a specified frequency. The *frequency equation* is a first-order guide. A second-order waveguide design is performed with computer calculations and modeling experiments.

Assume that δ, the spacing between irises, is small compared to the wavelength of the traveling wave; the boundary fields approximate a continuous function. The tube radius is R_0 and the aperture radius is R. The complete solution consists of standing waves in the volume between the irises and a traveling wave matched to the reactive boundary at $r = R_0$. The solution must satisfy the following boundary conditions:

$$E_z \text{ (standing wave)} = 0 \quad \text{at} \quad r = R_0, \tag{14.1}$$

$$E_z \text{ (traveling wave)} \cong E_z \text{ (standing wave)} \quad \text{at} \quad r = R, \tag{14.2}$$

$$B_\theta \text{ (traveling wave)} \cong B_\theta \text{ (standing wave)} \quad \text{at} \quad r = R. \tag{14.3}$$

The last two conditions proceed from the fact that **E** and **B** must be continuous in the absence of surface charges or currents.

Following Section 12.3, the solution for azimuthally symmetric standing waves in the space between the irises is

$$E_z(r, t) = AJ_0(\omega r/c) + BY_0(\omega r/c). \qquad (14.4)$$

The Y_0 term is retained because the region does not include the axis. Applying Eq. (14.1), Eq. (14.4) becomes

$$E_z = E_0[Y_0(\omega R_0/c)J_0(\omega r/c) - J_0(\omega R_0/c)Y_0(\omega r/c)]. \qquad (14.5)$$

The toroidal magnetic field is determined from Eq. (12.45) as

$$B_\theta = -(jE_0/c)[Y_0(\omega R_0/c)J_1(\omega r/c) - J_0(\omega R_0/c)Y_1(\omega r/c)]. \qquad (14.6)$$

The traveling wave has an electric field of the form

$$E_z = E_0\exp[j(kz - \omega t)].$$

We shall see in Section 14.4 that the axial electric field of the traveling wave is approximately constant over the aperture. Therefore, the net displacement current carried by a wave with phase velocity equal to c is

$$I_d = \pi R^2(\partial E_z/\partial t)/\mu_0 c^2$$

$$= -(j\omega/\mu_0 c^2)(\pi R^2)E_0\exp[j(kz - \omega t)]. \qquad (14.7)$$

The toroidal magnetic field of the wave at $r = R$ is

$$B_\theta = -(j\omega R/2c^2)E_0\exp[j(kz - \omega t)]. \qquad (14.8)$$

The frequency equation is determined by setting E_z/B_θ for the cavities and for the traveling wave equal at $r = R$ [Eqs. (14.2) and (14.3)]:

$$\omega R/c = \frac{2[Y_0(\omega R_0/c)J_1(\omega R/c) - J_0(\omega R_0/c)Y_1(\omega R/c)]}{Y_0(\omega R_0/c)J_0(\omega R/c) - J_0(\omega R_0/c)Y_0(\omega R/c)}. \qquad \boxed{14.9}$$

Equation (14.9) is a transcendental equation that determines ω in terms of R and R_0 to generate a traveling wave with phase velocity equal to the speed of light. A plot of the right-hand side of the equation is given in Figure 14.5. A detailed analysis shows that power flow is maximized and losses minimized when there are about four irises per wavelength. Although the assumptions underlying Eq. (14.9) are not well satisfied in this limit, it still provides a good first-order estimate.

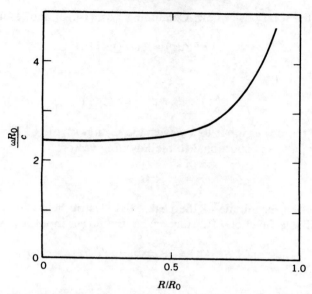

Figure 14.5 Frequency of TM_{01} mode with $\omega/k = c$ in iris-loaded cylindrical waveguide as function of R/R_0, where R is iris inner radius and R_0 is inner radius of waveguide.

E. Electromagnetic Energy Flow

Radio-frequency power is inserted into the waveguides periodically at locations separated by a distance l. For a given available total power P_t and accelerator length L, we can show that there is an optimum spacing l such that the final beam energy is maximized. In analogy with standing wave cavities, the quantity Q characterizes resistive energy loss in the waveguide according to

$$-(dP/dz) = \mathcal{U}\omega/Q. \tag{14.10}$$

In Eq. (14.10), dP/dz is the power lost per unit length along the slow-wave structure and $\mathcal{U}$ is the electromagnetic energy per unit length. Following the discussion of Section 12.10, the group velocity of the traveling waves is equal to

$$v_g = \frac{\text{energy flux}}{\text{electromagnetic energy density}}.$$

Multiplying the numerator and denominator by the area of the waveguide implies

$$\mathcal{U}v_g = P, \tag{14.11}$$

where P is the total power flow. Combining Eqs. (14.10) and (14.11),

$$-(dP/dz) = (\omega/Qv_g)P,$$

or

$$P(z) = P_0\exp(-\omega z/Qv_g), \tag{14.12}$$

where P_0 is the power input to a waveguide section at $z = 0$. The electromagnetic power flow is proportional to the Poynting vector,

$$\mathbf{S} = \mathbf{E} \times \mathbf{B} \sim E_z^2,$$

where E_z is the magnitude of the peak axial electric field. We conclude that electric field as a function of distance from the power input is described by

$$E_z(z) = E_{z0}\exp(-z/l_0), \tag{14.13}$$

where

$$l_0 = 2Qv_g/\omega.$$

An electron traveling through an accelerating section of length l gains an energy

$$\Delta T = e\int_0^l E_z(z)\,dz. \tag{14.14}$$

Substituting from Eq. (14.13) gives

$$\Delta T = eE_{z0}l\,[1 - \exp(-l/l_0)]/(l/l_0). \tag{14.15}$$

In order to find an optimum value of l, we must carefully define the following constraints:

1. The total rf power P_t and total accelerator length L are specified. The power input to an accelerating section of length l is $\Delta P = P_t(l/L)$.
2. The waveguide properties Q, v_g, and ω are specified.

The goal is to maximize the total energy, $T = \Delta T(L/l)$ by varying the number of power input points. The total power scales as

$$P_t \sim (v_g E_{z0}^2)(L/l),$$

where the first factor is proportional to the input power flux to a section and the second factor is the number of sections. Therefore, with constant power, E_{z0} scales as $l^{1/2}$. Substituting the scaling for E_{z0} in Eq. (14.15) and multiply-

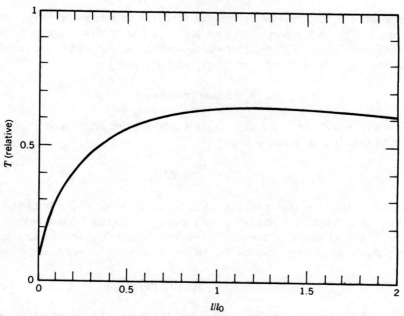

Figure 14.6 Relative output beam kinetic energy from traveling wave electron accelerator as function of l/l_0, where $l_0 = 2Qv_g/\omega$: Q is quality factor of waveguide, v_g is group velocity of traveling waves in slow-wave structure, and ω is rf angular frequency).

ing by L/l, we find that the beam output energy scales as

$$T \sim l^{1/2}\left[1 - \exp(-l/l_0)\right]/l$$

or

$$T \sim \left[1 - \exp(-l/l_0)\right]/\sqrt{l/l_0}. \qquad (14.16)$$

Inspection of Figure 14.6 shows that T is maximized when $l/l_0 = 1.3$; the axial electric field drops to 28% of its initial value over the length of a section. It is preferable from the point of view of particle dynamics to maintain a constant gradient along the accelerator. Figure 14.6 implies that l/l_0 can be reduced to 0.8 with only a 2% drop in the final energy. In this case, the output electric field in a section is 45% of the initial field.

Fields can also be equalized by varying waveguide properties over the length of a section. If the wall radius and the aperture radius are decreased consistent with Eq. (14.9), the phase velocity is maintained at c while the axial electric field is raised for a given power flux. Waveguides can be designed for constant axial field in the presence of decreasing power flux. In practice, it is difficult to fabricate precision waveguides with continuously varying geometry. A common compromise is to divide an accelerator section into subsections with varying geometry. The sections must be carefully matched so that there is no phase discontinuity between them. This configuration has the additional benefit of reducing the growth of beam breakup instabilities.

F. Shunt Impedance

The *shunt impedance* is a figure-of-merit quantity for electron and ion linear accelerators. It is defined by

$$P_t = V_0^2/(\mathscr{Z}L),   \boxed{14.17}$$

where P_t is the total power dissipated in the cavity walls of the accelerator, V_0 is the total accelerator voltage (the beam energy in electron volts divided by the particle charge), and L is the total accelerator length. The shunt impedance $\mathscr{Z}$ has dimensions of ohms per meter. An alternate form for shunt impedance is

$$\mathscr{Z} = \langle E_z^2 \rangle/(dP/dz),   \boxed{14.18}$$

where dP/dz is the resistive power loss per meter. The power loss of Eq. (14.17) has the form of a resistor of value $\mathscr{Z}L$ in parallel with the beam load. This is the origin of the term shunt impedance.

The efficiency of a linear accelerator is given by

$$\text{energy efficiency} = Z_b/(Z_b + \mathscr{Z}L),   (14.19)$$

where Z_b is the beam impedance,

$$Z_b = V_0/i_b.$$

The shunt impedance for most accelerator rf structures lies in the range of 25 to 50 MΩ/m. As an example, consider a 2.5-GeV linear electron accelerator with a peak on-axis gradient of 8 MV/m. The total accelerator length is 312 m. With a shunt impedance of 50 MΩ/m, the total parallel resistance is 1.6×10^{10} Ω. The power determined from Eq. (14.17) to maintain the high acceleration gradient is 400 MW.

14.2 LINEAR ION ACCELERATOR CONFIGURATIONS

Linear accelerators for ions differ greatly from electron machines. Ion accelerators must support traveling wave components with phase velocity well below the speed of light. In the energy range accessible to linear accelerators, ions are nonrelativistic; therefore, there is a considerable change in the synchronous particle velocity during acceleration. Slow-wave structures are not useful for

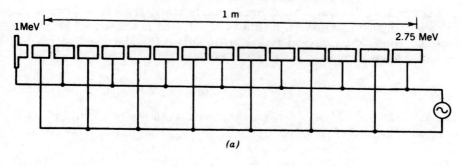

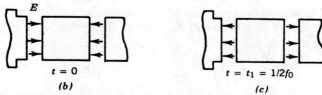

Figure 14.7 Wideröe linear accelerator for heavy ions. (*a*) Scale drawing of accelerator with following parameters: Ion species: Cs-137, $f = 10$ MHz, $T_i = 1$ MeV, $V_0 = 100$ kV, and $\phi_s = 60°$. (*b*) Electric fields in acceleration gaps 1 and 2 at ion injection ($t = 0$). (*c*) Electric fields at time $t = 1/2f$, where f is the rf frequency.

ion acceleration. An iris-loaded waveguide has small apertures for $\omega/k \ll c$. The conduction of electromagnetic energy via slow waves is too small to drive a multicavity waveguide. Alternative methods of energy coupling are used to generate traveling wave components with slow phase velocity.

An ion linear accelerator typically consists of a sequence of cylindrical cavities supporting standing waves. Cavity oscillations are supported either by individual power feeds or through intercavity coupling via magnetic fields. The theory of ion accelerators is most effectively carried out by treating cavities as individual oscillators interacting through small coupling terms.

Before studying rf linear ion accelerators based on microwave technology, we will consider the Wideröe accelerator[†] (Fig. 14.7*a*), the first successful linear accelerator. The Wideröe accelerator operates at a low frequency (1–10 MHz); it still has application for initial acceleration of heavy ions. The device consists of a number of tubes concentric with the axis connected to a high-voltage oscillator. At a particular time, half the tubes are at negative potential with respect to ground and half the tubes are positive. Electric fields are concentrated in narrow acceleration gaps; they are excluded from the interior of the tubes. The tubes are referred to as *drift tubes* since ions drift at constant velocity inside the shielded volume.

[†]R. Wideroe, *Arch. Elektrotechn.*, **21**, 387 (1928).

Assume that the synchronous ion crosses the first gap at $t = 0$ when the fields are aligned as shown in Figure 14.7b. The ion is accelerated across the gap and enters the zero-field region in the first drift tube. The ion reaches the second gap at time

$$\Delta t_1 = L_1/v_{s1}. \tag{14.20}$$

The axial electric fields at $t = t_1$ are distributed as shown in Figure 14.7c if t_1 is equal to half the rf period, or

$$\Delta t_1 = \pi/\omega. \tag{14.21}$$

The particle is accelerated in the second gap when Eq. (14.21) holds.

It is possible to define a synchronous orbit with continuous acceleration by increasing the length of subsequent drift tubes. The velocity of synchronous ions following the nth gap is

$$v_n = \left[2(T_0 + nqV_0\sin \phi_s)/m_i\right]^{1/2}, \tag{14.22}$$

where T_0 is the injection kinetic energy, V_0 is the peak gap voltage, and ϕ_s is the synchronous phase. The length of drift tube n is

$$L_n = v_n(\pi/\omega). \tag{14.23}$$

The drift tubes of Figure 14.7a are drawn to scale for the acceleration of Hg^+ ions injected at 2 MeV with a peak gap voltage of 100 kV and a frequency of 4 MHz.

The Wideroe accelerator is not useful for light-ion acceleration and cannot be extrapolated to produce high-energy heavy ions. At high energy, the drift tubes are unacceptably long, resulting in a low average accelerating gradient. The drift tube length is reduced if the rf frequency is increased, but this leads to the following problems:

1. The acceleration gaps conduct large displacement currents at high frequency, loading the rf generator.
2. Adjacent drift tubes act as dipole antennae at high frequency with attendant loss of rf energy by radiation.

The high-frequency problems are solved if the acceleration gap is enclosed in a cavity with resonant frequency ω. The cavity walls reflect the radiation to produce a standing electromagnetic oscillation. The cavity inductance in combination with the cavity and gap capacitance constitute an LC circuit. Displacement currents are supported by the electromagnetic oscillations. The power supply need only contribute energy to compensate for resistive losses and beam loading.

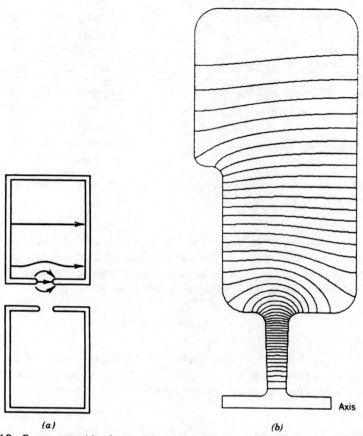

Figure 14.8 Resonant cavities for particle acceleration. (*a*) Electric fields in cylindrical cavity with "noses" to define beam drift space. (*b*) Computer calculation of electric fields of an acceleration cavity using the SUPERFISH code; $f = 454$ MHz. (Courtesy G. Boicourt, Los Alamos National Laboratory.)

A resonant cavity for ion acceleration is shown in Figure 14.8*a*. The TM_{010} mode produces good electric fields for acceleration. We have studied the simple cylindrical cavity in Section 12.3. The addition of drift tube extensions to the cylindrical cavity increases the capacitance on axis, thereby lowering the resonant frequency. The resonant frequency can be determined by a perturbation analysis or through the use of computer codes. The electric field distribution for a linac cavity computed by the program SUPERFISH is shown in Figure 14.8*b*.

Linear ion accelerators are composed of an array of resonant cavities. We discussed the synthesis of slow waves by independently phased cavities in Section 12.9. Two frequently encountered cases of cavity phasing are il-

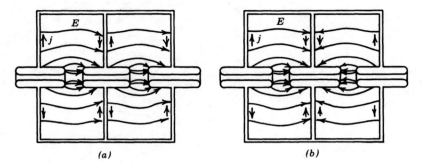

Figure 14.9 Electric field lines and wall currents of TM_{010} modes in two cavities of standing wave linear accelerator: (a) $\beta\lambda$ linac, (b) $\frac{1}{2}\beta\lambda$ linac.

lustrated in Figures 14.9a and 14.9b. In the first, the electric fields of all cavities are in phase, while in the second there is a phase change of 180° between adjacent cavities. The synchronous condition for the in-phase array is satisfied if ions traverse the intergap distance L_n in one rf period:

$$L_n = v_n(2\pi/\omega) = \beta\lambda, \qquad \boxed{14.24}$$

where $\beta = v_n/c$ and $\lambda = 2\pi c/\omega$. Hence, an accelerator with the phasing of Figure 14.9a is referred to as a $\beta\lambda$ linac. Similarly, the accelerator of Figure 14.9b is a $\frac{1}{2}\beta\lambda$ linac since the synchronous condition implies that

$$L_n = \frac{1}{2}\beta\lambda. \qquad \boxed{14.25}$$

In this notation, the Wideröe accelerator is a $\frac{1}{2}\beta\lambda$ structure.

The advantages of an individually phased array are that all cavities are identical and that a uniform accelerating gradient can be maintained. The disadvantage is technological; each cavity requires a separate rf amplifier and waveguide. The cost of the accelerator is reduced if a number of cavities are driven by a single power supply at a single feed point. Two geometries that accomplish this are the drift tube linac (Alvarez linac)[†] and the coupled cavity array. We shall study the drift tube accelerator in the remainder of this section. Coupled cavities are treated in Section 14.3.

The concept of the drift tube linac is most easily understood by following an evolution from the independently phased array. The $\beta\lambda$ cavity array of Figure 14.10a is an improvement over the independently phase array in terms of reduction of microwave hardware. There are separate power feeds but only one amplifier. Synchronization of ion motion to the rf oscillations is accomplished

[†] L. W. Alvarez, *Phys. Rev.*, **70**, 799 (1946).

by varying the drift lengths between cavities. The structure of Figure 14.10b is a mechanically simplified version in which the two walls separating cavities are combined. In the absence of the drift tubes, the cavities have the same resonant frequency, since ω_{010} does not depend on the cavity length (Table 12.1). This reflects the fact that the capacitance of a cylindrical cavity scales as $1/d$ while the inductance increases as d. The additional capacitance of the acceleration gap upsets the balance. It is necessary to adjust the gap geometry in different cavities to maintain a constant resonant frequency. The capacitance is determined by the drift tube diameter and the gap width. Figure 14.10b illustrates variation of drift tube diameter to compensate for increasing cavity length along the direction of acceleration. Resonant frequencies of individual cavities must be matched to within a factor of $1/Q$ so that all cavities are excited by the driving wave; a typical requirement is 1 part in 10^4. The design procedure for a cavity array often consists of the following stages

1. Approximate dimensions are determined by analytic or computer calculations.

2. Measurements are performed on a low-power model.

3. The final cavity array is tuned at low power. Small frequency corrections can be made by deforming cavity walls (dimpling) or by adjusting tuning slugs which change the capacitance or inductance of individual cavities.

The electric fields and wall currents for the TM_{010} mode in a $\beta\lambda$ structure are illustrated in Figure 14.9a. Note the distribution of electric field and current on the wall separating two cavities:

1. The currents in the two cavities are opposite and approximately equal; therefore, the wall carries zero net current.

2. Electric fields have equal magnitude and direction on both sides of the wall; therefore, the surface charge densities on the two sides of the wall have equal magnitude and opposite sign. There is zero net charge per area on the wall.

The field pattern is almost unchanged if the wall is removed (Fig. 14.10c). Eliminating the intervening walls leads to the drift tube accelerator of Figure 14.10c. Shaped drift tubes with increasing length along the direction of acceleration are supported by rods. The rods are located at positions of zero radial electric field; they do not seriously perturb the field distribution. An alternate view of the DTL is that it is a long cylindrical cavity with a single rf power feed to drive the TM_{010} mode; the variation of drift tube length and diameter maintains synchronization with accelerated particles and compensates the tube perturbations to maintain a constant axial electric field.

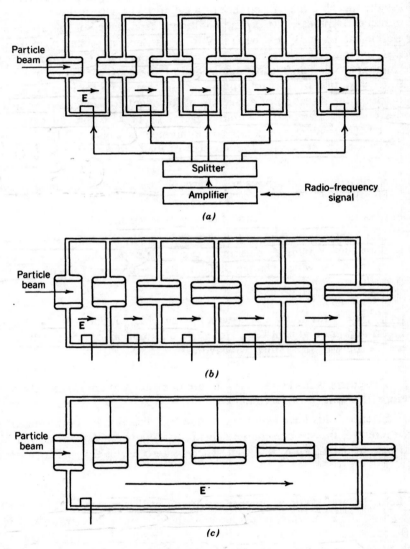

Figure 14.10 Evolution of drift tube linear accelerator. (*a*) Array of resonant cavities in the $\beta\lambda$ configuration with particle synchronization maintained by variation of distance between cavities. (*b*) Simplified $\beta\lambda$ structure with sychronization by varying cavity length; uniform resonant frequency maintained by variation of drift tube and acceleration gap geometry. (*c*) Alvarez linac tank. (*d*) Drift tube linac with postcouplers to shift the frequency of undesired rf modes. (Courtesy R. Jameson, Los Alamos National Laboratory.)

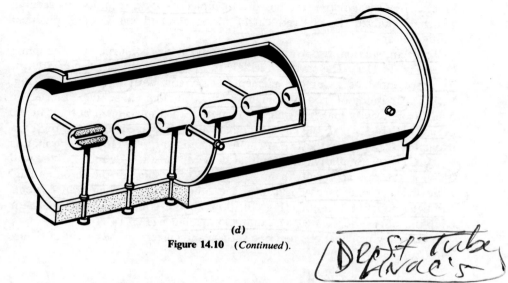

(d)
Figure 14.10 (*Continued*).

Drift Tube Linac's

Magnetic quadrupole lenses for beam focusing are located inside the drift tubes. Power and cooling water for the magnets enter along the tube supports. The development of strong permanent magnetic materials (such as orientated samarium-cobalt) has generated interest in adjustable permanent magnet quadrupole lenses. One of the main operational problems in DTLs is maintaining the TM_{010} mode in a complex structure with many competing modes. Contributions of modes with transverse electric fields are particularly dangerous since they lead to beam loss. An effective solution to stabilize the rf oscillations is to incorporate tuning elements in the structure. *Post couplers* are illustrated in Figure 14.10*d*. The posts are orthogonal to the drift tube supports. They have little effect on the fundamental acceleration mode which has only longitudinal electric fields. On the other hand, the combination of drift tube support and postcoupler causes a significant perturbation of other modes that have transverse electric fields. The effect is to shift the frequency of competing modes away from that of the fundamental so that they are less likely to be excited. A second purpose of the post couplers is to add periodic loading of the drift tube structures. Rotation of the post adds a small shunt capacitance to selected drift tubes. The variable loading is used to adjust the distribution of fundamental mode accelerating fields along the resonant cavity.

14.3 COUPLED CAVITY LINEAR ACCELERATORS

For a constrained frequency (set by rf power tube technology) and peak electric field (set by breakdown limits), a $\frac{1}{2}\beta\lambda$ linac has twice the average accelerating gradient as a $\beta\lambda$ structure such as the drift tube linac. For a given beam output energy, a $\frac{1}{2}\beta\lambda$ accelerator is half as long as a $\beta\lambda$ machine.

Practical $\frac{1}{2}\beta\lambda$ geometries are based on coupled cavity arrays. In this section, we shall analyze the coupled cavity formalism and study some practical configurations.

To begin, we treat two cylindrical resonant cavities connected by a coupling hole (Fig. 14.11a). The cavities oscillate in the TM$_{010}$ mode. Each cavity can be represented as a lumped element LC circuit with $\omega_0 = 1/\sqrt{LC}$ (Fig. 14.11b). Coupling of modes through an on-axis hole is capacitive. The electric field of one cavity makes a small contribution to displacement current in the other (Fig. 14.11c). In the circuit model, we can represent the coupling by a capacitor C_c between the two oscillator circuits (Fig. 14.11b). If coupling is weak, $C_c \ll C$. Similarly, an azimuthal slot near the outer diameter of the wall between the cavities results in magnetic coupling. Some of the toroidal magnetic field of one cavity leaks into the other cavity, driving wall currents through inductive coupling (Fig. 14.10d). In the circuit model, a magnetic coupling slot is represented by a mutual inductance (Fig. 14.11e).

The following equations describe voltage and current in the circuit of Figure 14.11b:

$$-C\dot{V}_1 = I_1, \tag{14.26}$$

$$V_1 = L(\dot{I}_1 - i), \tag{14.27}$$

$$-C\dot{V}_2 = I_2, \tag{14.28}$$

$$V_2 = L(\dot{I}_2 + i), \tag{14.29}$$

$$i = C_c(\dot{V}_1 - \dot{V}_2) = (C_c/C)(-I_1 + I_2). \tag{14.30}$$

When coupling is small, voltages and currents oscillate at frequency $\omega \cong \omega_0$ and the quantity i is much smaller than I_1 or I_2. In this case, Eq. (14.30) is approximated as

$$i \cong (-C_c\omega_0^2)(V_1 - V_2). \tag{14.31}$$

Assuming solutions of the form $V_1, V_2 \sim \exp(j\omega t)$, Eqs. (14.26)–(14.31) can be combined to give

$$V_1(1 - LC\omega^2 - LC_c\omega_0^2) + V_2(LC_c\omega_0^2) = 0, \tag{14.32}$$

$$V_1(LC_c\omega_0^2) + V_2(1 - LC\omega^2 - LC_c\omega_0^2) = 0. \tag{14.33}$$

Substituting $\Omega = \omega/\omega_0$ and $\kappa = C_c/C$, Eqs. (14.32) and (14.33) can be written in matrix form:

$$\begin{bmatrix} 1 - \Omega^2 - \kappa & \kappa \\ \kappa & 1 - \Omega^2 - \kappa \end{bmatrix}\begin{pmatrix} V_1 \\ V_2 \end{pmatrix} = 0. \tag{14.34}$$

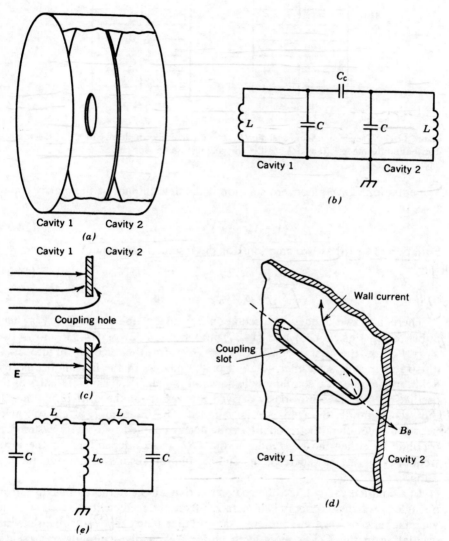

Figure 14.11 Coupled cavities. (*a*) Two resonant cavities with TM_{010} modes coupled capacitively through hole on-axis. (*b*) Equivalent circuit model of two electrically coupled cavities. (*c*) Electric field distribution near coupling hole with cavity 1 excited and cavity 2 unexcited. (*d*) Magnetic field distribution from TM_{010} mode at azimuthal slot near outer cavity radius; cavity 1 excited, cavity 2 unexcited. (*e*) equivalent circuit model for two magnetically coupled cavities.

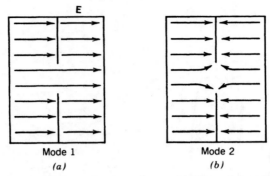

Figure 14.12 TM_{010} modes of oscillation for two capacitively coupled cavities. (*a*) Electric field distribution for mode 1 (0 mode). (*b*) Electric field distribution for mode 2 (π mode).

The equations have a nonzero solution if the determinant of the matrix equals zero, or

$$(1 - \Omega^2 - \kappa)^2 - \kappa^2 = 0. \qquad (14.35)$$

Equation (14.35) has two solutions for the resonant frequency:

$$\Omega_1 = \omega_1/\omega_0 = 1, \qquad (14.36)$$

$$\Omega_2 = \omega_2/\omega_0 = \sqrt{1 - 2\kappa}. \qquad (14.37)$$

There are two modes of oscillation for the coupled two-cavity system. Substituting Eqs. (14.36) and (14.37) into Eq. (14.32) or (14.33) shows that $V_1 = V_2$ for the first mode and $V_1 = -V_2$ for the second. Figure 14.12 illustrates the physical interpretation of the modes. In the first mode, electric fields are aligned; the coupling hole does not influence the characteristics of the oscillation. We have previously derived this result for the drift tube linac. In the second mode, the fields are antialigned. The interaction of electric fields near the hole alters the resonant frequency. A coupled two-cavity system can oscillate in either the $\beta\lambda$ mode or the $\frac{1}{2}\beta\lambda$ mode, depending on the input frequency of the rf generator. A similar solution results with magnetic coupling.

In a coupled cavity linac, the goal is to drive a large number of cavities from a single power feed. Energy is transferred from the feed cavity to other cavities via magnetic or electric coupling. Assume that there are N identical cavities oscillating in the TM_{010} mode with uniform capacitive coupling, represented by C_c. Figure 14.13 illustrates current and voltage in the circuit model of the nth cavity. The equations describing the circuit are

$$-C\ddot{V}_n = \dot{I}_n, \qquad (14.38)$$

$$V_n = L(\dot{I}_n - i_n - i_{n-1}), \qquad (14.39)$$

$$i_n \cong C_c\omega_0^2(V_{n+1} - V_n). \qquad (14.40)$$

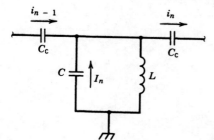

Figure 14.13 Equivalent circuit model for array of uniform, capacitively coupled resonant cavities.

The assumption of small coupling is inherent in Eq. (14.40). Taking time variations of the form $\exp(j\omega t)$, Eqs. (14.38)–14.40 can be combined into the single finite difference equation

$$V_{n+1} + \left[(1 - \Omega^2 - 2\kappa)/\kappa\right]V_n + V_{n-1} = 0, \qquad (14.41)$$

where κ and Ω are defined as above.

We have already solved a similar equation for the thin-lens array in Section 8.5. Again, taking a trial solution with amplitude variations between cells of the form

$$V_n = V_0\cos(n\mu + \phi), \qquad (14.42)$$

we find that

$$\cos\mu = -(1 - \Omega^2 - 2\kappa)/2\kappa. \qquad (14.43)$$

The resonant frequencies of the coupled cavity system can be determined by combining Eq. (14.43) with appropriate boundary conditions. The cavity oscillation problem is quite similar to the problem of an array of unconstrained, coupled pendula. The appropriate boundary condition is that the displacement amplitude (voltage) is maximum for the end elements of the array. Therefore, the phase term in Eq. (14.42) is zero. Applying the boundary condition in the end cavity implies that

$$\cos(N\mu) = \pm 1. \qquad (14.44)$$

Equation (14.44) is satisfied if

$$\mu_m = \pi m/N - 1, \qquad m = 0, 1, 2, \ldots, N - 1. \qquad (14.45)$$

The quantity m has a maximum value $N - 1$ since there can be at most N different values of V_n in the coupled cavity system.

A coupled system of N cavities has N modes of oscillation with frequencies given by

$$\Omega_m = \omega_m/\omega_0 = \left\{1 - 2\kappa\left[1 - \cos(2\pi m/N)\right]\right\}^{1/2}. \qquad (14.46)$$

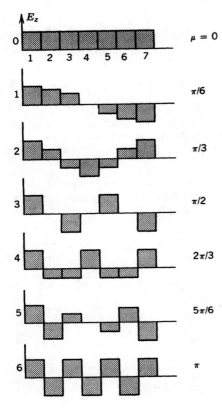

Figure 14.14 Amplitudes of axial electric fields for allowed modes of array of seven coupled cavities.

The physical interpretation of the allowed modes is illustrated in Figure 14.14. Electric field amplitudes are plotted for the seven modes of a seven-cavity system. In microwave nomenclature, the modes are referenced according to the value of μ. The 0 mode is equivalent to a $\beta\lambda$ structure while the π mode corresponds to $\frac{1}{2}\beta\lambda$.

At first glance, it appears that the π mode is the optimal choice for a high-gradient accelerator. Unfortunately, this mode cannot be used because it has a very low energy transfer rate between cavities. We can demonstrate this by calculating the group velocity of the traveling wave components of the standing wave. In the limit of a large number of cavities, the positive-going wave can be represented as

$$V_+(z,t) = \exp[\,j(\mu z/d - \omega t)]. \tag{14.47}$$

The wavenumber, k, is equal to μ/d. The phase velocity is

$$\omega/k = \omega_0 \Omega d/\mu, \tag{14.48}$$

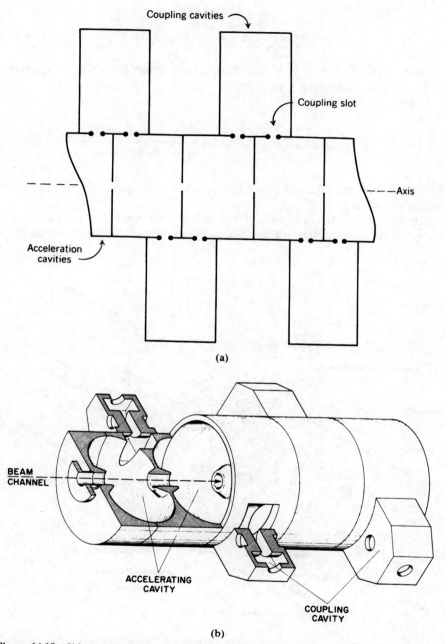

Figure 14.15 Side-coupled linear accelerator with resonant coupling a) Schematic diagram. b) Scale cross section of LAMF accelerator cavities (Courtesy O. B. van Dyck, Los Alamos National Laboratory.)

465

where ω_0 is the resonant frequency of an uncoupled cavity. For the π mode, Eq. (14.48) implies

$$d = (\omega/k)\pi/\omega_0\Omega = (\beta\lambda/2)/\Omega. \qquad (14.49)$$

Equation (14.49) is the $\frac{1}{2}\beta\lambda$ condition adjusted for the shift in resonant oscillation caused by cavity coupling.

The group velocity is

$$d\omega/dk = (\omega_0 d)\, d\Omega/d\mu = -(\omega_0 d)\frac{\kappa \sin \mu}{(1 - 2\kappa + 2\kappa \cos \mu)^{1/2}}. \qquad (14.50)$$

Note that v_g is zero for the 0 and π modes, while energy transport is maximum for the $\frac{1}{2}\pi$ mode.

The $\frac{1}{2}\pi$ mode is the best choice for rf power coupling, but it has a relatively low gradient since half of the cavities are unexcited. An effective solution to

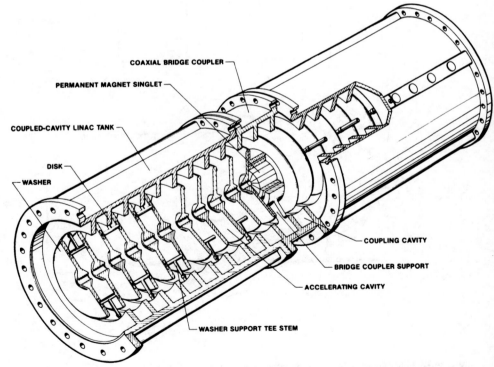

Figure 14.16 Cutaway view of a disk-and-washer accelerator structure. (Courtesy D. Swenson, Los Alamos National Laboratory and Texas A & M University).

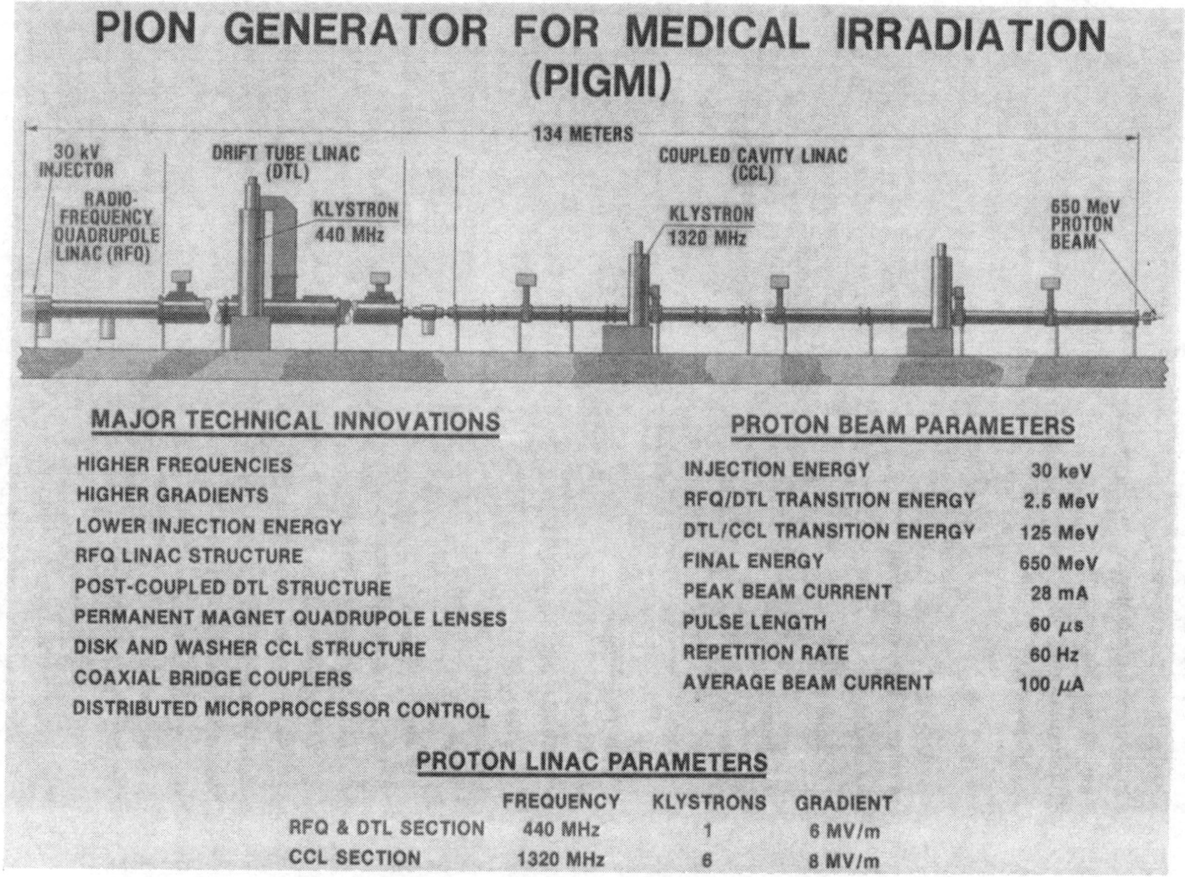

PION GENERATOR FOR MEDICAL IRRADIATION (PIGMI)

134 METERS

30 kV INJECTOR

RADIO-FREQUENCY QUADRUPOLE LINAC (RFQ)

DRIFT TUBE LINAC (DTL)

KLYSTRON 440 MHz

COUPLED CAVITY LINAC (CCL)

KLYSTRON 1320 MHz

650 MeV PROTON BEAM

MAJOR TECHNICAL INNOVATIONS

HIGHER FREQUENCIES
HIGHER GRADIENTS
LOWER INJECTION ENERGY
RFQ LINAC STRUCTURE
POST-COUPLED DTL STRUCTURE
PERMANENT MAGNET QUADRUPOLE LENSES
DISK AND WASHER CCL STRUCTURE
COAXIAL BRIDGE COUPLERS
DISTRIBUTED MICROPROCESSOR CONTROL

PROTON BEAM PARAMETERS

INJECTION ENERGY	30 keV
RFQ/DTL TRANSITION ENERGY	2.5 MeV
DTL/CCL TRANSITION ENERGY	125 MeV
FINAL ENERGY	650 MeV
PEAK BEAM CURRENT	28 mA
PULSE LENGTH	60 μs
REPETITION RATE	60 Hz
AVERAGE BEAM CURRENT	100 μA

PROTON LINAC PARAMETERS

	FREQUENCY	KLYSTRONS	GRADIENT
RFQ & DTL SECTION	440 MHz	1	6 MV/m
CCL SECTION	1320 MHz	6	8 MV/m

Figure 14.17 Diagram and parameters of PIGMI accelerator. (Courtesy D. Swenson, Los Alamos National Laboratory and Texas A & M University).

467

TABLE 14.2   Parameters of the LAMPF Accelerator

Accelerator length	800 m
Output beam energy	800 MeV
Output beam current	15 mA
Macropulselength	1 ms
Repetition rate	120 Hz
Duty cycle	12%

Injector

Ion species	H^+, H^-
Maximum output current	30 mA, H^+
Voltage	750 kV
Voltage generator	Cockcroft–Walton generator
Bunchers	201.25 MHz, 4-kV prebuncher, 10-kV main buncher

Drift Tube Linac

Energy variation	0.75–100 MeV
Length	61.7 m
Operating frequency	201.25 MHz
Cavity Q	5×10^4
rf filling time	200 μs
Number of tanks	4
Tank diameter	0.9 m
Number of drift tubes	165
Drift tube outer diameter	0.16 m
Drift tube bore	0.75–1.5 cm
Shunt impedance	42 MΩ/m
Average axial field	1.6–2.4 MV/m peak
Maximum surface field	12 MV/m
Synchronous phase	64°
rf power units	Triode power amplifiers
Number of rf units	4
rf power rating/unit	2.7 MW
Number of focusing quadrupoles	135
Focusing magnetic field gradient	8 to 0.8 kG/cm
Focusing mode	FDFD
System normalized acceptance	7π mm · mrad

Side-Coupled Linac

Energy variation	100–800 MeV
Total length	726.9 m
Operating frequency	805 MHz
Cavity Q	1.6 to 2.4 × 10⁴
rf filling time	0.15 ms
Number of tanks	104
Tank length	2.9 to 7.8 m
Number of cavities	5000
Bore diameter	3.2–3.8 cm
Shunt impedance	30–42 MΩ/m
Average axial field	1.1 MV/m
Synchronous phase	64–70°
rf power units	Klystrons
Number of rf power units	44
rf power rating/unit	1.25 MW
Number of focusing quadrupoles	204
Focusing magnetic field gradient	2.2 to 3.2 kG/cm
Focusing mode	Doublets
Normalized acceptance	17π mm · mrad

this problem is to displace the unexcited cavities to the side and pass the ion beam through the even-numbered cavities. The result is a $\frac{1}{2}\beta\lambda$ accelerator with good power coupling. The *side-coupled linac*[†] is illustrated in Figure 14.15*a*. Intermediate cavities are coupled to an array of cylindrical cavities by magnetic coupling slots. Low-level electromagnetic oscillations in the side cavities act to transfer energy along the system. There is little energy dissipation in the side cavities. Figure 14.15*b* illustrates an improved design. The side cavities are reentrant to make them more compact (see Section 12.2). The accelerator cavity geometry is modified from the simple cylinder to reduce shunt impedance. The simple cylindrical cavity has a relatively high shunt impedance because wall current at the outside corners dissipates energy while making little contribution to the cavity inductance.

The disk and washer structure (Fig. 14.16) is an alternative to the side-coupled linac. It has high shunt impedance and good field distribution stability. The accelerating cavities are defined by "washers." The washers are suspended by supports connected to the wall along a radial electric field null.

[†]See B. C. Knapp, E. A. Knapp, G. J. Lucas, and J. M. Potter, *IEEE Trans. Nucl. Sci.*, **NS-12**, 159 (1965).

LOS ALAMOS MESON PHYSICS FACILITY

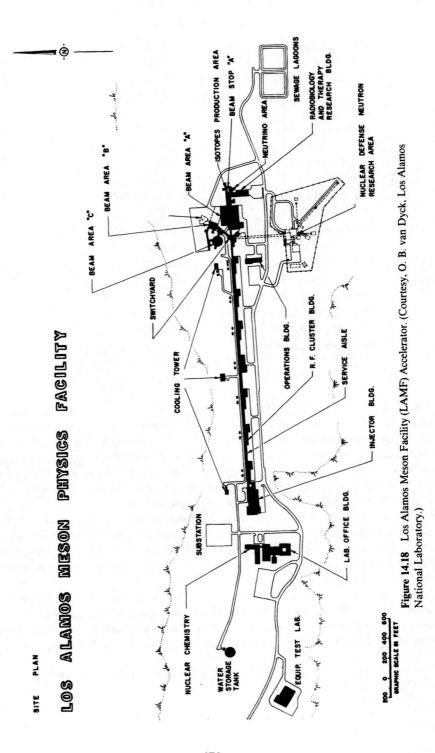

Figure 14.18 Los Alamos Meson Facility (LAMF) Accelerator. (Courtesy, O. B. van Dyck, Los Alamos National Laboratory.)

TABLE 14.3 Parameters of the UNILAC[a]

Accelerator length	125 m
Particle species	Variety of heavy ions, oxygen through uranium
Number of isotopes available	~ 50
Output energy (U)[b]	~ 4 GeV
Energy/nucleon (U)	17 MeV
Average output current (U)	0.05 μA
Macropulselength	5 ms
Duty cycle factor	0.25, 5 ms each 20 ms
Ion source	Penning discharge
Typical charge state (U)	$+10$
Typical Z/A (U)	0.042
Electrostatic accelerator voltage	300 kV
Species selection	Magnetic mass separator
Wideröe accelerator, number of cavities	4
Matched entrance β	0.05
Bore diameter	0.03 m
Number of accelerating gaps	120
Total accelerating voltage	30 MV
rf frequency	27 MHz
rf power	3 MW
Energy/nucleon at exit (U)	1.4 MeV
Average charge state after stripper	$+40$
Alvarez accelerator, number of cavities	4
Total accelerating voltage	100 MeV
rf frequency	108 MHz = 4 × 27 MHz
rf power	5 MW
Number of cavities and rf amplifiers, independently phased array for beam energy variation	17
Accuracy of final beam energy	0.1%
Micropulselength	0.2 to 4 ns

[a]Gesellschaft für Schwerionenforschung.
[b]Parameters quoted for uranium ions.

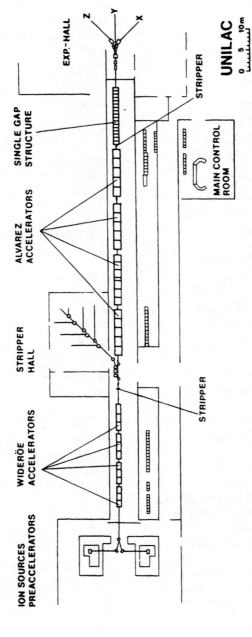

Figure 14.19 UNILAC heavy-ion linear accelerator. (Courtesy D. Böhne. Gesellschaft für Schwerionenforschung).

The coupling cavities extend around the entire azimuth. The individual sections of the disk-and-washer structure are strongly coupled. The perturbation analysis we used to treat coupled cavities is inadequate to determine the resonant frequencies of the disk-and-washer structure. The development of strongly coupled cavity geometries results largely from the application of digital computers to determine normal modes.

In contrast to electron accelerators, ion linear accelerators may be composed of a variety of acceleration structures. Many factors must be considered in choosing the accelerating components, such as average gradient, field stability, shunt impedance, fabrication costs, and beam throughput. Energy efficiency has become a prime concern; this reflects the rising cost of electricity as well as an expansion of interest in the accelerator community from high-energy physics to commercial applications. Figure 14.17 shows an accelerator designed for medical irradiation. Three types of linear accelerators are used. Notice that the factor of 4 increase in frequency between the low- and high-energy sections. Higher frequency gives higher average gradient. The beam microbunches are compressed during acceleration in the drift-tube linac (see Section 13.4) and are matched into every fourth bucket of the coupled cavity linac.

Parameters of the Los Alamos Meson Facility (LAMF) accelerator are listed in Table 14.2. The machine, illustrated in Figure 14.18, was designed to accelerate high-current proton beams for meson production. Parameters of the UNILAC are listed in Table 14.3. The UNILAC, illustrated in Figure 14.19, accelerates a wide variety of highly ionized heavy ions for nuclear physics studies.

14.4 TRANSIT-TIME FACTOR, GAP COEFFICIENT, AND RADIAL DEFOCUSING

The diameter of accelerator drift tubes and the width of acceleration gaps cannot be chosen arbitrarily. The dimensions are constrained by the properties of electromagnetic oscillations. In this section, we shall study three examples of rf field properties that influence the design of linear accelerators: the transit-time factor, the gap coefficient, and the radial defocusing forces of traveling waves.

The *transit-time factor* applies mainly to drift tube accelerators with narrow acceleration gaps. The transit-time factor is important when the time for particles to cross the gap is comparable to or longer than the half-period of an electromagnetic oscillation. If d is the gap width, this condition can be written

$$d/v_s \geq \pi/\omega, \qquad (14.51)$$

where v_s is the synchronous velocity. In this limit, particles do not gain energy $eE_0 d \sin \omega t$. Instead, they are accelerated by a time-averaged electric field smaller than $E_0 \sin \omega t$.

Assume that the gap electric field has time variation

$$E_z(r, z, t) = E_0 \cos(\omega t + \phi). \tag{14.52}$$

The longitudinal equation of motion for a particle crossing the gap is

$$dp_z/dt = qE_0 \sin(\omega t + \phi). \tag{14.53}$$

Two assumptions simplify the solution of Eq. (14.53).

1. The time $t = 0$ corresponds to the time that the particle is at the middle of the gap.
2. The change in particle velocity over the gap is small compared to v_s.

The quantity ϕ is equivalent to the particle phase in the limit of a gap of zero thickness (see Fig. 13.1). The change in longitudinal motion is approximately

$$\Delta p_z \cong qE_0 \int_{-(d/2v_s)}^{(d/2v_s)} \cos(\omega t + \phi)\, dt$$

$$= qE_0 \int_{-(d/2v_s)}^{(d/2v_s)} (\cos \omega t \sin \phi - \sin \omega t \cos \phi)\, dt. \tag{14.54}$$

Note that the term involving $\sin \omega t$ is an odd function; its integral is zero. The total change in momentum is

$$\Delta p \cong (2qE_0/\omega)\sin(\omega d/2v_s)\sin \phi. \tag{14.55}$$

The momentum gain of a particle in the limit $d \to 0$ is

$$\Delta p_0 = eE_0 \sin \phi (d/v_s). \tag{14.56}$$

The ratio of the momentum gain for a particle in a gap with nonzero width to the ideal thin gap is defined as the transit-time factor:

$$\mathcal{T} = \Delta p/\Delta p_0 = \sin(\omega d/2v_s)/(\omega d/2v_s). \boxed{14.57}$$

The transit-time factor is also approximately equal to the ratio of energy gain in a finite-width gap to that in a zero-width gap.

Defining a particle transit time as $\Delta t = d/v_s$, Eq. (14.57) can be rewritten

$$\mathcal{T} = \sin(\omega \Delta t/2)/(\omega \Delta t/2). \tag{14.58}$$

The transit-time factor is plotted in Figure 14.20 as a function of $\omega \Delta t$.

As an application example, consider acceleration of 5 MeV Cs^+ ions in a Wideröe accelerator operating at $f = 2$ MHz. The synchronous velocity is

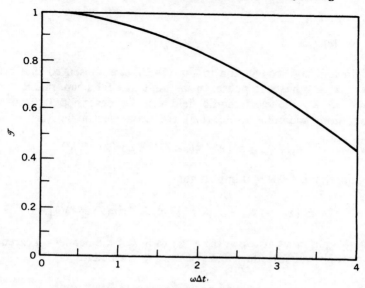

Figure 14.20   Transit-time factor as function of $\omega \Delta t$.

2.6×10^6 m/s. The transit time across a 2-cm gap is $\Delta t = 7.5$ ns. The quantity $\omega \Delta t$ equals 0.95; the transit-time factor is 0.963. If the synchronous phase is $60°$ and the peak gap voltage is 100 kV, the cesium ions gain an average energy of $(100)(0.963)(\sin 60°) = 83$ keV per gap.

The *gap coefficient* parametrizes the radial variation of accelerating fields across the dimension of the beam. Variations in E_z lead to a spread in beam energy; particles with large-amplitude transverse oscillations gain a different energy than particles on the axis. Large longitudinal velocity spread is undesirable for research applications and may jeopardize longitudinal confinement in rf buckets. We shall first perform a nonrelativistic derivation because the gap coefficient is primarily of interest in linear ion accelerators.

The slow-wave component of electric field chiefly responsible for particle acceleration has the form

$$E_z(0, z, t) = E_0 \sin(\omega t - \omega z/v_s). \qquad (14.59)$$

As discussed in Section 13.3, a slow wave appears to be an electrostatic field with no magnetic field when observed in a frame moving at velocity v_s. The magnitude of the axial electric field is unchanged by the transformation. The on-axis electric field in the beam rest frame is

$$E_z(0, z') = -E_0 \sin(2\pi z'/\lambda'), \qquad (14.60)$$

where λ' is the wavelength in the rest frame. In the nonrelativistic limit,

$\lambda' = \lambda$, so that

$$\lambda' = 2\pi v_s/\omega. \tag{14.61}$$

The origin and sign convention in Eq. (14.60) are chosen so that a positive particle at $z' = 0$ has zero phase. In the limit that the beam diameter is small compared to λ', the electrostatic field can be described by the paraxial approximation. According to Eq. (6.5), the radial electric field is

$$E_r(r', z') \cong (r'/2)(2\pi/\lambda')E_0\cos(2\pi z'/\lambda'). \tag{14.62}$$

The equation $\nabla \times \mathbf{E} = 0$ implies that

$$E_z(r', z') \cong -E_0\left[1 - (\pi r'/\lambda')^2\sin(2\pi z'/\lambda')\right]. \tag{14.63}$$

The energy gain of a particle at the outer radius of the beam (r_b) is reduced by a factor proportional to the square of the gap coefficient:

$$\Delta T/T \cong -(\pi r_b/\lambda')^2. \qquad \boxed{14.64}$$

The gap coefficient must be small compared to unity for a small energy spread. Equation (14.64) sets a limit on the minimum wavelength of electromagnetic waves in terms of the beam radius and allowed energy spread:

$$\lambda > \pi r_b/\sqrt{\Delta T/T}. \tag{14.65}$$

As an example, consider acceleration of a 10-MeV deuteron beam of radius 0.01 m. To obtain an energy spread less than 1%, the wavelength of the slow wave must be greater than 0.31 m. Using a synchronous velocity of 3×10^7 m/s, the rf frequency must be lower than $f < 100$ MHz.

This derivation can also be applied to demonstrate radial defocusing of ion beams by the fields of a slow wave. Equation (14.62) shows that slow waves must have radial electric fields. Note that the radial field is positive in the range of phase $0 < \phi < 90°$ and negative in the range $90° < \phi < 180°$. Therefore, the rf fields radially defocus particles in regions of axial stability. The radial forces must be compensated in ion accelerators by transverse focusing elements, usually magnetic quadrupole lenses. The stability properties of a slow wave are graphically illustrated in Figure 14.21. The figure shows three-dimensional variations of the electrostatic confinement potential (see Section 13.3) of an accelerating wave viewed in the wave rest frame. It is clear that there is no position in which particles have stability in both the radial and axial directions.

The problems of the gap coefficient and radial defocusing are reduced greatly for relativistic particles. For a relativistic derivation, we must include the fact that the measured wavelength of the slow wave is not the same in the stationary frame and the beam rest frame. Equation (2.23) implies that the

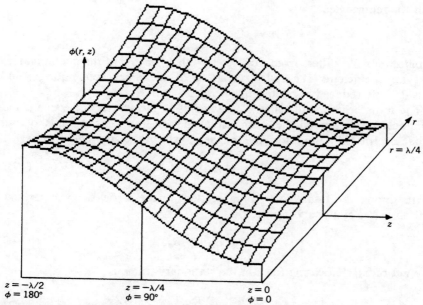

Figure 14.21 Three-dimensional view of variations of the electrostatic potential of a slow wave viewed in the wave rest frame.

measurements are related by

$$\lambda = \lambda'/\gamma, \tag{14.66}$$

where γ is the relativistic factor,

$$\gamma = 1/\left[1 - (v_s/c)^2\right]^{1/2}.$$

Again, primed symbols denote the synchronous particle rest frame.

The radial and axial fields in the wave rest frame can be expressed in terms of the stationary frame wavelength:

$$E_r'(r', z') \cong E_0'(-r'/2)(2\pi/\gamma\lambda)\cos(2\pi z'/\gamma\lambda), \tag{14.67}$$

$$E_z'(r', z') \cong E_0'\left[1 - (\pi r'/\gamma\lambda)^2\right]\sin(2\pi z'/\gamma\lambda). \tag{14.68}$$

Note that the peak value of axial field is unchanged in a relativistic transformation ($E_0' = E_0$). Transforming Eq. (14.68) to the stationary frame, we find that

$$E_z(r, z) = E_0\left[1 - (\pi r/\gamma\lambda)^2\right]\sin(2\pi z/\lambda), \tag{14.69}$$

with the replacement

$$r = r', \qquad z = z'/\gamma.$$

Equation (14.69) differs from Eq. 14.63 by the γ factor in the denominator of the gap coefficient. The radial variation of the axial accelerating field is considerably reduced at relativistic energies.

The transformation of radial electric fields to the accelerator frame is more complicated. A pure radial electric field in the rest frame corresponds to both a radial electric field and a toroidal magnetic field in the stationary frame:

$$E_r' = \gamma(E_r + v_z B_\theta). \tag{14.70}$$

Furthermore, the total radial force exerted by the rf fields on a particle is written in the stationary frame as

$$F_r = q(E_r + v_z B_\theta). \tag{14.71}$$

The net radial defocusing force in the stationary frame is

$$F_r = [E_0(r/2)(2\pi/\lambda)\cos(2\pi z/\lambda)]/\gamma^2. \tag{14.72}$$

Comparison with Eq. (14.62) shows that the defocusing force is reduced by a factor of γ^2. Radial defocusing by rf fields is negligible in high-energy electron linear accelerators.

14.5   VACUUM BREAKDOWN IN RF ACCELERATORS

Strong electric fields greater than 10 MV/m can be sustained in rf accelerators. This results partly from the fact that there are no exposed insulators in regions of high electric field. In addition, rf accelerators are run at high duty cycle, and it is possible to condition electrodes to remove surface whiskers. The accelerators are operated for long periods of time at high vacuum, minimizing problems of surface contamination on electrodes.

Nonetheless, there are limits to the voltage gradient set by resonant particle motion in the oscillating fields. The process is illustrated for electrons in an acceleration gap in Figure 14.22. An electron emitted from a surface during the accelerating half-cycle of the rf field can be accelerated to an opposing electrode. The electron produces secondary electrons at the surface. If the transit time of the initial electron is about one-half that of the rf period, the electric field will be in a direction to accelerate the secondary electrons back to the first surface. If the secondary electron coefficient δ is greater than unity, the electron current grows. Table 14.4 shows maximum secondary electron coefficients for a variety of electrode materials. Also included are the incident electron energy corresponding to peak emission and to $\delta = 1$. Emission falls

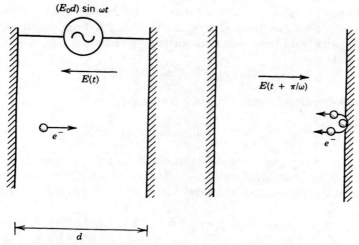

Figure 14.22 Geometry for calculating electron multipactoring. Electron field indicated at times of electron emission and electron collision with opposing electrode.

off at a higher electron energy. Table 14.4 gives values for clean, outgassed surfaces. Surfaces without special cleaning may have a δ as high as 4.

The resonant growth of electron current is called *multipactoring*, implying multiple electron impacts. Multipactoring can lead to a number of undesirable effects. The growing electron current absorbs rf energy and may clamp the magnitude of electric fields at the multipactoring level. Considerable energy can be deposited in localized regions of the electrodes, resulting in outgassing or evaporation of material. This often leads to general cavity breakdown.

TABLE 14.4 Secondary Electron Coefficient

Metal	d (electrons per incident electron)	Electron Energy at δ_{max} (eV)	Electron Energy for $\delta = 1$ (eV)
Al	1.9	220	35
Au	1.1	330	160
Cu	1.3	240	100
Fe	1.3	350	120
Mo	1.3	360	120
Ni	1.3	460	160
W	1.45	700	200

Conditions for electron multipactoring can be derived easily for the case of a planar gap with electrode spacing d. The electric field inside the gap is assumed spatially uniform with time variation given by

$$E(x, t) = -E_0 \sin(\omega t + \phi). \tag{14.73}$$

The nonrelativistic equation of motion for electrons is

$$m_e (d^2 x / dt^2) = eE_0 \sin(\omega t + \phi). \tag{14.74}$$

The quantity ϕ represents the phase of the rf field at the time an electron is produced on an electrode. Equation (14.74) can be integrated directly. Applying the boundary conditions that $x = 0$ and $dx/dt = 0$ at $t = 0$, we find that

$$x = -(eE_0 / m_e \omega^2)[\omega t \cos \phi + \sin \phi - \sin(\omega t + \phi)]. \tag{14.75}$$

Resonant acceleration occurs when electrons move a distance d in a time interval equal to an odd number of rf half-periods. When this condition holds, electrons emerging from the impacted electrode are accelerated in the $-x$ direction; they follow the same equation of motion as the initial electrons. The resonance condition is

$$\Delta t = (2n + 1)(\pi/\omega), \tag{14.76}$$

for $n = 0, 1, 2, 3, \ldots$. Combining Eqs. (14.75) and (14.76), the resonant condition can be rewritten

$$d = -(eE_0 / m_e \omega^2)[(2n + 1)\pi \cos \phi + 2 \sin \phi], \tag{14.77}$$

since $\sin(\omega \Delta t + \phi) = -\sin \phi$. Furthermore, we can use Eq. (14.75) to find the velocity of electrons arriving at an electrode:

$$v_x(x = d) = -(2eE_0 / m_e \omega)\cos \phi. \tag{14.78}$$

The solution of Eq. (14.74) is physically realizable only for particles leaving the initial electrode within a certain range of ϕ. First, the electric field must be negative to extract electrons from the surface at $t = 0$, or $\sin \phi > 0$. A real solution exists only if electrons arrive at the opposite electrode with positive velocity, or $\cos \phi > 0$. These two conditions are met in the phase range

$$0 < \phi < \tfrac{1}{2}\pi. \tag{14.79}$$

Figure 14.23 is a plot of Eq. (14.77) showing the breakdown parameter $(eE_0 / m_e \omega^2 d)$ versus the rf phase at which electrons leave the surface. Electron resonance is possible, in principle, over a range of gap voltage from 0 to $V = m_e \omega^2 d^2 / 2e$.

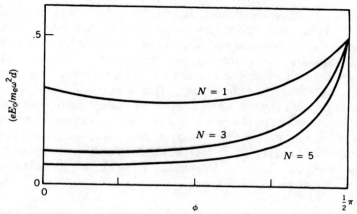

Figure 14.23 Conditions for electron multipactoring in planar gap of width d. Normalized resonant electric field, $eE_0/m_e\omega^2 d$, versus rf phase at time of electron emission: E_0, peak electric field; ω, rf angular frequency; N, number of rf half-periods during electron transit.

Electron multipactoring is a significant problem in the low-energy sections of linear ion accelerators. Consider, for example, an acceleration gap for 2-MeV protons. Assume that the proton transit time Δt is such that $\omega \Delta t = 1$. This implies that

$$\omega d = \beta c.$$

Substituting the above condition in Eq. (14.78) and assuming that $n = 0$ resonance condition, the electron energy at impact is

$$E_e = (2/\pi^2)(\beta \cos \phi)^2 (m_e c^2).$$

The quantity β equals 0.065 for 2-MeV protons. The peak electron energy occurs when $\cos \phi = 1$ ($\phi = 0$); for the example, it is 440 eV. Table 14.4 shows that this value is close to the energy of peak secondary electron emission. Electrons emitted at other phases arrive at the opposing electron with lower energy; therefore, they are not as likely to initiate a resonant breakdown. For this reason, the electron multipactoring condition is often quoted as

$$V_0 = (d\omega)^2 m_e/\pi e = (2\pi d/\lambda)^2 (m_e c^2)/\pi e. \qquad \boxed{14.80}$$

Equation (14.80) is expressed in terms of λ, the vacuum wavelength of the rf oscillations.

Electron multipactoring for the case quoted is probably not significant for values of n greater than zero since the peak electron energy is reduced by a factor of about $2n^2$. Therefore, breakdowns are usually not observed until the gap reaches a voltage level near that of Eq. (14.80). For an rf frequency $f = 400$ MHz, an acceleration gap 0.8 cm in width has $\omega \Delta t = 1$ for 2-MeV protons. This corresponds to a peak voltage of 730 V. At higher field levels, the resonance condition can be met over longer pathlengths at higher field stresses. This corresponds to high-energy electrons, which generally have secondary emission coefficients less than unity. Therefore, with clean surfaces it is possible to proceed beyond multipactoring by raising the rf electric field level rapidly. This may not be the case with contaminated electrodes; surface effects contribute much of the mystery and aggravation associated with rf breakdown.

The ultimate limits for rf breakdown in clean acceleration gaps were investigated experimentally by Kilpatrick.[†] The following formula is consistent with a wide variety of observations:

$$V_K = (2\pi d/\lambda)^2 (m_p c^2)/\pi e. \qquad \boxed{14.81}$$

Note that Eq. (14.81) is identical to Eq. 14.80 with the replacement of the electron mass by that of the proton. The Kilpatrick voltage limit is about a factor of 2000 times the electron multipactoring condition. The similarity of the equations suggests proton multipactoring as a mechanism for high-voltage rf breakdown. The precise mechanisms of proton production on electrode surfaces are unknown. Proton production may be associated with thin surface coatings. Present research on extending rf systems past the Kilpatrick limit centers on the use of proton-free electrodes.

14.6 RADIO-FREQUENCY QUADRUPOLE

The rf quadrupole[‡] is an ion accelerator in which both acceleration and transverse focusing are performed by rf fields. The derivations of Section 14.4 (showing lack of absolute stability in an rf accelerator) were specific to a cylindrical system; the fields in an RFQ are azimuthally asymmetric. There is no moving frame of reference in which RFQ fields can be represented as an electrostatic distribution. We shall see that the electric fields on the RFQ consist of positive and negative traveling waves; the positive wave continually accelerates ions in the range of stable phase. The beam is focused by oscillating transverse electric field components. These fields provide net beam focusing if the accelerating fields are not too high.

The major application of the RFQ is in low-energy ion acceleration. In the past, low-velocity ion acceleration presented one of the main technological

[†] W. D. Kilpatrick, "Criterion for Vacuum Sparking to Include Both RF and DC," University of California Radiation Laboratory, UCRL-2321, 1953.

[‡] I. M. Kapchinskii and V. A. Teplyakov, *Pribory i Teknika Eksperimenta* **2**, 19 (1970); R. H. Stokes, K. R. Crandall, J. E. Stovall, and D. A. Swenson, *IEEE Trans. Nucl. Sci.* **NS-26**, 3469 (1979).

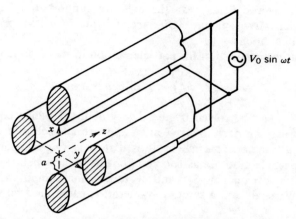

Figure 14.24 Radio-frequency electrostatic quadrupole for transverse confinement of charged particles.

difficulties for high-flux accelerators. A conventional ion beam injector consists of an ion source floating at high voltage and an electrostatic acceleration column. Space charge forces are strong for low-velocity ion beams; this fact motivates the choice of a high injection voltage, typically greater than 1 MV. The resulting system with adequate insulation occupies a large volume. The extracted beam must be bunched for injection into an rf accelerator. This implies long transport sections with magnetic quadrupole lenses. Magnetic lenses are ineffective for focusing low-energy ion beams, so that flux limits are low.

In contrast, the RFQ relies on strong electrostatic focusing in a narrow channel; this allows proton beam current in the range of 10 to 100 mA. An additional advantage of the RFQ is that it can combine the functions of acceleration and bunching. This is accomplished by varying the geometry of electrodes so that the relative magnitudes of transverse and longitudinal electric fields vary through the machine. A steady-state beam can be injected directly into the RFQ and reversibly bunched while it is being accelerated.

The quadrupole focusing channel treated in Chapter 8 has static fields with periodically alternating field polarity along the beam axis. In order to understand the RFQ, we will consider the geometry illustrated in Figure 14.24. The quadrupole electrodes are axially uniform but have time-varying voltage of the form $V_0\sin\omega t$. It is valid to treat the fields near the axis in the electrostatic limit if $a \ll c/\omega$, where a is the distance between the electrodes and the axis. In this case, the electric fields are simply the expressions of Eqs. (4.22) and (4.23) multiplied by $\sin\omega t$:

$$E_x(x, y, z, t) = (E_0 x/a)\sin\omega t, \qquad (14.82)$$

$$E_y(x, y, z, t) = -(E_0 y/a)\sin\omega t. \qquad (14.83)$$

The oscillating electric fields near the axis are supported by excitation of surrounding microwave structures. Off-axis fields must be described by the full set of Maxwell equations.

The nonrelativistic equation for particle motion in the x direction is

$$m(d^2x/dt^2) = (eE_0x/a)\sin \omega t. \qquad (14.84)$$

Equation (14.84) can be solved by the theory of the Mathieu equations. We will take a simpler approach to arrive at an approximate solution. Assume that the period for a transverse particle orbit oscillation is long compared to $2\pi/\omega$. In this limit, particle motion has two components; a slow betatron oscillation (parametrized by frequency Ω) and a rapid small-amplitude motion at frequency ω. We shall seek a solution by iteration using the trial solution

$$x(t) = x_0\sin(\Omega t) + x_1\sin(\omega t). \qquad (14.85)$$

where

$$x_1 \ll x_0, \qquad (14.86)$$

$$\Omega \ll \omega, \qquad (14.87)$$

$$\Omega^2 x_0 \ll \omega^2 x_1. \qquad (14.88)$$

Substituting Eq. (14.85) into Eq. (14.84), we find that

$$-\Omega^2 x_0\sin \Omega t - \omega^2 x_1\sin \omega t = (qE_0/ma)(\sin \omega t)(x_0\sin \Omega t + x_1\sin \omega t). \qquad (14.89)$$

The first term on the left-hand side and the second term on the right-hand side of Eq. (14.89) are dropped according to Eqs. (14.88) and (14.86). The result is an equation for the high-frequency motion:

$$-\omega^2 x_1 \cong (qE_0 x_0\sin \Omega t/a)/m$$

or

$$x_1 \cong -(qE_0 x_0\sin \Omega t/a)/m\omega^2. \qquad (14.90)$$

The second step is to substitute Eq. (14.90) into eq. (14.84) and average over a fast oscillation period to find the long-term motion. Terms containing $\sin \omega t$ average to zero. The remaining terms imply the following approximate equation for x_0:

$$-\Omega^2 x_0\sin \Omega t \cong -(qE_0/ma)^2(x_0\sin \Omega t)(\langle\sin^2\omega t\rangle)/\omega^2, \qquad (14.91)$$

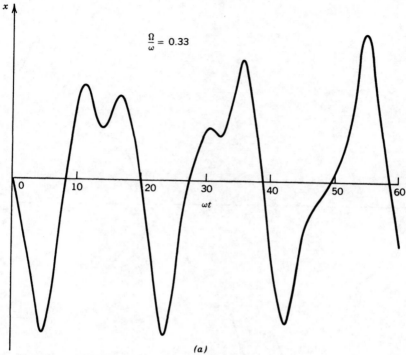

$$\frac{\Omega}{\omega} = 0.33$$

(a)

Figure 14.25 Numerical solutions for transverse motion of non-relativistic particle in oscillating electrostatic quadrupole field: $\Omega = (qE_0/m_0 a\omega)/\sqrt{2}$; E_0, peak electric field at electrode tip; a, distance from axis to electrode tip; m_0, particle rest mass; ω, rf angular frequency. *(a)* $\Omega/\omega = 0.33$. *(b)* $\Omega/\omega = 0.1$.

where $\langle \sin^2 \omega t \rangle$ denotes the average over a time $2\pi/\omega$. Equation (14.91) implies that Ω has the real value

$$\Omega = (qE_0/ma\omega)/\sqrt{2} . \qquad \boxed{14.92}$$

The long-term motion is oscillatory; the time-varying quadrupole fields provide net focusing. Numerical solutions to Eq. (14.84) are plotted in Figure 14.25 for $\omega = 3\ \Omega$ and $\omega = 10\ \Omega$. The phase relationship of Eq. (14.90) guarantees that particles are at a larger displacement when the fields are focusing. This is the origin of the average focusing effect. Orbit solutions in the y direction are similar.

The quadrupole lens of Figure 14.24 is useful only for focusing. It exerts no longitudinal force on ions. Axial field components are introduced if the shape of the electrodes is modified to that of Figure 14.26. The distance between the horizontal electrodes and the axis is modulated with spatial period D. There is

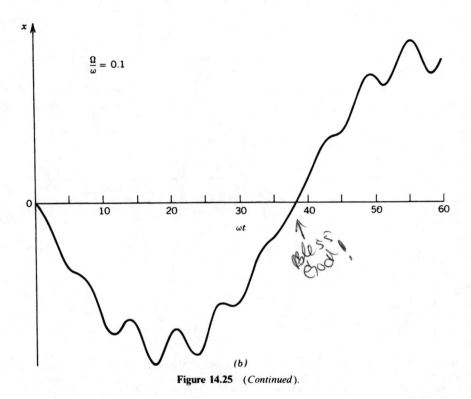

$\frac{\Omega}{\omega} = 0.1$

ωt

Bless
God!

(b)

Figure 14.25 (*Continued*).

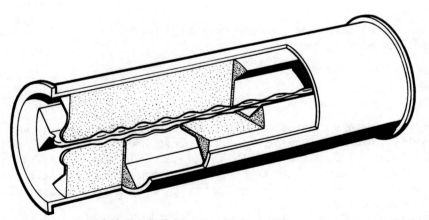

Figure 14.26 General configuration of rf quadrupole. (Courtesy, A. Wadlinger, Los Alamos National Laboratory).

a similar modulation of the vertical electrodes 90° out of phase. We postulate transverse fields of the form

$$E_x(x, y, z, t) = (E_0 x/a)[1 + \varepsilon \sin(2\pi z/D)]\sin \omega t, \qquad (14.93)$$

$$E_y(x, y, z, t) = -(E_0 y/a)[1 - \varepsilon \sin(2\pi z/D)]\sin \omega t. \qquad (14.94)$$

Again, the electrostatic approximation is invoked near the axis. Following the discussion of Section 4.4, Eqs. (14.93) and (14.94) are valid if (1) they are consistent with the Laplace equation and (2) the generating electrode surfaces lie on an equipotential. We shall show that both conditions can be satisfied.

Assume that a particle enters the system at the origin near time $t \leq \pi/2\omega$. The electric fields in the x–z plane are plotted in Figure 14.27a. The particle experiences a defocusing, quadrupolelike transverse field but also sees an accelerating component of field. Assume further than the particle moves a distance $\frac{1}{4}D$ in the time interval $\pi/2\omega$. The particle position and field configuration are sketched at $t = 3\pi/2\omega$ in Figure 14.27c. Transverse fields are focusing, while the axial component of the electric field is still positive. A synchronous particle orbit can be defined for the system.

We can find the synchronous orbit by determining the axial electric fields and solving the longitudinal equation of motion. If the electrostatic potential field pattern satisfies the Laplace equation, then

$$\partial E_z/\partial z = -(\partial E_x/\partial x + \partial E_y/\partial y). \qquad (14.95)$$

Substituting from Eqs. (14.93) and (14.94) and integrating, we find

$$E_z(x, y, z, t) = \boldsymbol{+}(2\varepsilon E_0 D/2\pi a)\cos(2\pi z/D)\sin \omega t. \qquad (14.96)$$

The standing wave pattern of Eq. (14.96) can be resolved into two traveling waves,

$$E_z = (\varepsilon E_0 D/2\pi a)[\sin(2\pi z/D + \omega t) - \sin(2\pi z/D - \omega t)]. \qquad (14.97)$$

The negative-going wave in the first term can be neglected. The positive-going component will interact strongly with particles moving at the synchronous velocity,

$$v_s = D\omega/2\pi. \qquad (14.98)$$

Assume that the synchronous particle enters the system at the origin of Figure 14.27 with velocity v_s at time $t = 0 + \phi/\omega$. Subsequently, the synchronous particle experiences a constant accelerating axial electric field of magnitude

$$E_{zs} = (\varepsilon E_0 D/2\pi a)\sin \phi. \qquad (14.99)$$

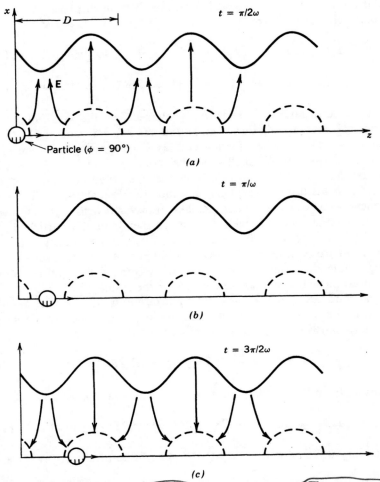

Figure 14.27 Particle motion and electric fields (projected in $x-z$ plane) in beam transport region of rf quadrupole. Profile of vertical electrode (vane) designated as solid line; location of horizontal electrode extensions (toward axis) indicated by dashed lines. Particle position corresponds to synchronous particle injected with $\phi_s = 90°$; time measured from zero crossing (positive slope) of vertical vane voltage. (a) $t = \pi/2\omega$, particle injected at $z = 0$. (b) $t = \pi\omega$. (c) $t = 3\pi/2\omega$.

The choice of axial origin and rf field phase illustrated in Figure 14.27 makes ϕ synonymous with the particle phase defined in Figure 13.3. As we found in Section 13.1, particle bunches have longitudinal stability if the synchronous phase is in the range $0 < \phi < 90°$. In contrast to a drift tube accelerator, an RFQ can be designed with only two traveling wave components. An alternate view is that the RFQ provides almost continuous acceleration. The equation of motion for the synchronous particle is

$$m(dv_s/dt) = (\varepsilon q E_0 D/2\pi a)\sin\phi. \tag{14.100}$$

Substituting from Eq. (14.98), Eq. (14.100) can be rewritten as

$$dD/dz = (2\pi\varepsilon q E_0 \sin\phi/m\omega^2 a). \tag{14.101}$$

If the field modulation factor ε is constant, Eq. (14.101) indicates that the length of modulations should increase linearly moving from entrance to exit of the RFQ.

The transverse equation of motion for a particle passing $z = 0$ at time ϕ/ω is

$$(d^2x/dt^2) = (qE_0x/ma)[1 + \varepsilon\sin(2\pi z/D)]\sin(\omega t + \phi)$$

$$= (qE_0x/ma)\sin(\omega t + \phi)$$

$$+ (\varepsilon q E_0 x/2ma)[\cos(2\pi z/D - \omega t - \phi)$$

$$- \cos(2\pi z/D + \omega t + \phi)]. \tag{14.102}$$

Again, we retain only the part of the force resonant with synchronous particles. Applying the synchronous condition [Eq. (14.98)], Eq. (14.102) becomes

$$(d^2x/dt^2) = (eE_0x/ma)\sin(\omega t + \phi) + (\varepsilon q E_0 x/2ma)\cos\phi. \tag{14.103}$$

The first term on the right-hand side represents the usual transverse focusing from the rf quadrupole. This component of motion is solved by the same method as the axially uniform oscillating quadrupole. The second term represents a defocusing force arising from the axial modulation of the quadrupole electrodes. The origin of this force can be understood by inspection of Figure 14.27. A sequence of particle position and electrode polarities is shown for a particle with a phase near 90°. On the average, the electrode spacing in the x direction is smaller during transverse defocusing and larger during the focusing phase for $\phi < 90°$. This brings about a reduction of the average focusing force.

The solution for average betatron oscillations of particles is

$$x(t) = x_0\cos(\Omega t), \tag{14.104}$$

where

$$\Omega = \left\{ \tfrac{1}{2}\left[(qE_0/ma\omega)^2\right] - (\varepsilon q E_0/2ma)\cos\phi \right\}^{1/2}.$$

The same result is determined for motion in the y direction. There is net transverse focusing if Ω is a real number, or

$$\varepsilon \le qE_0/(ma\omega^2\cos\phi).$$ $\boxed{14.105}$

The longitudinal electric field is proportional to ε. Therefore, there are limits on the accelerating gradient that can be achieved while preserving transverse focusing:

$$E_z \le (qE_0^2D\tan\phi)/(2\pi a^2m\omega^2).$$ (14.106)

Note that high longitudinal gradient is favored by high pole tip field (E_0) and a narrow beam channel (a).

The following parameters illustrate the results for the output portion of a 2.5-MeV RFQ operating at 440 MHz. The channel radius is $a = 0.0025$ m, the synchronous phase is $\phi = 60°$, the cell length is 0.05 m, and the pole tip field is 10 MV/m, well below the Kilpatrick limit. The limiting longitudinal gradient is 4.4 MV/m. The corresponding modulation factor is $\varepsilon = 0.05$. A typical RFQ design accelerates protons to 2.5 MeV in a length less than 2 m.

Equations (14.93), (14.94), and (14.96) can be used to find the electrostatic potential function following the same method used in Section 4.4. The result is

$$\Phi(x, y, z) = (E_0x^2/2a)[1 + \varepsilon\sin(2\pi z/D)]$$

$$-(E_0y^2/2a)[1 - \varepsilon\sin(2\pi z/D)] + (\varepsilon E_0D^2/4\pi^2a)(\sin 2\pi z/D).$$ $\boxed{14.107}$

The equipotential surfaces $\Phi = \pm\frac{1}{2}E_0a$ determine the three-dimensional electrode shape. The equation for the minimum displacement of the vertical vanes from the axis is

$$x_{min}^2 = [a^2 - (\varepsilon D^2/2\pi^2)\sin(2\pi z/D)]/[1 + \varepsilon\sin(2\pi z/D)].$$

This function is plotted in Figure 14.28. For a modulation factor $\varepsilon = 0.02$ and an average minimum electrode displacement of 0.0025 m, the distance from the electrode to the axis varies between 0.0019 and 0.0030 m.

The design of RFQ electrodes becomes more complex if the modulation factor is varied to add bunching capability. The design procedure couples results from particle orbit computer codes into a computer-controlled mill to generate complex electrodes such as that illustrated in Figure 14.29. The structure transports an incoming 30-mA, 30-keV proton beam. Electrode modulations increase gradually along the direction of propagation, adding longitudinal field components. The synchronous phase rises from 0 to the final value. Note the increasing modulation depth and cell length along the direction of acceleration.

A cross section of a complete RFQ is illustrated in Figure 14.30. The volume outside the transport region is composed of four coupled cylindrical

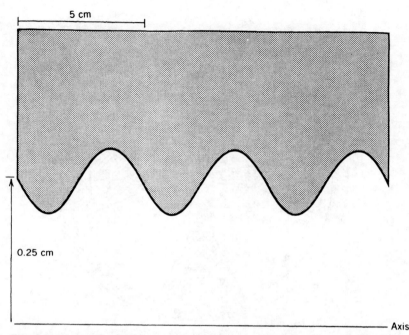

5 cm

0.25 cm

Axis

Figure 14.28 Profile of RFQ electrode. Parameters: $D = 0.05$ m, $a = 0.0025$ m, $\varepsilon = 0.02$.

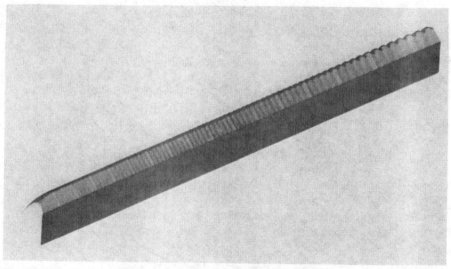

Figure 14.29 Computer-generated view of RFQ electrode designed for adiabatic bunching and acceleration of low-energy protons. Particles injected at left side. (Courtesy R. Jameson, Los Alamos National Laboratory.)

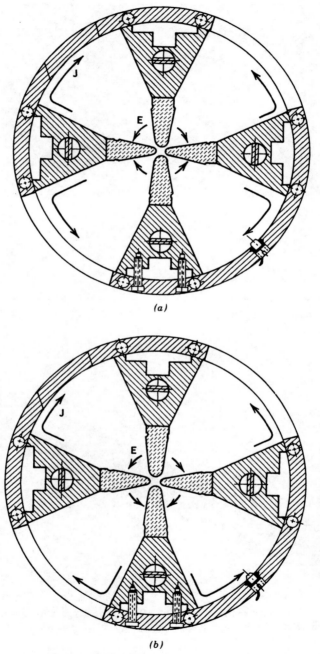

Figure 14.30 Diagram of rf modes in RFQ resonant structure; electric fields and wall currents outside beam transport region. (*a*) Quadrupole model (*b*) Dipole mode.

section cavities. The desired excitation modes for the cavities have axial magnetic fields and properties which are uniform along the longitudinal direction. Field polarities and current flows are illustrated for the quadrupole mode. A 440-MHz cavity has a radius of about 0.2 m. The mode for the quadrupole oscillation in all four cavities is designated TE_{210}. This terminology implies that

1. electric fields are transverse to the z direction in the rf portion of the cavity;
2. field quantities vary in azimuth according to $\cos(2\theta)$;
3. the electric field is maximum on axis and decreases monotonically toward the wall; and
4. there is no axial variation of field magnitudes in the standing wave.

Coupling of the four lines through the narrow transport region is not strong; equal distribution of energy demands separate drives for each of the lines. The usual procedure is to surround the RFQ with an annular resonator (*manifold*) driven at a single feed point. The manifold symmetrizes the rf energy; it is connected to the transmission lines by multiple coupling slots.

Other modes of oscillation are possible in an RFQ cavity. The dipole mode illustrated in Figure 14.30b is particularly undesirable since it results in electrostatic deflections and beam loss. The dipole mode frequency does not differ greatly from that of the quadrupole. Another practical problem is setting end conditions on the electrodes to maintain a uniform electric field magnitude over the length. Problems of mode coupling and field uniformity multiply as the length of the RFQ increases. This is the main reason why RFQ applications are presently limited to low-energy acceleration. The RFQ has been studied as a preaccelerator for heavy ions. In this case, the frequency is low. Low-frequency RFQs are sometimes fabricated as a nonresonant structure driven by an oscillator like the Wideröe accelerator.

14.7 RACETRACK MICROTRON

The extensive applications of synchrotron radiation to atomic and solid-state physics research has renewed interest in electron accelerators in the GeV range. The microtron is one of the most promising electron accelerators for research. Its outstanding feature is the ability to generate a continuous beam of high-energy electrons with average current approaching 100 μA. The time-average output of a microtron is much higher than a synchrotron or high-electron linac, which produce pulses of electrons at relatively low repetition rates.

The racetrack microtron[†] is illustrated in Figure 14.31a. Electrons are accelerated in a short linac section. Uniform field sector magnets at each end

[†] V. Veksler, *Compt. Rend. Acad. Sci. U.S.S.R.*, **43**, 444 (1944).

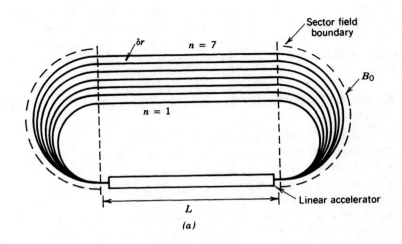

Sector field boundary

δr

$n = 7$

B_0

$n = 1$

L

Linear accelerator

(a)

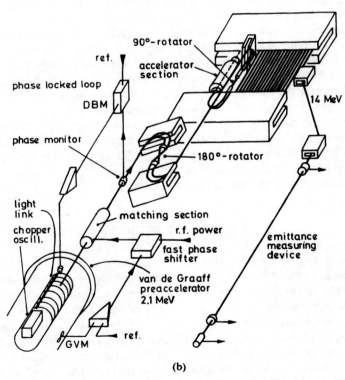

90°-rotator

ref.

accelerator section

phase locked loop

DBM

14 MeV

phase monitor

180°-rotator

light link

matching section

chopper oscill.

r.f. power

fast phase shifter

emittance measuring device

van de Graaff preaccelerator 2.1 MeV

GVM

ref.

(b)

Figure 14.31 Racetrack microtron. (a) Section view of components and electron orbits after from one to seven passes through linear accelerator. (b) Isometric view of microtron components and injection system for MAMI I. (Courtesy H. Herminghaus, Universität Mainz.)

of the accelerator confine the electrons. Electrons at a variety of energy levels are contained simultaneously in the machine. Electron orbits in the magnets are half-circles. It is easily demonstrated that electrons return to the linear accelerator axis after each revolution, independent of their energy. The size of the orbit increases as the electron energy increases.

The microtron combines linear accelerator technology with circular accelerator particle dynamics. Beam recirculation allows more efficient utilization of the linac. In contrast to the high-energy electron linear accelerators of Section 14.1 (where the machine length is a major constraint), acceleration gradient is not the primary concern in the microtron. This means that the accelerator need not operate at high field stress; therefore, power dissipation is a factor of 25–100 times lower than a high-gradient electron linac. This accounts for the capability of CW steady-state operation. The phase velocity of traveling wave components in the microtron linac is equal to the speed of light. In contrast to high-energy electron linacs, microtrons have phase stability. The orbit size (and, hence, the time to return to the linac) depends on electron energy. Therefore, electrons can be longitudinally confined during acceleration, even at low values of accelerating gradient.

Some parameters of a medium-energy microtron are listed in Table 14.5. The machine is designed as a preaccelerator in a three-microtron facility to produce an 840-MeV beam. The 14-MeV microtron with its associated injection and extraction system is illustrated in Figure 14.31b. The injected beam is generated by a 2-MV Van de Graaff accelerator. A beam chopper in the terminal of the electrostatic accelerator produces short pulses of electrons phased-matched to the linac. The complex series of lenses and deflection magnets matches the transverse and longitudinal distributions of the electron beam to the acceptance of the microtron. The origin of the parameters in Table 14.5 will be evident after we develop the theory of microtron equilibrium orbits.

In order to describe the microtron analytically, assume that the sector fields have sharp boundaries and a uniform field magnitude, B_0. Electrons are injected with initial total energy U_0 and gain an energy ΔU in each pass through the linear accelerator. Assume, further, that $U_0 \gg m_e c^2$, so that the electron velocity is always approximately equal to the speed of light. We have shown that acceleration in linear accelerators arises from a traveling wave component of the form

$$E_z = E_0 \sin(\omega t - \omega z/c).$$

The energy gain of a relativistic electron traversing an accelerator of length L is

$$\Delta U = e E_0 L \sin \phi,$$

where ϕ is the phase of the particle with respect to the traveling wave. The

TABLE 14.5   Parameters of MAMI, Stages I and II[a]

	Stage I	Stage II
General		
Input energy	2.1 MeV	14 MeV
Output energy	14 MeV	175 MeV
Number of traversals	20	51
Power consumption (total)	280 kW	
Design current	100 μA	
Magnet System		
Magnet separation	1.66 m	5.59 m
Magnetic field	0.10 T	0.54 T
Maximum orbit diameter	0.97 m	2.17 m
Magnet weight (each)	1.3 tonne	43 tonne
Gap width	6 cm	7 cm
rf System		
Linac length	0.80 m	3.55 m
Numbers of klystrons	2	
Frequency	2.449 GHz	
rf power	9 kW	64 kW
Beam load	1.2 kW	16 kW
Energy gain per turn	0.59 MeV	3.16 MeV

[a] Nuclear Physics Institute, University of Mainz.

energy gain for synchronous electrons passing through the linac is independent of their total energy.

The index n designates the number of times that an electron has passed through the linear accelerator; the orbits in Figure 14.31 are labeled accordingly. The time for an electron to traverse the microtron is

$$\Delta t_n = 2L/c + 2\pi r_{gn}/c, \qquad (14.108)$$

where

$$r_{gn} = \gamma_n m_e c / e B_0 = U_n / e B_0 c.$$

The quantity U_n is the total energy of an electron on the nth orbit:

$$U_n = U_0 + n \, \Delta U. \qquad (14.109)$$

The condition for synchronous electrons is that they arrive at the entrance to the linac at the same phase of the rf period. In other words, the traversal time must be an integer multiple of the rf period. Letting ω be the frequency of rf oscillations in the linac, this condition is written

$$\Delta t_n = 2L/c + 2\pi U_n/eB_0 c^2 = m(2\pi/\omega). \tag{14.110}$$

As electrons gain energy, the particle velocity is constant but the orbit size increases. The traversal time of high-energy particles is longer than that of low-energy particles. The difference in traversal times between particles on the n and $n-1$ orbits is

$$\Delta t_n - \Delta t_{n-1} = 2\pi \Delta U/eB_0 c^2. \tag{14.111}$$

Clearly, for synchronization Δt_n must also equal an integer multiple of the rf period:

$$2\pi \Delta U/eB_0 c^2 = q(2\pi/\omega). \qquad \boxed{14.112}$$

As an example, we pick $q = 1$. This means that electrons take one extra rf period for a traversal with each energy increment. Following Table 14.5, assume a 20-turn microtron with an injection energy of 2 MeV. The bending magnetic field is 0.1 T, and the linear accelerator length is 0.8 m. The energy gain per turn is $\Delta U = 0.6$ MeV, implying an average acceleration gradient of 0.75 MV/m. Substituting into Eq. (14.112), the matched frequency is $\omega = 1.5 \times 10^{10}$ Hz, or $f = 2.4$ GHz. Electrons injected at 2 MeV are boosted to 2.6 MeV in their first passage through the linac. The initial gyroradius in the bending field is 0.174 m; the total distance around the system on the first orbit is 2.7 m. Equation 14.111 implies that the time for the first traversal equals 22 rf periods. This is a high number; the particle must return to the linac entrance with equal phase after an interval of $22(2\pi/\omega)$. Synchronization requires excellent bending field uniformity and a constant energy input beam with little velocity dispersion. The problem becomes more acute as electrons are accelerated. Electrons on the highest-energy orbit take 42 rf periods to traverse the system. The synchronization problem limits the practical number of turns in a single microtron. A choice of $q > 1$ worsens the problem.

The separation between adjacent orbits on the side opposite the linac is

$$\delta r \cong 2\,\Delta r_g \cong \Delta U/eB_0 c.$$

For the parameters of the example, $\delta r = 0.04$ m. The large orbit separation makes extraction of high-energy electrons relatively easy.

The two main problems of microtrons are beam steering and beam breakup instabilities. Regarding the first problem, the uniform magnetic field of the microtron has horizontal focusing but no vertical focusing. Lenses must be added to each beam line on the straight sections opposite the accelerator. Even with the best efforts to achieve bending field uniformity, it is necessary to add beam steering magnets with active beam sensing and compensation to meet the synchronization condition. The beam breakup instability is severe in the microtron because the current of all beams is concentrated in the high-Q resonant cavities of the linear accelerator. The beam breakup instability is the main reason why microtron average currents are limited to less than 1 mA. It has also impeded the development of microtrons with superconducting linear accelerator cavities. These cavities have extremely high values of Q for all modes.

Phase stability is an interesting feature of microtrons. In contrast to high-energy electron linear accelerators, variations of electron energy lead to phase shifts because of the change in orbit pathlength. For instance, a particle with energy greater than that of the synchronous particle has a larger gyroradius; therefore, it enters the linac with increased phase. For longitudinal

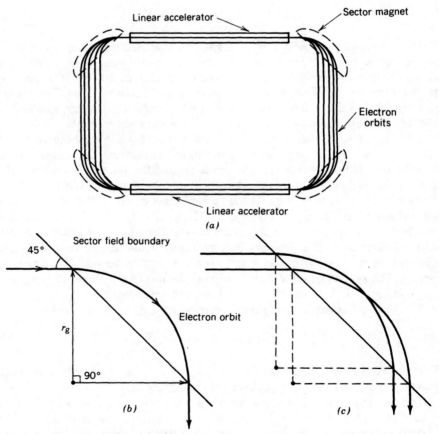

Figure 14.32 Double-sided microtron. (*a*) Sectional view of components and particle orbits. (*b*) Specular reflection of particle orbit incident on 45° sector magnet. (*c*) Neutral focusing property of 45° sector magnet.

stability, the higher-energy electrons must receive a reduced energy increment in the linac. This is true if the synchronous phase is in a region of decreasing field, $90° \leq \phi_s \leq 180°$. Particle phase orbits are the inverse of those in a linear ion accelerator.

The double-sided microtron (DSM) illustrated in Figure 14.32*a* is an alternative to the racetrack microtron. The DSM has linear accelerators in both straight sections. Beam deflection is performed by four 45° sector magnets. The major advantage, compared to the racetrack microtron, is that approximately double the electron energy can be achieved for the same magnet mass. The 45° sector magnet has the feature that the orbits of electrons of any energy are reflected at exactly 90° (see Fig. 14.32*b*).

Unfortunately, the DSM has unfavorable properties for electron focusing. Figure 14.32*c* shows a particle trajectory on the main orbit compared to an orbit displaced horizontally off-axis. Note that there is no focusing; the DSM has neutral stability in the horizontal direction. Furthermore, the sector magnets contribute defocusing forces in the vertical direction. There are edge-focusing effects because the magnets boundaries are inclined 45° to the particle orbits. Reference to Section 6.9 shows that the inclination gives a negative focal length resulting in defocusing.

15

Cyclotrons and Synchrotrons

The term *circular accelerator* refers to any machine in which beams describe a closed orbit. All circular accelerators have a vertical magnetic field to bend particle trajectories and one or more gaps coupled to inductively isolated cavities to accelerate particles. Beam orbits are often not true circles; for instance, large synchrotrons are composed of alternating straight and circular sections. The main characteristic of resonant circular accelerators is synchronization between oscillating acceleration fields and the revolution frequency of particles.

Particle recirculation is a major advantage of resonant circular accelerators over rf linacs. In a circular machine, particles pass through the same acceleration gap many times (10^2 to $> 10^8$). High kinetic energy can be achieved with relatively low gap voltage. One criterion to compare circular and linear accelerators for high-energy applications is the energy gain per length of the machine; the cost of many accelerator components is linearly proportional to the length of the beamline. Dividing the energy of a beam from a conventional synchrotron by the circumference of the machine gives effective gradients exceeding 50 MV/m. The gradient is considerably higher for accelerators with superconducting magnets. This figure of merit has not been approached in either conventional or collective linear accelerators.

There are numerous types of resonant circular accelerators, some with specific advantages and some of mainly historic significance. Before beginning a detailed study, it is useful to review briefly existing classes of accelerators. In the following outline, a standard terminology is defined and the significance of each device is emphasized.

500

Most resonant circular accelerators can be classed as either cyclotrons or synchrotrons. One exception is the microtron (Section 14.7), which is technologically akin to linear accelerators. The microtron may be classified as a cyclotron for relativistic electrons, operating well beyond the transition energy (see Section 15.6). The other exception is the synchrocyclotron (Section 15.4).

A. Cyclotron

A cyclotron has constant magnetic field magnitude and constant rf frequency. Beam energy is limited by relativistic effects, which destroy synchronization betwen particle orbits and rf fields. Therefore, the cyclotron is useful only for ion acceleration. The virtue of cyclotrons is that they generate a continuous train of beam micropulses. Cyclotrons are characterized by large-area magnetic fields to confine ions from zero energy to the output energy.

1. Uniform-Field Cyclotron

The uniform-field cyclotron has considerable historic significance. It was the first accelerator to generate multi-MeV particle beams for nuclear physics research. The vertical field is uniform in azimuth. The field magnitude is almost constant in the radial direction, with small positive field index for vertical focusing. Resonant acceleration in the uniform-field cyclotron depends on the constancy of the nonrelativistic gyrofrequency. The energy limit for light ion beams is about 15–20 MeV, determined by relativistic mass increase and the decrease of magnetic field with radius. There is no synchronous phase in a uniform-field cyclotron.

2. Azimuthally-Varying-Field (AVF) Cyclotron

The AVF cyclotron is a major improvement over the uniform-field cyclotron. Variations are added to the confining magnetic field by attaching wedge-shaped inserts at periodic azimuthal positions of the magnet poles. The extra horizontal-field components enhance vertical focusing. It is possible to tolerate an average negative-field index so that the bending field increases with radius. With proper choice of focusing elements and field index variation, the magnetic field variation balances the relativistic mass increase, resulting in a constant-revolution frequency. An AVF cyclotron with this property is called an isochronous cyclotron. An additional advantage of AVF cyclotrons is that the stronger vertical focusing allows higher beam intensity. AVF machines have supplanted the uniform-field cyclotron, even in low-energy applications.

3. Separated-Sector Cyclotron

The separated-sector cyclotron is a special case of the AVF cyclotron. The azimuthal field variation results from splitting the bending magnet into a number of sectors. The advantages of the separated sector cyclotron are (1) modular magnet construction and (2) the ability to locate rf feeds and acceleration gaps between the sectors. The design of separated-sector cyclotrons is complicated by the fact that particles cannot be accelerated from low energy. This feature can be used to advantage; beams with lower emittance (better coherence) are achieved if an independent accelerator is used for low-energy acceleration.

4. Spiral Cyclotron

The pole inserts in a spiral cyclotron have spiral boundaries. Spiral shaping is used in both standard AVF and separated-sector machines. In a spiral cyclotron, ion orbits have an inclination at the boundaries of high-field regions. Vertical confinement is enhanced by edge focusing (Section 6.9). The combined effects of edge focusing and defocusing lead to an additional vertical confinement force.

5. Superconducting Cyclotron

Superconducting cyclotrons have shaped iron magnet poles that utilize the focusing techniques outlined above. The magnetizing force is supplied by superconducting coils, which consume little power. Superconducting cyclotrons are typically compact machines because they are operated at high fields, well above the saturation level of the iron poles. In this situation, all the magnetic dipoles in the poles are aligned; the net fields can be predicted accurately.

B. Synchrocyclotron

The synchrocyclotron is a precursor of the synchrotron. It represents an early effort to extend the kinetic energy limits of cyclotrons. Synchrocyclotrons have a constant magnetic field with geometry similar to the uniform-field cyclotron. The main difference is that the rf frequency is varied to maintain particle synchronization into the relativistic regime. Synchrocyclotrons are cyclic machines with a greatly reduced time-averaged output flux compared to a cyclotron. Kinetic energies for protons to 1 GeV have been achieved. In the sub-GeV energy range, synchrocyclotrons were supplanted by AVF cyclotrons, which generate a continuous beam. Synchrocyclotrons have not been extended to higher energy because of technological and economic difficulties in fabricating the huge, monolithic magnets that characterize the machine.

C. Synchrotron

Synchrotrons are the present standard accelerators for particle physics research. They are cycled machines. Both the magnitude of the magnetic field and the rf frequency are varied to maintain a synchronous particle at a constant orbit radius. The constant-radius feature is very important; bending and focusing fields need extend over only a small ring-shaped volume. This minimizes the cost of the magnets, allowing construction of large-diameter machines for ion energies of up to 800 GeV. Synchrotrons are used to accelerate both ions and electrons, although electron machines are limited in energy by emission of synchrotron radiation. The only limit on achievable energy for ions is the cost of the machine and availability of real estate. Cycling times are long in the largest machines, typically many seconds. Electron synchrotrons and proton boosters cycle at frequencies in the range of 15 to 60 Hz.

1. Weak Focusing Synchrotron

Early synchrotrons used weak focusing. The bending magnets were shaped to produce a field with index in the range $0 < n < 1$. With low focusing force, the combined effects of transverse particle velocity and synchrotron oscillations (see Section 15.6) resulted in beams with large cross section. This implies costly, large-bore magnets.

2. Strong Focusing Synchrotron

All modern synchrotrons use transverse focusing systems composed of strong lenses in a focusing–defocusing array. Strong focusing minimizes the beam cross section, reducing the magnet size. Beam dynamics are more complex in a strong focusing synchrotron. The magnets must be constructed and aligned with high precision, and care must be taken to avoid resonance instabilities. Advances in magnet technology and beam theory have made it possible to overcome these difficulties.

Alternating Gradient Synchrotron (AGS). The bending field in an alternating gradient synchrotron is produced by a ring of wedge-shaped magnets which fit together to form an annular region of vertical field. The magnets have alternate positive and negative field gradient with $n \gg 1$. The combination of focusing and defocusing in the horizontal and vertical directions leads to net beam confinement.

Separated Function Synchrotron. Most modern synchrotrons are configured as separated function synchrotrons. The bending field is provided by sector magnets with uniform vertical field. Focusing is performed by quadrupole magnetic lens set between the bending magnets. Other magnets may be included for correction of beam optics.

3. Storage Ring

A storage ring usually has the same focusing and bending field configuration as a separated function synchrotron, but provides no acceleration. The magnetic fields are constant in time. An rf cavity may be included for longitudinal beam manipulations such as stacking or, in the case of electrons, maintaining kinetic energy in the presence of radiation loss. A storage ring contains energetic particles at constant energy for long periods of time. The primary applications are for colliding beam experiments and synchrotron radiation production.

4. Collider

A collider is a synchrotron, storage ring, dual synchrotron, or dual storage ring with special geometry to allow high-energy charged particles moving in opposite directions to collide head-on at a number of positions in the machine. The use of colliding beams significantly increases the amount of energy available to probe the structure of matter for elementary particle physics. Colliders have been operated (or are planned) for counterrotating beams of protons (pp collider), electrons and positrons (e^-e^+), and protons and antiprotons ($p\bar{p}$).

Section 15.1 introduces the uniform-field cyclotron and the principles of circular resonant accelerators. The longitudinal dynamics of the uniform-field cyclotron is reviewed in Section 15.2. The calculations deal with an interesting application of the phase equations when there is no synchronous particle. The model leads to the choice of optimum acceleration history and to limits on achievable kinetic energy. Sections 15.3 and 15.4 are concerned with AVF, or isochronous, cyclotrons. Tranverse focusing is treated in the first section. Section 15.4 summarizes relationships between magnetic field and rf frequency to preserve synchronization in fixed-field, fixed-frequency machines. There is also a description of the synchrocyclotron.

Sections 15.5–15.7 are devoted to the synchrotron. The first section describes general features of synchrotrons, including focusing systems, energy limits, synchrotron radiation, and the kinematics of colliding beams. The longitudinal dynamics of synchrotrons is the subject of Section 15.6. Material includes constraints on magnetic field and rf frequency variation for synchronization, synchrotron oscillations, and the transition energy. To conclude, Section 15.7 summarizes the principles and benefits of strong focusing. Derivations are given to illustrate the effects of alignment errors in a strong focusing system. Forbidden numbers of betatron wavelengths and mode coupling are discussed qualitatively.

15.1 PRINCIPLES OF THE UNIFORM-FIELD CYCLOTRON

The operation of the uniform-field cyclotron[†] is based on the fact that the gyrofrequency for nonrelativistic ions [Eq. (3.39)] is independent of kinetic energy. Resonance between the orbital motion and an accelerating electric field can be achieved for ion kinetic energy that is small compared to the rest

---

[†] E. O. Lawrence, *Science*, **72**, 376 (1930).

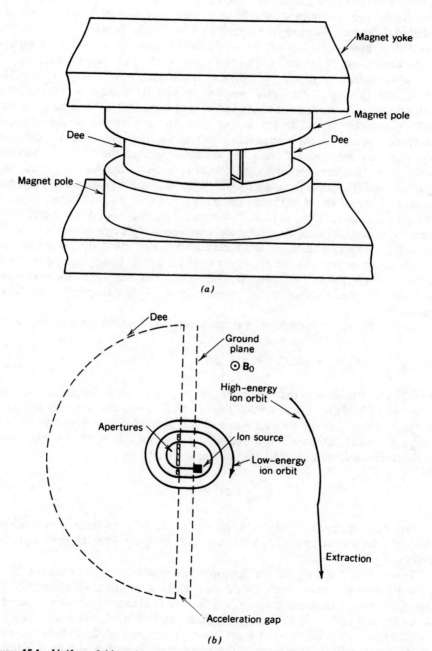

(a)

(b)

Figure 15.1 Uniform-field cyclotron. (*a*) General layout of beam acceleration region. (*b*) Overhead sectional view of acceleration region, showing a cyclotron with one dee and a ground plane. A single dee facilitates injection and extraction.

energy. The configuration of the uniform-field cyclotron is illustrated in Figure
15.1a. Ions are constrained to circular orbits by a vertical field between the
poles of a magnet. The ions are accelerated in the gap between two D-shaped
metal structures (dees) located within the field region. An ac voltage is applied
to the dees by an rf resonator. The resonator is tuned to oscillate near ω_g.

The acceleration history of an ion is indicated in Figure 15.1b. The
accelerator illustrated has only one dee excited by a bipolar waveform to
facilitate extraction. A source, located at the center of the machine continuously generates ions. The low-energy ions are accelerated to the opposite
electrode during the positive-polarity half of the rf cycle. After crossing the
gap, the ions are shielded from electric fields so that they follow a circular
orbit. When the ions return to the gap after a time interval π/ω_{g0}, they are
again accelerated since the polarity of the dee voltage is reversed. An aperture
located at the entrance to the acceleration gap limits ions to a small range of
phase with respect to the rf field. If the ions were not limited to a small phase
range, the output beam would have an unacceptably large energy spread. In
subsequent gap crossings, the ion kinetic energy and gyroradius increase until
the ions are extracted at the periphery of the magnet. The cyclotron is similar
to the Wideröe linear accelerator (Section 14.2); the increase in the gyroradius
with energy is analogous to the increase in drift-tube length for the linear
machine.

The rf frequency in cyclotrons is relatively low. The ion gyrofrequency is

$$f_0 = qB_0/2\pi m_i = (1.52 \times 10^7)B_0 \text{ (tesla)}/A, \qquad \boxed{15.1}$$

where A is the atomic mass number, m_i/m_p. Generally, frequency is in the
range of 10 MHz for magnetic fields near 1 T. The maximum energy of ions in
a cyclotron is limited by relativistic detuning and radial variations of the
magnetic field magnitude. In a uniform-field magnet field, the kinetic energy
and orbit radius of nonrelativistic ions are related by

$$T_{max} = 48(Z^*RB)^2/A, \qquad \boxed{15.2}$$

where T_{max} is given in MeV, R in meters, and B in tesla. For example,
30-MeV deuterons require a 1-T field with good uniformity over a 1.25-m
radius.

Transverse focusing in the uniform-field cyclotron is performed by an
azimuthally symmetric vertical field with a radial gradient (Section 7.3). The
main differences from the betatron are that the field index is small compared to
unity ($\nu_r \cong 1$ and $\nu_z \ll 1$) and that particle orbits extend over a wide range of
radii. Figure 15.2 diagrams magnetic field in a typical uniform-field cyclotron
magnet and indicates the radial variation of field magnitude and field index, n.
The field index is not constant with radius. Symmetry requires that the field
index be zero at the center of the magnet. It increases rapidly with radius at the

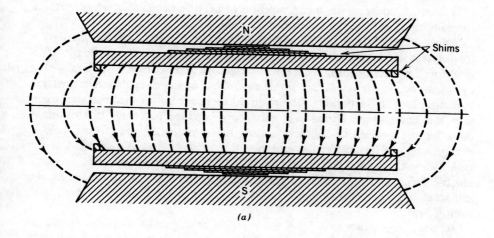

(a)

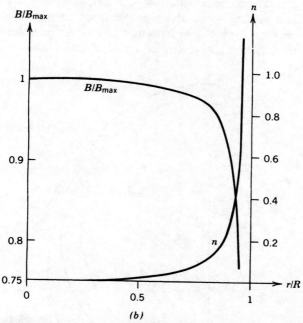

(b)

Figure 15.2 Magnetic fields of uniform-field cyclotron. (*a*) Sectional view of cyclotron magnetic poles showing shims for optimizing field distribution. (*b*) Radial variation of vertical field magnitude and field index. (M. S. Livingston and J. P. Blewett, *Particle Accelerators*, used by permission, McGraw-Hill Book Co.)

edge of the pole. Cyclotron magnets are designed for small n over most of the field area to minimize desynchronization of particle orbits. Therefore, vertical focusing in a uniform-field cyclotron is weak.

There is no vertical magnetic focusing at the center of the magnet. By a fortunate coincidence, electrostatic focusing by the accelerating fields is effective for low-energy ions. The electric field pattern between the dees of a cyclotron act as the one-dimensional equivalent of the electrostatic immersion lens discussed in Section 6.6. The main difference from the electrostatic lens is that ion transit-time effects can enhance or reduce focusing. For example, consider the portion of the accelerating half-cycle when the electric field is rising. Ions are focused at the entrance side of the gap and defocused at the exit. When the transit time is comparable to the rf half-period, the transverse electric field is stronger when the ions are near the exit, thereby reducing the net focusing. The converse holds in the part of the accelerating half-cycle with falling field.

In order to extract ions from the machine at a specific location, deflection fields must be applied. Deflection fields should affect only the maximum energy ions. Ordinarily, static electric (magnetic) fields in vacuum extend a distance comparable to the spacing between electrodes (poles) by the properties of the Laplace equation (Section 4.1). Shielding of other ions is accomplished with a *septum* (separator), an electrode or pole that carries image charge or current to localize deflection fields. An electrostatic septum is illustrated in Figure 15.3. A strong radial electric field deflects maximum

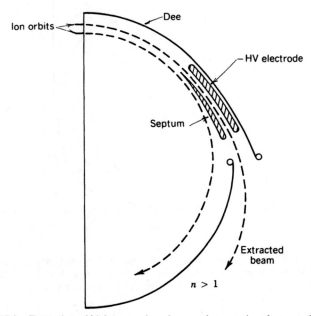

Figure 15.3 Extraction of high-energy ions from cyclotron using electrostatic septum.

energy ions to a radius where $n > 1$. Ions spiral out of the machine along a well-defined trajectory. Clearly, a septum should not intercept a substantial fraction of the beam. Septa are useful in the cyclotron because there is a relatively large separation between orbits. The separation for nonrelativistic ions is

$$\Delta R \cong (R/2)(2qV_0 \sin \phi_s/T). \qquad (15.3)$$

For example, with a peak dee voltage $V_0 = 100$ kV, $\phi_s = 60°$, $R = 1$ m, and $T = 20$ MeV, Eq. (15.3) implies that $\Delta R = 0.44$ cm.

15.2 LONGITUDINAL DYNAMICS OF THE UNIFORM-FIELD CYCLOTRON

In the uniform-field cyclotron, the oscillation frequency of gap voltage remains constant while the ion gyrofrequency continually decreases. The reduction in ω_g with energy arises from two causes: (1) the relativistic increase in ion mass and (2) the reduction of magnetic field magnitude at large radius. Models of longitudinal particle motion in a uniform-field cyclotron are similar to those for a traveling wave linear electron accelerator (Section 13.6); there is no synchronous phase. In this section, we shall develop equations to describe the phase history of ions in a uniform-field cyclotron. As in the electron linac, the behavior of a pulse of ions is found by following individual orbits rather than performing an orbit expansion about a synchronous particle. The model predicts the maximum attainable energy and energy spread as a function of the phase width of the ion pulse. The latter quantity is determined by the geometry of the aperture illustrated in Figure 15.1. The model indicates strategies to maximize beam energy.

The geometry of the calculation is illustrated in Figure 15.4. Assume that the voltage of dee1 relative to dee2 is given by

$$V(t) = V_0 \sin \omega t, \qquad (15.4)$$

where ω is the rf frequency. The following simplifying assumptions facilitate development of a phase equation:

1. Effects of the gap width are neglected. This is true when the gap width divided by the ion velocity is small compared to $1/\omega$.
2. The magnetic field is radially uniform. The model is easily extended to include the effects of field variations.
3. The ions circulate many times during the acceleraton cycle, so that it is sufficient to approximate kinetic energy as a continuous variable and to identify the centroid of the particle orbits with the symmetry axis of the machine.

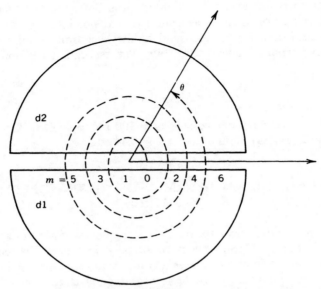

Figure 15.4 Geometry for treating longitudinal dynamics of uniform-field cyclotron.

The phase of an ion at azimuthal position θ and time t is defined as

$$\phi = \omega t - \theta(t). \tag{15.5}$$

Equation (15.5) is consistent with our previous definition of phase (Chapter 13). Particles crossing the gap from dee1 to dee2 at $t = 0$ have $\phi = 0$ and experience zero accelerating voltage. The derivative of Eq. (15.5) is

$$\dot{\phi} = \omega - \dot{\theta} = \omega - \omega_g, \tag{15.6}$$

where

$$\omega_g = qB_0/\gamma m_i = qc^2 B_0/E. \tag{15.7}$$

The quantity E in Eq. (15.7) is the total relativistic ion energy,

$$E = T + m_i c^2.$$

In the limit that $T \ll m_i c^2$, the gyrofrequency is almost constant and Eq. (15.6) implies that particles have constant phase during acceleration. Relativistic effects reduce the second term in Eq. (15.6). If the rf frequency equals the nonrelativistic gyrofrequency $\omega = \omega_{g0}$, then $d\phi/dt$ is always positive. The limit of acceleration occurs when ϕ reaches 180°. In this circumstance, ions arrive at the gap when the accelerating voltage is zero; ions are trapped at a particular energy and circulate in the cyclotron at constant radius.

Equation (15.4), combined with the assumption of small gap width, implies that particles making their mth transit of the gap with phase ϕ_m gain an energy.

$$\Delta E_m = qV_0 \sin \phi_m. \qquad (15.8)$$

In order to develop an analytic phase equation, it is assumed that energy increases continually and that phase is a continuous function of energy, $\phi(E)$. The change of phase for a particle during the transit through a dee is

$$\Delta\phi = \dot\phi(\pi/\omega_g) = \pi\left[(\omega E/c^2 qB_0) - 1\right]. \qquad (15.9)$$

Dividing Eq. (15.9) by Eq. (15.8) gives an approximate equation for $\phi(E)$:

$$\Delta\phi/\Delta E \cong d\phi/dE \cong (\pi/qV_0\sin\phi)\left[(\omega E/c^2 qB_0) - 1\right]. \qquad (15.10)$$

Equation (15.10) can be rewritten

$$\sin\phi\, d\phi = (\pi/qV_0)\left[(\omega E/c^2 qB_0) - 1\right] dE. \qquad (15.11)$$

Integration of Eq. (15.11) gives an equation for phase as a function of particle energy:

$$\cos\phi = \cos\phi_0 - (\pi/qV_0)\left[(\omega/2c^2 qB_0)(E^2 - E_0^2) - (E - E_0)\right], \qquad (15.12)$$

where ϕ_0 is the injection phase. The cyclotron phase equation is usually expressed in terms of the kinetic energy T. Taking $T = E - m_0 c^2$ and $\omega_{g0} = qB_0/m_i$, Eq. (15.12) becomes

$$\cos\phi = \cos\phi_0 + (\pi/qV_0)(1 - \omega/\omega_{g0})T - (\pi/2qV_0 m_i c^2)(\omega/\omega_{g0})T^2.$$

$$\boxed{15.13}$$

During acceleration, ion phase may traverse the range $0° < \phi < 180°$. The content of Eq. (15.13) can be visualized with the help of Figure 15.5. The quantity $\cos\phi$ is plotted versus T with ϕ_0 as a parameter. The curves are parabolas. In Figure 15.5a, the magnetic field is adjusted so that $\omega = \omega_{g0}$. The maximum kinetic energy is defined by the intersection of the curve with $\cos\phi = -1$. The best strategy is to inject the particles in a narrow range near $\phi_0 = 0$. Clearly, higher kinetic energy can be obtained if $\omega < \omega_{g0}$ (Fig. 15.5b). The particle is injected with $\phi_0 > 0$. It initially gains on the rf field phase and then lags. A particle phase history is valid only if $\cos\phi$ remains between -1 and $+1$. In Figure 15.5b, the orbit with $\phi_0 = 45°$ is not consistent with acceleration to high energy. The curve for $\phi_0 = 90°$ leads to a higher final energy than $\phi_0 = 135°$.

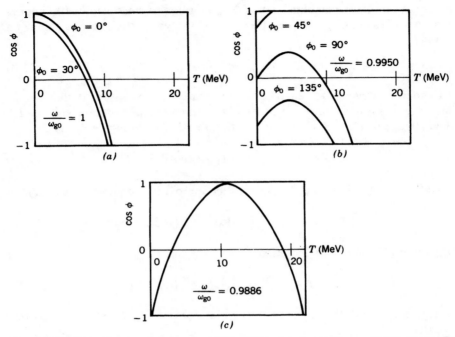

Figure 15.5 Phase histories of protons in uniform-field cyclotron; $\cos \phi$ versus kinetic energy (T) for different injection phase (ϕ_0). (*a*) $\omega/\omega_{g0} = 1$, where ω is rf angular frequency and ω_{g0} is nonrelativistic gyrofrequency. (*b*) $\omega/\omega_{g0} = 0.9950$. (*c*) $\omega/\omega_{g0} = 0.9896$, $\phi_0 = 180°$ (parameters for maximum kinetic energy).

The curves of Figure 15.5 depend on V_0, m_i, and ω/ω_{g0}. The maximum achievable energy corresponds to the curve illustrated in Figure 15.5c. The particle is injected at $\phi_0 = 180°$. The rf frequency is set lower than the nonrelativistic ion gyrofrequency. The two frequencies are equal when ϕ approaches $0°$. The curve of Figure 15.5c represents the maximum possible phase excursion of ions during acceleration and therefore the longest possible time of acceleration. Defining T_{max} as the maximum kinetic energy, Figure 15.5c implies the constraints

$$\cos \phi = -1 \quad \text{for } T = T_{max} \tag{15.14}$$

and

$$\cos \phi = +1 \quad \text{for } T = \tfrac{1}{2}T_{max}. \tag{15.15}$$

The last condition proceeds from the symmetric shape of the parabolic curve. Substitution of Eqs. (15.14) and (15.15) in Eq. (15.13) gives two equations in two unknowns for T_{max} and ω/ω_{g0}. The solution is

$$\omega/\omega_{g0} = 1/\left(1 + T_{max}/2m_ic^2\right) \tag{15.16}$$

and

$$T_{max} \cong \left(16qV_0m_ic^2/\pi\right)^{1/2}. \qquad \boxed{15.17}$$

Equation (15.17) is a good approximation when $T_{max} \ll m_ic^2$.

Note that the final kinetic energy is maximized by taking V_0 large. This comes about because a high gap voltage accelerates particles in fewer revolutions so that there is less opportunity for particles to get out of synchronization. Typical acceleration gap voltages are ± 100 kV. Inspection of Eq. (15.17) indicates that the maximum kinetic energy attainable is quite small compared to m_ic^2. In a typical cyclotron, the relativistic mass increase amounts to less than 2%. The small relativistic effects are important because they accumulate over many particle revolutions.

To illustrate typical parameters, consider acceleration of deuterium ions. The rest energy is 1.9 GeV. If $V_0 = 100$ kV, Eq. (15.17) implies that $T_{max} = 31$ MeV. The peak energy will be lower if radial variations of magnetic field are included. With $B_0 = 1.5$ T, the nonrelativistic gyrofrequency is $f_{g0} = 13.6$ MHz. For peak kinetic energy, the rf frequency should be about 13.5 MHz. The ions make approximately 500 revolutions during acceleration.

15.3 FOCUSING BY AZIMUTHALLY VARYING FIELDS (AVF)

Inspection of Eqs. (15.6) and (15.7) shows that synchronization in a cyclotron can be preserved only if the average bending magnetic field increases with radius. A positive field gradient corresponds to a negative field index in a magnetic field with azimuthal symmetry, leading to vertical defocusing. A positive field index can be tolerated if there is an extra source of vertical focusing. One way to provide additional focusing is to introduce azimuthal variations in the bending field. In this section, we shall study particle orbits in azimuthally varying fields. The intent is to achieve a physical understanding of AVF focusing through simple models. The actual design of accelerators with AVF focusing[†] is carried out using complex analytic calculations and, inevitably, numerical solution of particle orbits. The results of this section will be applied to isochronous cyclotrons in Section 15.4. In principle, azimuthally varying fields could be used for focusing in accelerators with constant particle orbit radius, such as synchrotrons or betatrons. These configurations are usually referred to as FFAG (fixed-field, alternating-gradient) accelerators. In

[†] K. R. Symon et al., *Phys. Rev.*, **103**, 1837 (1956); F. T. Cole et al., *Rev. Sci. Instrum.*, **28**, 403 (1957).

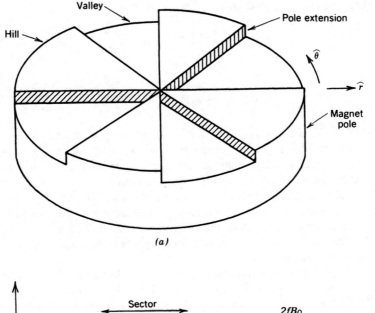

(a)

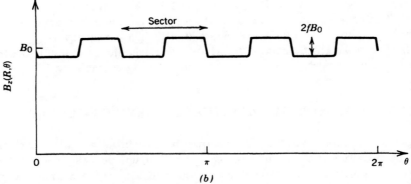

(b)

Figure 15.6 Magnetic fields in AVF cyclotron. (*a*) Magnet pole of AVF cyclotron, no spiral angle. (*b*) Vertical field amplitude as function of azimuth at constant radius.

practice, the cost of magnets in FFAG machines is considerably higher than more conventional approaches, so AVF focusing is presently limited to cyclotrons.

Figure 15.6*a* illustrates an AVF cyclotron field generated by circular magnet poles with wedge-shaped extensions attached. We begin by considering extentions with boundaries that lie along diameters of the poles; more general extension shapes, such as sections with spiral boundaries, are discussed below. Focusing by fields produced by wedge-shaped extensions is usually referred to as *Thomas focusing*.[†] The raised regions are called hills, and the recessed

[†] L. H. Thomas, *Phys. Rev.*, **54**, 580 (1938).

regions are called valleys. The magnitude of the vertical magnetic field is approximately inversely proportional to gap width; therefore, the field is stronger in hill regions. An element of field periodicity along a particle orbit is called a *sector*; a sector contains one hill and one valley. The number of sectors equals the number of pole extensions and is denoted N. Figure 15.6a shows a magnetic field with $N = 3$. The variation of magnetic with azimuth along a circle of radius R is plotted in Figure 15.6b. The definition of sector (as applied to the AVF cyclotron) should be noted carefully to avoid confusion with the term *sector magnet*.

The terminology associated with AVF focusing systems is illustrated in Figure 15.6b. The azimuthal variation of magnetic field is called *flutter*. Flutter is represented as a function of position by

$$B_z(R, \theta) = B_0(R)\Phi(R, \theta), \tag{15.18}$$

where $\Phi(R, \theta)$ is the *modulation function* which parametrizes the relative changes of magnetic field with azimuth. The modulation function is usually resolved as

$$\Phi(R, \theta) = 1 + f(R)g(\theta), \tag{15.19}$$

where $g(\theta)$ is a function with maximum amplitude equal to 1 and an average value equal to zero. The modulation function has a θ-averaged value of 1. The function $f(R)$ in Eq. (15.19) is the *flutter amplitude*.

The modulation function illustrated in Figure 15.6b is a step function. Other types of variation are possible. The magnetic field corresponding to a sinusoidal variation of gap width is approximately

$$B_z(R, \theta) = B_0(R)[1 + f(R)\sin N\theta], \tag{15.20}$$

so that

$$\Phi(\theta) = 1 + f(R)\sin N\theta. \tag{15.21}$$

The *flutter function* $F(R)$ is defined as the mean square relative azimuthal fluctuation of magnetic field along a circle of radius R:

$$F(R) = \langle \{[B_z(R, \theta) - B_0(R)]/B_0(R)\}^2 \rangle$$
$$= (1/2\pi) \int_0^{2\pi} [\Phi(R, \theta) - 1]^2 \, d\theta. \tag{15.22}$$

For example, $F(R) = f(R)^2$ for a step-function variation and $F(R) = \frac{1}{2}f(R)^2$ for the sinusoidal variation of Eq. (15.21).

Particle orbits in azimuthally varying magnetic bending fields are generally complex. In order to develop an analytic orbit theory, simplifying assumptions will be adopted. We limit consideration to a field with sharp transitions of

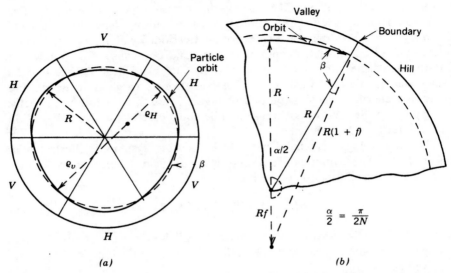

Figure 15.7 Thomas focusing with sharp field region boundaries in the limit of small flutter amplitude. (*a*) Main orbits: dashed line, uniform field; solid line, field with flutter. (*b*) Geometry for calculating focusing effect of fringing fields at valley–hill transition.

magnitude between hills and valleys (Fig. 15.6*b*). The hills and valleys occupy equal angles. The step-function assumption is not too restrictive; similar particle orbits result from continuous variations of gap width. Two limiting cases will be considered to illustrate the main features of AVF focusing: (1) small magnetic field variations ($f \ll 1$) and (2) large field variations with zero magnetic field in the valleys. In the latter case, the bending field is produced by a number of separated sector magnets. Methods developed in Chapters 6 and 8 for periodic focusing can be applied to derive particle orbits.

To begin, take $f \ll 1$. As usual, the strategy is to find the equilibrium orbit and then to investigate focusing forces in the radial and vertical directions. The magnetic field magnitude is assumed independent of radius; effects of average field gradient will be introduced in Section 15.4. In the absence of flutter, the equilibrium orbit is a circle of radius $R = \gamma m_i c / q B_0$. With flutter, the equilibrium orbit is changed from the circular orbit to the orbit of Figure 15.7*a*. In the sharp field boundary approximation, the modified orbit is composed of circular sections. In the hill regions, the radius of curvature is reduced, while the radius of curvature is increased in the valley regions. The main result is that the equilibrium orbit is not normal to the field boundaries at the hill–valley transitions.

There is strong radial focusing in a bending field with zero average field index; therefore, flutter has little relative effect on radial focusing in the limit $f \ll 1$. Focusing in a cyclotron is conveniently characterized by the dimensionless parameter ν (see Section 7.2), the number of betatron wavelengths during a particle revolution. Following the discussion of Section 7.3, we find that

$$\nu_r^2 \cong 1 \qquad (15.23)$$

for a radially uniform average field magnitude.

In contrast, flutter plays an important part in vertical focusing. Inspection of Figure 15.7a shows that the equilibrium orbit crosses between hill and valley regions at an angle to the boundary. The vertical forces acting on the particle are similar to those encountered in edge focusing (Section 6.9). The field can be resolved into a uniform magnetic field of magnitude $B_0[1 - f(R)]$ superimposed on fields of magnitude $2B_0 f(R)$ in the hill regions. Comparing Figure 15.7a to Figure 6.20, the orbit is inclined so that there is focusing at both the entrance and exit of a hill region. The vertical force arises from the fringing fields at the boundary; the horizontal field components are proportional to the change in magnetic field, $2B_0 f(R)$. Following Eq. 6.30, the boundary fields act as a thin lens with positive focal length

$$\text{focal length} \cong \left(\gamma m_i c/q[2B_0 f]\right)/|\tan \beta| = -R/2f|\tan \beta|, \qquad (15.24)$$

where β is the angle of inclination of the orbit to the boundary. The ray transfer matrix corresponding to transit across a boundary is

$$\mathbf{A}_b = \begin{bmatrix} 1 & 0 \\ -2f \tan \beta/R & 1 \end{bmatrix}. \qquad (15.25)$$

The inclination angle can be evaluated from the geometric construction of Figure 15.7b. The equilibrium orbit crosses the boundary at about $r = R$. The orbit radii of curvature in the hill and valley regions are $R(1 \pm f)$. To first order, the inclination angle is

$$|\beta| = \pi f/2N, \qquad (15.26)$$

where N is the number of sectors. The ray transfer matrix for a boundary is expressed as

$$\mathbf{A}_b = \begin{bmatrix} 1 & 0 \\ -\pi f^2/NR & 1 \end{bmatrix} \qquad (15.27)$$

for small β. Neglecting variations in the orbit length through hills and valleys caused by the flutter, the transfer matrix for drift is

$$\mathbf{A}_d = \begin{bmatrix} 1 & \pi R/N \\ 0 & 1 \end{bmatrix}. \qquad (15.28)$$

A focusing cell, the smallest element of periodcity, consists of half a sector (a drift region and one boundary transition). The total ray transfer matrix is

$$\mathbf{A} = \begin{bmatrix} 1 & \pi R/N \\ -\pi f^2/NR & 1 - (\pi f/N)^2 \end{bmatrix}. \tag{15.29}$$

The phase advance in the vertical direction is

$$\cos \mu \cong 1 - \tfrac{1}{2}\mu^2 = \tfrac{1}{2}\text{Tr}\,\mathbf{A} = 1 - \tfrac{1}{2}(\pi f/N)^2, \tag{15.30}$$

or

$$\mu \cong \pi f/N. \tag{15.31}$$

The net phase advance during one revolution is equal to $2N\mu$. The number of betatron oscillations per revolution is therefore

$$\nu_z = 2N\mu/2\pi = f. \tag{15.32}$$

The final form is derived by substituting from Eq. (15.31). The vertical number of betatron wavelengths can also be expressed in terms of a flutter function as

$$\nu_z^2 = F. \qquad \boxed{15.33}$$

Equation (15.33) is not specific to a step-function field. It applies generally for all modulation functions.

Stronger vertical focusing results if the hill–valley boundaries are modified from the simple diametric lines of Figure 15.6. Consider, for instance, spiral-shaped pole extensions, as shown in Figure 15.8. At a radius R, the boundaries between hills and valleys are inclined at an angle $\zeta(R)$ with respect to a diameter. Spiral-shaped pole extensions lead to an additional inclination of magnitude $\zeta(R)$ between the equilibrium particle orbit and the boundary. The edge fields from the spiral inclination act to alternately focus and defocus particles, depending on whether the particle is entering or leaving a hill region. For example, the spiral of Figure 15.8 is defocusing at a hill-to-valley transition. A focusing–defocusing lens array provides net focusing.

The effect of boundary inclination can easily be derived in the limit that $f \ll 1$ and combined with Thomas focusing for a total ν_z. A focusing cell extends over a sector; a cell consists of a drift region of length $\pi R/N$, a thin lens of focal length $+2f \tan \zeta/R$, a second drift region, and a lens with focal length $-2f \tan \zeta/R$. The total ray transfer matrix for a sector is

$$\mathbf{A} = \begin{bmatrix} \left(1 + 2f\pi \tan \zeta/N - (2\pi f \tan \zeta/N)^2\right) & \left(2\pi R/N - 2f\pi^2 R \tan \zeta/N^2\right) \\ -4\pi f^2 \tan^2\zeta/NR & \left(1 - 2\pi f \tan \zeta/N\right) \end{bmatrix}. \tag{15.34}$$

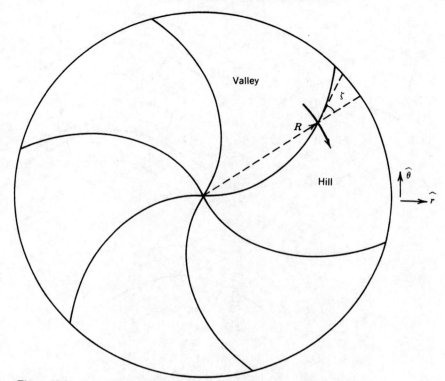

Figure 15.8 Geometry of magnet pole of spiral cyclotron showing inclination angle $\zeta(R)$.

Again, identifying $\frac{1}{2}\mathrm{Tr}\,\mathbf{A}$ with $\cos\mu$, we find that

$$\mu \cong \sqrt{2}\,\pi f \tan\zeta/N. \qquad (15.35)$$

Following the method used above, the number of vertical betatron oscillations per revolution is expressed simply as

$$\nu_z^2 = f^2(1 + 2\tan\zeta) = F(1 + 2\tan\zeta). \qquad \boxed{15.36}$$

Vertical focusing forces can be varied with radius through the choice of the spiral shape. The Archimedean spiral is often used; the boundaries of the pole extensions are defined by

$$r = A[\theta + J(2\pi)/2N], \qquad (15.37)$$

where $J = 0, 1, 2, \ldots, 2N - 1$. The corresponding inclination angle is

$$\tan\zeta(r) = d(r\theta)/dr = 2r/A. \qquad (15.38)$$

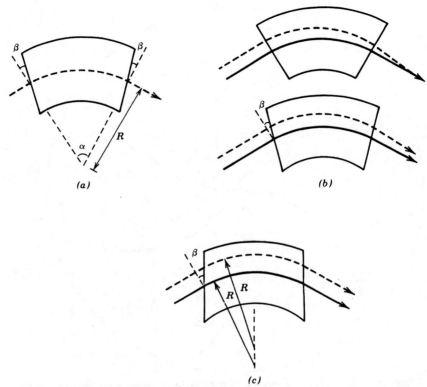

Figure 15.9 Focusing properties of a uniform field sector magnet with inclined boundaries. (*a*) Definition of angular extent of sector magnet (*α*) and boundary inclination angle (*β*) with respect to entering or exiting main particle orbit. (*b*) Horizontal trajectories of initially parallel particle orbits on main orbit (solid line) and displaced from main orbit (dashed line) with (bottom) and without (top) boundary inclination ($\beta < \frac{1}{2}\alpha$). (*c*) Horizontal orbits when $\beta = \frac{1}{2}\alpha$.

Archimedean spiral pole extensions lead to vertical focusing forces that increase with radius.

An analytical treatment of AVF focusing is also possible for a step-function field with $f = 1$. In this case (corresponding to the separated sector cyclotron), the bending field consists of regions of uniform magnetic field separated by field-free regions. Focusing forces arise from the shape of the sector magnet boundaries. As an introduction, consider vertical and radial focusing in a single-sector magnet with inclined boundaries (Fig. 15.9*a*). The equilibrium orbit in the magnetic field region is a circular section of radius R centered vertically in the gap. The circular section subtends an angle α. Assume that the boundary inclinations, β, are equal at the entrance and exit of the magnet.

In the vertical direction, the ray transfer matrix for the magnet is the product of matrices representing edge focusing at the entrance, a drift distance

αR, and focusing at the exit. We can apply Eqs. (15.27) and (15.28) to calculate the total ray transfer matrix. In order to calculate focusing in the radial direction, we must include the effect of the missing sector field introduced by the inclination angle β. For the geometry of Figure 15.9a, the inclination reduces radial focusing in the sector magnet. Orbits with and without a boundary inclination are plotted in Figure 15.9b. Figure 15.9c shows the equilibrium particle orbit and an off-axis parallel orbit in a sector magnet with $\beta = \frac{1}{2}\alpha$. The boundary is parallel to a line through the midplane of the magnet; the gyrocenters of both orbits also lie on this line. Therefore, the orbits are parallel throughout the sector and there is no focusing. A value of inclination ($\beta < \frac{1}{2}\alpha$) moves the gyrocenter of the off-axis particle to the left; the particle emerges from the sector focused toward the axis. The limit on β for radial focusing in a uniform-field sector magnet is

$$\beta < \tfrac{1}{2}\alpha. \tag{15.39}$$

We now turn our attention to the AVF sector field with diametric boundaries shown in Figure 15.10. The equilibrium orbits can be constructed with compass and straightedge. The orbits are circles in the sector magnets and straight lines in between. They must match in position and angle at the boundaries. Figures 15.10a, b show solutions with $N = 2$ and $N = 3$ for hills and valleys occupying equal azimuths ($\alpha = \pi/N$). Note that in all cases the inclination angle of the orbit at a boundary is one-half the angular extent of the sector, $\beta = \alpha/2$. Figures 15.10c and d illustrate the geometric construction of off-axis horizontal orbits for conditions corresponding to stability ($\alpha > \pi/N$, $\beta < \alpha/2$) and instability ($\alpha < \pi/N$, $\beta > \alpha/2$). The case of $N = 2$ is unstable for all choices of α. This arises because particles are overfocused when $\alpha > \pi/N$. This effect is clearly visible in Figure 15.10e. It is generally true that particle orbits are unstable in any type of AVF field with $N = 2$.

Spiral boundaries may also be utilized in separated sector fields. Depending on whether the particles are entering or leaving a sector, the edge-focusing effects are either focusing or defocusing in the vertical direction. Applying matrix algebra and the results of Section 6.9, it is easy to show that ν_z is

$$\nu_z^2 = (1 + 2\tan\zeta) \tag{15.40}$$

for $\alpha = \pi/N$. Spiral boundaries contribute alternate focusing and defocusing forces in the radial direction that are 180° out of phase with the axial forces. For $\alpha = \pi/N$, the number of radial betatron oscillations per revolution is approximately

$$\nu_r^2 \cong 2\tan\zeta. \tag{15.41}$$

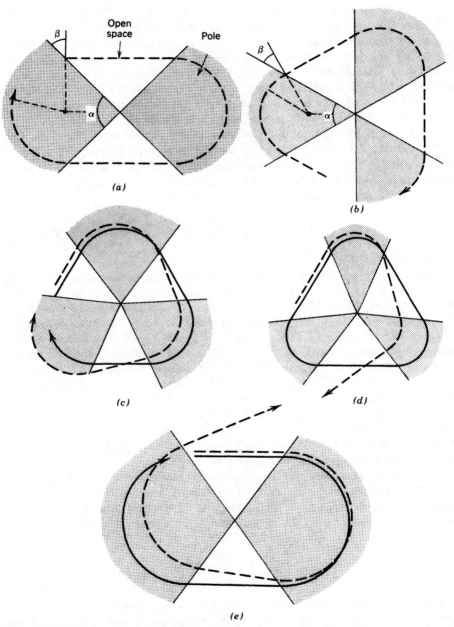

Figure 15.10 Horizontal particle orbits constructed with compass and straightedge in separated sector cyclotron with sector boundaries on diameter showing sector angular extent (α) and boundary inclination angle (β). (*a*) Main orbit with hills and valleys of equal angular extent ($\beta = \frac{1}{2}\alpha$), $N = 2$. (*b*) Main orbit with $\beta = \frac{1}{2}\alpha$, $N = 3$. (*c*) Stable horizontal orbits in three-sector field with $\beta < \frac{1}{2}\alpha$. Solid line, main orbit; dashed line, off-axis orbit. (*d*) Unstable horizontal orbits when $\beta > \frac{1}{2}\alpha$. (*e*) Illustration of instability in two-sector field with $\beta < \frac{1}{2}\alpha$.

15.4 THE SYNCHROCYCLOTRON AND THE AVF CYCLOTRON

Following the success of the uniform-field cyclotron, efforts were made to reach higher beam kinetic energy. Two descendants of the cyclotron are the synchrocyclotron and the AVF (isochronous) cyclotron. The machines resolve the problem of detuning between particle revolutions and rf field in quite different ways. Synchrocyclotrons have the same geometry as the SF cyclotron. A large magnet with circular poles produces an azimuthally symmetric vertical field with positive field index. Ions are accelerated from rest to high energy by an oscillating voltage applied between dees. The main difference is that the frequency is varied to preserve synchronism.

There are a number of differences in the operation of synchrocyclotrons and cyclotrons. Synchrocyclotrons are cycled, rather than continuous; therefore, the time-average beam current is much lower. The longitudinal dynamics of particles in a synchrocyclotron do not follow the model of Section 15.2 because there is a synchronous phase. The models for phase dynamics developed in Chapter 13 can be adapted to the synchrocyclotron. The machine can contain a number of confined particle bunches with phase parameters centered about the bunch that has ideal matching to the rf frequency. The beam bunches are distributed as a group of closely spaced turns of slightly different energy. The acceptance of the rf buckets decreases moving away from the ideal match, defining a range of time over which particles can be injected into the machine. In research applications, the number of bunches contained in the machine in a cycle is constrained by the allowed energy spread of the output beam.

There are technological limits on the rate at which the frequency of oscillators can be swept. These limits were particularly severe in early synchrocyclotrons that used movable mechanical tuners rather than the ferrite tuners common on modern synchrotrons. The result is that the acceleration cycle of a synchrocyclotron extends over a longer period than the acceleration time for an ion in a cyclotron. Typically, ions perform between 10,000 and 50,000 revolutions during acceleration in a synchrocyclotron. The high recirculation factor implies lower voltage between the dees. The cycled operation of the synchrocyclotron leads to different methods of beam extraction compared to cyclotrons. The low dee voltage implies that orbits have small separation (< 1 mm), ruling out the use of a septum. On the other hand, all turns can be extracted at the same time by a pulsed field since they are closely spaced in radius. Figure 15.11 illustrates one method of beam extraction from a synchrocyclotron. A pulsed electric field is used to deflect ions on to a perturbed orbit which leads them to a magnetic shield. The risetime of voltage on the kicker electrodes should be short compared to the revolution time of ions. Pulsed extraction is characteristic of cycled machines like the synchrocyclotron and synchrotron. In large synchrotrons with relatively long revolution time, pulsed magnets with ferrite cores are used for beam deflection.

Containment of high-energy ions requires large magnets. For example, a 600-MeV proton has a gyroradius of 2.4 m in a 1.5-T field. This implies a pole

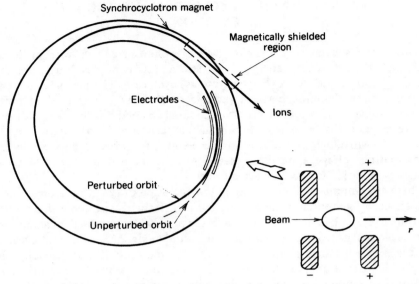

Figure 15.11 Beam extraction from a synchrocyclotron by a pulsed radial electric field.

diameter greater than 15 ft. Synchrocyclotron magnets are among the largest monolithic, iron core magnets ever built. The limitation of this approach is evident; the volume of iron required rises roughly as the cube of the kinetic energy. Two synchrocyclotrons are still in operation: the 184-in. machine at Lawrence Berkeley Laboratory and the CERN SC.

The AVF cyclotron has fixed magnetic field and rf frequency; it generates a continuous-beam pulse train. Compensation for relativistic mass increase is accomplished by a magnetic field that increases with radius. The vertical defocusing of the negative field index is overcome by the focusing methods described in Section 15.3.

We begin by calculating the radial field variations of the θ-averaged vertical field necessary for synchronization. The quantity $B(R)$ is the averaged field around a circle of radius R and B_0 is the field at the center of the machine. Assume that flutter is small, so that particle orbits approximate circles of radius R, and let $B(R)$ represent the average bending field at R. Near the origin ($R = 0$), the AVF cyclotron has the same characteristics as a uniform field cyclotron; therefore, the rf frequency is

$$\omega = qB_0/m_i, \tag{15.42}$$

where m_i is the rest energy of the ion. Synchronization with the fixed frequency at all radii implies that

$$B(R) = \gamma(R)m_i\omega/q, \tag{15.43}$$

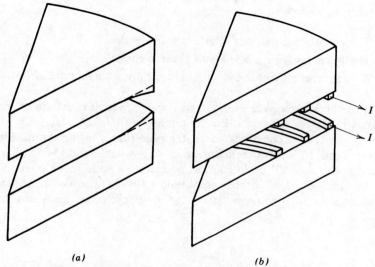

Figure 15.12 Methods for generating vertical magnetic field with negative field index (positive radial gradient). (*a*) Radial variation of gap width. (*b*) Trim coils.

or

$$B(R)/B_0 = \gamma(R). \tag{15.44}$$

The average magnetic field is also related to the average orbit radius and ion energy through Eq. (3.38):

$$R = \gamma(R)\beta m_i c/qB(R) = [m_i c/qB(R)]\sqrt{\gamma^2 - 1}$$

$$= (m_i c/qB_0)\sqrt{\gamma^2 - 1}/\gamma. \tag{15.45}$$

Combining Eqs. (15.44) and (15.45), we find

$$B(R)/B_0 = \gamma(R) = \left[1 - (qB_0 R/m_i c)^2\right]^{-1/2}. \tag{15.46}$$

Equation (15.46) gives the following radial variation of field index:

$$n(R) = -[R/B(R)][dB(R)/dR] = -(\gamma^2 - 1). \tag{15.47}$$

Two methods for generating a bending field with negative field index (positive radial gradient) are illustrated in Figure 15.12. In the first, the distance between poles decreases as a function of radius. This method is useful mainly in small, low-energy cyclotrons. It has the following drawbacks for large research machines:

1. The constricted gap can interfere with the dees.
2. The poles must be shaped with great accuracy.
3. A particular pole shape is suitable for only a single type of ion.

A better method to generate average radial field gradient is the use of *trimming coils*, illustrated in Figure 15.12*b*. Trimming coils (or *k* coils) are a set of adjustable concentric coils located on the pole pieces inside the magnet gap. They are used to shift the distribution of vertical field. With adjustable trimming coils, an AVF cyclotron can accelerate a wide range of ion species.

In the limit of small flutter amplitude ($f \ll 1$), the radial and vertical betatron oscillations per revolution in an AVF cyclotron are given approximately by

$$\nu_r^2 \cong 1 - n + F(R)n^2/N^2 + \cdots, \qquad \boxed{15.48}$$

$$\nu_z^2 \cong n + F(r) + 2F(r)\tan^2\zeta + F(R)n^2/N^2 + \cdots. \qquad \boxed{15.49}$$

Equations (15.48) and (15.49) are derived through a linear analysis of orbits in an AVF field in the small flutter limit. The terms on the right-hand side represent contributions from various types of focusing forces. In Eq. (15.48), the terms have the following interpretations:

Term 1:   Normal radial focusing in a bending field.
Term 2:   Contribution from an average field gradient ($n < 0$ in an AVF cyclotron).
Term 3:   Alternating-gradient focusing arising from the change in the actual field index between hills and valleys. Usually, this is a small effect.

A term involving the spiral angle ζ is absent from the radial equation. This comes about because of cancellation between the spiral term and a term arising from differences of the centrifugal force on particles between hills and valleys.

The terms on the right-hand side of Eq. (15.49) for vertical motion represent the following contributions:

Term 1:   Defocusing by the average radial field gradient.
Term 2:   Thomas focusing.
Term 3:   FD focusing by the edge fields of a spiral boundary.
Term 4:   Same as the third term of Eq. (15.48).

Symmetry considerations dictate that the field index and spiral angle near the center of an AVF cyclotron approach zero. The flutter amplitude also approaches zero at the center since the effects of hills and valleys on the field cancel out at radii comparable to or less than the gap width between poles. As in the conventional cyclotron, electrostatic focusing at the acceleraton gaps plays an important role for vertical focusing of low-energy ions. At large radius, there is little problem in ensuring good radial focusing. Neglecting the third term, Eq. (15.48) may be rewritten as

$$\nu_r \cong \gamma. \tag{15.50}$$

using Eq. (15.47). The quantity ν_r is always greater than unity; radial focusing is strong. Regarding vertical focusing, the combination of Thomas focusing and spiral focusing in Eq. (15.49) must increase with radius to compensate for the increase in field index. This can be accomplished by a radial increase of $F(R)$ or $\zeta(R)$. In the latter case, boundary curves with increasing ζ (such as the Archimedean spiral) can be used. Isochronous cyclotrons have the property that the revolution time is independent of the energy history of the ions. Therefore, there are n phase oscillations, and ions have neutral stability with respect to the rf phase. The magnet poles of high-energy isochronous cyclotrons must be designed with high accuracy so that particle synchronization is maintained through the acceleration process.

In addition to high-energy applications, AVF cyclotrons are well suited to low-energy medical and industrial applications. The increased vertical focusing compared to a simple gradient field means that the accelerator has greater transverse acceptance. Higher beam currents can be contained, and the machine is more tolerant to field errors (see Section 15.7). Phase stability is helpful, even in low-energy machines. The existence of a synchronous phase implies higher longitudinal acceptance and lower beam energy spread. The AVF cyclotron is much less expensive per ion produced than a uniform-field cyclotron.

In the range of kinetic energy above 100 MeV, the separated sector cyclotron is a better choice than the single-magnet AVF cyclotron. The separated sector cyclotron consists of three or more bending magnets separated by field-free regions. It has the following advantages:

1.   Radio-frequency cavities for beam acceleration can be located between the sectors rather than between the magnet poles. This allows greater latitude in designing the focusing magnetic field and the acceleration system. Multiple acceleration gaps can be accommodated, leading to rapid acceleration and large orbit separation.

2.   The bending field is produced by a number of modular magnets rather than a single larger unit. Modular construction reduces the problems of fabrication and mechanical stress. This is particularly important at high energy.

The main drawback of the separated sector cyclotron is that it cannot

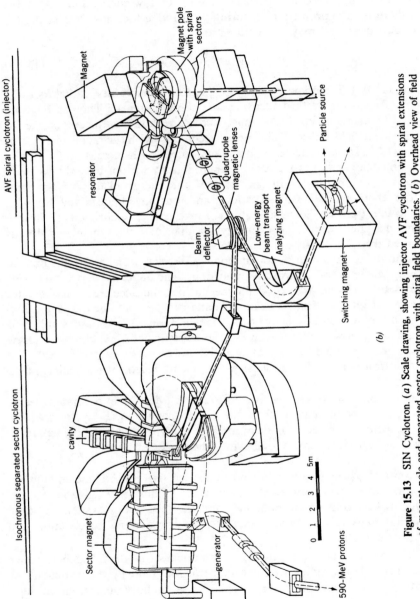

Figure 15.13 SIN Cyclotron. (*a*) Scale drawing, showing injector AVF cyclotron with spiral extensions of magnet pole and separated sector cyclotron with spiral field boundaries. (*b*) Overhead view of field boundaries in separated sector cyclotron with calculated proton orbits at 75, 177, 279, 381, 483, and 585 MeV. (Courtesy W. Joho, Swiss Institute for Nuclear Studies.)

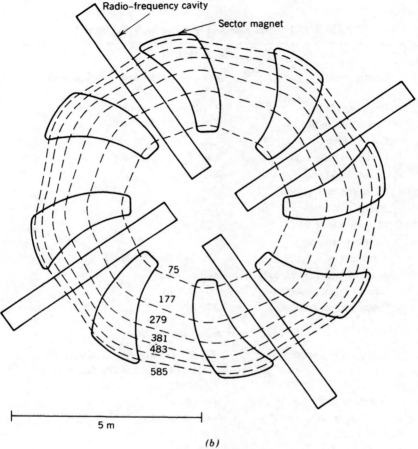

Figure 15.13 (*Continued*).

accelerate ions from zero energy. The beam transport region is annular since structures for mechanical support of the individual magnet poles must be located on axis. Ions are preaccelerated for injection into a separated sector cyclotron. Preacceleration can be accomplished with a low-energy AVF cyclotron or a linac. The injector must be synchronized so that micropulses are injected into the high-energy machine at the proper phase.

Figure 15.13*a* shows the separated sector cyclotron at the Swiss Nuclear Institute. Parameters of the machine are summarized in Table 15.1. The machine was designed for a high average flux of light ions to generate mesons for applications to radiation therapy and nuclear research. The accelerator has eight spiral sector magnets with a maximum hill field of 2.1 T. Large waveguides connect rf supplies to a four acceleration gaps. In operation, the

TABLE 15.1   Parameters of SIN 590 MeV Cyclotron[a]

Injector Cyclotron

Configuration	AVF cyclotron with spiral-shaped sectors
Injection radius	1.5 cm
Extraction radius	105 cm
Pole diameter	250 cm
Peak magnetic field	1.65 T
Number of sectors	4
Maximum spiral angle	55°
Number of trimming coils	12
Magnet power	500 kW
Number of dees	1
rf frequency	4.6–17 MHz (tunable)
Maximum energy gain/turn	160 kV/turn
Beam energy (protons)	10–72 MeV
Beam current (protons)	170 μA

590-MeV Separated Sector Cyclotron

Main applications	Research in nuclear physics, solid-state physics and chemistry, pion therapy
Injection radius	210 cm
Extraction radius	445 cm
Peak magnetic field	2.09 T
Averaged field at exit radius	0.87 T
Number of sector magnets	8
Maximum spiral angle	35°
Number of trimming coils	18
Magnet power	670 MW
Number of acceleration cavities	4
rf frequency	50.63 MHz
Maximum energy gain/turn	2200 kV/turn
rf power (maximum)	800 MW
Extraction system	Electrostatic septum

[a]Swiss Nuclear Research Institute.

machine requires 0.5 MW of rf input power. The peak acceleration gap voltage is 500 kV. The maximum orbit diameter of the cyclotron is 9 m for a maximum output energy of 590 MeV (protons). The time-averaged beam current is 200 μA. A standard AVF cyclotron with four spiral-shaped sectors is used as an injector. An increase of average beam current to 1 mA is expected with the addition of a new injector. The injector is a spiral cyclotron with four sectors. The injector operates at 50.7 MHz and generates 72-MeV protons. Figure 15.13*b* is an overhead view of the magnets and rf cavities in the separated sector cyclotron. Six selected orbits are illustrated at equal energy intervals from 72 to 590 MeV. Note that the distance an ion travels through the sector field increases with orbit radius (negative effective field index). The diagram also indicates the radial increase of the inclination angle between sector field boundaries and the particle orbits.

15.5 PRINCIPLES OF THE SYNCHROTRON

Synchrotrons are resonant circular particle accelerators in which both the magnitude of the bending magnetic field and the rf frequency are cycled. An additional feature of most modern synchrotrons is that focusing forces are adjustable independent of the bending field. Independent variation of the focusing forces, beam-bending field, and rf frequency gives synchrotrons two capabilities that lead to beam energies far higher than those from other types of circular accelerators:

1.   The betatron wavelength of particles can be maintained constant as acceleration proceeds. This makes it possible to avoid the orbital resonances that limit the output energy of the AVF cyclotron.

2.   The magnetic field amplitude is varied to preserve a constant particle orbit radius during acceleration. Therefore, the bending field need extend over only a small annulus rather than fill a complete circle. This implies large savings in the cost of the accelerator magnets. Furthermore, the magnets can be fabricated as modules and assembled into ring accelerators exceeding 6 km in circumference.

The main problems of the synchrotron are (1) a complex operation cycle and (2) low average flux.

The components of a modern separated function synchrotron are illustrated in Figure 15.14. An ultra-high-vacuum chamber for beam transport forms a closed loop. Circular sections may be interrupted by straight sections to facilitate beam injection, beam extraction, and experiments. Acceleration takes place in a cavity filled with ferrite cores to provide inductive isolation over a broad frequency range. The cavity is similar to a linear induction accelerator cavity. The two differences are (1) an ac voltage is applied across the gap and (2) the ferrites are not driven to saturation to minimize power loss.

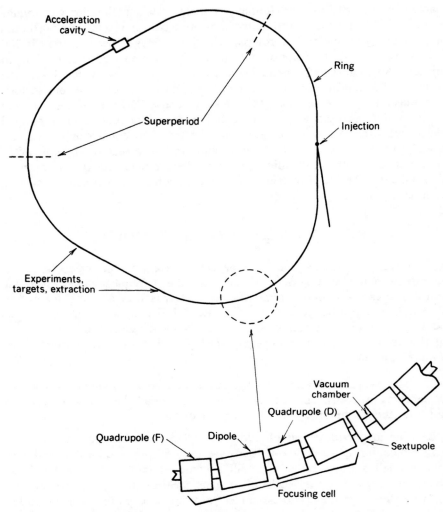

Figure 15.14 Major components and definition of terms in separated function synchrotron.

Beam bending and focusing are accomplished with magnetic fields. The separated function synchrotron usually has three types of magnets, classified according to the number of poles used to generate the field. *Dipole magnets* (Fig. 15.15a) bend the beam in a closed orbit. *Quadrupole magnets* (Fig. 15.15b) (grouped as quadrupole lens sets) focus the beam. *Sextupole magnets* (Fig. 15.15c) are usually included to increase the tolerance of the focusing system to beam energy spread. The global arrangement of magnets around the synchrotron is referred to as a *focusing lattice*. The lattice is carefully designed to maintain a stationary-beam envelope. In order to avoid resonance instabilities, the lattice design must not allow betatron wavelengths to equal a char-

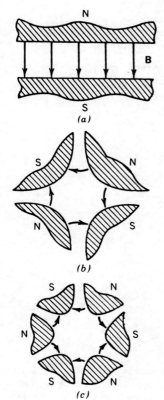

Figure 15.15 Classification of synchrotron magnets. (*a*) Dipole magnet (bending particle orbits). (*b*) Quadrupole (transverse focusing). (*c*) Sextupole (chromatic correction; assuring that particles in a range of energy about the mean have the same ν).

acteristic dimension of the machine (such as the circumference). Resonance conditions are parametrized in terms of *forbidden values* of ν_r and ν_z.

A focusing cell is stricty defined as the smallest element of periodicity in a focusing system. A period of a noncircular synchrotron contains a large number of optical elements. A cell may encompass a curved section, a straight section, focusing and bending magnets, and transition elements between the sections. The term *superperiod* is usually used to designate the minimum periodic division of a synchrotron, while *focusing cell* is applied to a local element of periodicity within a superperiod. The most common local cell configuration is the *FODO* cell. It consists of a focusing quadrupole (relative to the r or z direction), a dipole magnet, a defocusing quadrupole, and another dipole. Horizontal focusing forces in the bending magnets are small compared to that in the quadrupoles. For transverse focusing, the cell is represented as a series of focusing and defocusing lenses separated by drift (open) spaces.

The alternating-gradient synchrotron (AGS) is the precursor of the separated function synchrotron. The AGS has a ring of magnets which combine the functions of beam bending and focusing. Cross sections of AGS magnets

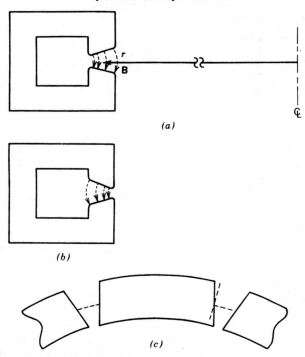

Figure 15.16 Synchrotron Magnets. (*a*) Magnet in alternating-gradient synchrotron with strong radial focusing and vertical defocusing. (*b*) AGS magnet with strong vertical focusing and radial defocusing. (*c*) Overhead view of uniform-field sector magnet boundaries in zero-gradient synchrotron.

are illustrated in Figure 15.16. A strong positive or negative radial gradient is superimposed on the bending field; horizontal and vertical focusing arises from the transverse fields associated with the gradient (Section 7.3). The magnet of Figure 15.16*a* gives strong radial focusing and horizontal defocusing, while the opposite holds for the magnet of Figure 15.16*b*.

Early synchrotrons utilized simple gradient focusing in an azimuthally symmetric field. They were constructed from a number of adjacent bending magnets with uniform field index in the range $0 < n < 1$. These machines are now referred to as *weak focusing synchrotrons*, since the betatron wavelength of particles was larger than the machine circumference. The zero-gradient synchrotron (ZGS) (Fig. 15.16*c*) was an interesting variant of the weak focusing machine. Bending and focusing were performed by sector magnets with uniform-field magnitude (zero gradient). The sector field boundaries were inclined with respect to the orbits to give vertical focusing [via edge focusing (Section 6.9)] and horizontal focusing [via sector focusing (Section 6.8)]. The advantage of the ZGS compared to other weak focusing machines was that higher bending fields could be achieved without local saturation of the poles.

The limit on kinetic energy in an ion synchrotron is set by the bending magnetic field magnitude and the area available for the machine. The ring radius of relativistic protons is given by

$$R = 3.3E/\langle B \rangle \text{ (m)}, \tag{15.51}$$

where $\langle B \rangle$ is the average magnetic field (in tesla) and E is the total ion energy in GeV. Most ion synchrotrons accelerate protons; protons have the highest charge-to-mass ratio and reach the highest kinetic energy per nucleon for a given magnetic field. Synchrotrons have been used for heavy-ion acceleration. In this application, ions are preaccelerated in a linear accelerator and directed through a thin foil to strip electrons. Only ions with high charge states are selected for injection into the synchrotron.

The maximum energy in an electron synchrotron is set by emission of *synchrotron radiation*. Synchrotron radiation results from the continuous transverse acceleration of particles in a circular orbit. The total power emitted per particle is

$$P = 2cE^4 r_0 / 3R^2 (m_0 c^2)^3 \text{ (watts)}, \qquad \boxed{15.52}$$

where E is the total particle energy and R is the radius of the circle. Power in Eq. (15.52) is given in electron volts per second if all energies on the right-hand side are expressed in electron volts. The quantity r_0 is the classical radius of the particle,

$$r_0 = q^2 / 4\pi\varepsilon_0 m_0 c^2. \tag{15.53}$$

The classical radius of the electron is

$$r_e = 2.82 \times 10^{-15} \text{ m.} \tag{15.54}$$

Inspection of Eqs. (15.52) and (15.53) shows that synchrotron radiation has a negligible effect in ion accelerators. Compared to electrons, the power loss is reduced by a factor of $(m_e / m_i)^4$. To illustrate the significance of synchrotron radiation in electron accelerators, consider a synchrotron in which electrons gain an energy eV_0 per turn. The power input to electrons (in eV/s) is

$$P = cV_0 / 2\pi R. \tag{15.55}$$

Setting Eqs. (15.52) and (15.55) equal, the maximum allowed total energy is

$$E \leq \left[3V_0 (m_0 c^2)^3 R / 4\pi r_0 \right]^{0.25}. \tag{15.56}$$

For example, with $R = 20$ m and $V_0 = 100$ kV, the maximum energy is $E = 2.2$ GeV. Higher energies result from a larger ring radius and higher power input to the accelerating cavities, but the scaling is weak. The peak energy achieved in electron synchrotrons is about 12 GeV for $R = 130$ m. Linear accelerators are the only viable choice to reach higher electron energy for particle physics research. Nonetheless, electron synchrotrons are actively employed in other areas of applied physics research. They are a unique source of intense radiation over a wide spectral range via synchrotron radiation. New synchrotron radiation facilities are planned as research tools in atomic and solid-state physics.

Synchrotron radiation has some advantageous effects on electron beam dynamics in synchrotrons. The quality of the beam (or the degree to which particle orbit parameters are identical) is actually enhanced by radiation. Consider, for instance, the spread in longitudinal energy in a beam bunch. Synchrotron radiation is emitted over a narrow cone of angle

$$\Delta \theta = \left(m_e c^2 / E \right) \tag{15.57}$$

in the forward direction relative to the instantaneous electron motion. Therefore, the emission of photons slows electrons along their main direction of motion while making a small contribution to transverse motion. According to Eq. (15.52), higher-energy electrons lose more energy; therefore, the energy spread of an electron bunch decreases. This is the simplest example of *beam cooling*. The process results in a reduction of the random spread of particle orbits about a mean; hence, the term *cooling*.

The highest-energy accelerator currently in operation is located at the Fermi National Accelerator Laboratory. The 2-km-diameter proton synchrotron consists of two accelerating rings, built in two stages. In the main ring (completed in 1971), beam focusing and bending are performed by conventional magnets. Beam energies up to 450 GeV have been achieved in this ring. After seven years of operation, an additional ring was added in the tunnel beneath the main ring. This ring, known as the energy doubler, utilizes superconducting magnets. The higher magnetic field makes it possible to generate beams with 800 GeV kinetic energy. The total experimental facility, with beam transport elements and experimental areas designed to accommodate the high-energy beams, is known as the Tevatron. A scale drawing of the accelerator and experimental areas is shown in Figure 15.17a. Protons, extracted from a 750-kV electrostatic accelerator, are accelerated in a 200-MeV linear accelerator. The beam is then injected into a rapid cycling booster synchrotron which increases the energy to 8 GeV. The booster synchrotron cycles in 33 ms. The outputs from 12 cycles of the booster synchrotron are used to fill the main ring during a constant-field initial phase of the main ring acceleration cycle. The booster synchrotron has a circumference equal to 1/13.5 that of the main ring. The 12 pulses are injected head to tail to fill most of the main ring circumference.

A cross section of a superconducting bending magnet from the energy doubler is shown in Figure 15.17b. It consists of a central bore tube of average

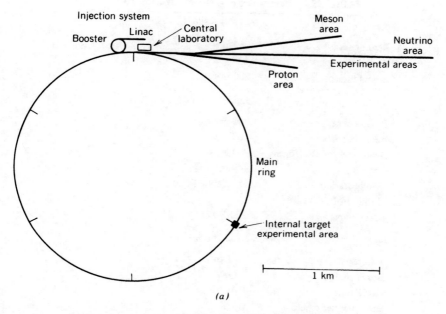

(a)

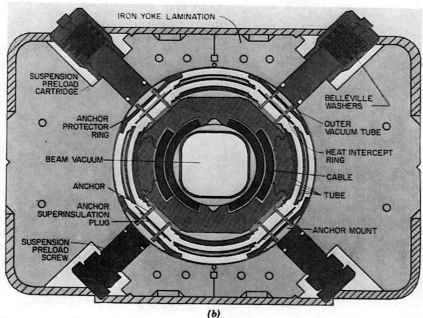

IRON YOKE LAMINATION

SUSPENSION
PRELOAD
CARTRIDGE

ANCHOR
PROTECTOR
RING

BEAM VACUUM

ANCHOR

ANCHOR
SUPERINSULATION
PLUG

SUSPENSION
PRELOAD
SCREW

BELLEVILLE
WASHERS

OUTER
VACUUM TUBE

HEAT INTERCEPT
RING

CABLE

TUBE

ANCHOR MOUNT

(b)

Figure 15.17 FNAL Synchrotron. (a) Scale over head view showing injector accelerators, main accelerator ring, and experimental areas for elementary particle physics research. (b) Cross-sectional view energy doubler superconducting bending magnet. (Courtesy F. Cole, Fermi National Accelerator Laboratory.)

537

TABLE 15.2 Parameters of 400-GeV Synchrotron[a]

Injector	
Beam energy	750 keV
Particle	p^+
Beam current	200 mA
Linac	
Beam energy	200 MeV
Beam current	100 mA
Beam macropulselength	3–12 μs
Length	140 m
Configuration	Drift tube linac
Number of drift tubes	295
Frequency	200 MHz
Booster Synchrotron	
Radius	74.1 m
Beam energy	8 GeV
Cycle time	0.067 s
Peak magnetic field	0.7 T
Energy gain/turn	0.7 MeV/turn
Configuration	Alternating gradient, H-type magnets
Number of magnets	96
Pulse rate	15 Hz
Number of rf cavities	16
Frequency excursion	30–53 MHz
Transition energy	4.2 GeV
Peak intensity	10^{12} protons/pulse
Main Ring	
Configuration	Separated function, bending dipole magnets, and quadrupole focusing magnets
Number of magnets	1014
Peak magnetic field	2.0 T
Magnet alignment accuracy	< ±1 mm
Injection pulses	12
Radius	1000 m
Energy gain rate	100–125 GeV/s
Cycling time (400 GeV)	12 s
Energy gain/turn	2.5 MeV
ν	19.4

[a] Fermi National Accelerator Laboratory.

radius 7 cm surrounded by superconducting windings with a spatial distribution calculated to give a highly uniform bending field. The windings are surrounded by a layer of stainless steel laminations to clamp the windings securely to the tube. The assembly is supported in a vacuum cryostat by fiberglas supports, surrounded by a thermal shield at liquid nitrogen temperature. A flow of liquid helium maintains the low temperature of the magnet coils. Bending magnets in the energy doubler are 6.4 m in length. A total of 774 units are necessary. Quadrupoles are constructed in a similar manner; a total of 216 focusing magnets are required. The parameters of the FNAL accelerator are listed in Table 15.2.

Storage rings consist of bending and focusing magnets and a vacuum chamber in which high-energy particles can be stored for long periods of time. The background pressure must be very low to prevent particle loss through collisions. Storage rings are filled with particles by a high-energy synchrotron or a linear accelerator. Their geometry is almost identical to the separated function synchrotron. The main difference is that the particle energy remains constant. The magnetic field is constant, resulting in considerable simplification of the design. A storage ring may have one or more acceleration cavities to compensate for radiative energy loss of electrons or for longitudinal bunching of ions.

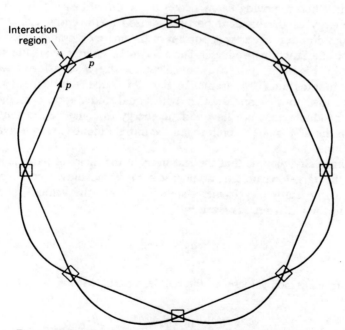

Figure 15.18 Arrangement of Intersecting Storage Ring, CERN.

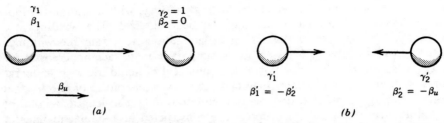

Figure 15.19 Collision of high-energy particle with target particle of equal rest mass. (*a*) View in stationary frame. (*b*) View in center-of-momentum frame.

One of the main applications of storage rings is in colliding beam facilities for high-energy particle physics. A geometry used in the ISR (intersecting storage ring) at CERN is shown in Figure 15.18. Two storage rings with straight and curved sections are interleaved. Proton beams circulating in opposite directions intersect at small angles at eight points of the ring. Proton–proton interactions are studied by detectors located near the intersection points.

Colliding beams have a significant advantage for high-energy physics research. The main requisite for probing the nature of elementary particles is that a large amount of energy must be available to drive reactions with a high threshold. When a moving beam strikes a stationary target (Fig. 15.19*a*), the kinetic energy of the incident particle is used inefficiently. Conservation of momentum dictates that a large portion of the energy is transformed to kinetic energy of the reaction products. The maximum energy available to drive a reaction in Figure 15.19*a* can be calculated by a transformation to the center-of-momentum (CM) frame. In the CM frame, the incident and target particles move toward one another with equal and opposite momenta. The reaction products need not have kinetic energy to conserve momentum when viewed in the CM frame; therefore, all the initial kinetic energy is available for the reaction.

For simplicity, assume that the rest mass of the incident particle is equal to that of the target particle. Assume the CM frame moves at a velocity $c\beta_u$ relative to the stationary frame. Using Eq. (2.30), the velocity of the target particle in the CM frame is given by

$$c\beta_2' = -c\beta_u, \tag{15.58}$$

and the transformed velocity of the incident particle is

$$c\beta_1' = c(\beta_1 - \beta_u)/(1 - \beta_u\beta_1). \tag{15.59}$$

Both particles have the same value of γ' in the CM frame; the condition of

**TABLE 15.3 Stationary Target: Available Reaction Energy (T_{cm})
Versus Incident Particle Kinetic Energy (T)**

γ	γ'	T_{cm}/T $2(\gamma'-1)/(\gamma-1)$	T (protons) (GeV)	T_{cm} (protons) (GeV)
2.00	1.22	0.449	0.94	0.42
4.00	1.58	0.387	2.81	1.09
8.00	2.12	0.320	6.57	2.10
16.00	2.96	0.255	14.07	3.59
32.00	4.06	0.198	29.08	5.76
64.00	5.70	0.149	59.09	8.81
128.00	8.03	0.111	119.13	13.22
256.00	11.34	0.081	239.19	19.37
512.00	16.14	0.059	479.32	28.27
1024.00	23.41	0.044	959.57	42.22

equal and opposite momenta implies

$$-\beta_2' = \beta_1'. \tag{15.60}$$

Combining Eqs. (15.58), (15.59), and (15.60), we find that

$$\beta_u = (1/\beta_1) - \left[(1/\beta_1^2) - 1\right]^{0.5}. \tag{15.61}$$

Equation (15.61) allows us to compare the energy invested in the incident particle,

$$T = (\gamma_1 - 1)m_0c^2, \tag{15.61}$$

to the maximum energy available for particle reactions,

$$T_{cm} = 2(\gamma_1' - 1)m_0c^2, \tag{15.62}$$

where

$$\gamma_1 = 1/\sqrt{(1 - \beta_1^2)}, \qquad \gamma_1' = 1/\sqrt{(1 - \beta_u^2)}.$$

Table 15.3 shows T_{cm}/T as a function of γ_1 along with equivalent kinetic energy values for protons. In the nonrelativistic range, half the energy is available. The fraction drops off at high kinetic energy. Increasing the kinetic energy of particles striking a stationary target gives diminishing returns.

The situation is much more favorable in an intersecting storage ring. The stationary frame is the CM frame. The CM energy available from ring particles with γ_1 is

$$T_{cm} = 2(\gamma_1 - 1)m_0c^2. \tag{15.63}$$

For example, a 21-GeV proton accelerator operated in conjunction with an intersecting storage ring can investigate the same reactions as a 1000-GeV accelerator with a stationary target. The price to pay for this advantage is reduction in the number of measurable events for physics experiments. A

TABLE 15.4   High-Energy Accelerators and Storage Rings

Machine	Location	Particle	Energy
Synchrotrons			
KEK	Tsukuba, Japan	p^+	12
ZGS (Zero-gradient synchrotron)	Argonne, Illinois, U.S.A.	p^+	12
CERN PS	Geneva, Switzerland	p^+	28
AGS (Alternating-gradient synchrotron)	Upton, New York, U.S.A.	p^+	32
Serpukhov	Serpukhov, USSR.	p^+	76
CERN SPS	Geneva, Switzerland	p^+	400
Fermilab	Batavia, Illinois, U.S.A.	p^+	800
Cornell Electron Synchrotron	Ithaca, New York, U.S.A.	e^-	12
DESY	Hamburg, Germany	e^-	7
Linear Accelerator			
SLAC	Stanford, California, U.S.A.	e^-	25
Colliding Beam Storage Rings			
ADONE	Frascati, Italy	$e^- e^+$	1.5 + 1.5
DCI	Orsay, France	$e^- e^+$	1.8 + 1.8
SLAC-SPEAR	Stanford, California, U.S.A.	$e^- e^+$	4 + 4
DESY-DORIS	Hamburg, Germany	$e^- e^+$	5 + 5
VEPP-4	Novosibirsk, USSR	$e^- e^+$	7 + 7
CESR	Ithaca, New York, U.S.A.	$e^- e^+$	8 + 8
SLAC-PEP	Stanford, California,	$e^- e^+$	18 + 18
DESY-PETRA	Hamburg, Germany	$e^- e^+$	19 + 19
CERN-ISR	Geneva, Switzerland	$p\bar{p}$	30 + 30

stationary target is usually at solid density. The density of a stored beam is more than 10 orders of magnitude lower. A major concern in intersecting storage rings is *luminosity*, a measure of beam density in physical space and velocity space. Given a velocity-dependent cross section, the luminosity determines the reaction rate between the beams. The required luminosity depends on the cross section of the reaction and the nature of the event detectors.

A list of accelerators and storage rings with the most energetic beams is given in Table 15.4. The energy figure is the kinetic energy measured in the accelerator frame. The history of accelerators for particle physics during the last 50 years has been one of exponential increase in the available CM energy. Although this is attributable in part to an increase in the size of equipment, the main reason for the dramatic improvement has been the introduction of new acceleration techniques. When a particular technology reached the knee of its growth curve, a new type of accelerator was developed. For example, proton accelerators evolved from electrostatic machines to cyclotrons. The energy limit of cyclotrons was resolved by synchrocyclotrons which lead to the weak focusing synchrotron. The development of strong focusing made the construction of large synchrotrons possible. Subsequently, colliding beam techniques brought about a substantial increase in CM energy from existing machines. At present, there is considerable activity in converting the largest synchrotrons to colliding beam facilities.

In the continuing quest for high-energy proton beams for elementary particle research, the next stated goal is to reach a proton kinetic energy of 20 TeV (20×10^{12} eV). At present, the only identified technique to achieve such an extrapolation is to build an extremely large machine. A 20-TeV synchrotron with conventional magnets operating at an average field of 1 T has a radius of 66 km and a circumference of 414 km. The power requirements of conventional magnets in such a large machine are prohibitive; superconducting magnets are essential. Superconducting magnets can be designed in two ranges. Superconducting coils can be combined with a conventional pole assembly for fields below saturation. Since superconducting coils sustain a field with little power input, there is also the option for high-field magnets with completely saturated poles. A machine with 6-T magnets has a circumference of 70 km.

Studies have recently been carried out for a superconducting super collider (SSC).[†] This machine is envisioned as two interleaved 20-TeV proton synchrotrons with counterrotating beams and a number of beam intersection regions. Estimates of the circumference of the machine range from 90 to 160 km, depending on details of the magnet design. The CM energy is a factor of 40 higher than that attainable in existing accelerators. If it is constructed, the SSC may mark the termination point of accelerator technology in terms of particle energy; it is difficult to imagine a larger machine. Considerations of cost versus

[†]See, M. Tigner, Ed., *Accelerator Physics Issues for a Superconducting Super Collider*, University of Michigan, UM HE 84-1, 1984).

rewards in building the SSC raise interesting intellectual questions about economic limits to our knowledge of the universe.

15.6 LONGITUDINAL DYNAMICS OF SYNCHROTRONS

The description of longitudinal particle motion in synchrotrons has two unique aspects compared to synchrocyclotrons and AVF cyclotrons. The features arise from the geometry of the machine and the high energy of the particles:

1. Variations of longitudinal energy associated with stable phase confinement of particles in an rf bucket result in horizontal particle oscillations. The *synchrotron oscillations* sum with the usual betatron oscillations that arise from spreads in transverse velocity. Synchrotron oscillations must be taken into account in choosing the size of the "good field" region of focusing magnets.

2. The range of stable synchronous phase in a synchrotron depends on the energy of particles. This effect is easily understood. At energies comparable to or less than m_0c^2, particles are nonrelativistic; therefore, their velocity depends on energy. In this regime, low-energy particles in a beam bunch take a longer time to complete a circuit of the accelerator and return to the acceleration cavity. Therefore, the accelerating voltage must rise with time at ϕ_s for phase stability ($0 < \phi_s < \frac{1}{2}\pi$). At relativistic energies, particle velocity is almost independent of energy; the particle orbit circumference is the main determinant of the revolution time. Low-energy particles have smaller orbit radii and therefore take less time to return to the acceleration gap. In this case, the range of stable phase is $\frac{1}{2}\pi < \phi < \pi$. The energy that divides the regimes is called the *transition energy*. In synchrotrons that bridge the transition energy, it is essential to shift the phase of the rf field before the bunched structure of the beam is lost. This effect is unimportant in electron synchrotrons since electrons are always injected above the transition energy.

Models are developed in this section to describe the longitudinal dynamics of particles in synchrotrons. We begin by introducing the quantity γ_t, the transition gamma factor. The parameter characterizes the dependence of particle orbit radius in the focusing lattice to changes in momentum. We shall see that γ_t corresponds to the relativistic mass factor at the transition energy. After calculating examples of γ_t in different focusing systems, we shall investigate the equilibrium conditions that define a synchronous phase. The final step is to calculate longitudinal oscillations about the synchronous particle.

The transition gamma factor is defined by

$$\gamma_t^2 = \frac{\delta p / p_s}{\delta S / S},$$

15.64

where p_s is the momentum of the synchronous particle and S is the pathlength of its orbit around the machine. In a circular accelerator with no straight sections, the equilibrium radius is related to pathlength by $S = 2\pi R$; therefore,

$$\gamma_t^2 = \frac{\delta p/p_s}{\delta r/R}. \tag{15.65}$$

The transition gamma factor must be evaluated numerically for noncircular machines with complex lattices. We will develop simple analytic expressions for γ_t in ideal circular accelerators with weak and strong focusing.

In a weak focusing synchrotron, momentum is related to vertical magnetic field and position by Eq. (3.38),

$$p = qrB,$$

so that

$$\delta p/p_s = (\delta r/R) + (\delta B/B_0). \tag{15.66}$$

for $\delta r \ll R$ and $\delta B \ll B_0$. The relative change in vertical field can be related to the change in radius though Eq. (7.18), so that

$$\boxed{\gamma_t^2 = \frac{\delta p/p_s}{\delta r/R} = (1 - n) = \left(\frac{\omega_r}{\omega_{g0}}\right)^2. \quad 15.67}$$

The requirement of stable betatron oscillations in a weak focusing machine limits γ_t to the range $0 < \gamma_t < 1$.

We can also evaluate γ_t for an ideal circular machine with uniform bending field and a strong focusing system. Focusing in the radial direction is characterized by ν_r, the number of radial betatron oscillations per revolution. For simplicity, assume that the particles are relativistic so that the magnetic forces are almost independent of energy. The quantity R is the equilibrium radius for particles of momentum $\gamma_0 m_0 c$. The radial force expanded about R is

$$F_r \cong -\gamma_0 m_0 \omega_r^2 \, \delta r - qB_0 c, \tag{15.68}$$

where $\delta r = r - R$. The equilibrium radius for momentum $(\gamma_0 + \delta\gamma) m_0 c$ is determined by the balance of magnetic forces with centrifugal force, $(\gamma_0 + \delta\gamma) m_0 c^2/r$. Neglecting second-order terms, we find that

$$\gamma_0 m_0 \omega_r^2 \, \delta r + qB_0 c \cong \gamma_0 m_0 c^2/R + \delta\gamma \, m_0 c^2/R - \gamma_0 m_0 c^2 \, \delta r/R^2. \tag{15.69}$$

Zero-order terms cancel, leaving

$$(\delta r/R)\left(\omega_r^2 + \omega_{g0}^2\right) \cong (\delta\gamma/\gamma_0)\omega_{g0}^2 \cong (\delta p/p)\omega_{g0}^2,$$

or

$$\gamma_t^2 \cong 1 + \left(\omega_r/\omega_{g0}\right)^2 = 1 + \nu_r^2.$$

$$\boxed{15.70}$$

Note that $\gamma_t \gg 1$ in a strong focusing system with high ν_r. Therefore, particle position in a strong focusing system is much less sensitive to momentum errors than in a weak focusing system.

Both the magnetic field and frequency of accelerating electric fields must vary in a synchrotron to maintain a synchronous particle with constant radius R. There are a variety of possible acceleration histories corresponding to different combinations of synchronous phase, cavity voltage amplitude, magnetic field strength, and rf frequency. We shall derive equations to relate the different quantities.

We begin by calculating the momentum of the synchronous particle as a function of time. Assume the acceleration gap has narrow width δ so that transit-time effects can be neglected. The electric force acting on the synchronous particle in a gap with peak voltage V_0 is

$$qE = (qV_0\sin\phi_s/\delta).$$ (15.71)

The momentum change passing through the gap is the electric force times the transit time, or

$$\Delta p_s = (qV_0\sin\phi_s/\delta)(\delta/v_s),$$ (15.72)

where v_s is the synchronous particle velocity. Acceleration occurs over a large number of revolutions; it is sufficient to approximate p_s as a continuous function of time. The smoothed derivative of p_s is found by dividing both sides of Eq. (15.72) by the revolution time

$$\tau_0 = 2\pi R/v_s.$$ (15.73)

The result is

$$dp_s/dt \cong qV_0\sin\phi_s/2\pi R.$$ (15.74)

If V_0 and ϕ_s are constant, Eq. (15.74) has the solution

$$p_s = p_{s0} + (qV_0\sin\phi_s/2\pi R)t.$$ (15.75)

Either Eq. (15.74) or (15.75) can be used to find $p_s(t)$. Equation (2.37) can then be used to determine $\gamma_s(t)$ from $p_s(t)$. The time history of the frequency is then constrained. The revolution frequency is

$$\omega_{g0} = v_s/R = (c/R)\left(1 - 1/\gamma_s^2\right)^{0.5}$$ (15.76)

through Eq. (2.21). The rf frequency must be an integer multiple of the revolution frequency, $\omega = M\omega_{g0}$. In small synchrotrons, M may equal 1 to minimize the rf frequency. In larger machines, M is usually greater than unity. In this case, there are M circulating beam bunches contained in the ring. The rf frequency is related to the particle energy by

$$\omega = (Mc/R)(1 - 1/\gamma_s^2)^{0.5}. \tag{15.77}$$

Similarly, the equation $B_0 = \gamma_0 m_0 v_s/qR$ implies that the magnetic field magnitude is

$$B_0 = (m_0 c/qR)(\gamma_s^2 - 1)^{0.5}. \tag{15.78}$$

The rf frequency and magnetic field are related to each other by

$$\omega = \frac{MqB_0/m_0}{\left[1 + (qB_0R/m_0c)^2\right]^{0.5}}. \qquad \boxed{15.79}$$

As an example of the application of Eqs. (15.75), (15.77), and (15.78), consider the parameters of a moderate-energy synchrotron (the Bevatron). The injection energy and final energy for protons is 9.8 MeV and 6.4 GeV, respectively. The machine radius is 18.2 m and $M = 1$. The variations of F and B_0 during an acceleration cycle are plotted in Figure 15.20. The magnetic field rises from 0.025 to 1.34 T and the frequency ($\omega/2\pi$) increases from 0.37 to 2.6 MHz.

The reasoning that leads to Eq. (15.74) can also be applied to derive a momentum equation for a nonsynchronous particle. Again, averaging the momentum change around one revolution,

$$dp/dt \cong (qV_0/2\pi R)\sin\phi, \tag{15.80}$$

where R is the average radial position of the particle. Substituting $\delta p = p - p_s$, we find (as in Section 13.3) that

$$\delta\dot{p} = (qV_0/2\pi R)(\sin\phi - \sin\phi_s) = (qV_0\omega_{g0}/2\pi\beta_s c)(\sin\phi - \sin\phi_s). \tag{15.81}$$

Applying Eq. (15.6), changes of phase can be related to the difference between the orbital frequency of a nonsynchronous particle to the rf frequency,

$$\dot{\phi} = \omega - M\omega_g. \tag{15.82}$$

The orbital frequency must be related to variations of relativistic momentum in order to generate a closed set of equations. The revolution time for a nonsynchronous particle is $\tau = 2\pi r/v = 2\pi/\omega_g$. Differential changes in τ arise from

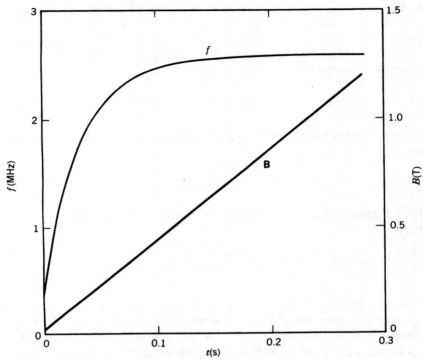

Figure 15.20 Variation of magnetic field and rf frequency during acceleration cycle of protons from 9.4 MeV to 5.6 GeV (Bevatron).

variations in particle velocity and changes in orbit radius. The following equations pertain to small changes about the parameters of the synchronous particle orbit:

$$\delta\tau/\tau_0 = -\delta\omega_g/\omega_{g0} = (\delta r/R) - (\delta v/v_s) = (\delta r/R) - (\delta\beta/\beta_s). \quad (15.83)$$

The differential change in momentum ($p = \gamma m_0\beta c$) is

$$(\delta p/p_s) = \delta\gamma/\gamma_0 + \delta\beta/\beta_s = (\delta\beta/\beta_s)/(1 - \beta_s^2). \quad (15.84)$$

The final form is derived from Eq. (2.22) with some algebraic manipulation. Noting that

$$\delta\beta/\beta_s = (1 - \beta_s^2)(\delta p/p_s)$$

we find that

$$-\delta\omega_g/\omega_{g0} = (\delta r/R) - (\delta p/p_s)/\gamma_s^2 = [(1/\gamma_t^2) - (1/\gamma_s^2)](\delta p/p_s). \quad (15.85)$$

Equation (15.85) implies that

$$\dot{\omega}_g = -\delta \dot{p}(\omega_{g0}/p_s)\left[(1/\gamma_t^2) - (1/\gamma_s^2)\right]. \tag{15.86}$$

Equations (15.81), (15.82), and (15.86) can be combined into a single equation for phase in the limit that the parameters of the synchronous particle and the rf frequency change slowly compared to the time scale of a phase oscillation. This is an excellent approximation for the long acceleration cycle of synchrotrons. Treating ω as a constant in Eq. (15.82), we find

$$\ddot{\phi} = -M\dot{\omega}_g. \tag{15.87}$$

Combining Eqs. (15.85), (15.86), and (15.87), the following equation describes phase dynamics in the synchrotron:

$$\ddot{\phi} = \left(M\omega_{g0}^2/\gamma_0 m_0 c^2 \beta_s^2\right)(eV_0/2\pi)\left[(1/\gamma_t^2) - (1/\gamma_s^2)\right](\sin\phi - \sin\phi_s).$$

$$\boxed{15.88}$$

Equation (15.88) describes a nonlinear oscillator; it is similar to Eq. (13.21) with the exception of the factor multiplying the sine functions. We discussed the implications of Eq. (13.21) in Section 13.3, including phase oscillations, regions of acceptance for longitudinal stability, and compression of phase oscillations. Phase oscillations in synchrotrons have two features that are not encountered in linear accelerators:

1. Phase oscillations lead to changes of momentum about p_s and hence to oscillation of particle orbit radii. These radial oscillations are called *synchrotron oscillations*.
2. The coefficient of the sine terms may be either positive or negative, depending on the average particle energy.

In the limit of small phase excursion ($\Delta\phi \ll 1$), the angular frequency for phase oscillations in a synchrotron is

$$\omega_s = \omega_{g0}\left[-\frac{M\cos\phi_s}{2\pi\beta_s^2}\frac{eV_0}{\gamma_s m_0 c^2}\left(\frac{1}{\gamma_t^2} - \frac{1}{\gamma_s^2}\right)\right]^{1/2}. \tag{15.89}$$

Note that the term in brackets contains dimensionless quantities and a factor proportional to the ratio of the peak energy gain in the acceleration gap divided by the particle energy. This is a very small quantity; therefore, the synchrotron oscillation frequency is small compared to the frequency for particle revolutions or betatron oscillations. The radial oscillations occur at angular frequency ω_s. In the range well beyond transition ($\gamma_s \gg \gamma_t^2$), the amplitude of radial oscillations can be expressed simply as

$$\delta r \cong R(\Delta\phi/M)(\omega_s/\omega_{g0}),\qquad(15.90)$$

where $\Delta\phi$ is the maximum phase excursion of the particle from ϕ_s.

The behavior of the expression $[(1/\gamma_t^2) - (1/\gamma_s^2)]$ determines the range of stable phase and the transition energy. For large γ_t or small γ_s, the expression is negative. In this case, the stability range is the same as in a linear accelerator, $0 < \phi_s < \frac{1}{2}\pi$. At high values of γ_s, the sign of the expression is positive, and the stable phase regime becomes $\frac{1}{2}\pi < \phi_s < \pi$.

In a weak focusing synchrotron, γ_t is always less than unity; therefore, particles are in the post-transition regime at all values of energy. Transition is a problem specific to strong focusing synchrotrons. The transition energy in a strong focusing machine is given approximately by

$$E_t = (m_0c^2)\nu_r.\qquad\boxed{15.91}$$

15.7 STRONG FOCUSING

The strong focusing principle,[†] originated in 1950, was in large part responsible for the subsequent development of synchrotrons with output beam kinetic energy exceeding 10 GeV. Strong focusing leads to a reduction in the dimensions of a beam for a given transverse velocity spread and magnetic field strength. In turn, the magnet gap and transverse extent of the "good field region" can be reduced, bringing about significant reductions in the overall size and cost of accelerator magnets.

Weak focusing refers to beam confinement systems in circular accelerators where the betatron wavelength is longer than the machine circumference. The category includes the gradient-type field of betatrons and uniform-field cyclotrons. Strong focusing accelerators have $\lambda_b < 2\pi R$, a consequence of the increased focusing forces. Examples are the alternating-gradient configuration and FD or FODO combinations of quadrupole lenses. Progress in rf linear accelerators took place largely in the early 1950s after the development of high-power rf equipment. Although some early ion linacs were built with solenoidal lenses, all modern machines use strong focusing quadrupoles, either magnetic or electric.

The advantage of strong focusing can be demonstrated by comparing the vertical acceptance of a weak focusing circular accelerator to that of an alternating-gradient (AG) machine. Assume that the AG field consists of FD focusing cells of length l (along the beam orbit) with field index $\pm n$, where $n \gg 1$. The vertical position of a particle at cell boundaries is given by

$$z = z_0\cos(M\mu + \phi),\qquad(15.92)$$

[†] N. C. Christofilos, U.S. Patent No. 2,736,799 (1950).

where

$$\mu = \cos^{-1}\left[\cos\left(\sqrt{n}\,\omega_{g0}/v_s\right)\cosh\left(\sqrt{n}\,\omega_{g0}/v_s\right)\right]$$

and M is the cell number. For $\mu \leq 1$, the orbit consists of a sinusoidal oscillation extending over many cells with small-scale oscillations in individual magnets. Neglecting the small oscillations, the orbit equation for particles on the beam envelope is

$$z \cong z_0\cos(\mu S/l + \phi), \tag{15.93}$$

where S, the distance along the orbit, is given by $S = Ml$. The angle of the orbit is approximately

$$z' \cong -(z_0\mu/l)\sin((\mu S/l + \phi). \tag{15.94}$$

Combining Eqs. (15.93) and (15.94), the vertical acceptance is

$$A_v = \pi z_0 z_0' = \pi z_0^2\mu/l. \tag{15.95}$$

In a weak focusing system, vertical orbits are described by

$$z = z_0\cos\left(\sqrt{n}\,s/R + \phi\right). \tag{15.96}$$

Following the same development, the vertical acceptance is

$$A_v = \pi z_0^2\sqrt{n}\,/R. \tag{15.97}$$

In comparing Eqs. (15.95) and (15.97), note that the field index for weak focusing must be less than unity. In contrast, the individual field indices of magnets in the alternating gradient are made as large as possible, consistent with practical magnet design. Typically, the field indices are chosen to give $\mu \sim 1$. For the same field strength, the acceptance of the strong focusing system is therefore larger by a factor on the order of R/l or $N/2\pi$, where N is the number of focusing cells. The quantity N is a large number. For example, $N = 60$ in the AGS accelerator at Brookhaven National Laboratory.

The major problem of strong focusing systems is that they are sensitive to alignment errors and other perturbations. The magnets of a strong focusing system must be located precisely. We shall estimate the effects of alignment error in a strong focusing system using the transport matrix formalism (Chapter 8). The derivation gives further insight into the origin of resonant instabilities introduced in Section 7.2.

For simplicity, consider a circular strong focusing machine with uniformly distributed cells. Assume that there is an error of alignment in either the horizontal or vertical direction between two cells. The magnets may be displaced a distance ε, as shown in Figure 15.21a. In this case, the position

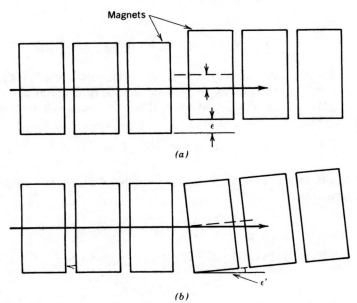

Figure 15.21 Alignment errors between magnets in strong focusing system. (*a*) Error in position. (*b*) Error in angle.

component of an orbit vector is transformed according to

$$x \rightarrow x + \varepsilon \tag{15.98}$$

when the particle crosses the boundary. An error in magnet orientation by an angle ε' (Fig. 15.21*b*) causes a change in the angular part of the orbit vector:

$$x' \rightarrow x' + \varepsilon'. \tag{15.99}$$

The general transformation at the boundary is

$$\mathbf{u}_{n+1} = \mathbf{u}_n + \boldsymbol{\varepsilon}, \tag{15.100}$$

where

$$\boldsymbol{\varepsilon} = (\varepsilon, \varepsilon').$$

Let **A** be the transfer matrix for a unit cell of the focusing system and assume that there are N cells distributed about the circle. The initial orbit vector of a particle is $\mathbf{u}_0$. For convenience, $\mathbf{u}_0$ is defined at a point immediately following the imperfection. After a revolution around the machine and traversal of the field error, the orbit vector becomes

$$\mathbf{u}_N = \mathbf{A}^N \mathbf{u}_0 + \boldsymbol{\varepsilon}. \tag{15.101}$$

The orbit vector after two revolutions is

$$\mathbf{u}_{2N} = \mathbf{A}^N \mathbf{u}_N + \boldsymbol{\varepsilon} = \mathbf{A}^{2N} \mathbf{u}_0 + (\mathbf{A}^N + \mathbf{I})\boldsymbol{\varepsilon}, \tag{15.102}$$

where $\mathbf{I}$ is the identity matrix. By induction, the transformation of the orbit matrix for n revolutions is

$$\mathbf{u}_{nN} = \mathbf{A}^{nN} \mathbf{u}_0 + \mathbf{D}_n \boldsymbol{\varepsilon}, \tag{15.103}$$

where

$$\mathbf{D}_n = (\mathbf{A}^{(n-1)N} + \mathbf{A}^{(n-2)N} + \cdots + \mathbf{A}^N + I). \tag{15.104}$$

We found in Chapter 8 that the first term on the right-hand side of Eq. (15.103) corresponds to bounded betatron oscillations when stability criteria are satisfied. The amplitude of the term is independent of the perturbation. Particle motion induced by the alignment error is described by the second term. The expression for $\mathbf{D}_n$ can be simplified using the eigenvectors (Section 8.6) of the matrix $\mathbf{A}$, $\boldsymbol{v}_1$ and $\boldsymbol{v}_2$. The eigenvectors form a complete set; any two-dimensional vector, including $\boldsymbol{\varepsilon}$, can be resolved into a sum of eigenvectors:

$$\boldsymbol{\varepsilon} = a_1 \boldsymbol{v}_1 + a_2 \boldsymbol{v}_2. \tag{15.105}$$

We found in Section 8.6 that the eigenvalues for a transfer matrix $\mathbf{A}$ are

$$\lambda_1 = \exp(j\mu), \qquad \lambda_2 = \exp(-j\mu), \tag{15.106}$$

where μ is the phase advance in a cell. Substituting Eq. (15.106) in Eq. (15.103), we find

$$\mathbf{D}_n \boldsymbol{\varepsilon} = a_1 \boldsymbol{v}_1 \{ \exp[j(n-1)N\mu] + \exp[j(n-2)N\mu] + \cdots + 1 \}$$

$$+ a_2 \boldsymbol{v}_2 \{ \exp[-j(n-1)N\mu] + \exp[-j(n-2)N\mu] + \cdots + 1 \}. \tag{15.107}$$

The sums of the geometric series can be rewritten as

$$\mathbf{D}_n \boldsymbol{\varepsilon} = \frac{\exp(jnN\mu) - 1}{\exp(jN\mu) - 1} a_1 \boldsymbol{v}_1 + \frac{\exp(-jnN\mu) - 1}{\exp(-jN\mu) - 1} a_2 \boldsymbol{v}_2 \tag{15.108}$$

or, alternately,

$$\mathbf{D}_n \boldsymbol{\varepsilon} = \left[\sin(\tfrac{1}{2}nN\mu) / \sin(\tfrac{1}{2}N\mu) \right]$$

$$\times \left\{ \exp[j(n-1)\tfrac{1}{2}N\mu] a_1 \boldsymbol{v}_1 + \exp[-j(n-1)\tfrac{1}{2}N\mu] a_2 \boldsymbol{v}_2 \right\}. \boxed{15.109}$$

The second term in curly brackets is always bounded; it has a magnitude on the order of ε. The first term in brackets determines the cumulative effect of many transitions across the alignment error. The term becomes large when the denominator approaches zero; this condition occurs when

$$\mu = 2\pi M/N, \tag{15.110}$$

where M is an integer. Equation (15.110) can be rewritten in terms of ν, the number of betatron wavelengths per revolution:

$$\nu = M. \tag{15.111}$$

This is the condition for an orbital resonance. When there is a resonance, the effects of an alignment error sum on successive revolutions. The amplitude of oscillatory motion grows with time. The motion induced by an error when $\nu \neq M$ is an oscillation superimposed on betatron and synchrotron oscillations. The amplitude of the motion can be easily estimated. For instance, in the case of a position error of magnitude ε, it is $\varepsilon/\sin(\frac{1}{2}N\mu)$.

An alternate view of the nature of resonant instabilities, mode coupling, is useful for general treatments of particle instabilities. The viewpoint arises from conservation of energy and the second law of thermodynamics. The second law implies that there is equipartion of energy between the various modes of oscillation of a physical system in equilibrium. In the treatment of resonant instabilities in circular accelerators, we included two modes of oscillation: (1) the revolution of particles at frequency ω_{g0} and (2) betatron oscillations. There is considerable longitudinal energy associated with particle revolution and, under normal circumstances, a small amount of energy in betatron oscillations.

TABLE 15.5   Forbidden ν Values[a]

Condition	Description
$\nu_r, \nu_z = M$	Magnet misalignment, error in field strength
$\nu_r, \nu_z = M/2$	Error in field index of bending magnets.
$\nu_r, \nu_z = N$ or $N/2$	Particularly strong special case of conditions 1 or 2, where N is the number of focusing cells in the circular machine.
$\nu_r, \nu_z = N'$ or $N'/2$	Another special case of 1 or 2, where N' is the number of superperiods in a machine with curved and straight sections.
$\nu_r + \nu_z = M$	Linear coupling between horizontal and vertical betatron oscillations. Induced by rotational error in a quadrupole focusing magnet.

[a]M, N, and N' are integers.

In a linear analysis, there is no exchange of energy between the two modes. A field error introduces a nonlinear coupling term, represented by $\mathbf{D}_n \varepsilon$ in Eq. (15.103). This term allows energy exchange. The coupling is strong when the two modes are in resonance. The second law implies that the energy of the betatron oscillations increases. A complete nonlinear analysis predicts that the system ultimately approaches an equilibrium with a thermalized distribution of particle energy in the transverse and longitudinal directions. In an accelerator, the beam is lost on vacuum chamber walls well before this state is reached.

In a large circular accelerator, there are many elements of periodicity that can induce resonance coupling of energy to betatron oscillations. In synchrotrons, where particles are contained for long periods of time, all resonance conditions must be avoided. Resonances are categorized in terms of forbidden numbers of betatron wavelengths per revolution. The physical bases of some forbidden values are listed in Table 15.5.

Bibliography

L. L. Alston (Ed.), *High Voltage Technology*, Oxford University Press, Oxford, 1968.

R. Bakish, *Introduction to Electron Beam Technology*, Wiley, New York, 1962.

A. P. Banford, *The Transport of Charged Particle Beams*, Spon, London, 1966.

A. H. W. Beck, *Space Charge Waves and Slow Electromagnetic Waves*, Pergamon Press, London, 1958.

M. Y. Bernard, "Particles and Fields: Fundamental Equations," in A. Septier, Ed., *Focusing of Charged Particles*, Vol. 1, Academic, New York, 1967.

P. Bonjour, "Numerical Methods for Computing Electrostatic and Magnetic Fields," in A. Septier, Ed., *Applied Charged Particle Optics*, *Part A*, Academic, New York, 1980.

M. Born and E. Wolf, *Principles of Optics*, Pergamon Press, Oxford, 1965.

D. Boussard, "Focusing in Linear Accelerators," in A. Septier, Ed., *Focusing of Charged Particles*, Vol. 2, Academic, New York, 1967.

H. Brechner, *Superconducting Magnet Systems*, Springer-Verlag, Berlin, 1973.

L. Brillouin, *Wave Propagation in Periodic Structures*, Dover, New York, 1953.

G. Brewer, *Ion Propulsion—Technology and Applications*, Gordon and Breach, New York, 1970.

H. Bruck, *Accelerateurs Circulaires de Particules*, Presses Universitaires de France, Paris, 1966.

R. A. Carrigen, F. P. Huson, and M. Month, *Summer School on High Energy Particle Accelerators*, American Institute of Physics, New York, 1981.

J. D. Cobine, *Gaseous Conductors*, Dover, New York, 1958.

R. E. Collin, *Foundations for Microwave Engineering*, McGraw-Hill, New York, 1966.

T. Collins, "Concepts in the Design of Circular Accelerators," in *Physics of High Energy Particle Accelerators* (SLAC Summer School, 1982), American Institute of Physics, New York, 1983.

J. S. Colonias, *Particle Accelerator Design—Computer Programs*, Academic, New York, 1974.

V. E. Coslett, *Introduction to Electron Optics*, Oxford University Press, Oxford, 1950.

P. Dahl, *Introduction to Electron and Ion Optics*, Academic, New York, 1973.

H. A. Enge, "Deflecting Magnets," in A. Septier, Ed., *Focusing of Charged Particles*, Vol. 2, Academic, New York, 1967.

C. Fert and P. Durandeau, "Magnetic Electron Lenses," in A. Septier, Ed., *Focusing of Charged Particles*, Vol. 1, Academic, New York, 1967.

J. F. Francis, *High Voltage Pulse Techniques*, Air Force Office of Scientific Research, AFOSR-74-2639-5, 1974.

J. C. Francken, "Analogical Methods for Resolving Laplace's and Poisson's Equation," in A. Septier, Ed., *Focusing of Charged Particles*, Vol. 1, Academic, New York, 1967.

A. Galejs and P. H. Rose, "Optics of Electrostatic Accelerator Tubes," in A. Septier, Ed., *Focusing of Charged Particles*, Vol. 2, Academic, New York, 1967.

C. Germain, "Measurement of Magnetic Fields," in A. Septier, Ed., *Focusing of Charged Particles*, Vol. 1, Academic, New York, 1967.

G. N. Glasoe and J. V. Lebacqz, *Pulse Generators*, Dover, New York, 1965.

M. Goldsmith, *Europe's Giant Accelerator, the Story of the CERN 400 GeV Proton Synchrotron*, Taylor and Francis, London, 1977.

H. Goldstein, *Classical Mechanics*, Addison-Wesley, Reading, Mass., 1950.

P. Grivet and A. Septier, *Electron Optics*, Pergamon Press, Oxford, 1972.

K. J. Hanszen and R. Lauer, "Electrostatic Lenses," in A. Septier, Ed., *Focusing of Charged Particles*, Vol. 1, Academic, New York, 1967.

E. Harting and F. H. Read, *Electrostatic Lenses*, Elsevier, Amsterdam, 1976.

W. V. Hassenzahl, R. B. Meuser, and C. Taylor, "The Technology of Superconducting Accelerator Dipoles," in *Physics of High Energy Particle Accelerators* (SLAC Summer School, 1982), American Institute of Physics, New York, 1983.

P. W. Hawkes, *Electron Optics and Electron Microscopy*, Taylor and Francis, London, 1972.

P. W. Hawkes (Ed.), *Magnetic Electron Lens Properties*, Springer-Verlag, Berlin, 1980.

P. W. Hawkes, "Methods of Computing Optical Properties and Combatting Aberrations for Low-Intensity Beams," in A. Septier, Ed., *Applied Charged Particle Optics*, Part A, Academic, New York, 1980.

P. W. Hawkes, *Quadrupoles in Electron Lens Design*, Academic, New York, 1970.

P. W. Hawkes, *Quadrupole Optics*, Springer-Verlag, Berlin, 1966.

R. Hutter, "Beams with Space-charge," in A. Septier, Ed., *Focusing of Charged Particles*, Vol. 2, Academic, New York, 1967.

J. D. Jackson, *Classical Electrodynamics*, Wiley, New York, 1975.

I. M. Kapchinskii, *Dynamics in Linear Resonance Accelerators*, Atomizdat, Moscow, 1966.

S. P. Kapitza and V. N. Melekhin, *The Microtron* Harwood Academic, New York, 1978. (I. N. Sviatoslavsky, (trans.))

E. Keil, "Computer Programs in Accelerator Physics," in *Physics of High Energy Particle Accelerators* (SLAC Summer School, 1982), American Institute of Physics, New York, 1983.

O. Klemperer and M. E. Barnett, *Electron Optics*, Cambridge University Press, London, 1971.

A. A. Kolomensky and A. N. Lebedev, *Theory of Cyclic Accelerators* (trans. from Russian by M. Barbier), North-Holland, Amsterdam, 1966.

R. Kollath (Ed.), *Particle Accelerators*, (trans. from 2nd German edition by W. Summer), Pittman and Sons, London, 1967.

P. M. Lapostolle and A. Septier (Eds.), *Linear Accelerators*, North Holland, Amsterdam, 1970.

L. J. Laslett, "Strong Focusing in Circular Particle Accelerators," in A. Septier, Ed., *Focusing of Charged Particles*, Vol. 2, Academic, New York, 1967.

J. D. Lawson, *The Physics of Charged-Particle Beams*, Clarendon Press, Oxford, 1977.

B. Lehnert, *Dynamics of Charged Particles*, North-Holland, Amsterdam, 1964.

A. J. Lichtenberg, *Phase Space Dynamics of Particles*, Wiley, New York, 1969.

R. Littauer, "Beam Instrumentation," in *Physics of High Energy Particle Accelerators* (SLAC Summer School, 1982), American Institute of Physics, New York, 1983.

J. J. Livingood, *Principles of Cyclic Particle Accelerators*, Van Nostrand, Princeton, New Jersey, 1961.

J. J. Livingood, *The Optics of Dipole Magnets*, Academic, New York, 1969.

M. S. Livingston (Ed.), *The Development of High Energy Particle Accelerators*, Dover, New York, 1966.

M. S. Livingston, *High Energy Accelerators*, Interscience, New York, 1954.

M. S. Livingston, *Particle Accelerators, A Brief History*, Harvard University Press, Cambridge, Mass., 1969.

M. S. Livingston and J. P. Blewett, *Particle Accelerators*, McGraw-Hill, New York, 1962.

G. A. Loew and R. Talman, "Elementary Principles of Linear Accelerators," in *Physics of High Energy Particle Accelerators* (SLAC Summer School, 1982), American Institute of Physics, New York, 1983.

W. B. Mann, *The Cyclotron*, Methuen, London, 1953.

J. W. Mayer, L. Eriksson, and J. A. Davies, *Ion Implantation in Semiconductors*, Academic, New York, 1970.

N. W. McLachlan, *Theory and Application of Matheiu Functions*, Oxford University Press, Oxford, 1947.

A. H. Maleka, *Electron-Beam Welding—Principles and Practice*, McGraw-Hill, New York, 1971.

R. B. Miller, *Intense Charged Particle Beams*, Plenum Press, New York, 1982.

M. Month (Ed.), *Physics of High Energy Particle Accelerators* (SLAC Summer School, 1982), American Institute of Physics, New York, 1983.

R. B. Neal (Ed.), *The Stanford Two-mile Accelerator*, Benjamin, Reading, Mass., 1968.

T. J. Northrup, *The Adiabatic Motion of Charged Particles*, Interscience, New York, 1963.

H. Patterson, *Accelerator Health Physics*, Academic, New York, 1973.

E. Perisco, E. Ferrari, and S. E. Segre, *Principles of Particle Accelerators*, Benjamin, New York, 1968.

J. R. Pierce, *Theory and Design of Electron Beams*, Van Nostrand, Princeton, New Jersey 1954.

D. Potter, *Computational Physics*, Wiley-Interscience, New York, 1973

R. E. Rand, *Recirculating Electron Accelerators*, Harwood Academic, New York, 1984.

S. Ramo, J. R. Whinnery, and T. Van Duzer, *Fields and Waves in Communication Electronics*, Wiley, New York, 1965.

E. Regenstreif, "Focusing with Quadrupoles, Doublets and Triplets," in A. Septier, Ed., *Focusing of Charged Particles*, Vol. 1, Academic, New York, 1967.

J. Rosenblatt, *Particle Accelerators*, Methuen, London, 1968.

W. Scharf, *Particle Accelerators and Their Uses*, Harwood Academic, New York, 1985.

S. Schiller, U. Heisig, and S. Panzer, *Electron Beam Technology*, Wiley, New York, 1982.

R. W. Southwell, *Relaxation Methods in Theoretical Physics*, Oxford University Press, Oxford, 1946.

A. Septier (Ed.), *Applied Charged Particle Optics, Part A*, Academic, New York, 1980.

A. Septier (Ed.), *Applied Charged Particle Optics, Part B*, Academic, New York, 1980.

A. Septier (Ed.), *Applied Charged Particle Optics, Part C, Very-High Density Beams*, Academic, New York, 1983.

A. Septier (Ed.), *Focusing of Charged Particles*, Academic, New York, 1967.

J. C. Slater, *Microwave Electronics*, Van Nostrand, Princeton, New Jersey, 1950.

K. G. Steffen, *High Energy Beam Optics*, Wiley-Interscience, New York, 1965.

E. Stuhlinger, *Ion Propulsion for Space Flight*, McGraw-Hill, New York, 1964.

P. Sturrock, *Static and Dynamic Electron Optics*, Cambridge University Press, London, 1955.

M. Tigner and H. Padamsee, "Superconducting Microwave Cavities in Accelerators for Particle Physics," in *Physics of High Energy Particle Accelerators* (SLAC Summer School, 1982), American Institute of Physics, New York, 1983.

A. D. Vlasov, *Theory of Linear Accelerators*, Atomizdat, Moscow, 1965.

C. Weber, "Numerical Solutions of Laplace's and Poisson's Equations and the Calculation of Electron Trajectories and Electron Beams," in A. Septier, Ed., *Focusing of Charged Particles*, Vol. 1, Academic, New York, 1967.

R. G. Wilson and G. R. Brewer, *Ion Beams with Applications to Ion Implantation*, Wiley, New York, 1973.

H. Wollnik, "Electrostatic Prisms," in A. Septier, Ed., *Focusing of Charged Particles*, Vol. 2, Academic, New York, 1967.

H. Wollnik, "Mass Spectrographs and Isotope Separators," in A. Septier, Ed., *Applied Charged Particle Optics*, Part B, Academic, New York, 1980.

O. C. Zienkiewicz, *The Finite Element Method in Engineering Science*, McGraw-Hill, New York, 1971.

J. F. Ziegler, *New Uses of Ion Accelerators*, Plenum, New York, 1975.

V. K. Zworykin, G. A. Morton, E. G. Ramberg, J. Hillier, and A. W. Vance, *Electron Optics and the Electron Microscope*, Wiley, New York, 1945.

Index

About the Author

Stanley Humphries spent the first part of his career as an experimentalist in the fields of plasma physics, controlled fusion and charged particle acceleration. A notable contribution was the creation and demonstration of methods to generate and to transport intense pulsed ion beams. His current work centers on simulations of electromagnetic fields, biomedical processes and material response at high pressure and temperature. Dr. Humphries is the author of over 150 journal publications and the textbooks *Principles of Charged Particle Acceleration* (John Wiley, New York, 1986), *Charged Particle Beams* (John Wiley, New York, 1990) and *Field Solutions on Computers* (CRC Press, Boca Raton, 1997). He received a B.S. in Physics from the Massachusetts Institute of Technology and a Ph.D. in Nuclear Engineering from the University of California at Berkeley. He was elected a Fellow of the American Physical Society and the Institute of Electrical and Electronic Engineers. Dr. Humphries is a professor emeritus in the Department of Electrical and Computer Engineering at the University of New Mexico and President of Field Precision LLC, an engineering software company.

CPSIA information can be obtained at www.ICGtesting.com
Printed in the USA
LVOW08s1736160616

492919LV00016B/354/P

9 780486 498188